Michael A. Taylor

STRUCTURAL STEEL DESIGN

STRUCTURAL STEEL DESIGN

Joseph E. Bowles
Professor of Civil Engineering
Bradley University

McGraw-Hill Book Company

New York   St. Louis   San Francisco   Auckland   Bogotá   Hamburg
Johannesburg   London   Madrid   Mexico   Montreal   New Delhi
Panama   Paris   São Paulo   Singapore   Sydney   Tokyo   Toronto

STRUCTURAL STEEL DESIGN

Copyright © 1980 by McGraw-Hill, Inc. All rights reserved.
Printed in the United States of America. No part of this publication
may be reproduced, stored in a retrieval system, or transmitted, in any
form or by any means, electronic, mechanical, photocopying, recording, or
otherwise, without the prior written permission of the publisher.

1234567890 DODO 89876543210

This book was set in Times Roman by Science Typographers, Inc.
The editors were Julienne V. Brown and Madelaine Eichberg;
the cover was designed by Anne Canevari Green;
the production supervisor was Dominick Petrellese.
The drawings were done by J & R Services, Inc.
R. R. Donnelley & Sons Company was printer and binder.

Library of Congress Cataloging in Publication Data

Bowles, Joseph E
    Structural steel design.

    Bibliography: p.
    Includes index.
    1. Building, Iron and steel. 2. Steel,
Structural. 3. Structures, Theory of. I. Title.
TA684.B78   624'.1821   79-18155
ISBN 0-07-006765-1

CONTENTS

Preface ix

Chapter 1 General Design Considerations 3
- 1-1 Types of Structures 3
- 1-2 Design Procedures 4
- 1-3 Steel as a Structural Material 4
- 1-4 Steel Products 7
- 1-5 Steel Strength 10
- 1-6 Temperature Effects on Steel 12
- 1-7 Structural Design Codes 16
- 1-8 Building Loads 18
- 1-9 Highway and Railroad Bridge Loads 26
- 1-10 Impact Loads 32
- 1-11 Earthquake Loads 33
- 1-12 Fatigue 39
- 1-13 Steel Structures 42
- 1-14 Accuracy of Computations and Electronic Calculators 47
- 1-15 Structural Engineering Computations in SI 48

Chapter 2 Elements of Frame, Truss, and Bridge Design 53
- 2-1 Methods of Analysis 53
- 2-2 Beam Analysis 57
- 2-3 Determinate Structures 59
- 2-4 Truss Analysis 59
- 2-5 Rigid Frame Analysis 62
- 2-6 Bridge Analysis 63

2-7	The Computer Program Furnished in the Appendix	64
2-8	The P Matrix	66
2-9	Load Conditions	70
2-10	Checking Computer Output	71
2-11	Design Examples	72

Chapter 3 Elastic, Plastic, and Buckling Behavior of Structural Steel — 113

3-1	Introduction	113
3-2	Elastic versus Plastic Design Theory	114
3-3	Safety Factors in Elastic and Plastic Design	122
3-4	Elastic versus Plastic Design Deflections	125
3-5	Length of Plastic Hinge	125
3-6	Elastic versus Plastic Design	126
3-7	Load Resistance Factor Design	132
3-8	Local Buckling of Plates	133
3-9	Post-Buckling Strength of Plates	138

Chapter 4 Design of Beams for Bending — 143

4-1	General Considerations	143
4-2	Design of Beams by the Elastic Method	147
4-3	Design of Continuous Beams	154
4-4	Web Buckling and Crippling	156
4-5	Shear Criteria	159
4-6	Strong versus Weak-Axis Bending	161
4-7	Deflections	161
4-8	Biaxial Bending and Bending on Unsymmetrical Sections	164
4-9	Shear Center of Open Sections	171
4-10	Design of Laterally Unsupported Beams	174
4-11	Beams with Nonparallel Flanges	182
4-12	Design of Bridge Stringers and Floor Beams	183
4-13	Composite Beams	189
4-14	Beam Design Using Load Resistance Factor Design (LRFD)	208

Chapter 5 Design of Tension Members — 217

5-1	Types of Tension Members	217
5-2	Allowable Tension Stresses	219
5-3	General Design Characteristics	220
5-4	Stresses Due to Axial Load on the Net Section	228
5-5	Design of AISC Tension Rods	230
5-6	Net Sections	232
5-7	Design of Tension Members	237
5-8	Design of Bridge Tension Members	240
5-9	Cable Design	242
5-10	Design of Tension Members Using LRFD	248

Chapter 6 Axially Loaded Columns and Struts 255

6-1	Introduction	255
6-2	The Euler Column Formula	256
6-3	Columns with End Conditions	258
6-4	Allowable Stresses in Steel Columns	260
6-5	Design of Built-up Compression Members	272
6-6	Column Base Plates	284
6-7	Lateral Bracing of Columns	291
6-8	Column and Strut Design Using LRFD	292

Chapter 7 Beam–Column Design 297

7-1	Introduction	297
7-2	General Considerations of Axial Load with Bending	300
7-3	Effective Lengths of Columns in Building Frames	303
7-4	Developing the Beam–Column Design Formulas	312
7-5	Determination of the Interaction Reduction Coefficient C_m	319
7-6	AASHTO and AREA Beam–Column Design Formulas	321
7-7	Beam–Column Design Using Interaction Equations	322
7-8	Stepped Columns and Columns with Intermediate Axial Load	329
7-9	Control of Sidesway	336
7-10	Beam–Column Design Using LRFD	338

Chapter 8 Bolted and Riveted Connections 347

8-1	Introduction	347
8-2	Rivets and Riveted Connections	351
8-3	High-Strength Bolts	355
8-4	Factors Affecting Joint Design	359
8-5	Rivets and Bolts Subjected to Eccentric Loading	375
8-6	Beam Framing Connections	384
8-7	Fasteners Subjected to Tension	388
8-8	Connections Subjected to Combined Shear and Tension	396
8-9	Moment (Type 1) Connections	399
8-10	Load Resistance Factor Design (LRFD) for Connections	405

Chapter 9 Welded Connections 411

9-1	General Considerations	411
9-2	Welding Electrodes	414
9-3	Types of Joints and Welds	414
9-4	Lamella Tearing	420
9-5	Orientation of Welds	421
9-6	Welded Connections	422
9-7	Eccentrically Loaded Welded Connections	430
9-8	Welded Column Base Plates	439
9-9	Welded End Plate Connections	442
9-10	Welded Corner Connections	445
9-11	Fillet Weld Design Using LRFD	451

Chapter 10 Plate Girders 455

10-1	General	455
10-2	Loads	460
10-3	Proportioning Flanges and Webs of Girders and Built-up Sections	463
10-4	Partial-Length Cover Plates	466
10-5	General Proportions of Plate Girders	469
10-6	Plate Girder Design Theory—AISC	470
10-7	Plate Girder Design Theory—AASHTO and AREA	490

Appendix Selected Computer Programs 509

A-1	Frame Analysis Program	509
A-2	Load Matrix Generator for AASHTO Truck Loading on a Truss Bridge	521
A-3	Load Matrix Generator for AREA Cooper's E-80 Loading on a Truss Bridge	524

Index 529

PREFACE

The primary purpose of this textbook is to provide the basic material for the first course in structural steel design. The text contains elements of both building and bridge design for use in the structural engineering sequence of civil engineering programs. If the instructor wishes to emphasize building frames, the text is also suitable for an introduction to structural steel design in architectural programs.

Approximately equal emphasis is given to fps and SI units. In the discussion material both systems of units are used; the examples and homework problems are in either fps or SI. This format was arrived at through discussions with a number of interested faculty members and people in industry. The consensus was that the text discussion should continue to use both systems of units because transition to metric is not occurring as rapidly in the construction industry as in other areas of engineering. Dual usage seems necessary to provide both student and instructor with a feeling for what is a reasonable member size (number), deflection, or other design parameter in both systems of units.

Practical SI instruction requires use of design data, and since none were readily available, I have assembled a set of computer-generated rolled steel section data tables as a supplement to the text. These tables are in general agreement with the AISC and ASTM A-6 specifications. This bound set of data also includes *edited* material from the AISC, AASHTO, and AREA specifications. It is intended that the textbook, together with the supplemental Structural Steel Design Data (SSDD) manual, will provide adequate material for a steel design course without the need for any other reference material. The course material should be sufficient to enable students to design routine (and some not-so-routine) structural members in either fps or SI units and by any one of the three steel design specifications which are most likely to control the design—

at least in American practice. Specialized problems are not generally addressed in a classroom environment, and for these (as well as for design office practice and other nonacademic work) the reader should obtain a copy of the latest specifications from the appropriate agency.

I use the digital computer as a design aid in a somewhat interactive mode (via batch processing) for the design portion of the steel design course. I have found that the use of the computer in the steel design course is one of the better academic experiences for students, because it helps them rapidly gain experience in structural behavior. This may be by accident (from mispunching data on the modulus of elasticity, cross-sectional area, or moment of inertia of a member) or by iteration of a design problem in which member sizes are changed as indicated by the computer output. In either case, students readily see the effects of member section properties on structural behavior. Using the computer programs permits this with only a modest amount of work on the part of the student—and no program writing.

Several computer programs are listed in the Appendix to the text for those who are not already using the computer as a design aid. These programs are relatively simple, but efficient, and can easily be punched on cards for use on a local computer system. The band matrix reduction method is used so that computer cpu requirements are minimal. I can furnish these programs on tape at the cost of tape, reproduction, and mailing for anyone who is using the text in the classroom.

I have not attempted to cite, or promote the use of, desktop programmable calculators for simple tasks such as beam or column designs because of the variety of devices available (e.g., HP, TI, Sharp, Casio, etc.), each requiring a different programming method, and because of continuing rapid change in the state of the art. Listing of the multiple programs necessary for use of the various calculators would take too much text space, at the expense of more important topics.

The text attempts to strike a balance between theory and "how to." The topical treatment is not so exhaustive as to obscure the fundamentals but is of sufficient depth that the reader is aware of the source of the design equations in the various specifications. A number of the equations are partially to completely derived so that the reader can be aware of the limitations. A reasonably detailed explanation is given of the basic design problems, and the illustrative examples are essentially step by step. With this format students should be able to cope with the more complex design problems on the professional level, and to obtain design solutions for the assigned home problems.

Appropriate references are cited directly in the text for topics for which coverage is limited but which are sufficiently important that the reader may wish to study the subject in greater depth. The inclusion of references will generally be of more use to those in professional practice than to student users. My experience in teaching steel design for a number of years is that most students in the first design course are primarily interested in learning how to design the various types of structural members they will be assigned for home or laboratory work. At this point in their professional development they are not overly interested in the theoretical considerations and the extensive laboratory work by researchers and theoreticians that has produced the current design equations.

The complexity of semitheoretical and empirical design equations, coupled with the nature of structural design and its intimate association with design specifications and codes, makes it necessary to take a strong "how to" approach in teaching steel design. It is essential to present the user with a set of hypothetical (or real) data and by illustration produce a design. Students are presumed to have a sufficient background in the basic engineering and math sequence to appreciate what has been illustrated and are taught how to duplicate the steps with a similar problem to gain confidence, and, based on the illustrative problems, to extrapolate to a problem where the design parameters are considerably different, with a minimum of supervision.

Fabrication and practical considerations are introduced in the example problems as appropriate. Fastener spacing, edge distances, erection clearances, standard gage distances, thread runout, and maintenance are considered in various sections. This should give the user an appreciation of fabrication problems and other practical considerations. In conjunction with this, the text has a large number of photographs, supplemented with line drawings of structural elements and connections, which should be of particular aid to the novice. The reader should supplement these illustrations by observing steel frames under construction. The photographs were all taken especially for this text, to display individual structural features as appropriate to the development of the discussion.

Plastic design is introduced briefly in Chapter 3 together with the basics of plate theory. This is done so that the design equations with origins in plastic design or plate theory can be efficiently referenced back to Chapter 3, thereby saving text space. Plastic design methods are not emphasized, for two basic reasons: there is not enough time in a first course to adequately treat the subject, and elastic design seems to be preferred in professional practice.

I have deviated from the current textbook trend to reflect the format incorporated in some of the steel texts published in the 1950s. This format includes the use of simple illustrative examples where the design data are simply stated as well as more realistic design examples. These examples are analyzed in Chapter 2 using the computer, and selected members are subsequently designed in the later chapters. The use of simple examples gives the reader a quick grasp of the general objectives of the discussion. More detailed design examples are used to generate a sense of realism and to clearly indicate that steel design is not just a matter of manipulating numbers. The examples are accompanied with a reasonable amount of discussion of the analysis provided.

Within the framework of classroom time restraints, a steel design course should be as realistic as possible. For this reason the user is encouraged to carry any structural design problems assigned in Chapter 2 through succeeding chapters, redesigning members as necessary and recycling the problem one or more times for member sizing before the connections are designed in Chapters 8 and 9. A false sense of security regarding the actual complexity of structural design, and even how the design loads are finally arrived at, can be developed if the user is simply given the loads for each design problem. Admittedly, the more realistic design problems require more physical *and mental* effort on the part of the student and more grading effort on the part of the instructor. This extra effort can be offset somewhat by assigning fewer total problems, but including some in which loads are given, to build confidence, and some with design problems, to build design skill.

The following text sequence might be appropriate in the semester system:

3 semester hours Rapid coverage of Chapters 1 and 3, with Chapter 2 assigned for reading. Reasonable coverage of Chapters 4 to 10. Probably two weeks each on Chapters 4, 7, and 10.

4 semester hours Rapid coverage of Chapters 1 and 3. Two weeks on Chapters 2, 4, 7, and 10, followed by actual design of a building frame and highway bridge truss, or industrial building, based on the analysis in Chapter 2. One structure should be done in fps, the other in SI. A design notebook should be kept, showing computations and computer input/output. It is also suggested that this work be done in groups, each with no more than four students.

ACKNOWLEDGMENTS

Several persons and organizations have provided considerable encouragement and assistance in producing this textbook. First, I should like to express my sincere appreciation to Dr. Peter Z. Bulkeley, Dean of Engineering and Technology, Bradley University, who provided me with released teaching time.

I would also like to thank Mr. Andrew Lally and Mr. Frank Stockwell, Jr., of AISC, who provided me with a preliminary copy of the new AISC specifications and took the time to go over the major changes with me. Mr. Lally also provided useful information on making the SI conversions. Mr. Robert Lorenz of the Chicago Regional Office, AISC, was also helpful in providing me with last-minute corrections to the preliminary specification changes.

Both Bethlehem and US Steel corporations were most helpful in providing copies of their new steel section profiles, nearly a year in advance of their becoming official. This allowed work to proceed early on computer generation of the *Structural Steel Design Data Manual* tables. Particular appreciation is due to Mr. Roland Graham of US Steel, who carefully reviewed selected portions of the manuscript and the entire steel data manual and made some very useful suggestions.

Grateful acknowledgment is also made of the very considerable contributions of Dr. Eugene Chesson, Civil Engineering Department, University of Delaware, who carefully reviewed both the preliminary and final text manuscripts. Thanks are due Dr. T. V. Galambos, Civil Engineering Department, Washington University, St. Louis, who reviewed the load resistance factor design material.

I should also like to express appreciation to Mr. Gary Zika, construction engineer with Pittsburg-Des Moines Steel Company and a former student, who assisted me in obtaining a number of the photographs used in the text.

Finally thanks are due my wife, Faye, who helped type the manuscript, checked the figures and put them in order, and most important, tolerated me during this trying time.

Joseph E. Bowles

STRUCTURAL STEEL DESIGN

Figure I-1 The Eads bridge across the Mississippi River at St. Louis, Missouri. This railroad and highway bridge completed in 1874 represents one of the first uses of steel (and high-strength $F_y = 50$ to 55 ksi steel) in the United States for a major structure. The 192-m (630-ft) high St. Louis "gateway" arch, with an exterior skin of stainless steel, can be seen in the background.

CHAPTER
ONE

GENERAL DESIGN CONSIDERATIONS

1-1 TYPES OF STRUCTURES

The structural engineer will be concerned with the design of a variety of structures including, but not necessarily limited to, the following:

Bridges: for railroads, highways, and pedestrians.
Buildings: including rigid framed, simple connected frames, load-bearing wall, cable-stayed, and cantilevered. Numerous lateral bracing schemes, including trussed, staggered trussed, and rigid central core, may be considered or used. Buildings may be further classified as to occupancy or height as office, industrial, mill, high-rise, and so on.
Other structures: including power transmission towers, towers for radar and TV installations, telephone relay towers, water supply facilities, and transportation terminal facilities, including railroad, trucking, aviation, and marine.

In addition to the foregoing structures, the structural engineer is also engaged in the design of ships, airplanes, parts of various machines and other mechanical equipment, automobiles, and dams and other hydraulic structures, including water supply and waste disposal.

This text will focus primarily on structural design using metal, and in particular standard structural shapes as produced directly by the several steel producers or in a few cases use of members that are built up from steel plates and shapes and fabricated either by the steel producers or in local steel fabrication shops.

1-2 DESIGN PROCEDURES

Structural design involves application of engineering judgment to produce a structural system that will adequately satisfy the client/owner's needs. Next, this system is incorporated into a mathematical model to obtain the member forces. Since the mathematical model never accurately represents the real structure, engineering judgment is again required to assess the validity of the analysis so that adequate allowance can be made for uncertainties in both deformations and statics.

Based on material properties, structural function, environmental considerations and esthetics, geometrical modifications in the analysis model are made and the solution process iterated until a solution is obtained that produces a satisfactory balance among material selection, economics, client desires/financial ability, and various architectural considerations. Seldom, except possibly in the most elementary structure, will a unique solution be obtained—unique in the sense that two structural engineering firms would obtain exactly the same solution.

In structural engineering practice the designer will have available for possible use numerous structural materials, including steel, concrete, wood, and possibly plastics and/or other metals, such as aluminum and cast iron. Often the occupancy/use, type of structure, location, or other design parameter will dictate the structural material. In this text we will assume that the design has proceeded to the point where the structural form has been decided (i.e., as truss, girder, frame, dome, etc.) and the several possible alternative structural materials have all been eliminated in favor of using steel. We will then proceed with any additional analysis required, and make the member selection and connection design appropriate to the topic being studied.

Textbook space and classroom time limitations will of necessity reduce to the bare essentials the complexity of the design presentations. The reader should be aware that real design is considerably more complex, even with experience, than the simplifications presented in the following chapters.

Safety as a design concern takes precedence over all other design considerations. The "safety" of any structure, of course, depends on the subsequent loadings. Since the structure is always loaded after it is built and not always in the mode or manner used in the design, the selection of design loads is a problem in statistics and probability. This part of the problem would become rather subjective and produce extremely divergent designs had not building codes been developed (and in some form or another, almost universally used) which place minimum required/suggested bounds for loads where public safety is an important factor.

1-3 STEEL AS A STRUCTURAL MATERIAL

Steel is one of the most important structural materials. Properties of particular importance in structural usage are high strength, compared to any other available material, and ductility. *Ductility* is the ability to deform substantially in either tension or compression before failure. Other important considerations in the use of steel include widespread availability and durability, particularly with a modest amount of weather protection.

Steel is produced by refining iron ore and scrap metals together with appropriate fluxing agents, coke (for carbon), and oxygen in high-temperature furnaces to produce large masses of iron called "pigs" or "pig iron." The pig iron is further refined to remove excess carbon and other impurities and/or is alloyed with other metals, such as copper, nickel, chromium, manganese, molybdenum, phosphorus, silicon, sulfur, titanium, columbium, and vanadium, to produce the desired strength, ductility, welding, and corrosion-resistance characteristics.

The steel ingots obtained from this process are passed between two rolls revolving at the same speed in opposite directions to produce a semifinished, long, rectangular-shaped product called either a slab, bloom, or billet, depending on the cross-sectional area. From this point the product is sent to other rolling mills to produce the final section geometry, including structural shapes as well as bars, wire, strips, plates, and pipes. The process of rolling, in addition to producing the desired shape, tends to improve the material properties of toughness, strength, and malleability. From these rolling mills the structural shapes are shipped to steel fabricators or warehouses on order.

The steel fabricator works from the engineering or architectural drawings to produce shop detail drawings from which the required dimensions are obtained to shear, saw, or gas-cut the shapes to size and to accurately locate holes for drilling or punching. The original drawings also indicate the necessary surface finishes to cuts. In many cases the parts are assembled in the shop to determine if a proper fit has been obtained. The pieces are marked for ease of field identification and shipped in pieces or subassemblies to the job site for erection. The site erection is often done by the steel fabricator but may be done by a general contractor.

Some of the most important structural properties of steel are the following:

1. *Modulus of elasticity*, E. The typical range for all steels (relatively independent of yield strength) is 28 000 to 30 000 ksi or 193 000 to 207 000 MPa.[†] The value for design is commonly taken as 29 000 ksi or 200 000 MPa.
2. *Shear modulus*, G. The shear modulus of any elastic material is computed as

$$G = \frac{E}{2(1 + \mu)}$$

   where μ = Poisson's ratio taken as 0.3 for steel. Using $\mu = 0.3$ gives $G = 11\,000$ ksi or 77 000 MPa.
3. *Coefficient of expansion*, α. The coefficient of expansion may be taken as

$$\alpha = 11.25 \times 10^{-6} \text{ per °C}$$
$$\Delta L = \alpha(T_f - T_i)L \qquad \text{(ft or m depending on length } L\text{)}$$

[†] MPa, megapascal = 1×10^6 N/m².

Table 1-1 Structural steel shape data

Type	ASTM designation		F_y		$F_{ultimate}$		Plate and bar thickness		Group[a]
			ksi	MPa	ksi	MPa	in	mm	
Carbon	A-36		36	250	58–80	400–550	8	203	1 through 5
High-strength, low-alloy[b]	A-242		40	275	60	415	4 to 8	102 to 203	—
	A-440		42	290	63	435	1.5 to 4	38 to 102	4 and 5
			46	315	67	460	0.75 to 1.5	19 to 38	3
	A-441		50	345	70	485	0.75	19	1 and 2
High-strength, low-alloy, columbium–vanadium	A-572	grade 42	42	290	60	415	to 6	to 152.4	1 through 5
		grade 45	45	310	60	415	to 2	to 50.8	1 through 5
		grade 50	50	345	65	450	to 2	to 50.8	1 through 4
		grade 55	55	380	70	485	to 1.5	to 38.1	1, 2, 3, and 4 to 426 lb/ft (639 kg/m)
		grade 60	60	415	75	520	to 1.25	to 31.8	1 and 2
		grade 65	65	450	80	550	to 1.25	to 31.8	1
High strength, low-alloy	A-588	grade 42	42	290	63	435	5 to 8	127 to 203	—
		grade 46	46	315	67	460	4 to 5	102 to 127	—
		grade 50	50	345	70	485	to 4	to 102	1 through 5

[a] See steel section data tables (e.g., Table I-1 or V-1 of *Structural Steel Design Data* by J. E. Bowles).
[b] A-440 steel may be difficult to obtain because it is not weldable.

GENERAL DESIGN CONSIDERATIONS 7

In these equations the temperature is in degrees Celsius. To convert from Farenheit to Celsius, use

$$C = \tfrac{5}{9}(F - 32)$$

4. *Yield point and ultimate strength.* Table 1-1 gives the yield points of the several grades of steel of interest to the structural designer that are produced by the steel mills.
5. *Other properties of some interest.* These properties include the mass density of steel, which is 490 pcf or 7.850 t/m^3 (1 t = 1000 kg); or in terms of unit weight, the value for steel is 490 pcf or 76.975 kN/m^3. The specific gravity of steel is generally accepted as 7.85. The conversion of fps units of lb/ft to SI units of kN/m and kg/m is accomplished as follows.
   Given: lb/ft and required to convert to:

$$\text{kg/m:} \quad \frac{\text{lb}}{\text{ft}} \times 0.4535924 \frac{\text{kg}}{\text{lb}} \times 3.2808 \frac{\text{ft}}{\text{m}} = 1.488164 \frac{\text{lb}}{\text{ft}} = \text{kg/m}$$

$$\text{kN/m:} \quad \frac{\text{lb}}{\text{ft}} \times 0.4535924 \frac{\text{kg}}{\text{lb}} \times 3.2808 \frac{\text{ft}}{\text{m}} \times 0.009806650 \frac{\text{kN}}{\text{kg}}$$

$$= 0.0145939 \frac{\text{lb}}{\text{ft}} = \text{kN/m}$$

Note that lb mass and lb weight or force have been used interchangeably in the fps system because the acceleration-producing force is that of gravity. This cannot be done in the SI system, since the newton is a derived unit that defines the force necessary to accelerate a 1-kg mass 1 m/s^2. The acceleration due to gravity is approximately 9.807 m/s^2.

Example: Given: A rolled structural shape weighs 300 lb/ft (the largest, W36).
Required: mass/m and weight/m.
Solution: mass/m = kg/m = 1.488164(300) = 446.4 kg/m
weight/m = kN/m = 0.0145939(300) = 4.38 kN/m

The *mass* value would be used in making material cost estimates and ordering quantities from the steel mills. The *weight* value is used in computing loads, bending moments, and stresses.

1-4 STEEL PRODUCTS

Steel ingots from the refining of pig iron are rolled into plates of varying widths and thickness; several structural shapes: round, square, and rectangular bars; and pipes. Most of the rolling is done on hot steel, with the product termed "hot-rolled steel." Sometimes the thinner plates are further rolled or bent, after cooling, into cold-rolled or "cold-formed" steel products. Several of the most common steel shapes are described in the following sections.

1-4.1 W Shapes

The most commonly used structural shape is the wide-flange or W shape. This is a doubly symmetrical (symmetrical about both the x and y axes) shape consisting of two rectangular-shaped flanges connected by a rectangular web plate. The flange faces are essentially parallel with the inner flange distance for most of the groups, with a constant dimension.† There is some variation due to roll wear and other factors, but the distance is held constant within ASTM tolerances. The shape is produced as illustrated in Fig. 1-1.

The designation: W16 × 40 means a nominal overall section depth of 16 in with a weight of 40 lb/ft.

The designation: W410 × 59.5 is the same W16 as above with a nominal depth in mm (based on the approximate average depths of all the W16 sections and rounded to the nearest 5 mm) and with a mass of 59.5 kg/m.

Prior to 1978, at least one W section in a group designation was "exactly" the nominal depth given (i.e., one W16 was 16.00 in deep; one W18 was 18.00 in deep). Now the closest W16 is the W16 × 40, with a designated depth of 16.01 in. There can be substantial deviations between the nominal and actual depth (e.g., the W21 ranges from 20.66 to 22.06 in). For the W14 the SI equivalent is W360, but the actual range is 349 to 570 mm (in this case the "average" was too far from the nominal value and W360 was somewhat arbitrarily used).

It should be noted that the rolled product will contract on cooling and at a variable rate depending on the thickness at any point on the cross section. The rolls used to produce the shapes will undergo wear, and coupled with the enormous forces involved in the rolling process, only shapes of nominal dimension (varying from theoretical or design values) can be produced. American Society for Testing and Materials (ASTM) specification A-6 in Part 4 gives allowable rolling tolerances, including amount of flange and web warping and deviation of web depth permitted for the section to be satisfactory. Generally, the maximum permissible variation in depth as measured in the plane of the web is $\pm\frac{1}{8}$ in or 3 mm. Note, however, that the permissible difference in depth between two rolled beams with a theoretical depth of 16.01 can produce extreme depths of 15.885 to 16.135 in or a difference of $\frac{1}{4}$ in or 6 mm. These variations should be kept in mind, particularly when converting to SI dimensions for detailing, clearances, and mating of parts.

1-4.2 S Shapes

These are doubly symmetric shapes produced in accordance with dimensions adopted in 1896 and were formerly called I beams and American Standard

---

† The several sections with a constant nominal depth. Where a group consists of a large number of sections, a second inner flange distance may be used.

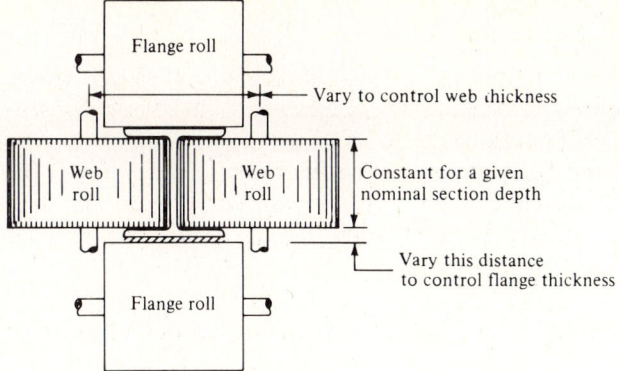

Figure 1-1 Method of rolling a steel ingot to produce a W shape. Note that varying of the rolls and roll configuration produces S, C, and angle shapes.

beams. There are three essential differences between S and W shapes:

1. The flange width of the S shape is less.
2. The inner face of the flange has a slope of approximately 16.7°.
3. The theoretical depth is the same as the nominal depth. An S510 × 111.6 is a shape of nominal depth 510 mm × 111.6 kg/m (S20 × 75).

1-4.3 M Shapes

These are doubly symmetrical shapes which are not classified as W or S shapes. There are about 20 lightweight shapes classified as M. An M360 × 25.6 is the largest M shape and is a section of nominal 360 mm depth with a mass of 25.6 kg/m (M14 × 17.2).

1-4.4 C Shapes

These are channel shapes produced in accordance with dimensional standards adopted in 1896. The inner flange slope is the same as for S shapes. These channels were formerly called Standard or American Standard Channels. The theoretical and nominal depths are identical (also for MC shapes following).

A C150 × 19.3 is a standard channel shape with a nominal depth of 150 mm and a mass of 19.3 kg/m (C6 × 13).

1-4.5 MC Shapes

These are channel shapes which are not classified as C shapes. These were formerly called Shipbuilding or Miscellaneous Channels.

10 STRUCTURAL STEEL DESIGN

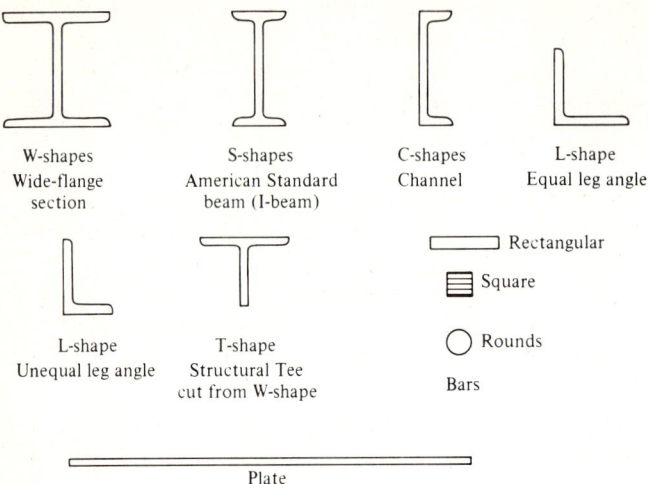

Figure 1-2 Structural shapes as directly produced by the steel producers.

1-4.6 L Shapes

These shapes are either equal or unequal leg angles. All angles have parallel flange faces. Angle leg dimensions can vary on the order of ± 1 mm in width.

An L6 × 6 × $\frac{3}{4}$ is an equal leg angle with nominal dimensions of 6 in and a thickness of $\frac{3}{4}$ in.

An L89 × 76 × 12.7 is an unequal leg angle with leg dimensions of 89 and 76 mm, respectively, and a leg thickness of 12.7 mm (L$3\frac{1}{2}$ × 3 × $\frac{1}{2}$).

1-4.7 T Shapes

Structural tees are structural members obtained by splitting W (for WT), S (for ST), or M (for MT) shapes. Generally, the splitting is such to produce a shape with one-half the area of the parent section, but offset splitting may be used if a deeper tee section is required. Published tables of T shapes are based on symmetrical splitting. No allowance is made for material loss from splitting the parent shape by sawing or flame cutting.

A WT205 × 29.8 is a structural tee with a nominal depth of 205 mm and mass of 29.8 kg/m and is obtained by splitting the W410 × 59.5 section (from W16 × 40).

Several rolled structural shapes are illustrated in Fig 1-2.

1-5 STEEL STRENGTH

All steel design takes into consideration the yield strength of the material. The yield strength of several grades of steel available for design is given in Table 1-1. The *yield strength* is that minimum value guaranteed by the steel producers and is based on statistical averaging and consideration of the minimum value of yield obtained from a large number of tests. Thus for A-36 steel the guaranteed F_y = 36 ksi or 250 MPa, but the most likely value is on the order of 43 to 48 ksi (300 to 330 MPa and refer to Figs 1-3a and 1-3b. Similarly, A-441 steel, with a yield point of 345 MPa, will have a yield strength on the order of 400 MPa. As the guaranteed yield increases to around 450 MPa (65 ksi), the actual and guaranteed values converge.

From about 1900 to 1960, the principal steel available in structural shapes was a grade called A-7 with F_y = 33 ksi. A grade A-373 for welding with F_y = 32 ksi was introduced in 1954 as welding became more popular following the shipbuilding activities of World War II. Where older buildings are being renovated, the structural engineer may be concerned with incorporating newer steels with those older grades.

Since 1960 both A-373 and A-7 grades of steel have been replaced by A-36 steel, which represented a 10 percent increase in yield strength over the A-7 grade. In the 1930s a high-strength corrosion-resistant steel began to be produced and designated A-272 (described in ASTM specification A-272). Specification ASTM A-440 was written in 1959 for another high strength steel with applicability to riveted and bolted construction, and A-441 steel was introduced in 1960 with application to welding. All three of these steels have a yield point that is dependent on the thickness of the metal, as shown in Table 1-1.

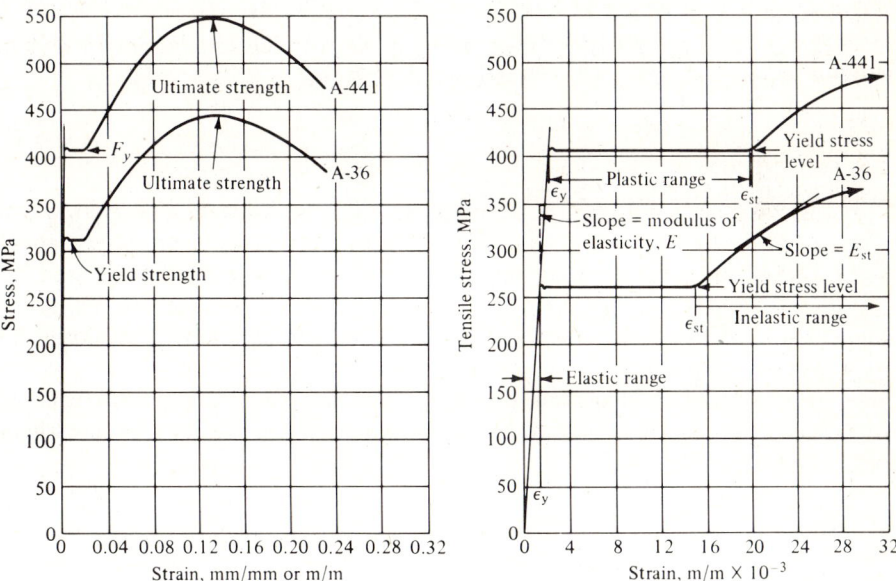

Figure 1-3a Typical stress–strain curves for structural steel.

Figure 1-3b Enlargement of initial part of stress–strain curve for two grades of structural steel. Note that the plastic range is on the order of $10\epsilon_y$.

Since about 1964, specifications for several other high-strength (low-alloy) steels have been incorporated into ASTM specifications as A-572 and A-588. Table 1-1 shows that the steel covered by the A-572 specification covers several yield strengths, termed grades, such as grades 42, 45, 50, 55, 60, and 65 for the corresponding guaranteed minimum yield stress in ksi. Generally, the yield strengths of these newer steels are also thickness-dependent, as shown in the table under the heading "plates and bar thickness." The steel producers have designated the several W shapes into five groups, depending on flange thickness (and as shown in Tables I-1, I-2, V-1, and V-2)†, compatible with the steel grade. The designer merely has to check these tables to see if the shape is available in the required/desired yield-stress grade. For example, in the 450-MPa grade, only shapes in group 1 qualify from flange thickness. W18 shapes are available in group 1 only from 35 to 60 lb/ft inclusive (the five smallest sections and with a maximum part thickness of 0.695 in).

Specification ASTM A-588 allows $F_y = 345$ MPa for a high-strength low-alloy steel which may be up to 100 mm (4 in) thick. The steel covered in this specification is primarily for welding and is corrosion-resistant.

In terms of cost/unit of mass, the A-36 steel is most economical. High-strength steels have principal application where the stresses are primarily tensile. High-strength steel beams may deflect excessively, owing to reduced section modulus. The high-strength steel columns may be less economical than A-36 steel if the slenderness ratio (KL/r) is large. Hybrid girders that use high-strength steel in the flanges, or built-up columns using high strength steels, may provide better solutions where member sizes are restricted. In a given case it is necessary to perform an economic and availability analysis to determine the suitability of using high-strength steel.

1-6 TEMPERATURE EFFECTS ON STEEL

1-6.1 High-Temperature Effects

Steel is not a flammable material; however, the strength is heavily temperature-dependent, as illustrated in Fig. 1-4. Both the yield and tensile strength at 1000°F is about 60 to 70 percent of that at room (about 70°F) temperature. The drop in strength is rather marked at higher temperatures, as shown on the figure, where the strength at 1600°F is only about 15 percent of that at room temperature.

Steel frames enclosing materials that are flammable will require fire protection to control the temperature of the metal for a sufficient time for the occupants to seek safety or for the fire to either consume the flammables or be extinguished before the building collapses. In many cases the building does not

---

† See footnote *a* of Table I-1 in J. E. Bowles, *Structural Steel Design Data Manual*, McGraw-Hill, New York, 1980.

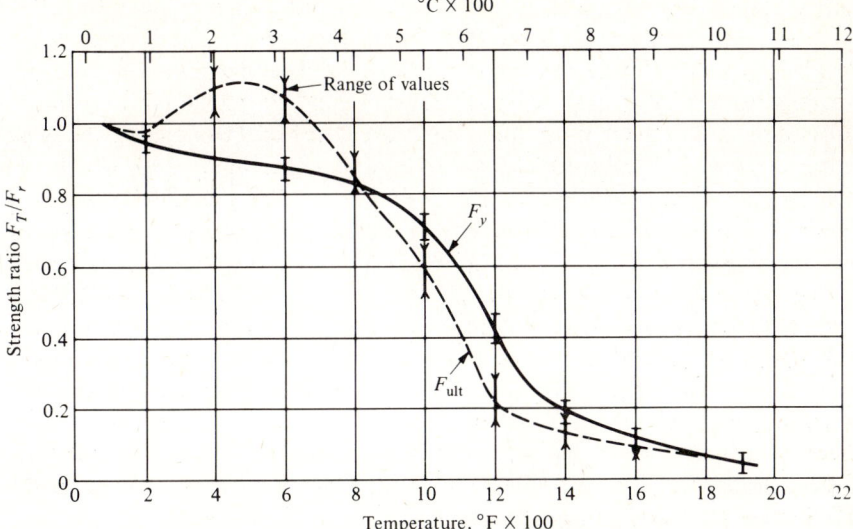

Figure 1-4 Effect of elevated temperatures on either yield or ultimate tensile strength expressed as a ratio of strength at room temperature of approximately 70°F.

collapse, even at high temperatures; rather, the deformations of the overheated members is too large to be acceptable and those members must be removed.

Fire protection ratings are established for the several materials and thickness that may be applied to a structural member to control the temperature. These include gypsum-based products or lightweight concrete that can be sprayed onto the member (see Fig. 1-5) or fiber insulation boards that are placed and banded to protect the steel. Where the steel member is hollow, a liquid with an antifreeze agent may be provided for high-temperature control. The fire rating is based on the number of hours it takes the steel to reach an average temperature of 540 to 650°C for the given thickness of fireproofing material using a standard test procedure and flame as given by ASTM E-119 (in Part 18). A 2-h rating, commonly used, designates that it takes 2 h for the steel temperature to reach the designated level in the standard test. Insurance rates are based on the fire rating in hours and are less for the higher time fire ratings because of the reduced likelihood of critical fire damage.

1-6.2 Low-Temperature Effects

Brittle fracture is a failure often associated with low temperatures. Essentially, brittle fracture is failure that takes place without material yielding. The stress–strain curves of Fig. 1-3 indicate that in the usual failure of a tensile specimen, considerable elongation takes place. As a matter of fact, a minimum percent elongation is specified for steel in the ASTM standard tensile test. Implicit in steel design is the resultant deformation (yielding) of the material

14 STRUCTURAL STEEL DESIGN

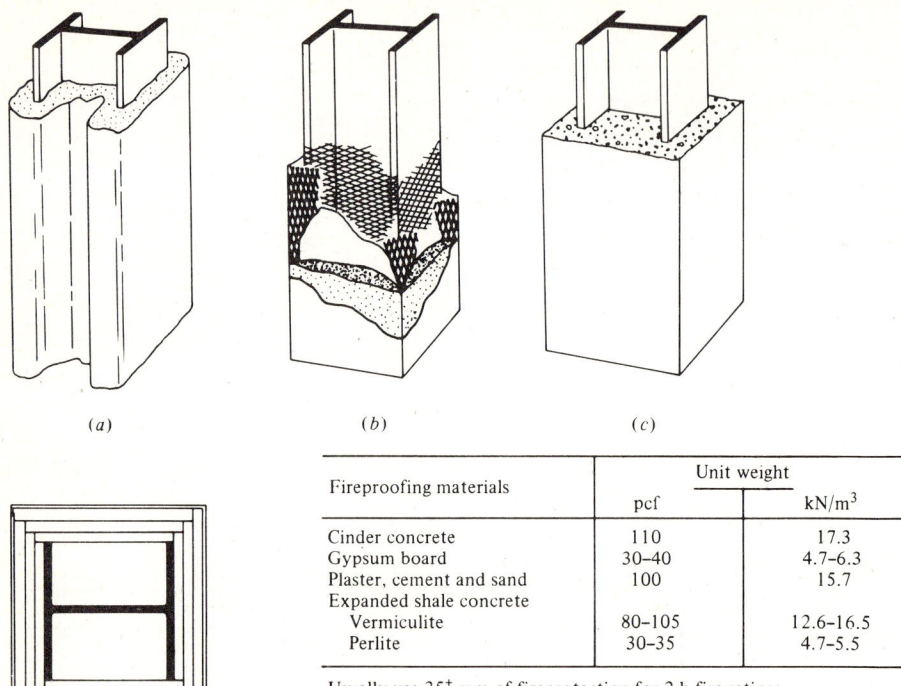

Fireproofing materials	Unit weight	
	pcf	kN/m³
Cinder concrete	110	17.3
Gypsum board	30–40	4.7–6.3
Plaster, cement and sand	100	15.7
Expanded shale concrete		
Vermiculite	80–105	12.6–16.5
Perlite	30–35	4.7–5.5

Usually use 35⁺ mm of fireprotection for 2-h fire rating; obtain specific thickness values from either tests or from the producers of gypsum, perlite, etc.

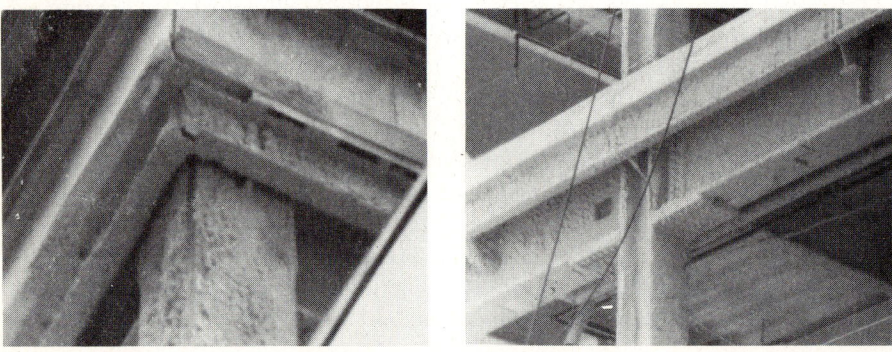

Figure 1-5 Methods of producing fireproofing of structural steel members. (*a*) Sprayed fiber. (*b*) Lath and plaster. (*c*) Lightweight concrete (formed). (*d*) Gypsum board—use boards to build thickness. (*e*) Corner detail of sprayed on fireproofing. Thickness is built in several sprayings. (*f*) Beam detail of sprayed fireproofing.

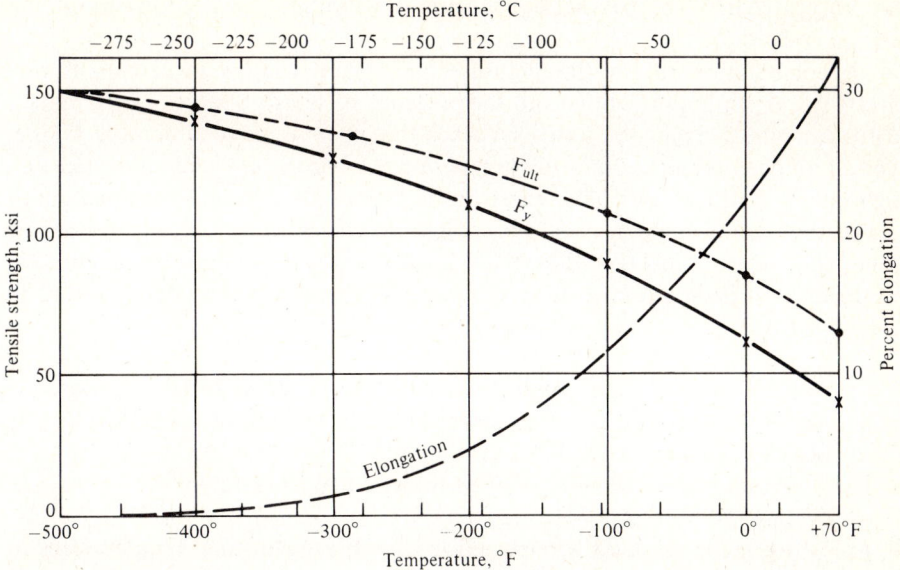

Figure 1-6 Effect of low temperature on yield and ultimate tensile strength and on elongation of a specimen in tension.

under high local stress. When the material elongates, the lateral dimensions contract, owing to the Poisson ratio effect. If the lateral dimensions are fully (or even partially) restrained, the steel will pull apart without fully developing the yield potential. This type of failure is a "brittle fracture." Figure 1-6 illustrates that the elongation decreases in any case with a lowered temperature. Figure 1-7 illustrates the stress concentration effect that occurs at a notch—a notch being

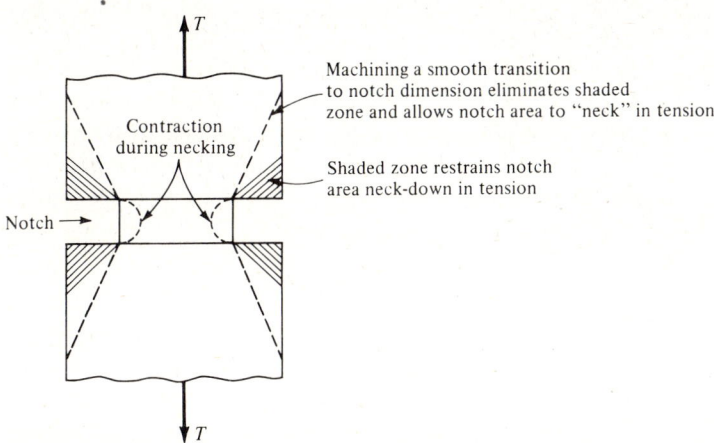

Figure 1-7 A notched bar in tension. The notched section attempts to contract under tension, as shown by dashed semicircles, and is resisted by the shaded zones, which is equivalent to applying a lateral force.

16 STRUCTURAL STEEL DESIGN

any abrupt change in cross-sectional area—to inhibit lateral contraction in a tension situation.

A combination of low temperature, an abrupt change in section (notch effect) size or an imperfection, and the presence of tensile stresses can initiate a brittle fracture. This *may* initiate as a crack that propagates to a member failure. Not all notched members in a low-temperature environment and subjected to a high rate of tensile strains fail—there has to be exactly the right combination of strain, strain rate, temperature, and notch effects. However, if a catastrophic failure does occur, the fact that a low probability of the failure combination occurring is of little aid in settling the resulting damage claims that are sure to follow. Brittle fracture can be controlled in several ways:

1. Detail members and their connections to minimize stress concentrations.
2. Specify the fabrication and assembly sequence to minimize residual tensile stresses.
3. Use steels that are especially alloyed for low-temperature environments (see ASTM A-633).
4. Apply a pretension stress at elevated (preferably) or normal temperatures to the notched zone to cause a local yield before the service loads and temperatures are encountered.
5. If possible, machine (or grind) the notch into a smooth transition.
6. Reduce the rate of tensile strain application.

1-7 STRUCTURAL DESIGN CODES

Local building departments almost always require structural designs under their jurisdiction to be designed in accordance with some code. The larger cities—New York, Atlanta, Chicago, San Francisco, and others—have their own codes (e.g., Chicago Building Code, New York Building Code, etc.). Smaller cities generally use one of the several national or regional codes listed later.

The various state departments of transportation (highway departments) generally use the specifications put forth by AASHTO, also listed below. The several railroads generally use specifications put forth by AREA, listed below. The structural designer doing highway or railroad work will have to follow closely the design specifications of these publications, particularly if the federal government is involved with any of the financing.

Similarly, the designer doing structural engineering will have to follow closely the design requirements in the appropriate building code/specifications of the local governing body as a *minimum* requirement. Special requirements of the owner/client may require a more stringent design than the building code criteria. Only in rare cases can the designer get a variance from the local governing body to deviate in a less conservative manner from the code. Variances usually require extensive documentation and bringing in additional consultants to gain approval. There are both good and bad features associated with this aspect of structural design. On the one hand, it sometimes takes considerable time for approval of new materials and methods; on the other hand, there is some advantage in not "going too fast." If the local building code is carefully followed and minimum design requirements met, or exceeded, and a catastrophe results, proof exists that good engineering practice has been followed. Finally, local building codes are supposed to reflect that part of the structural practice which is unique for that locale, such as temperatures, earthquakes, snow and rain quantity, frost depth, and average wind velocities.

The following list gives several design codes and/or specifications which the structural designer may have occasion to use:

National Building Code, published by, and available from, the American Insurance Association, 85 John Street, New York, N.Y. 10038.

Uniform Building Code by International Conference of Building Officials, 5360 South Workman Mill Road, Whittier, California 90601.

Basic Building Code, Building Officials and Code Administrators International, 1313 East 60th Street, Chicago, Illinois 60637 (formerly BOCA).

American National Standards Institute (ANSI), *Minimum Design Loads in Buildings and Other Structures*, ANSI 58-1, 1430 Broadway, New York, N.Y. 10018.

American Institute of Steel Construction (Specifications), *Steel Construction Manual*, 8th ed. (1979), 101 Park Avenue, New York, N.Y. 10017.

American Welding Society (AWS), *Structural Welding Code*, 2501 N.W., 7th Street, Miami, Florida 33125.

American Iron and Steel Institute (AISI), 1000 Sixteenth Street, Washington, D.C. 20036. Publishes various specifications for using iron and steel products.

American Association of State Highway and Transportation Officials (AASHTO), *Specifications for Highway Bridges*, 341 National Press Building, Washington, D.C. 20004.

American Railway Engineering Association (AREA), *Specifications for Steel Railway Bridges*, 59 East Van Buren Street, Chicago, Illinois 60605.

All the national and city codes use specification standards as applicable from specifications writing organizations, such as AWS, AISC, and AISI, for metals and other appropriate agencies for the other construction materials.

Organizations such as AISI and AISC, as well as the steel producers, provide tables of structural shape design data as well as data on other steel products, such as plates, bars, wire, and bolts. Certain of these data together with edited specifications from AISC, AASHTO, and AREA, have been prepared by the author in the *Structural Steel Design Data* (SSDD) manual published by McGraw-Hill Book Company and used as a supplement with this text. This manual was developed so that both fps and SI units would be available to the reader.

1-8 BUILDING LOADS

In the design of a building, it is necessary to determine the loads the building skeleton, or frame, must carry. These loads are considered as *dead* and *live loads*. This distinction is made because the frame and foundation must always carry the dead loads, or loads due to self-weight of the various building components and including the weight of the frame. Typically dead loads include:

Floor and roof materials, including floor and roof joists (beams).
Ceiling materials, including duct work for environmental control and power supplies.
Exterior walls supported by the frame, including windows, doors, and balconies.
Interior walls that are permanently placed.
Mechanical equipment (heating, air conditioning, ventilation, and electrical (such as elevators, including cage, cables, motors).
Fireproofing.
Beams, girders, and columns, including the footings making up the building frame.

From this list it is evident that any part of the building which is permanently installed contributes to the total dead load. Dead loads can be reasonably well established after the building elements are dimensioned, using the shape geometry and tables of unit weights of building materials, such as Table IV-3.†

Live loads are any loads that the building component must carry and is considered as somewhat transient. That is, the load may be applied for several hours to several years, but is variable in magnitude. Live loads tend to be prescribed by building codes based on occupancy and location, where wind, snow, and earthquake loads are considered. In addition to these latter loads, live loads include:

People, as in auditoriums, assembly halls, and classrooms.
Movable room partitions.
Office equipment and production machines if they are moved on occasion.
Warehouse products.
Furniture.

Building code values of live loads tend to be based on statistically conservative values, and while the designer may petition the local building department for a code variance, this should be done only after carefully assessing the proposed values. Table IV-4 gives typical values of live load based on occupancy as found in several building codes. Loads are given in force/area as a design/computational convenience and because the actual building plan geometry is not known.

† In *Structural Steel Design Data* (SSDD). Any tables referred to hereinafter will be from this manual unless it is evident that they are in the text.

Most building codes allow a reduction in the tabulated live loads when the contributing area is very large, because of the statistical probability of the entire area not being maximum loaded. Commonly,† the reduction is the lesser of the following factors:

fps: $R = 0.0008 \times \text{area}$ (when area $> 150 \text{ ft}^2$)

SI: $R = 0.0086 \times \text{area}$ (area $> 11.2 \text{ m}^2$)

$$R = \frac{D + L}{4.33 L} \tag{1-1}$$

$R \leq 0.60$ (some codes limit $R \leq 0.40$ for horizontal members such as beams and girders and $R \leq 0.60$ for vertical members)

where $R =$ reduction factor used as $(1 - R) \times L_{\text{table}}$
$D =$ dead load, psf or kPa (kilonewtons/m)
$L =$ live load, psf or kPa, but L is limited to not over 100 psf or 5 kPa; generally, values larger than this are not reduced

No live-load reductions are allowed for the following occupancies: places of public assembly (such as auditoriums), garages, and roofs.

It is necessary to determine the framing so that the load flow through the building can be made to obtain the area contributing load to any member.

Example 1-1 A portion of an office (multistory) floor plan is shown in Fig. E1-1. The floor is 4-in concrete on a metal deck over steel bar joists. What is the reduced live load for the floor beams and for an exterior column three floors down from the top floor?

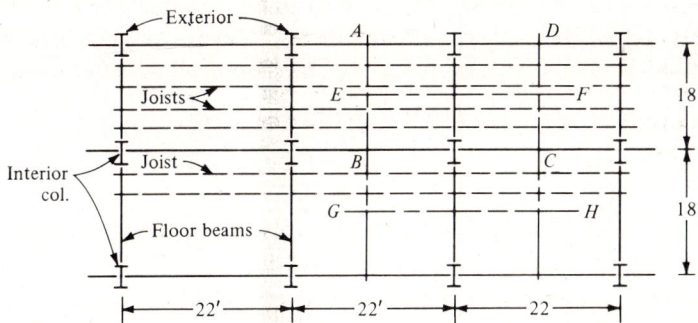

Figure E1-1

SOLUTION Estimate the dead load on the contributory (centered on member) floor area as:

Concrete floor and finish: $4 \times 144/12 = 48$ psf

Ceiling, metal deck, steel bar joists $= 12$ psf

Total $= 60$ psf

† As given in the several national building codes cited earlier.

20 STRUCTURAL STEEL DESIGN

The live load for offices, obtained from Table IV-4 (SSDD) = 80 psf. Compute R for the floor beam; the contributory area is, as shown in Fig. E1-1, 18×22 (area $ABCD$):

$$R = 0.0008(18 \times 22) = 0.32 < 0.60$$

$$R = \frac{D + L}{4.33L} = \frac{60 + 80}{4.33(80)} = 0.40 < 0.60$$

Use the smaller value of R computed, 0.32. The reduced live load L' is

$$L' = (1 - 0.32)80 = 54.4 \quad \text{say 55 psf}$$

Compute R for the column; the contributory area is centered on the column of 9×22 ($AEFD$); but for the accumulation of three stories, we have

$$R = 0.0008(3 \times 9 \times 22) = 0.475 < 0.60 \quad \text{O.K.}$$

$$R = \frac{60 + 80}{4.33(80)} = 0.40 < 0.60$$

Using the smaller value of R, 0.40, the reduced live load on the column is

$$L' = (1 - 0.4) \times 80 = 48 \text{ psf}$$

We note that 0.40 is the maximum R for the column and is much less at the uppermost floor level. ///

Example 1-2 A meeting/banquet room in a hotel has a floor plan of 22×27 m. The floor is 125 mm of concrete with a tile surfacing on a metal deck placed on steel bar joists (refer to Fig. E1-2). What is the reduced live load on floor beam C2 using not more than a 40 percent reduction?

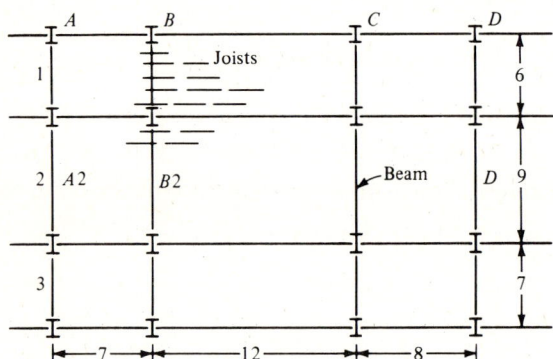

Figure E1-2

SOLUTION Note that a public room is not the same as an assembly hall, where the loading is primarily seating in fixed or movable seats. With this definition we may use a live-load reduction factor.

First, estimate the dead load using Table IV-3 of SSDD to obtain unit weight of concrete = 23.5 kN/m³.

Weight of concrete: 0.125×23.5	$= 2.940$ kPa
Tile at 10 mm	$= 0.188$ kPa
Steel deck, ceiling, joists (estimated pressure)	$= 0.575$ kPa
Total	$= 3.703$ kPa

From Table IV-4 of SSDD, the live load = 5.00 kPa. The reduction factor for a girder based on a contributory area as shown is

$$R = 0.0086[(12 + 8)/2 \times 9] = 0.774 > 0.40 \text{ (and also 0.60)}$$

$$R = \frac{D + L}{4.33 L} = \frac{3.703 + 5.00}{4.33(5.00)} = 0.402 < 0.60$$

Since the problem statement limits live-load reduction to not more than 0.40, the reduced live load is

$$L' = (1 - 0.4) \times 5.00 = 3.00 \text{ kPa} \qquad ///$$

1-8.1 Wind Loads

Wind loads have been extensively studied in recent years, particularly for larger high-rise structures. Generally, for tall structures wind-tunnel studies should be made to determine the wind forces on the structure. For smaller regular-shaped structures with heights on the order of 100 ft or about 30 m, the wind pressure stipulated in the appropriate building code is satisfactory to use. The National Building Code (NBC) for wind is as follows:

Height		Wind pressure	
ft	m	fps, psf	SI, kPa
< 30	<9	15	0.75
30 to 49	9.1 to 14.9	20	1.0
50 to 99	15 to 30	25	1.25
100 to 499	30.1 to 150	30	1.50

Tanks, chimneys, and similar structures must withstand this pressure times the following factor:

Shape of structure	Factor
Rectangular or square	1.00
Hexagonal or octagonal	0.80
Round or elliptical	0.60

No allowance is commonly made for the shielding effect of adjacent structures or from ground cover.

The building height is measured above the average ground level at the building base to obtain the NBC wind pressure.

The wind pressure is commonly computed between floor levels and prorated to the adjacent floors using simple beam theory if the vertical distance compared

22 STRUCTURAL STEEL DESIGN

to the lateral distance to adjacent columns is small enough that "one-way" bending action can be assumed. This assumption does not produce a bending moment due to the wind pressure along the column. When the column spacing is small compared to height, one may apply the wind pressure as a uniform loading on the column and producing a wind moment of $wL^2/8$ (if considered as a simple beam).

Aerodynamic effects are considered when wind is applied to buildings with sloping roofs. For roof slopes less than 30° (which includes flat roofs) the NBC roof wind pressure is a suction acting *outward* normal to the roof surface, with a value of 1.25 × pressure, previously given.

The wind load on roofs with slopes greater than 30° is obtained from the wind pressure acting normal to the roof surface, with the basic value previously given depending on the height. The height to determine the roof pressure is measured as the difference between average ground and average roof elevation.

We note that since wind can come from either side of a building, the frame is symmetrical even though the wind analysis is only from one direction.

Example 1-3 Given the gable roof shown in Fig. E1-3a and using NBC wind pressure, what are the lateral and roof forces due to wind from the left?

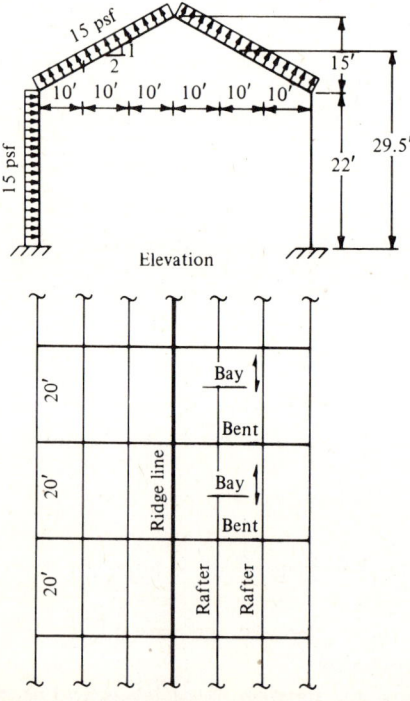

Figure E1-3a

GENERAL DESIGN CONSIDERATIONS 23

SOLUTION With a 22-ft column we will assume column bending, but even so this will result in eave and foundation shears as shown in Fig. E1-3b.

Average height = 22 + 7.5 = 29.5 ft

Wind pressure for this height = 15 psf = 0.015 ksf

Wind on the windward column = $P = \left(\dfrac{22}{2}\right)(20)(0.015)$

= 3.3 kips at roof line and into foundation

The roof wind is applied to the rafter points as shown. The slope distance is readily computed as 33.54 ft. Dividing into six parts, obtain 5.59 ft.

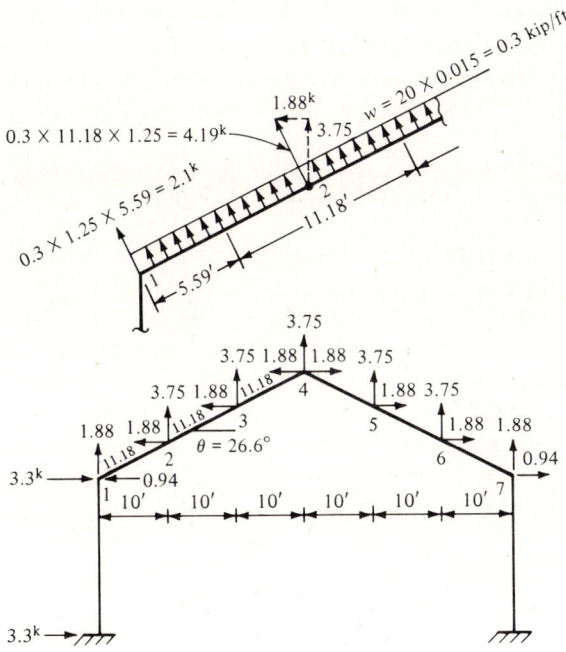

Figure E1-3b

At point 1 the wind is 5.59(20)(0.015 × 1.25) = 2.097 kips (roof slope < 30°). This force is broken into horizontal and vertical components of

$$\dfrac{1}{\sqrt{5}}(2.097) = 0.94 \text{ kip}$$

$$\dfrac{2}{\sqrt{5}}(2.097) = 1.88 \text{ kips}$$

At points 2 and 3 the contributory wind area is 2(5.59) = 11.18 ft. The wind components are 3.75 and 1.88 kips.

At point 4 the wind is also 3.75, but the horizontal components cancel.

The several wind values are shown in Fig. E1-3b. The data display is convenient for computer programming for frame stresses using the computer program discussed in Chap. 2. ///

Wind pressures can be approximately computed as

$$q = 0.00256 V^2 \quad \text{(psf)} \tag{1-2}$$

$$q = 0.0000473 V^2 \quad \text{(kPa)} \tag{1-2a}$$

where V is mi/h or km/h. This equation is readily derived as $q = \frac{1}{2}mv^2$, where the mass density of air is approximately 0.00238 lb · s^2/ft^4.

Since wind is a transient load, the building codes usually allow a one-third increase in the allowable design stresses with wind included as a part of the load condition as long as the required section is not less than required in the load condition of dead + live loads alone. For example, if a stress of 20 ksi is allowed, then with the wind load condition a stress of 20 × 1.33 = 26.6 ksi could be used.

1-8.2 Snow Loads

Snow loads are live loads acting on roofs. Snow and any other live loads are taken with respect to the horizontal projection of the roof, as illustrated in Fig. 1-8. Figure 1-9 is a map illustrating snow loads that may be used in the absence of specific load building code requirements. Even in areas where snow loads are minimal, a minimum roof live load should be used. The NBC stipulates the larger of the snow load or 20 psf or 1.0 kPa. Since 10 in of snow approximates 1 in of water, a 20-psf snow load corresponds to a roof snow depth of nearly 40 in —easily obtained where snow drifting occurs. When rain later falls on snow, however, the saturated snow weighs considerably more and the unit weight can approach that of water.

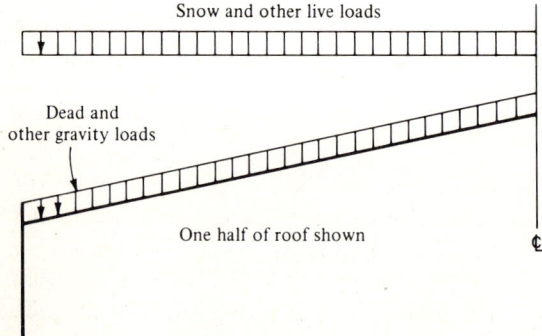

Figure 1-8 Snow and other roof live and dead loads.

GENERAL DESIGN CONSIDERATIONS **25**

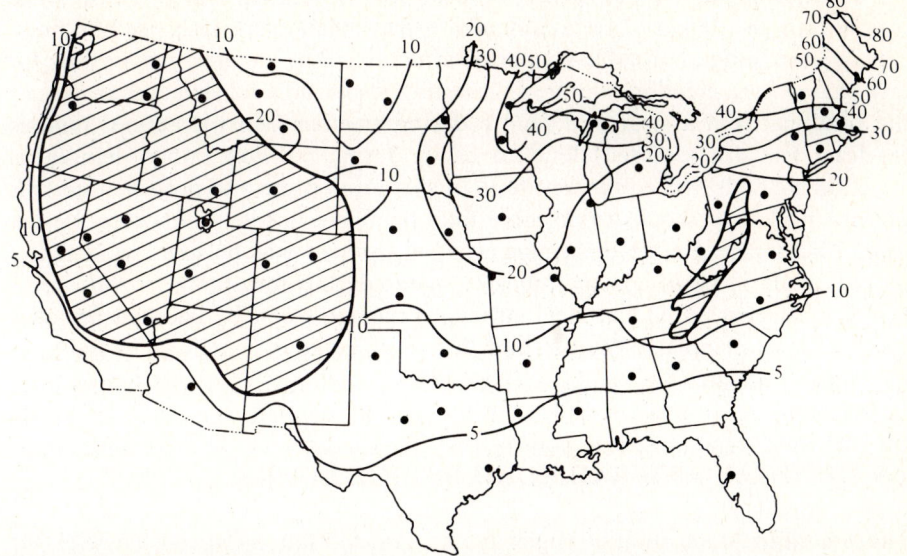

The ground snow loads as shown were derived from data reported by the U.S. Weather Bureau "Distribution of Maximum Annual Water Equivalent of Snow on the Ground" Monthly Weather Review. Vol. 94, No. 4, 1966.

The basic roof loads may be taken as 80 percent of the ground snow load.

Ground snow loads shown on this map are intended to be used in locations where there is no building code establishing snow load requirements, and are not intended to supersede state or local code requirements.

//// Special consideration shall be given to regions where no design loads are shown and where unusually high accumulations of snow may occur. The variation of ground snow loads with elevation and exposure is not yet completely understood and local differences in mountain regions are usually very significant.

Figure 1-9 Ground snow load in psf for 50-year recurrence interval. *(After Uniform Building Code, 1976).*

1-8.3 Other Building Loads

In addition to the types of pressure or area loads noted, building codes may stipulate checking for a concentrated load of some magnitude which may be placed anywhere on the floor or roof. Where roofs are used as recreational areas or sun decks, the live loads must be adjusted to values based on occupancy in addition to considering snow and/or wind.

Ponding is a special roof load that may require investigation. Ponding is a condition where water collects on a flat roof which has deflected locally (possibly due to an overload, poor construction, foundation settlement, or plugged roof drain), causing a concentration of water which in turn increases the load and deflection, causing a further concentration of water. Noting that a water depth of 1 in results in a live-load pressure of 5.2 psf, loads are readily developed which can locally fail roof members. Through progressive failure, the entire roof may collapse. Ponding design is considered in some detail by Marino in the July 1966 *AISC Engineering Journal.*

26 STRUCTURAL STEEL DESIGN

Erection loads are not directly considered in building codes. These loads may control the design of certain members, particularly very high rise buildings, cantilevered bridges, or cable-supported structures. The engineer responsible for any phase of the erection may be held legally responsible for damages or loss of life resulting from a structural failure during erection. Most structural failures (at least that are reported) tend to occur during erection rather than later. Erection methods and equipment tend to vary from project to project; thus it is not practical in textbooks to do more than point out this very important design area. The engineer must determine what equipment will be used, where it is placed, loads to be lifted, quantities of material, and the storage locations, so that the affected individual steel structural members can be checked for adequacy using principles of statics and design procedures contained in the later chapters on design.

1-9 HIGHWAY AND RAILROAD BRIDGE LOADS

The American Association of Highway and Transportation Officials (AASHTO) has designated standard bridge loadings as shown in Fig. 1-10. Generally, the standard truck loading produces the largest bending-moment values for bridge spans up to about 60 ft for H and up to 145 ft for HS trucks. The bending moment can be readily computed by referring to Fig 1-10. The maximum shear is obtained with either a wheel or the concentrated lane load at a reaction. For the HS truck and span lengths, shears are as follows:

Bridge span	Effect due from HS truck[a]
33.8 to 145.6 ft	$M_{max} = \dfrac{W}{L}[(0.9L + 4.206)(0.5L + 2.33) - 11.2L]$ ft · kips
10.3 to 44.38 m	$M_{max} = \dfrac{W}{L}[(0.9L + 1.282)(0.5L + 0.71) - 3.41L]$ kN · m
28 to 127.5 ft	$V_{max} = W\left(\dfrac{L - 16.8}{L} + 0.8\right)$ kips (truck loading)
8.53 to 38.86 m	$V_{max} = W\left(\dfrac{L - 5.1}{L} + 0.8\right)$ kN (truck loading)
Over 127.5 ft	$V_{max} = V + \dfrac{wL}{2}$ kips (lane loading)
Over 38.86 m	$V_{max} = V + \dfrac{wL}{2}$ kN (lane loading)

[a] W = 40, 30, and 20 kips or equivalent in kN (and is the basic truck load, not the total).

Railroad bridges are designed based on live loads devised by Theodore Cooper (ca. 1890), a railroad engineer, designated "Cooper's loading." The load consists of two "typical" railroad locomotives pulling a string of freight cars.

GENERAL DESIGN CONSIDERATIONS 27

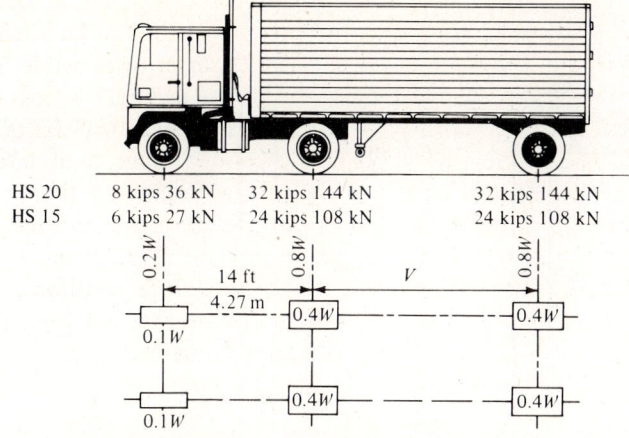

HS 20	8 kips 36 kN	32 kips 144 kN	32 kips 144 kN
HS 15	6 kips 27 kN	24 kips 108 kN	24 kips 108 kN

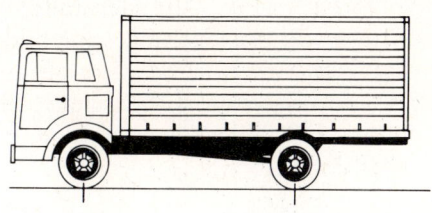

W = combined weight on the first two axles which is the same as for the corresponding H truck (4.27 to 9.14 m)
V = variable spacing — 14 to 30 ft inclusive spacing to be used is that which produces maximum stresses.

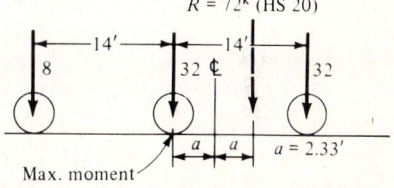

$R = 72^k$ (HS 20)

Max. moment

$a = 2.33'$

General location of maximum moment due to truck loadings is as shown (with HS 20 values)

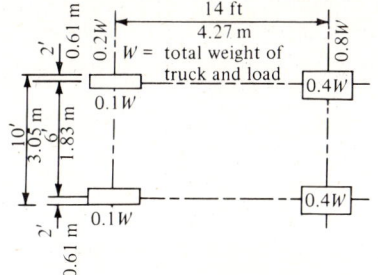

HS 20	8 kips 36 kN	32 kips 144 kN
HS 15	6 kips 27 kN	24 kips 108 kN
HS 10	4 kips 18 kN	16 kips 72 kN

W = total weight of truck and load

	W/length		Moment		Shear	
Truck	kip/ft	kN/m	kip	kN	kip	kN
HS 20	0.640	9.3	18	80	26	116
H 20	0.640	9.3	18	80	26	116
HS 15	0.480	7.0	13.5	60	19.5	87
H 15	0.480	7.0	13.5	60	19.5	87
H 10	0.320	4.7	9	40	13	58

V placed anywhere on span to produce maximum effect

W/length and covers full lane width

Figure 1-10 Standard AASHTO truck loadings for bridges.

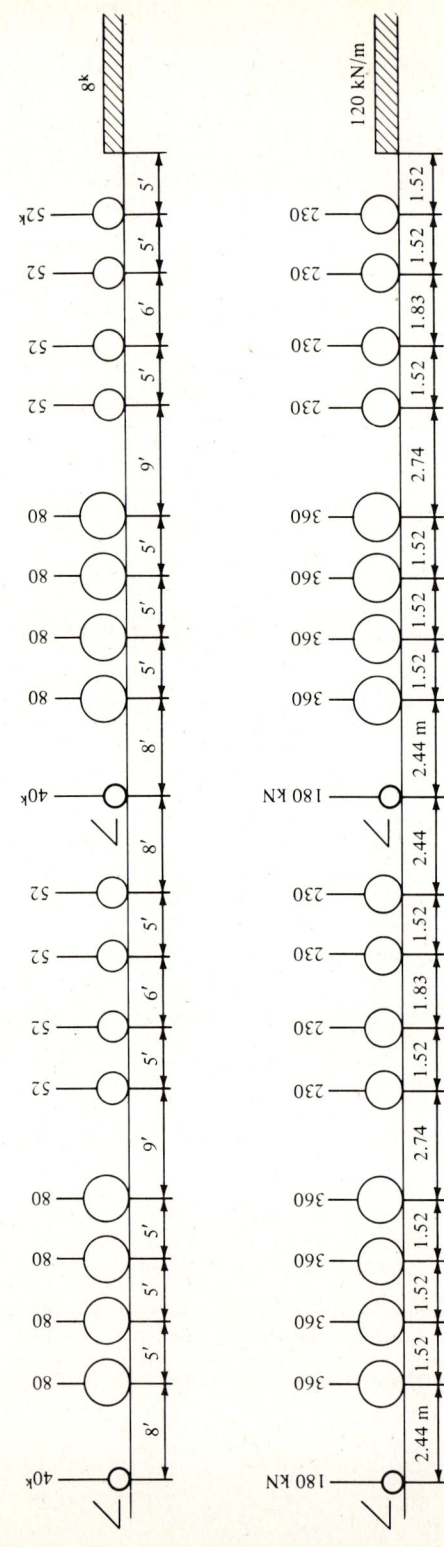

Figure 1-11 AREA standard Cooper E-80 loading. Other loadings obtained by direct proportion (i.e, E-60 = 0.75 × E-80).

Table 1-2 Moments, shears, and floor-beam reactions for Cooper's E-80 loading[a]

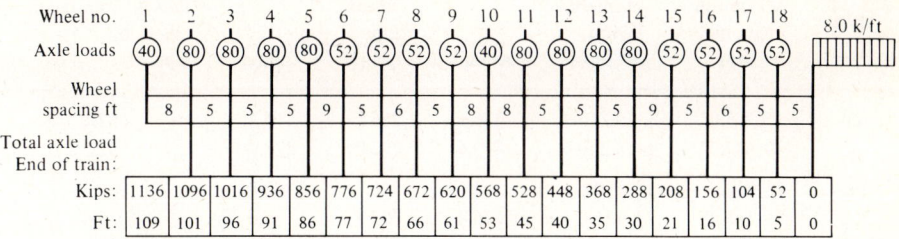

Span ft	Maximum moment, ft · kips	Moment, 1/4 pt, ft · kips	Shear, kips			Pier or floor beam reaction, kips
			At end	1/4 pt	Center	
5	50.0	37.5	40.0	30.0	20.0	40.0
6	60.0	45.0	46.9	30.0	20.0	53.3
7	70.0	55.0	51.4	31.4	20.0	62.9
8	80.0	70.0	55.0	35.0	20.0	70.0
9	90.0	85.0	57.6	37.8	20.0	75.6
10	112.5	100.0	60.0	40.0	20.0	80.0
11	131.8	115.0	65.5	41.8	21.8	87.3
12	160.0	130.0	70.0	43.3	23.3	93.3
13	190.0	145.0	73.8	44.6	24.6	98.5
14	220.0	165.0	77.1	47.1	25.7	104.3
15	250.0	188.0	80.0	49.8	26.6	109.2
16	280.0	210.0	85.0	52.5	27.5	113.7
18	340.0	255.0	93.3	56.7	28.9	121.3
20	412.5	300.0	100.0	60.0	28.7	131.1
24	570.4	420.0	110.8	70.0	31.8	147.9
28	731.0	555.0	120.9	77.1	34.3	164.6
30	820.8	623.0	126.0	80.1	35.8	172.4
32	910.9	692.5	131.4	83.1	37.5	181.9
36	1097.3	851.5	141.1	88.9	41.4	199.1
40	1311.3	1010.5	150.8	93.6	44.0	215.9
45	1601.2	1233.6	163.4	100.3	46.9	237.2
50	1901.8	1473.0	174.4	106.9	49.7	257.5
52	2030.4	1602.7	180.0	110.2	51.2	266.8
55	2233.1	1732.3	185.3	113.6	52.7	280.7
60	2597.8	2010.0	196.0	120.2	55.7	306.4
70	3415.0	2608.2	221.0	131.9	61.4	354.1
80	4318.9	3298.0	248.4	143.4	67.4	397.7
90	5339.1	4158.0	274.5	157.5	73.5	437.2
92	5553.2	4338.5	279.6	160.6	74.4	444.8
94	5768.0	4519.0	284.6	163.7	75.6	452.4
96	5989.2	4699.5	289.6	166.9	76.6	460.0
98	6210.9	4880.0	294.8	170.0	77.7	467.2
100	6446.3	5060.5	300.0	173.1	78.7	474.2
120	9225.4	7098.0	347.4	202.2	88.9	544.1
140	12 406.0	9400.0	392.6	230.2	101.6	614.9

Table 1-2 (Continued)

Span ft	Maximum moment, ft · kips	Moment, 1/4 pt, ft · kips	Shear, kips			Pier or floor beam reaction, kips
			At end	1/4 pt	Center	
160	15 908.0[b]	11 932.0	436.5	256.5	115.2	687.5
180	19 672.0[b]	14 820.0	479.6	282.0	128.1	762.2
200	23 712.0[b]	17 990.0	522.0	306.8	140.8	838.0
250	35 118.0[b]	27 154.0	626.4	367.3	170.0	1030.4
300	48 800.0[b]	38 246.0	729.3	426.4	197.9	1225.3

[a] All values shown are for one rail (one-half track load). Axle loads shown in diagram. Obtain values for other E loads by proportion.

[b] At center of span; other moment values are usually close to center of span, so one may obtain the total moment as the sum of $wL^2/8$ for dead load + live load value shown in the table.

Based on the locomotive weight, the Cooper load is designated as E-40, E-50, E-60, E-75, E-80, or E-110 and is directly proportional (i.e., E-60 = $\frac{6}{8}$ xE-80). The current AREA design criteria are based on the E-80 (sometimes E-110) loading shown in Fig. 1-11. Table 1-2 can be used to obtain the bending moments and shears at selected locations for girder bridges, with values given for a single rail loading (based on one-half the axle load shown in Fig. 1-11).

Where multitracks or road lanes are carried by the bridge, the live load is as follows:

Lane or track	Percent of live load	
	AASHTO	AREA
1	100	100
2	100	100
3	90	2 × 100 + 1 × 50
4	75	2 × 100 + 1 × 50 + 1 × 25
More than 4	75	As specified by designer

Other bridge loadings that must be considered include impact, wind, and longitudinal forces. Impact and longitudinal forces allow for dynamic effects from rolling equipment going across as well as for starts and stops made on the bridge. Impact will be considered in the next section. The wind force is self-explanatory and in the case of a loaded railroad bridge, the wind against the train may be a substantial load.

1-9.1 Wind

1. AASHTO wind requirements for wind at right angles to the longitudinal axis:

Bridge type or member (on exposed area)	Pressure, force/area	
	fps, psf	SI, kPa
Trusses	75	3.6
Girders and beams	50	2.4

The total uniform loading, however, must not be less than the following:

Trusses, 300 lb/ft or 4.38 kN/m.
Girders and beams, 150 lb/ft or 2.19 kN/m.
For wind on live load at 6 ft or 1.83 m above deck, 100 lb/ft or 1.46 kN/m.

2. AREA wind requirements:

	Pressure, force/area			
	Unloaded span		Loaded span	
Bridge member	fps, psf	SI, kPa	fps, psf	SI, kPa
Trusses	50	2.5	30	1.5
Girders and beams	75	3.75	45	2.25

The total uniform loading on the loaded span must not be less than the following:

Loaded chord or flange, 200 lb/ft or 2.92 kN/m.
Unloaded chord or flange, 150 lb/ft or 2.19 kN/m.
Wind on live load at 8 ft or 2.4 m above top rail, 300 lb/ft or 4.38 kN/m.

1-9.2 Longitudinal Force

The longitudinal force (parallel to bridge axis) is as follows:

AASHTO: $0.05 \times$ live load (without impact).
AREA: $0.15 \times$ live load (without impact).

Other loadings that may require consideration include differential temperatures between top and bottom flanges or chords, ice and snow loads, possible overloads, and for continuous bridges, support (pier) settlements.

32 STRUCTURAL STEEL DESIGN

Railroad bridges make a distinction between those bridges which consist of girders and cross-ties (open-deck) and those where there is a floor system that contains 6 in or more of crushed aggregate (ballast). The latter structures are termed ballasted-deck bridges. Since ballast is assumed to weigh 120 pcf or 18.86 kN/m^3, a ballasted structure has a much larger inertia and dead weight/live weight ratio, which should reduce both fatigue and impact effects. This advantage is offset somewhat by the greater likelihood of corrosion in the deck plates beneath the ballast.

1-10 IMPACT LOADS

All of the design specifications refer to impact loads. An impact factor is computed based on a combination of experience and theory to obtain the impact load as

$$L_{\text{impact}} = L(1 + I_f) \tag{1-3}$$

The AISC and National Building Code factors for buildings are as follows:

Item	I_f
Elevator loads	1.00
Machinery and other moving loads	≥ 0.25

Crane runways are further considered to be loaded with a force perpendicular (lateral) to the runway of 0.25 × weight of crane capacity + crane, with one-half of this value applied to the top of each rail. A horizontal and longitudinal force of 0.125 × total wheel load is also applied to the top of the crane runway rail.

The AASHTO impact requirement is

fps	SI
$I_f = \dfrac{50}{L + 125} \leq 0.30$	$I_f = \dfrac{15}{L + 38} \leq 0.30$

where L is the length of span or portion of span that is loaded, in ft or m. The AREA impact specifications depend on the rolling equipment. For diesel and electric locomotives and tenders:

	fps			SI	
$L < 80$ ft	$I_f = \dfrac{100}{S} + 40 - \dfrac{3L^2}{1600}$		$L < 24$ m	$I_f = \dfrac{30.5}{S} + 40 - \dfrac{3L^2}{150}$	
$L > 80$ ft	$I_f = \dfrac{100}{S} + 16 + \dfrac{600}{L - 30}$		$L > 24$ m	$I_f = \dfrac{30.5}{S} + 16 + \dfrac{185}{L - 9.0}$	

where I_f = impact factor in percent; must be divided by 100 to use in design computations

S = distance between centers of single or groups of longitudinal stringers which frame into transverse floor beams or girders, or between main trusses or girders, ft or m

L = length between transverse floor beams or between supports as applicable, ft or m

Use 90 percent of I_f computed above for ballasted-deck bridges.

Example 1-4 Given a highway truss bridge with HS 20 loading. The truss panels are 7.5 m, and the distance between reactions is 37.5 m. The distance between trusses (width) is 14.1 m. What is the impact factor, I_f?

SOLUTION The impact factor will vary for the floor beams, stringers, and the truss, depending on their lengths. For the stringers the impact factor is

$$I_f = \frac{15}{7.5 + 38} = 0.330 > 0.30 \quad \text{therefore, use } I_f = 0.30 \quad ///$$

Example 1-5 An open-deck railroad bridge consists in two trusses spaced at 17.00 ft apart. The trusses are made up of seven panels at 27.60 ft/panel. What is the impact factor?

SOLUTION Since $L > 80$ ft,

$$L = 7(27.60) = 193.2$$

$$I_f = \frac{100}{17} + 16 + \frac{600}{193.2 - 30} = 25.6 \text{ percent} \quad ///$$

Example 1-6 What is the impact for the floor beams of the AREA truss of Example 1-5? Floor beams are transverse members connecting the two trusses at panel points.

SOLUTION $S = 27.6$ ft, $L = 17$ ft < 80.

$$I_f = \frac{100}{27.6} + 40 - \frac{3(17)^2}{1600} = 43.1 \text{ percent} \quad ///$$

1-11 EARTHQUAKE LOADS

Earthquake analyses follow two general trends. One is to attempt to model the structure as an assemblage of masses and springs and use a digital computer to develop response spectra for various assumed earthquake accelerations. The other method is to approximate the earthquake accelerations based on earthquake probability, a period of excitation based on building geometry, and apply the force equation $F = ma$ to obtain an acceleration force which is prorated in some fashion to the several stories such that $\Sigma F_i = F$. The semiempirical procedure of the Uniform Building Code (the NBC is almost identical) for earthquake analysis will be given in the following several paragraphs [see following Eq. (1-7) for identification of terms not identified immediately following equations]:

The lateral earthquake force acting on the building in any direction can be computed using a modification of $F = ma$:

$$F = ZIKCSW \quad (1\text{-}4)$$

The codes (UBC and NBC) permit the product of the factors C and S to be

$$CS \leq 0.14$$

The C factor is based on the fundamental building period, T:

$$C = \frac{1}{15\sqrt{T}} \leq 0.12 \quad (1\text{-}5)$$

The fundamental building period is obtained as the period of a vibrating spring which has the general equation

$$T = 2\pi\sqrt{\frac{m}{k}}$$

where $m =$ mass (of building)
$k =$ spring constant (and obtained for buildings as the summation of the frame element contributions)

Since the spring constants are difficult to determine, the building period is more commonly computed using an experience factor, to obtain

$$T = \frac{JH}{\sqrt{D}} \quad \text{(seconds)} \quad (1\text{-}6)$$

where $J = 0.05$ when H (height) and D (lateral dimension of building in direction of interest) are in feet, and $J = 0.91$ when H and D are in meters. When the frame is made of a ductile material such as steel, it is allowed to approximate the building period as

$$T = 0.10 \times \text{number of stories above base}$$

The lateral force from Eq. (1-4) is distributed as follows:

$$F_{\text{top}} = 0.07TF \leq 0.25F$$

but F_{top} may be zero when $T \leq 0.7$ s. The force at any story level is prorated as

$$F_n = (F - F_{\text{top}})\frac{W_n h_n}{\Sigma W_n h_n} \quad (1\text{-}7)$$

In the preceding equations, terms not previously identified are:

$D =$ building width in direction of interest, ft or m
$H =$ building height above ground level, ft or m

h_n = height to any floor from ground level
I = occupancy importance factor: I = 1.5 for essential facilities
    = 1.25 for assembly halls
    = 1 for all other occupancy
K = lateral force coefficient based on building type as in Table 1-3
S = numerical coefficient for site resonance; use S = 1.5 unless refined site geotechnical data are available
W = total dead load or dead + 25 percent of live load in storage or warehouse structures
W_n = load of the nth floor
Z = earthquake zone coefficient, as obtained from Fig. 1-12, based on geographic location of building:

For zone:	Z
1	0.187
2	0.375
3	0.75
4	1.00

Table 1-3 Horizontal force factor K for buildings and other structures[a]

Type of structure	K
1. All building with full framing systems except as below	1.00
2. Buildings with a box system that is without a complete vertical-load-carrying space frame, such as part of gravity load carried by load-bearing walls	1.33
3. Buildings with a dual bracing system consisting of a ductile moment resisting frame and shear walls or braced frames as follows: *a.* The frames and shear walls resist the total lateral force in accordance with relative rigidities with wall–frame interaction *b.* The shear walls act independently of the moment-resisting capacity of the ductile frame and resist total lateral force *c.* The ductile frame must be able to resist at least 25 percent of total lateral force	0.80
4. Buildings with a ductile moment-resisting frame which can resist the total lateral force	0.67
5. Elevated tanks plus fill contents supported by four or more cross-braced legs and not supported by a building	2.5[b]
6. Structures other than buildings	2.00

[a] Where wind load produces higher stresses, use wind load instead of earthquake.
[b] The minimum value of "$K \times C$" = 0.12 and the maximum need not exceed 0.25. The tower supporting tank is designed for a torsion load based on the lateral force with an eccentricity of 5 percent of the structure dimension at the point of force application.

36 STRUCTURAL STEEL DESIGN

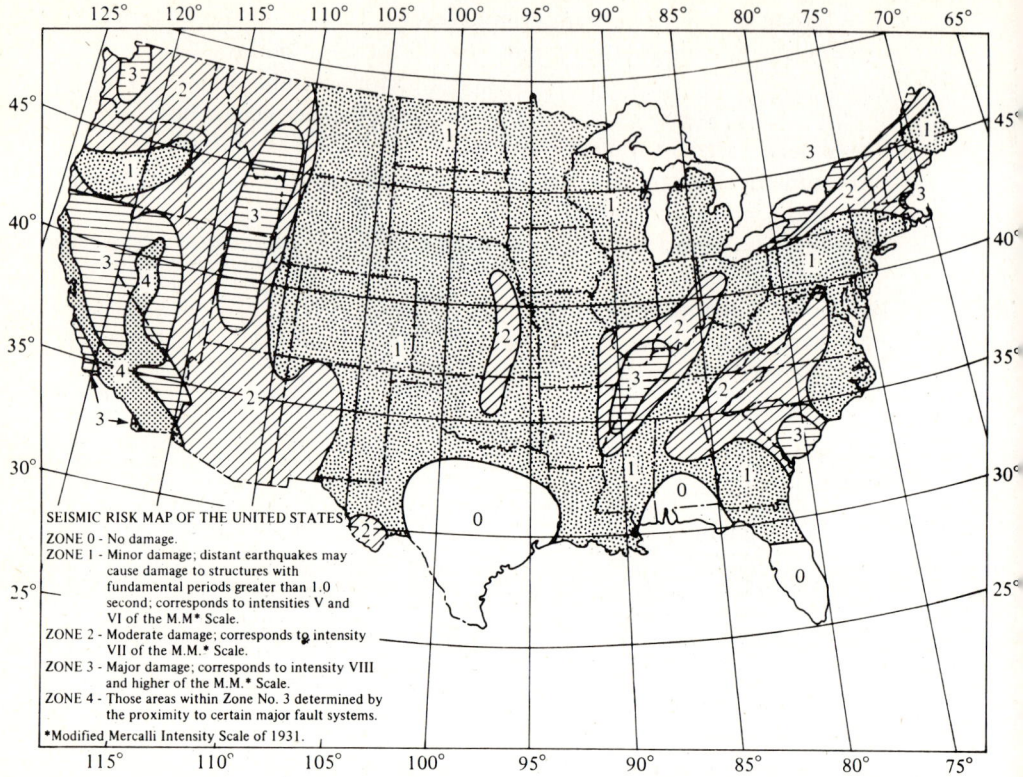

Figure 1-12 Earthquake zone map for the United States. *(After Uniform Building Code, 1976)*.

Example 1-7 A 10-story apartment building with basement is as shown in Fig. E1-7a. The exterior is insulated curtain walls and thermopane windows with an estimated weight of 15 psf. The interior walls are generally stud partitions plastered on both sides with insulation between apartments. Use 4-in concrete floors (tiled or carpeted) on corrugated metal pan supports carried by open web steel bar joists. The building site is in Memphis, Tennessee. Estimate the earthquake force and corresponding story load.

SOLUTION Estimate the roof and floor dead loads as follows:
Roof:

Wood sheathing	= 3 psf
5-ply felt roofing	= 7 psf
Ceiling and bar joists	= 11 psf
Total	21 psf

Total roof weight = $0.021(40 \times 90) = 75.6$ kips

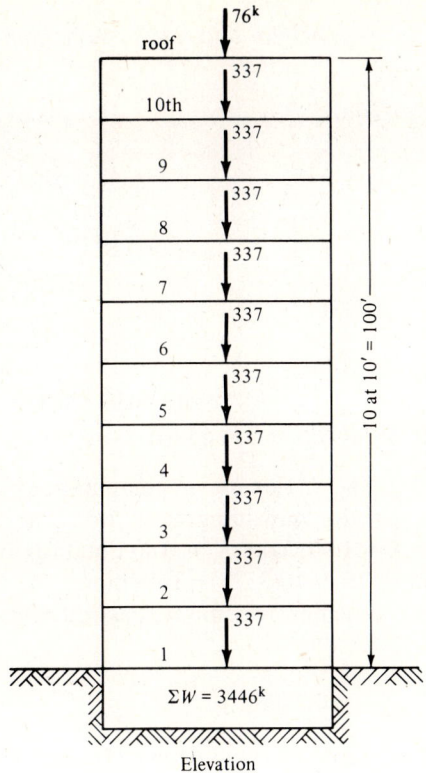
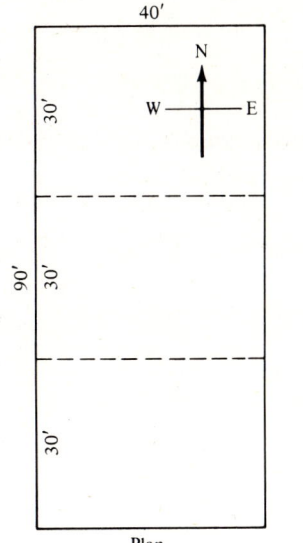

Figure E1-7a

Any floor:
Partitions in 40 × 30 apartment at 8-ft height and at $\frac{1}{2}$ perimeter + 2 cross-walls at 20 psf

$$\text{gives: } (40 \times 2 + 30 \times 2)(8)(20)/(40 \times 30) \qquad = 18.7 \text{ psf}$$

$$\text{Floor: } \frac{4}{12}(144) \text{ (concrete)} \qquad = 48.0 \text{ psf}$$

$$\text{Ceiling (estimated)} \qquad = 10 \text{ psf}$$

$$\text{Bar joists and metal pan} \qquad = 6 \text{ psf}$$

Exterior wall at 10-ft height
$$(2 \times 40 + 2 \times 90)(15)(10)/(40 \times 90) \qquad = 10.8 \text{ psf}$$
$$\text{Total } = 93.5 \text{ psf}$$
$$\text{Total floor weight} = 0.0935(40 \times 90) = 336.6 \text{ kips}$$

These weights are illustrated in Fig. E1-7a. For easier computations, use weights of 76 and 337 kips, respectively, for remaining work. For zone 3 of Fig 1-12, the Z factor is 0.75. Take $I = 1.00$; take $K = 0.80$ obtained from Table 1-3. The total building weight = 76 + 10(337) = 3446 kips.

The earthquake force in the E-W direction is computed as follows:
$$D = 40 \text{ ft}$$
$$T = \frac{0.05(100)}{\sqrt{40}} = 0.7906 \text{ s}$$

Since the frame is of steel, the alternative computation for period is
$$T = 0.1 \times \text{number of stories} = 0.1(10) = 1.0 \text{ s}$$

The author will average the two values of T to obtain $T = 0.895$ s.
$$C = \frac{1}{15\sqrt{T}} = \frac{1}{15\sqrt{0.895}} = 0.0705$$

Substitution of this accumulation of factors/weight into Eq. (1-4) gives
$$F = 0.75(1.0)(0.80)(0.0705)(1.5)(3446) = 218.6 \text{ kips}$$

The roof value is
$$F_{\text{top}} = 0.07TF = 0.07(0.895)(218.6) = 13.7 \text{ kips } (T > 0.7 \text{ s})$$

The story loads are found using h_n = distance ground to nth floor as follows: $\Sigma W_n h_n = 337(90) + 337(80) + 337(60) + \cdots + 337(10) = 151\,650 \text{ ft} \cdot \text{kips}$

$$F = (F - F_{\text{top}})\frac{W_n h_n}{\Sigma W_n h_n} = (218.6 - 13.7)\frac{W_n h_n}{151\,650}$$

For tenth floor
$$F_{10} = 0.001351(337 \times 90) = 40.98 \text{ kips} \quad \text{(shown in Fig. E1-7b)}$$
$$F_9 = 0.001351(337 \times 80) = 36.42 \text{ kips}$$
$$F_8 = 0.001351(337 \times 70) = 31.87 \text{ kips}$$
$$F_2 = 0.001351(337 \times 10) = 4.55 \text{ kips}$$

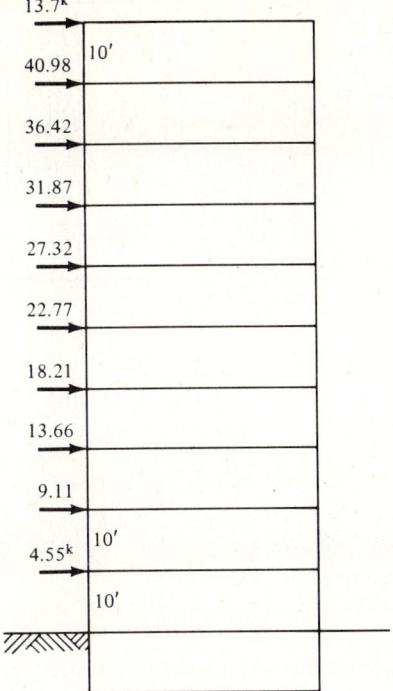

Figure E1-7b

Summing the horizontal floor loads and including the top value of 13.7 gives 218.59 versus 218.6 kips as a check. These lateral floor loads have to be further prorated to the several bays in the E-W direction for the frame analysis load condition(s), which includes earthquake forces.   ///

1-12 FATIGUE

There are numerous recorded structural failures which have been attributed to fatigue. Fatigue failure is a material fracture caused by a sufficiently large number of stress repetitions, cyclic or pulsating stresses, or stress reversals. The stress applications tend to cause a fracture of the material at a location where a small imperfection (which may be microscopic in size) exists. A crack forms, and depending on stress level, rapidly or slowly (sometimes so slowly that the crack can be visually detected before failure) progresses to failure of the part of member.

Most metals tested under repeated or cyclic loadings display stress range versus number of cycles to failure as qualitatively illustrated in Fig. 1-13. Formerly, these curves were commonly displayed as stress level versus cycles. At present, the stress range is used as the parameter of interest. The stress range F_{sr}

40 STRUCTURAL STEEL DESIGN

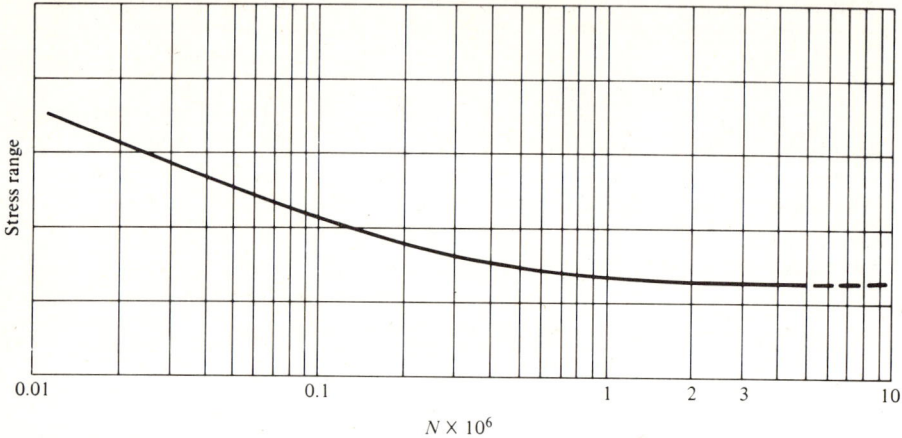

Figure 1-13 Qualitative plot of stress range F_{sr} versus number of cycles to failure.

can be defined as

$$F_{sr} = F_{max} - F_{min} \tag{1-8}$$

where F_{max} is the maximum stress and F_{min} the minimum stress the member is subjected to under service. The fatigue limit (stress range) is often defined as the stress range requiring at least 2×10^6 stress cycles to cause failure.

The AISC, AASHTO, and AREA specifications are very nearly identical (where member failure is not catastrophic) in specifying the stress range and number of stress cycles. These specifications are based on a large number of fatigue tests performed (see Fisher, "Fatigue Strength of Steel Members with Welded Details," *AISC Engineering Journal*, No. 4, 1977). Typical stress range values which may be used for all three specifications are given in Table 1-4. The reader should consult the *Structural Welding Code*, Sec. 9-14, which is the origin of the material used in the three specifications or the appropriate design specification for a more complete presentation of fatigue cases and F_{sr}.

The largest value in each cycle category in Table 1-4 is generally applicable to buildings. Lesser values than shown are necessary for reduced sections, certain types of joints, type of joining material, and for certain members in industrial buildings. The AISC (Appendix B) manual, AWS *Structural Welding Code*, the AASHTO specifications (Sec. 1-7.2) or AREA should be consulted for those situations where fatigue must be considered. Note that fatigue is not usually considered with wind or earthquake loadings on buildings. Fatigue is usually not considered for routine building design, since 10 load cycles/day over a 20-year period is only

$$N = 10 \times 365 \times 20 = 73\ 000 \text{ cycles}$$

This is seldom enough cycles of whatever the stress range to require a reduction in allowable stresses to account for fatigue.

Table 1-4 Allowable stress range F_{sr} for several materials and types of stress

Material	Stress	Stress range, ksi (MPa)			
		Cycles = 20 to 100 ×10³	100 to 500 ×10³	500 to 2000 ×10³	Over 2000 ×10³
Base metal with rolled surfaces	T or Rev[a]	60 (415)	36 (250)	24 (165)	24 (165)
Built-up members					
Base metal and weld metal, built up of plates and shapes with full or partial penetration grove welds	T or Rev	45 (310)	27.5 (190)	18 (124)	16 (110)
Calculated bending stress at toe of girder webs or flanges adjacent to welded transverse stiffeners	T or Rev	32 (221)	19 (131)	13 (90)	10[b] (69)
Base metal at end pr partial length cover plates (square or tapered ends)	T or Rev	21 (145)	12.5 (86)	8 (55)	5 (34.5)
Mechanically fastened connections					
Base metal at section using friction connection with high strength bolts—but not with out-of-plane bending in joint	T or Rev	45 (310)	27.5 (190)	18 (124)	16 (110)
Base metal at net section of riveted and other mechanically fastened joints	T or Rev	27 (186)	16 (110)	10 (69)	7 (48.3)
Groove welds					
Base metal and weld metal at full penetration, ground flush, and inspected	T or Rev	45 (310)	27.5 (190)	18 (124)	16 (110)
Base metal and weld metal in full-penetration splices	T or Rev	45 (310)	27.5 (190)	18 (124)	16 (110)
Fillet-welded connections					
Base metal at intermittent fillet welds	T or Rev	21 (145)	12.5 (86)	8 (55)	5 (34.5)
Base metal at junction of axially loaded member with fillet-welded end connections with welds balanced about axis of member	T or Rev	21 (145)	12.5 (86)	8 (55)	5 (24.5)
Continuous or intermittent longitudinal or transverse fillet welds	Shear	15 (103)	12 (83)	9 (62)	8 (55)

[a] T, tension; Rev, reversal of stresses (both tension and compression).
[b] Use 12 ksi or 83 MPa for girder webs.

Example 1-8 A rolled section will undergo an expected 1×10^6 cycles of stress over the design period of the structure. The stress analysis gives $f_{max} = P_{max}/A = 16$ ksi; $f_{min} = -P_{min}/A = -12$ ksi. The basic stress for this member is $F_a = 22$ ksi (tension) and 16.5 ksi (compression). The structural configuration limits $F_{sr} = 24$ ksi in the base metal. Is the section satisfactory?

SOLUTION

$$f_{sr} = 16 - (-12) = 28 \text{ ksi} > 24 \text{ ksi} \qquad \text{N.G.}$$

The section is inadequate for this number of cycles; increase the section so that $f_{sr} < 24$ ksi. Note that the section is adequate for "allowable" static stresses. ///

1-13 STEEL STRUCTURES

Structures of steel include bridges, buildings, transmission towers, storage tanks, sign supports, and even art objects. The primary focus of this text is on bridges and buildings, since these are the most common projects involving engineers.

Buildings are commonly classified according to use as industrial and multistory buildings. Little use is made at present of steel in residential construction except in multistory apartments.

1-13.1 Industrial Buildings

Industrial buildings are commonly one- or two-story structures used primarily for industrial (such as manufacturing, storage, or retail/wholesale operations) and institutional (including schools, hospitals, hotels, apartments) purposes. Other structures may include gymnasiums, arenas, churches, restaurants, and transportation terminals (land, sea, and air). These buildings may include a steel frame, as illustrated in Fig. 1-14, or have a roof supported by steel members resting on load-bearing walls (see Fig. 2-4). The steel skeleton of the building may be rigid or pinned; may be a two or three-hinged arch, or may be a truss-on-column system. The truss may be rigid or pinned. Several building frames under construction are illustrated in Fig. 4-1.

A building frame is a three-dimensional skeleton but is usually taken as rigid in only one plane. Some buildings are rigid in both the XY and XZ planes, but this type of frame will not be considered in this text. The planer frame resulting from considering only the principal frame elements and/or rigidity is termed a *bent* and may be one or more stories in height (Figs. 1-14 and 1-15 illustrate terms defined here and later). The term "bent" is used for all frames whether rigid, truss-on-column, rafters-on-columns, or other members are used to span between columns in the principal plane. The spacing between bents in the third dimension is the *bay* spacing. Spandrel and floor beams are used to

GENERAL DESIGN CONSIDERATIONS   43

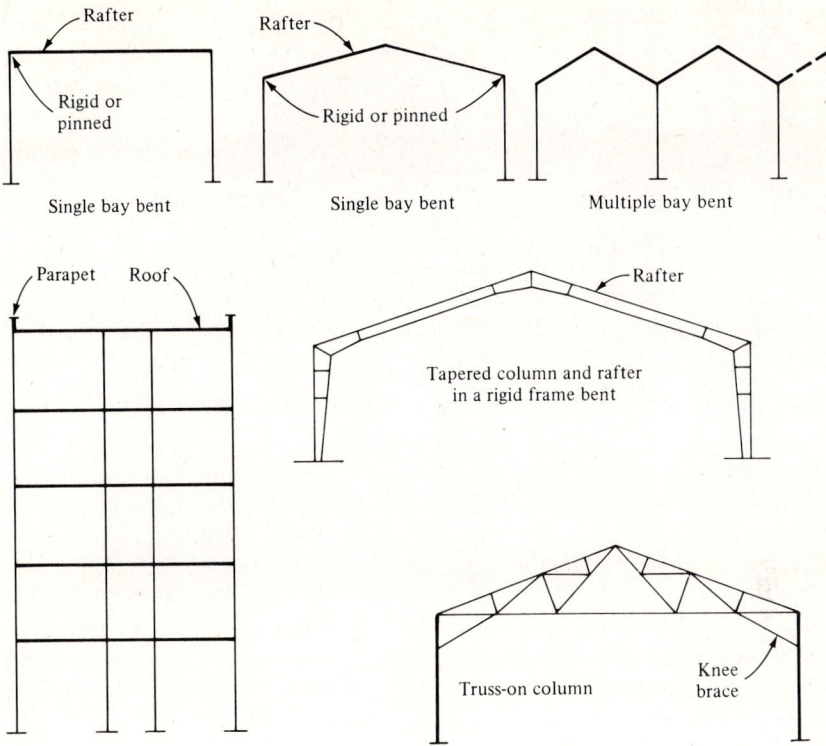

Figure 1-14 Several bents used in steel building frames.

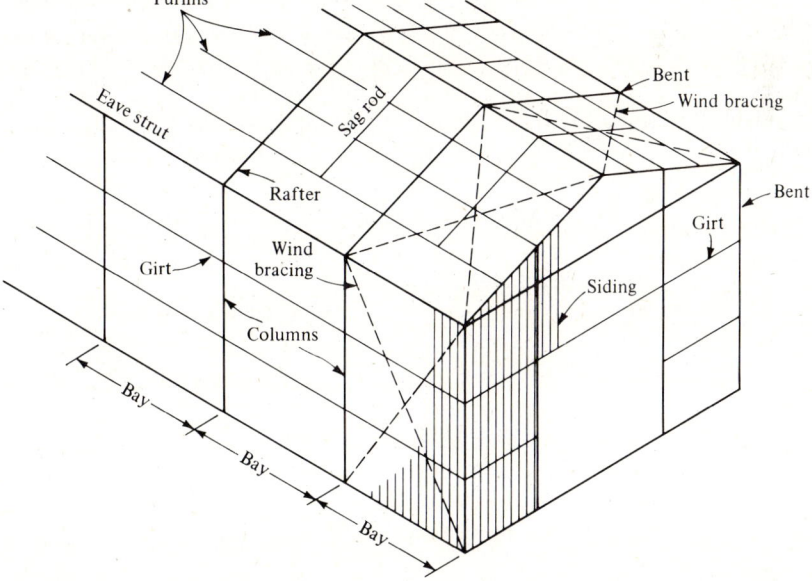

Figure 1-15 Additional terms used to identify structural members in industrial buildings.

44 STRUCTURAL STEEL DESIGN

Figure 1-16 Spandrel beams on a five-story office building. Note that the spandrels carry smaller girts to which the curtain walls will be attached.

span the bays in multistory buildings (see Figs. 1-16 and 4-1d) with girders (usually heavier members than floor beams) spanning between columns of bents. Figure 1-15 illustrates additional terms for one-story industrial buildings, where we note that lateral bracing (see Fig. 4-1d) is used in selected bays.

The roof system of all buildings consists of framing, some kind of decking, and a waterproof covering. The main roof framing consists of the rafters or the truss in any bent. Spanning the bay distance are *purlins* spaced 0.6 to 2 m (2 to 6 ft) or more on centers, depending on the type of roof decking used. Sag rods are provided as additional support for the purlins used on sloping roofs. Purlin design is rather complex for sloping roofs because of the unsymmetrical bending (see Example 4-10). The roof deck rests on the purlins and may be a metal deck, precast concrete slabs, wood planking, or asbestos or gypsum sheets.

Siding may be metal sheets, metal sandwich sheets (curtain walls) consisting of two metal sheets with some type of insulating filler, asbestos sheets, brick, concrete block, tile, or precast or poured concrete. Lightweight siding is carried by spandrel beams supplemented with girts for high-rise buildings. Lightweight siding is carried by the eave strut and girts for industrial buildings.

The spandrel beam previously mentioned is similar to a girt and is located at the floor line as the most exterior floor beam and carries a proportion of the floor load. It also carries a part of the siding. If the siding is heavy (brick, concrete block, tile, etc.), the spandrel may be "built up" using a channel or

GENERAL DESIGN CONSIDERATIONS **45**

Figure 1-17 Girts framed to columns using short segments of angle bolted to girt (small rectangular) tube and to column using high-strength bolts. Refer to Fig. 4-1b for girts partially installed for industrial building using tapered column and rafter sections. Note purlins resting on rafters.

Figure 1-18 Floor framing. Girders are deeper sections with floor beams framing into them. Metal decking is used over beams with concrete to be poured onto decking to produce a composite floor. Diagonal members shown is bracing for the frame. Direction of floor beams with respect to girders determines direction of load flow from any floor to the foundation.

46 STRUCTURAL STEEL DESIGN

Figure 1-19 Use of a deck-type truss for the approach span and a through truss for the main span. Note the lateral bracing at about midheight of the main truss used to reduce the unsupported length of the vertical truss members. The particular diagonal geometry used is to reduce the unsupported length of the diagonal members.

angle, depending on the load to be carried. Spandrel beams are illustrated in Figs. 1-5f, 1-16, and 4-1h. Girts are illustrated in Fig. 1-17.

The general floor framing system of a multistory building is illustrated in Figs. 1-18 and 4-1e and f. It is necessary to establish the floor framing system very early in the design so that the load flow can be made and the several members sized.

1-13.2 Bridges

Bridges may be classified as truss, plate girder (including rolled beam), arch, or suspension (using cables or rods as principal load-carrying members) types. The truss and plate girder bridges are the most commonly used and the only types considered in this text. The truss types include both deck (traffic on top of truss)

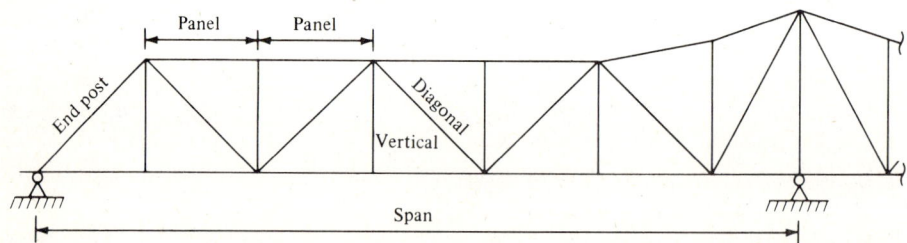

Figure 1-20 Truss terms. Continuous truss shown. It is not necessary that all panels be of the same length. When truss depth increases in region of high moment (over support), panel length should be reduced to reduce member lengths.

or through type, where traffic passes between the trusses. The deck-type truss is generally preferred if clearance beneath the truss is not a factor, because piers can be made shorter. Many truss bridges combine both types (see Fig. 1-19) of trusses. Combinations of a truss for the longer spans and girders for the shorter (or approach) spans is a common practice. This latter scheme is illustrated in Fig. 4-14a. Terms used with bridge trusses are shown in Fig. 1-20 (see also Fig. 4-13). A current trend in bridge design is to use girder structures, which require less fabrication over trusses. In all cases as much welding is used as possible, with most field connections either welded or fabricated using high-strength bolts.

1-14 ACCURACY OF COMPUTATIONS AND ELECTRONIC CALCULATORS

When the 10-in slide rule was the principal computational tool in the structural designer's office, the computations rarely exceeded three significant digits. This was (and is) generally satisfactory, for the reasons presented earlier in this chapter both directly and as implied in the example computations.

Presently, the electronic calculator and/or the digital computer are almost universally used for structural computations because of both the greater complexity of structural configurations and the greater speed of performing calculations and automatically setting the decimal. Now it is almost mandatory (both for ethical and economical reasons) to provide several iterations on a structure to greater optimize the design. This step almost always requires use of a digital computer.

These calculating devices can give a rather large number of digits to any computed result; the problem is how to treat this increased computing capacity. Certainly, the output is not any better than the input, but with a large number of apparently significant digits it looks very impressive. In nearly all design offices, the design computations are checked by a second person as a design precaution. If the original designer carries a large number of string calculations on an electronic calculator, the results will differ from those obtained where the checker truncates intermediate steps, then reenters these values and continues the computations. Where the discrepancies are not large, the question arises of whether the problem has been "checked" or whether one (or both) of the persons has made a design omission.

For these reasons it is suggested that regardless of the initial input data accuracy, computations should be carried to as many decimal places as is practical (0.001 perhaps) to obtain good checking convergence. The extra calculation effort is minimal. Any intermediate steps should be written to the same precision as they are used in subsequent calculations (e.g., do not write 106.1 and then use 106.153 in the following computations).

1-15 STRUCTURAL ENGINEERING COMPUTATIONS IN SI

1-15.1 Principal Units

The SI units of principal interest to the structural designer will be:

Unit weight, kN/m^3.
Unit mass (also mass density or simply "density"), kg/m^3 or t/m^3 (the metric ton = 1000 kg; 1 t/m^3 = 1 g/cm^3).
Pressure (over 1000 psi in fps units), MPa (MPa = megapascal = 10^6 N/m^2).
Pressure (less than 1000 psi in fps), kPa (kPa = kilopascal = 10^3 N/m^2).
Linear weight (as weight of beams for bending moments), kN/m.
Linear mass (as ordering mill quantities), kg/m.
Bending moments, $kN \cdot m$.

Tables in SSDD use SI units of mm for section dimensions and m^n for inertia $\times 10^{-6}$ and section modulus $\times 10^{-3}$. These exponents are multiples of 3, as in preferred usage. Also, the user should note that

$$f_b = \frac{M}{S} = \frac{M, kN \cdot m}{S, m^3 \times 10^{-3} \times 10^3} = MPa$$

$$M = f_b S = MPa \times m^3 \times 10^{-3} \times 10^3 = kN \cdot m$$

where the 10^{-3} does not have to be written after making the computation several times to obtain confidence.

Example 1-9 Given:

$$\text{Section modulus, } S = 6.33 \text{ m}^3 \ (\times 10^{-3})$$
$$F_b = 0.6 F_y = 0.6(250) = 150 \text{ MPa}$$
$$M = 1200 \text{ kN} \cdot \text{m}$$

Is the section adequate? If not, what S is required?

SOLUTION

$$f_b = \frac{M}{S} = \frac{1200}{6.33} = 189.6 \text{ MPa} > 150 \qquad \text{section not adequate}$$

The required S is

$$S = \frac{M}{F_b} = \frac{1200}{150} = 8.00 \text{ m}^3 \ (\times 10^{-3})$$

Note that it is not necessary to include the 10^3 with 150 MPa to produce 150 000 kPa for the division (after some practice), as the units will come out by simply shifting the decimal. ///

GENERAL DESIGN CONSIDERATIONS 49

1-15.2 Types and General Precision of SI Units

It is suggested that the following is of sufficient computational precision:

Item	To nearest	
Beam depths, flange widths, angle legs	1.0 mm	
Any member thickness, plate thickness, etc.	0.1 mm	
Moment of inertia ($m^4 \times 10^{-6}$), but depends on size	0.01	
Section modulus ($m^3 \times 10^{-3}$), but depends on size	0.01	
Radius of gyration	0.1 mm	
For weight items, use:	0.01 kN/m	
Mass:	0.1 kg/m	(ASTM A-6 currently uses nearest 1 kg/m for selected rolled shapes—it remains to be seen how the material will be purchased, since the mills are not likely to donate large masses to a purchaser who truncates excessively)
Pressure:	0.01 kPa	
Pressure:	0.1 MPa	
Unit weight:	0.01 kN/m³	

The reader should note that in the SSDD tables, some of the given dimensions and pressures vary slightly from the precision given above, because of generating them on the computer. The table values are acceptable as given, but any additionally computed values should conform to the above.

1-15.3 Selected SI Conversion Factors

Several of the more useful conversion factors are given next. Note that any derived force units can be developed using

1 ft³ of water weighs 62.4 pcf
1 m³ of water weighs 9.807 kN/m³
1 m³ of water has a mass density of 1000 kg/m³ = 1 g/cm³
1 ft = 0.3048 m
1 m = 3.2808 ft
$g = 32.17$ ft/s² = 9.8054 (use 9.807) m/s²

Example 1-10 Convert 120 lb/ft³ to SI.

$$\text{Unit weight} = \frac{120}{62.4}(9.807) = 18.86 \text{ kN/m}^3$$

What is 23.5 kN/m³ in fps?

$$\text{Unit weight} = \frac{23.5}{9.807}(62.4) = 149.5 \text{ pcf} \qquad ///$$

The reader should note that the several intermediate conversion steps that produce the end results 62.4 and 9.807 are not shown.

Several other useful conversion factors are as follows:

To convert	to	Multiply by
kilogram (kg)	kilonewton (kN)	0.009806650
pound (lb)	kg	0.4535924
pound	kN	0.004448222
kips (1000 lb)	kN	4.448222
lb/ft	kN/m	0.014593727
kips/ft	kN/m	14.593727
lb/ft	kg/m	1.48816404
psi	MPa	0.006894757
psi	kPa	6.8947577
ksi	MPa	6.8947577
ksi	kPa	6894.7577
psf	kPa	0.04788026
ksf	kPa	47.88026
kip · ft (moment)	kN · m	1.35584
kip · in	kN · m	0.1129862

PROBLEMS

1-1 What is ε_y for a steel with $F_y = 40$ ksi?

1-2 What is ε_y for a steel with $F_y = 365$ MPa?
Answer: 0.00182.

1-3 What is the increase in weight/ft of a W16 × 40 beam with 2 in of sprayed-on perlite (see Fig. 1-5) fireproofing as in Fig. P1-3? Note that the method of application is not likely to produce an exact geometric section; therefore, the computations should not be extremely precise.
Answer: $\Delta w \sim 30$ lb/ft.

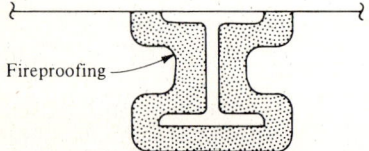

Figure P1-3

1-4 What is the increase in weight of a W410 × 46.1 rolled section with 55 mm of vermiculite fireproofing as in Fig. P1-3? (See the comment in Prob. 1-3 before starting this problem.)

1-5 What is the R factor for the interior column at the top floor and three floors down of Example 1-1?

1-6 What is the R factor for beam B1 and column B2 of Example 1-2?

1-7 What are the wind forces at points 1 through 9 of the roof shown in Fig. P1-7?
   Answer: $P(5) = -18.07$ kips.

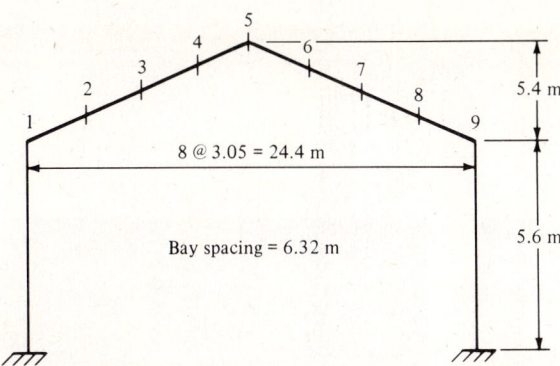

Figure P1-7

1-8 What is the snow load for the building in Fig. P1-7 if the site is near Chicago, Illinois, where the Chicago Building Code requires 1.25 kPa?

1-9 What is the maximum shear and bending moment in a bridge span of 92 ft for an HS 15 truck? What is the impact factor?
   Answer: $V = 48.5$ kips, $I = 0.23$.

1-10 What is the maximum shear and bending moment in a bridge span of 55.2 m for an HS 20 truck? What is the impact factor?

1-11 What is the maximum moment in a railroad bridge girder (one of two) carrying a single track. The span is 30.5 m and girders are spaced at $L/20$ apart. Use a E-72 loading. What is the impact factor if the track uses diesel locomotives and the bridge is a ballasted-through-deck type?
   Answer: $M = 7865$ kN · m; $I = 40.1$ percent.

1-12 What are the story shears in the N-S direction of the building of Example 1-7?

1-13 A structural element is subjected to 1×10^6 stress cycles of loading. The maximum axial force is 640 kN and the minimum is 320 kN. If the allowable axial stresses are 150 and 99 MPa for tension and compression, respectively, what area in m^2 is required to satisfy both allowable stresses and fatigue requirements? Assume that the third column of Table 1-4 is applicable to this member.

(a) (b)

Figure II-1 Two of world's tallest buildings using structural steel frameworks. (a) The 1450-ft (442-m) Sears Tower building in Chicago, Illinois—currently the world's tallest building. (b) The 337-m John Hancock Center building, also in Chicago, Illinois. View from the Sears Tower.

CHAPTER TWO

ELEMENTS OF FRAME, TRUSS, AND BRIDGE DESIGN

2-1 METHODS OF ANALYSIS

The design of any structure requires developing frame member forces (axial, shear, and bending stresses) based on the dead and critical live-load combinations.

The method of analysis depends on the complexity of the structure and whether it is rigid (indeterminate) or simply framed. An analysis may consider the structure as either two- or three-dimensional. Simple framed structures are generally determinate; that is, the three equations of statics (ΣF_h, ΣF_v, $\Sigma M = 0$) are sufficient to obtain the internal member forces. In any case, with simple framing the ends of the members are assumed to have no moment resistance transfer to adjacent members. The internal member forces of determinate structures are readily obtained by hand calculations and with considerable efficiency using pocket calculators.

Rigid framed structures are generally indeterminate since the member ends transfer shear forces and moments to the adjacent members. Indeterminate structures require deformation compatibility to supplement the equations of statics to determine the internal member forces. Digital computers are used to obtain solutions for all but the simplest indeterminate structures. Continuous beams and certain simple rigid frame structures have solutions that can be obtained from handbooks (or readily derived using mechanics-of-materials methods). A few beam solutions are presented in Part IV of SSDD, in the AISC manual, and in most engineering handbooks.

A fundamental part of structural design is to establish whether the structural frame is to be rigid or simple. A rigid frame generally gives smaller design moments but does not necessarily result in a more economical design. This is because:

1. Practical considerations of quality control/design of rigid versus simple joint connections.
2. Tendency to use the same depth beams across a bay even though certain spans are shorter or carry less load.
3. The use of a constant column size through at least two and often three or more floors to reduce splicing. Since the column must carry the largest load, it is over designed in the upper floor(s).

The *stiffness* (formerly called *slope deflection*) *method of analysis* is most commonly used for indeterminate structural analysis using the digital computer. This method is particularly adapted for computer use, as the matrix is always sparse and symmetrical. Advantage can be taken of symmetry to reduce the inversion effort so that rapid solutions of very large problems can be obtained economically.

All indeterminate structure solutions are iterative and generally nonunique in that the output depends on the input, or, stated differently, it is necessary to have member properties (area, A; moment of inertia, I) in order to obtain displacement compatibility. Since digital computer output can be obtained rapidly, one may initialize a problem using relative values of A and I, then using this output to select preliminary members and iterate as required. Where the input is considerable, with frames containing large numbers of members, it may be better to use some approximate methods of analysis to obtain preliminary member sizes. Approximate methods include simply assuming relative sizes of columns and beams (e.g., A = constant, $I_b = 1.2I_c$ or other relative values). Moment distribution (but limited to not more than two or three cycles) may be used together with relative values of A and I for preliminary sizing of members. The portal and cantilever methods of approximate frame analysis may be used for building frames of one or more story heights. The latter two methods are primarily to obtain the effects of lateral (wind) forces on a frame.

The *portal method* (refer to Fig. 2-1) makes the following assumptions:

1. Point of contraflexure occurs at midheight of all columns.
2. Sum of wind (lateral) load is distributed as shears in proportion to bay width.
3. Beam or girder moment is zero at midspan.
4. Interior columns carry no axial loads.

With these four assumptions and application of statics, the approximate joint moments can be obtained at any location.

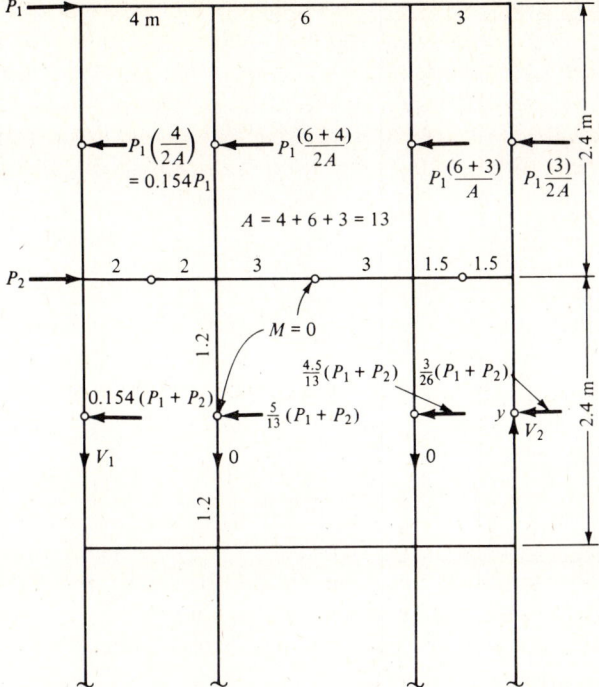

$\Sigma M_y = 0$ (assuming lateral loads do not introduce axial forces in interior cols.)

$$V_1 = \frac{1.2 P_2 + 3.6 P_1}{13}$$

$\Sigma F_v = 0 \quad \therefore V_2 = V_1$

Figure 2-1 Portal method of approximate frame analysis for a typical three-bay bent in a frame.

The *cantilever method* (Fig. 2-2) is an alternative method of approximate analysis for tall buildings with $H/W \leq 4$ to 5 with the following assumptions:

1. Points of contraflexure are the same as for the portal method—midpoint of beams and columns.
2. Axial loads are in all columns and are proportional to the distance from the neutral axis of the bent treated as a vertical cantilever beam. The location of the neutral axis is based on column areas (since these are not generally known, a value of $A = 1$ unit area is usually used). The equivalent moment of inertia of the vertical cantilever beam is computed as

$$I = \Sigma A d^2$$

The column loads are computed as

$$V_i = \frac{Mc}{I}$$

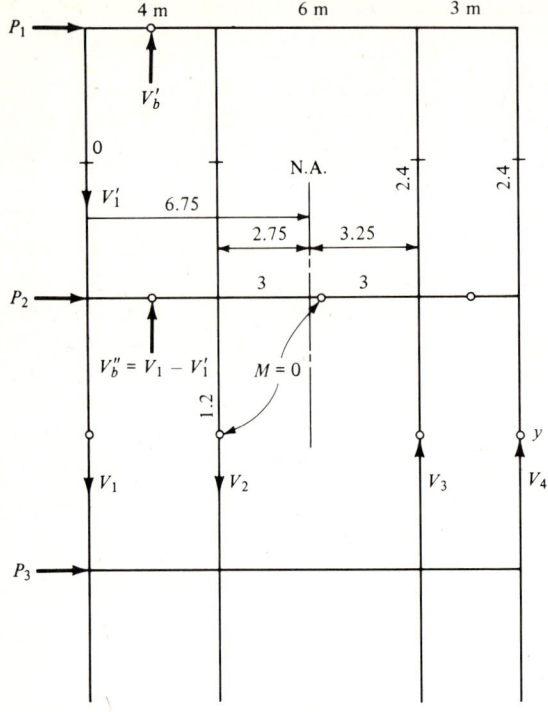

$\Sigma M_0 = A\bar{X}$ (assumes col. areas = constant)

$\bar{X} = \dfrac{4A + 10A + 13A}{A + A + A + A} = 6.75$ m

$I = 6.75^2 + 2.75^2 + 3.25^2 + 6.25^2 = 102.75$

$M_y = 3.6P_1 + 1.2P_2$

$V_i = \dfrac{M_y X}{I}$

$V_1 = \dfrac{M_y(6.75)}{102.75} = 0.0657 M_y \quad V_2 = 0.0268 M_y \quad V_3 = 0.0316 M_y$
$\hspace{22em} V_4 = 0.0608 M_y$

$\Sigma(V_1 + V_2 - V_3 - V_4) = 0$

Figure 2-2 Cantilever method of approximate frame analysis for a typical bent in a frame.

> where $M =$ sum of overturning moments above point of contraflexure
> $c =$ distance from neutral axis of bent to column i
> $I =$ moment of inertia of bent as $\Sigma A d^2$

Frames that are unsymmetrical in plan may be subjected to torsion effects from the lateral forces. A torsion analysis similar to the cantilever method can be made using a torsion constant, $J = A\Sigma(x^2 + y^2)$ instead of I with the torsion force obtained as

$$F_i = \dfrac{Tc}{J}$$

> where $T =$ torsion moment about neutral axis (plan view)
> $c =$ radial distance from neutral axis to ith column

2-2 BEAM ANALYSIS

Beams are horizontally (usually) positioned members carrying vertical load. The vertical load may be distributed across the span or applied over a small segment of the span. If the load is distributed over a short span segment, it is considered a point loading. To produce beam action, the l/d ratio (Fig. 2-3h) must be larger than 4 or 5, else the loads are transferred to the reactions more by arching than via shear. The w/d ratio must also be sufficiently large to develop bending rather than a twisting of the section. The floor framing scheme and directions for using floor joists will control the principal framing of the beams and/or girders making up the main frame. Floor joists carry concentrated end reactions to the floor beams, but it is customary where more than about three joists occur in a span to assume a uniform loading. The difference in bending moments for an assumed uniform load w instead of three equal loads P is obtained as follows:

$$w = \frac{3P}{L}$$

$$M_w = \frac{wL^2}{8} = \frac{3P}{L}\frac{L^2}{8} = \frac{3PL}{8}$$

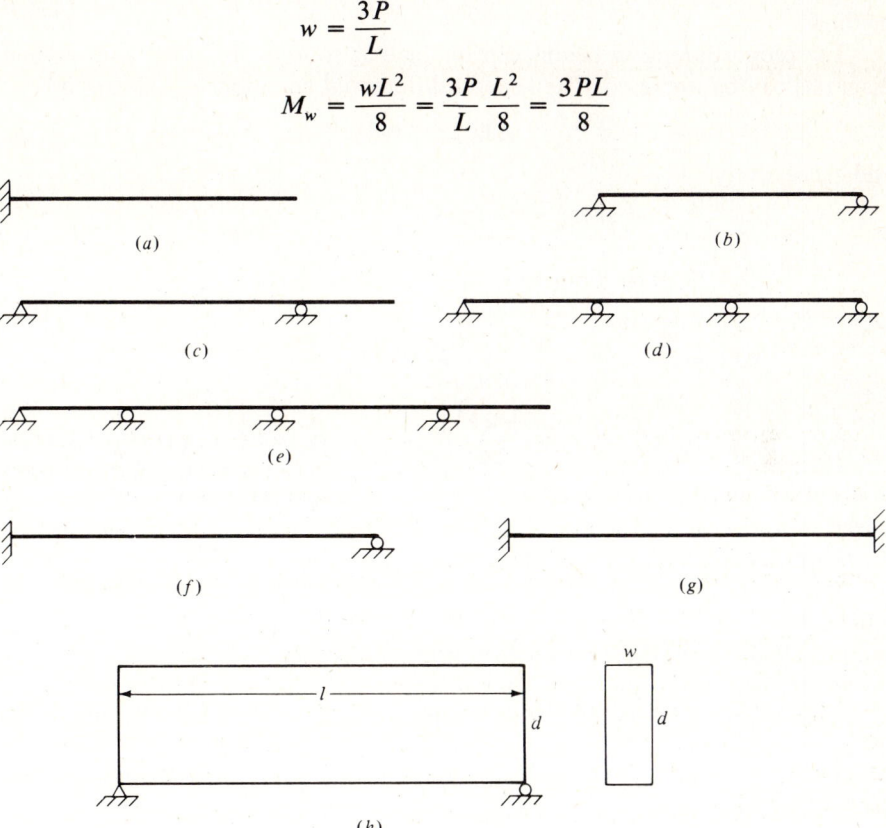

Figure 2-3 Beams. (*a*) Simple cantilever beam. (*b*) Simply supported beam. (*c*) Simply supported beam with overhang. (*d*) Continuous three-span beam. (*e*) Continuous beam with overhang. (*f*) Propped cantilever beam. (*g*) Fixed end beam. (*h*) Both L/d and w/d must be reasonable quantities for beam action to develop and to avoid twisting of thin, very deep beam sections.

The concentrated loads produce a moment of

$$M_c = \frac{3P}{2}\frac{L}{2} - \frac{PL}{4} = \frac{PL}{2}$$

The error is $[(0.5 - 0.375)/0.5](100) = 25$ percent (too small). The error rapidly decreases and reverses sign as the number of concentrated loads increases (noting that the situation is one where the total beam load is a constant and not increasing with the number of loads). For example, if the total load is constant $3P$ and placed at five equal spaces, the M_w value is still $3PL/8$; the M_c value is $9PL/25$.

Beams may be simply supported, overhanging, cantilevered, fixed, or continuous, as illustrated in Fig. 2-3. The designer must always make some assumptions involving point loads as well as how realistic the fixed, cantilever, or continuous-beam models will compare to the actual member geometry. It is seldom possible and never desirable to have actual point application of loads/reactions, although this is a common assumption made in all analysis methods.

The proportioning of beams can be done as soon as the shear and moment diagrams can be obtained. The general differential equation for a beam is

$$EIy^{iv} = -w$$

and successively:

$$EIy''' = V = \text{shear} = -wx + C_1$$

$$EIy'' = M = \text{moment} = \frac{-wx^2}{2} + C_1 x + C_2$$

$$EIy' = \text{slope} = \frac{-wx^3}{6} + \frac{C_1 x^2}{2} + C_2 x + C_3$$

$$EIy = \text{deflection} = \frac{-wx^4}{24} + \frac{C_1 x^3}{6} + \frac{C_2 x^2}{2} + C_3 x + C_4$$

The general equation for the elastic curve for a beam produces four constants of integration. Constants C_2 and $C_4 = 0$ for simple beams, and the w term is zero for beams loaded with concentrated loads.

The general beam equation is sometimes useful in that the appropriate beam information may be approximately obtained by replacing a series of concentrated loads with the equivalent uniform load value.

Beam deflections are often limited for both buildings and bridges. A value of $L/360$ to as low as $L/1000$ can be found in building codes and depending on occupancy and/or interior finish. Where deflection criteria are given, the approximate deflections should be computed before a design progresses very far. The reader should note that highly refined deflection computations are not possible because of the uncertainties in loading. It should also be noted that the deflection is heavily dependent on the moment of inertia—and E is nearly a constant for steel. Therefore, where a rigid deflection criterion is stipulated, the use of high-strength steel may not be economical, since the section size may be fixed by the moment of inertia rather than by the bending stress.

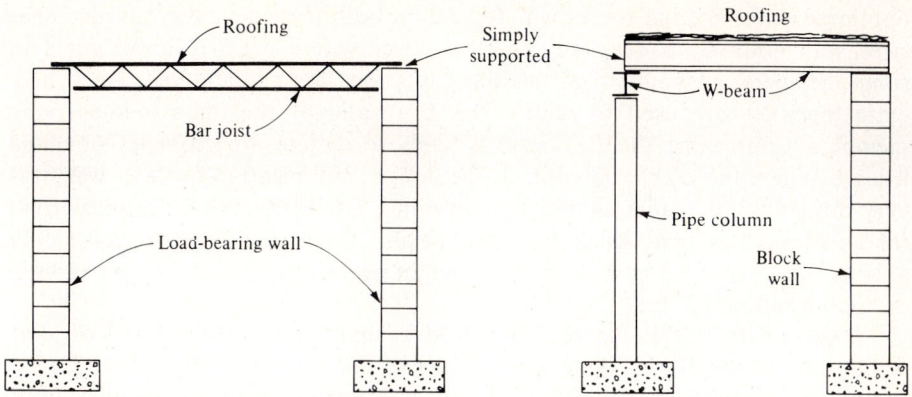

Figure 2-4 Load-bearing-wall systems for one-story buildings.

2-3 DETERMINATE STRUCTURES

Load-bearing-wall structures are widely used for warehouses, industrial buildings, shops, and other structures of small H/W ratios. A load-bearing-wall structure is one in which much of the roof load is carried by the perimeter (and sometimes interior partition) walls. Roof joists may span the entire covered space and rest on the exterior walls or be supported by the exterior wall and by an interior roof beam–column system, as illustrated in Fig. 2-4. In either case the joists are simply supported beams. The interior roof beam may be simply supported at the columns or continuous across them.

Beam–column connections cannot transmit any moment in simple frames. In a practical sense almost all beam–column connections can, and do, transmit both shear and moment. If the connection is made flexible, the usual assumption is that only shear is carried. Connections will be taken up in more detail in Chap. 8.

2-4 TRUSS ANALYSIS

Trusses may be either determinate or indeterminate, depending on whether more than enough bars are supplied to just make a series of triangles. In equation form, the number of bars to produce a determinate truss is

$$n = 2j - R$$

where $j =$ number of joints
$n =$ number of bar members
$R =$ number of reactions

If $n' > n$, the truss is indeterminate; $n' < n$ the truss is unstable.

Determinate trusses are readily optimized for least weight, since the bar forces depend only on truss geometry. Indeterminate trusses are not so readily

60 STRUCTURAL STEEL DESIGN

optimized, since the bar forces will depend on both truss geometry and member sizes. Indeterminate trusses are commonly used where stress reversals occur in some members, unsymmetrical loading conditions are encountered, and where extra members are used to reduce the L/r ratio of the main load-carrying members. Controlling the L/r ratio is likely to dictate the geometry of many trusses. Where the L/r ratio controls the design, optimization for least weight is very difficult. The reader should bear in mind, however, that the optimal truss (or any frame) is that which balances weight, fabrication, and overall safety against the total client cost. Minimum weight alone does not satisfy the criteria of "optimization."

Determinate trusses are readily solved by hand using the method of joint equilibrium or the alternative of equilibrium of the portion to the right or left of a cut section. Both determinate and indeterminate trusses are readily solved using one of the large number of computer programs currently available. The computer program in the Appendix can be used to solve either determinate or indeterminate trusses. This program uses the stiffness method of matrix analysis as outlined below. The user is expected to have taken or be taking a course in advanced structural analysis, so only a brief outline of the stiffness method as used in the computer program is given.

Referring to Fig. 2-5, the coding of a typical truss element is as shown, where the P–X code refers to nodal effects in a truss system of which only P_1 through P_4 are shown as affecting the ith member. The internal member force of the ith member is F_1, and for this force there is a corresponding axial displacement of e_1. Define the member slope as: $\cos \alpha = H/L$; $\sin \alpha = V/L$; and

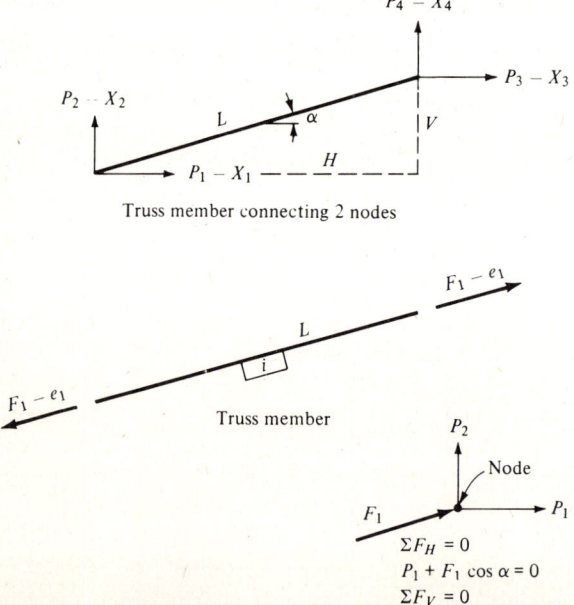

Figure 2-5 Coding of typical truss element.

$L = (H^2 + V^2)^{\frac{1}{2}}$. The nodal equilibrium in matrix notation is

$$P = AF \qquad (a)$$

Also, the member deformations (e's) are related to nodal displacements as

$$e = A^T X \qquad (b)$$

The member forces can be written as

$$F = Se = SA^T X \qquad (c)$$

Now combining Eqs. (c) and (a), we obtain

$$P = ASA^T X \qquad (d)$$

Equation (d) can be inverted to obtain

$$X = (ASA^T)^{-1} P \qquad (e)$$

The P matrix is based on the actual loads applied to the structure at the nodal points. Node points may be truss joints or points along a beam or frame. The S matrix is a member property matrix and the A matrix is the bridge between nodal and member forces and considers the truss (or frame) geometry. Inversion of Eq. (d) gives the nodal displacements, Eq. (e), and we can now use Eq. (c) to compute member forces:

$$F = SA^T X$$

The general A-matrix entries for any truss member are the direction cosines:

$$A_i = \begin{bmatrix} -\cos \alpha \\ -\sin \alpha \\ \cos \alpha \\ \sin \alpha \end{bmatrix}$$

The S matrix for any member i with cross-sectional area A, length L, and E is

$$S_i = \frac{AE}{L}$$

Performing matrix multiplication, we can obtain the SA^T as

$$SA_i^T = \frac{AE}{L}(-\cos \alpha \; -\sin \alpha \; +\cos \alpha \; +\sin \alpha)$$

The product of $A \times SA^T$ gives the element stiffness matrix as follows:

$$EASA_i^T = \frac{AE}{L}$$

P \	1	2	3	4
1	$\cos^2 \alpha$	$+\sin \alpha \cos \alpha$	$-\cos^2 \alpha$	$-\sin \alpha \cos \alpha$
2	$+\sin \alpha \cos \alpha$	$+\sin^2 \alpha$	$-\sin \alpha \cos \alpha$	$-\sin^2 \alpha$
3	$-\cos^2 \alpha$	$-\sin \alpha \cos \alpha$	$\cos^2 \alpha$	$\sin \alpha \cos \alpha$
4	$-\sin \alpha \cos \alpha$	$-\sin^2 \alpha$	$\sin \alpha \cos \alpha$	$\sin^2 \alpha$

This matrix is always symmetrical, as shown. The P–X entries are located at the correct location by noting the X and P coordinates, shown here as 1, 2, 3, and 4 in the total (also called global) truss matrix, and this is done for each truss member in turn. When all the truss members have been included, the matrix can be inverted to obtain the nodal displacements.

Node forces at reactions do not cause member forces or nodal displacements. To account for this, the coding for a truss is such that all the reactions are coded last, using the value NP + 1 = NPP1, where NP = number of P values that are included (at two per joint, with the first value being horizontal) up to the reaction value. Other nodes of known zero displacement or rotation should be coded NPP1 to reduce the computational effort.

The reader may note that reactions may be computed by inserting a single dummy member at a roller reaction and two dummy members at a hinge. Using a unit length and an area A of sufficient magnitude will give $\Delta \rightarrow 0$ and the axial force in the member is the reaction. The effect of support settlements can be evaluated by using a value of AE/L and axial node force whose effect on the dummy members computes

$$e = \frac{PL}{AE} = \text{given support settlement}$$

A support rotation requires inputting by hand the fixed-end moments (FEMs) and corresponding shears as P-matrix entries based on the known rotation. The computed member moments must be hand-corrected as required for the final member end moments, so the node is in static equilibrium.

2-5 RIGID FRAME ANALYSIS

Rigid frame analysis is similar to that of a truss. The essential differences are that each node has three degrees of freedom and the member has a moment on

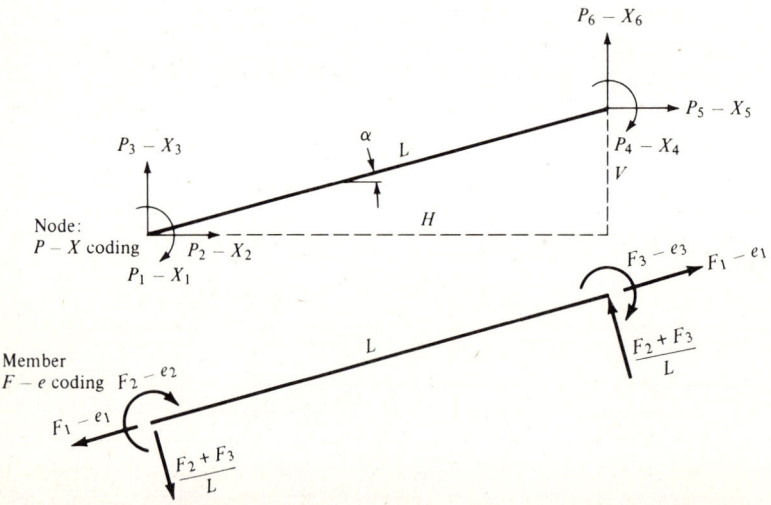

Figure 2-6 Element from rigid frame to develop general element A, S, and $EASA^T$ matrices.

each end as well as an axial force. The resulting element ASA^T is of size 6×6 instead of the 4×4 obtained with the truss element. Referring to Fig. 2-6, the A and S matrices for a typical member in a rigid frame are as follows:

$A =$

P \ F	1	2	3
1	0	1	0
2	$-\cos \alpha$	$\dfrac{\sin \alpha}{L}$	$\dfrac{\sin \alpha}{L}$
3	$-\sin \alpha$	$-\dfrac{\cos \alpha}{L}$	$-\dfrac{\cos \alpha}{L}$
4	0	0	1
5	$\cos \alpha$	$-\dfrac{\sin \alpha}{L}$	$-\dfrac{\sin \alpha}{L}$
6	$\sin \alpha$	$\dfrac{\cos \alpha}{L}$	$\dfrac{\cos \alpha}{L}$

$S =$

F \ e	1	2	3
1	EA/L	0	0
2	0	$4EI/L$	$2EI/L$
3	0	$2EI/L$	$4EI/L$

The computer produces the intermediate element matrices of SA^T and $EASA^T$. The total or global matrix is built by superposition from the element $EASA^T$. The P–X coding locates the ASA^T entries in the global matrix.

Figure 2-7 illustrates coding for joints that contain discontinuities or both truss and rigid frame members.

2-6 BRIDGE ANALYSIS

Bridge analysis is similar to that of a building truss or beam except that the live load is a moving value. With a moving load it is necessary to use influence lines, or alternatively, develop the load matrix for the load positioned at numerous locations along the span. It is generally as convenient with digital computers to develop the load matrix for several vehicle locations and directly obtain the member forces. The member is then designed for the critical member force obtained from inspection of the force matrix in the computer output. Note particularly, however, that impact loading as well as dead load must be used as a

64 STRUCTURAL STEEL DESIGN

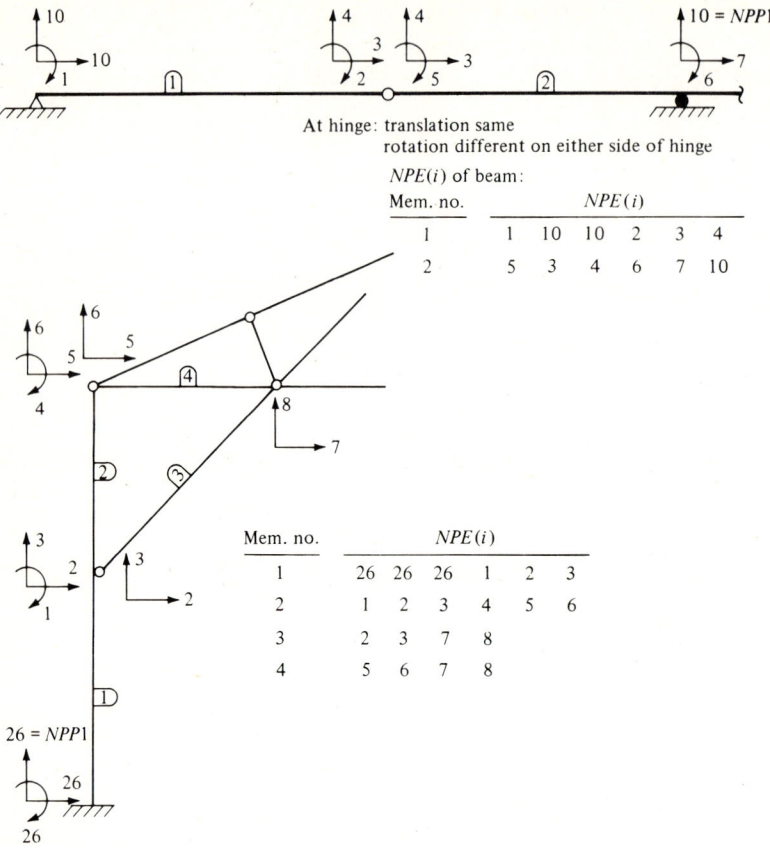

Figure 2-7 Coding of structures with discontinuities or with both beam and truss members.

part of the design. It is the designer's preference whether impact and/or dead load is directly incorporated into the live-load matrix to directly produce design forces in the output matrix or whether the designer combines the several effects by hand. Also note that the impact factor may depend on the member being designed.

2-7 THE COMPUTER PROGRAM FURNISHED IN THE APPENDIX

The computer program in the Appendix can be used to analyze any two-dimensional rigid frame or truss or combination of truss and rigid frame. It can be used for beam–column structures and is particularly efficient for continuous beams. It is necessary to draw the structure to approximate scale and properly code it (making allowance in coding for the type of structural element) to

ELEMENTS OF FRAME, TRUSS, AND BRIDGE DESIGN   65

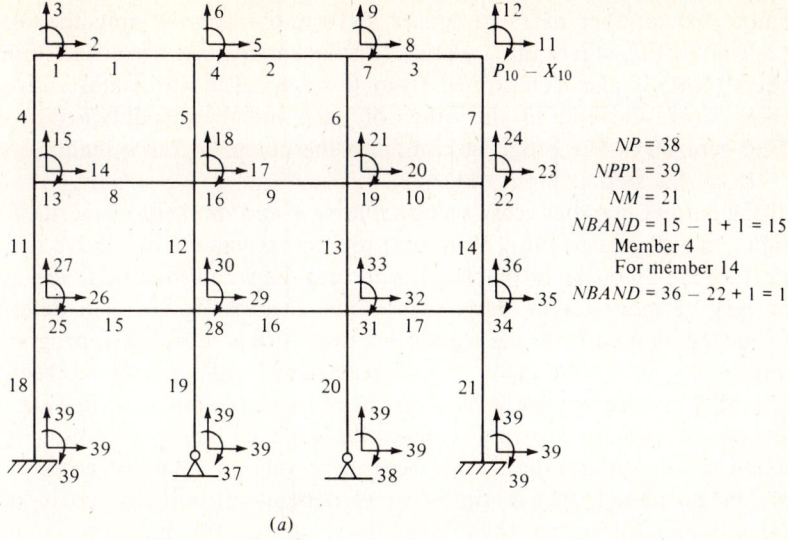

$NP = 38$
$NPP1 = 39$
$NM = 21$
$NBAND = 15 - 1 + 1 = 15$
Member 4
For member 14
$NBAND = 36 - 22 + 1 = 15$

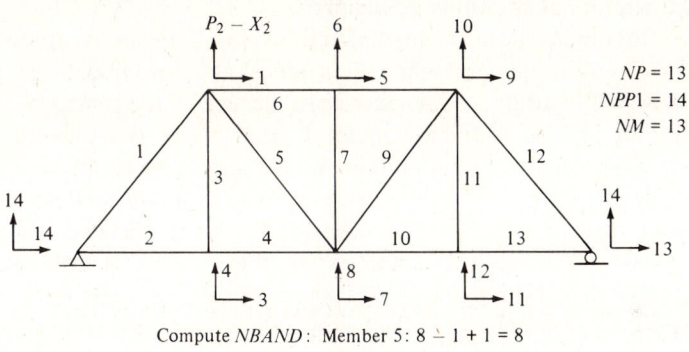

$NP = 13$
$NPP1 = 14$
$NM = 13$

Compute $NBAND$: Member 5: $8 - 1 + 1 = 8$

(b)

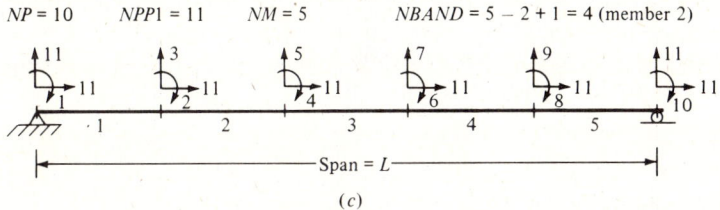

(c)

Figure 2-8 Coding of several structural frames. (a) Coding a rigid frame structure with column bases both pinned and fixed. (b) Coding a truss. (c) Coding of a beam across one span. Note that it is conventional to neglect changes in beam length, hence those $P-X$ values are made NPP1. This coding will give bending moments at each node as well as beam slope and vertical deflection. This configuration may be used for lower chords of bridge trusses or bridge beams with node locations located where transverse floor beams frame to the chord or beam/girder.

determine the number of *P–X* entries (NP) and those defining the reactions (NP + 1 = NPP1), where the *X* values (displacements) are zero. The number of members (NM) is also determined from this step. The horizontal and vertical distances from one end to the other of each member is obtained from the building geometry. The program computes the actual member length, using *H* and *V* from this step as input data.

If the correct member cross-sectional area *A* and moment of inertia *I* values are input, the displacements (*X* matrix) are actual values. If relative (or incorrect) values are used, the displacements are not the true values (also the member forces may be incorrect in many cases). The member forces of a determinate truss can be obtained from the use of any area (use area = 1.0 so program does not divide by zero), but again the displacements will only be correct if the correct member area is used. The moment of inertia is not used in determinate or indeterminate pinned truss computations. The member forces of indeterminate truss configurations are dependent on the member cross-sectional areas. The member forces in rigid frames depend on both the cross-sectional area and moment of inertia; thus this is always an iterative problem, as one must use some estimated intial values, obtain output, revise the initial input as required, and make additional solutions as required.

If it is desired to obtain floor beam deflections (or deflections along a highway bridge), it is only necessary to add nodes at those locations as illustrated in Fig. 2-8. While adding nodes increases the size of the matrix to be inverted, the routine (a banded reduction method) used in the computer program here only uses a part of the global matrix, so that very large problems can be very efficiently solved. To take full advantage of this reduction method, the coding should be such that the difference between the P_1 on the near end and P_4 (for trusses) or P_6 at the far end is as small as possible, since this difference [NBAND = NPE(4 or 6) − NPE(1) + 1] sets the size of the matrix to be reduced. Note that NBAND is only 4 for the beam coding illustrated in Fig. 2-8. This means that a very large number of beam segments can be used, or alternatively, a continuous bridge with five or six spans with 15 to 20 nodes/span can be easily solved using a very small amount of computer core, since the value of STIFF(I) is NBAND × NP = 4 × NP; with NP = 100, there are only 400 STIFF(I) entries versus 100 × 100 = 10,000 entries for a standard matrix inversion. The solution time is down to about $\frac{1}{10}$ that of the 100 × 100 matrix.

2-8 THE *P* MATRIX

The *P* matrix is developed from the structure loads considering each finite element as either a truss element that has just node forces (no bending) or a beam element based on using statics and the external forces applied to the member, such as wind, roof loads, and wheel loads. It should be noted that wheel loads between truss panels are prorated to adjacent nodes using simple beam analysis, as if that truss member is a beam with the reactions at the nodes.

The finite elements making up a beam are considered as a series of fixed-end beams; thus the P-matrix entries will include the fixed-end moments (FEMs) summed for the several beam elements meeting at a joint in addition to the shears as illustrated in Fig. 2-2.

In both beam and truss analyses, the node forces (equal and opposite to those acting on the connecting elements) are resolved into components parallel to the translation P_i's.

Where beam and beam–columns (axial force and bending) are analyzed, it is convenient to have the computer program compute the FEMs and shears for the several beam elements using the beam loading, and make the summing process as each element contribution to a node is found and then build the P matrix. The program should also be able to read in selected additional P-matrix entries, which can be used and/or added to the value(s) already in the P matrix. The computer program in the Appendix allows this procedure.

The user should note in using the computer program to develop the P-matrix entries that *all* dead and live loads are *applied along the length of the beam element* in the direction of gravity using a (+) sign. This allows the program to correctly compute the P-matrix entries (and signs) for sloping beam members.

Wind loads applied to sloping, flat, and vertical beam members are always applied along and normal to the member axis. If the slope is $0 \leq \theta \leq 30°$, the sign will usually be (−) because of the aerodynamic (uplift) effect. This allows the program to develop the P matrix for the wind NLC. Since gravity and wind loads along a member cannot be combined because of a horizontal component of wind on a sloping member, it is convenient to store the $D + L$ (or D only) analysis and treat the wind case totally separately, then combine wind with the stored $D + L$ analysis to obtain the design case for $D + L + W$. Specifications allow the allowable steel stresses to be increased one-third in any stress condition, including wind (as long as the resulting member is not smaller than for $D + L$ separately). This stress increase is equivalent to reducing the member forces, including wind (axial forces and moments), by 25 percent. The computer program also does this, so that the designer merely obtains the maximum axial force and/or moment from *any* of the NLC for design of that element. To take advantage of the program ability to do this, it is necessary to order the wind NLC (and using NLW) after the $D + L$ load NLC. This procedure is illustrated in the several examples displaying computer output. The reader should particularly inspect the input/output for Example 2-7, which uses wind pressure on a sloping beam element.

The coding of NP and noting NP + 1 = NPP1 for specifying zero displacements (including but not limited to) for reactions excludes those global matrix entries from consideration. Those P-matrix entries developed by the computer for P(NPP1) are not used in the analysis. The reader should note that these values will go directly into the reaction and therefore do not cause internal member forces (see the output check in Fig. E2-4*d*).

One of the very early decisions the structural designer must make is whether the superstructure (columns in particular) is pinned (allowing rotation) or rigidly

68 STRUCTURAL STEEL DESIGN

(no rotation) attached to the foundation and whether the foundation member can restrict rotation. There are no fixed rules to use for making this design decision—each structure is considered on its merits. If the structure is pinned, the method of attaching the member is usually different than for the rigid case, which does require sufficient design/analysis to develop the computed moment. It would appear reasonable for the office buildings in Examples 2-3 and 2-4 that a pinned connection to the basement wall (which would likely be pilastered at these locations) is appropriate. The pinned connections for the basement columns is more a matter of judgment, since rotational restraint could more easily be obtained at these locations with the footing geometry shown.

Example 2-1 Show the P-matrix entries for the 2 NLC (moving load) for the truss partially shown in Fig. E2-1a.

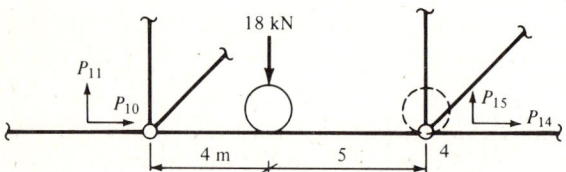

Figure E2-1a

SOLUTION Consider the fixed beam shown in Figure E2-1b:
For node 4:
Note that R_R is *opposite* to the direction of P_{15}, which has been defined as (+). Therefore, the P matrix (and for the other load condition also) is:

	NLC	
P	1	2
:		
10	0	0
11	−10	0
:		
14	0	0
15	−8	−18

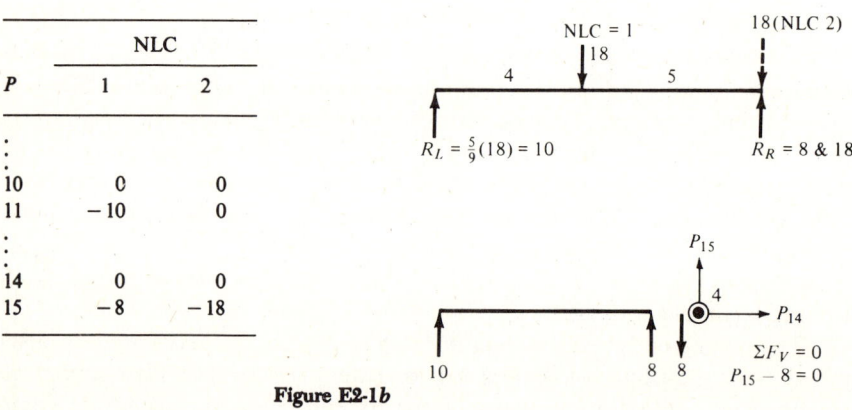

Figure E2-1b

///

Example 2-2 Show the P-matrix entries for the two-member frame of Fig. E2-2a.

ELEMENTS OF FRAME, TRUSS, AND BRIDGE DESIGN 69

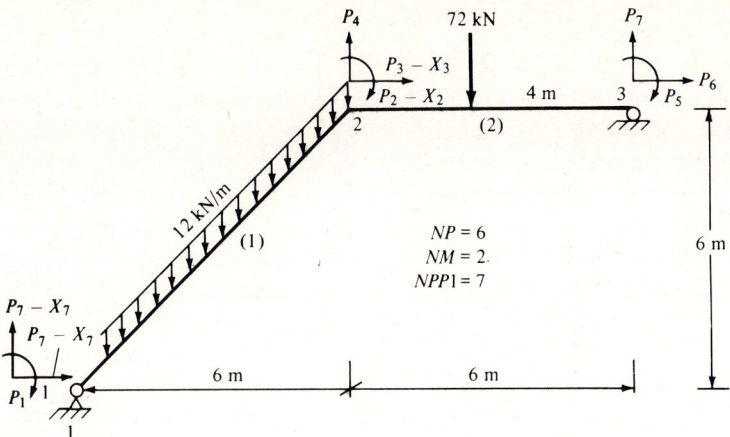

Figure E2-2a

SOLUTION It will be necessary to develop the necessary equations for FEM for sloping members. For this refer to Fig. E2-2b and note that loads normal to member cause FEM. Now, returning to the initial problem, for the sloping member $\theta = 45°$:

$$\text{FEM} = \frac{12(0.70711)}{12}(\frac{6}{0.70711})^2 = 50.91 \text{ kN} \cdot \text{m}$$

$$R_L = R_R = (12)\frac{6}{0.70711}(0.5) = 50.91 \text{ kN}$$

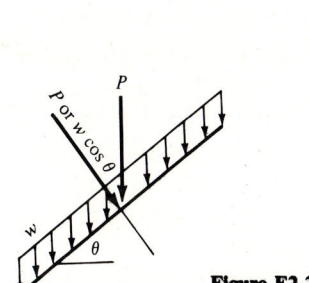

Figure E2-2b

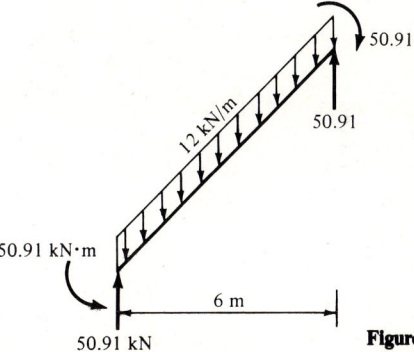

Figure E2-2c

For the horizontal member:

$$R'_L = \frac{4(72)}{6} = 48.0 \text{ kN} \quad R'_R = \frac{2(72)}{6} = 24.0 \text{ kN}$$

$$\text{FEM}'_L = \frac{Pab^2}{L^2} = \frac{72(2)(4)^2}{6^2} = 64 \text{ kN} \cdot \text{m}$$

$$\text{FEM}'_R = \frac{Pba^2}{L^2} = \frac{72(4)(2)^2}{6^2} = 32 \text{ kN} \cdot \text{m}$$

70 STRUCTURAL STEEL DESIGN

Now consider the following sketch for nodes 1, 2, and 3:
$$P_3 = 0$$
$$P_2 = 64 - 50.91 = 13.09 \text{ (note +)}$$
$$P_4 = -(50.91 + 48.0) = -98.91 \text{ kN}$$

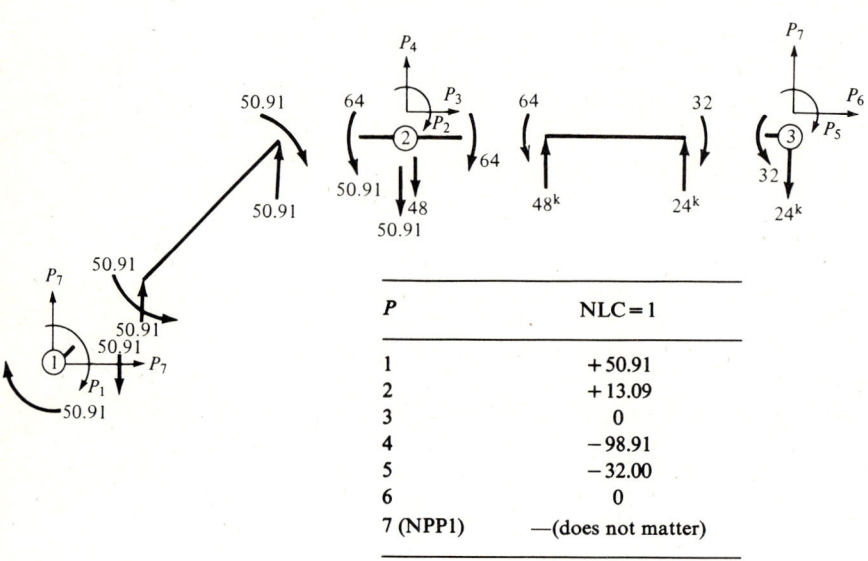

P	NLC = 1
1	+50.91
2	+13.09
3	0
4	−98.91
5	−32.00
6	0
7 (NPP1)	—(does not matter)

///

2-9 LOAD CONDITIONS

In most structural analysis there will be several combinations of loads which must be investigated to find that combination which produces the largest internal member forces. Each combination of dead or other loads is called a *load condition* (NLC in computer program). The load conditions produce the nodal loads making up the P matrix in the equation

$$X = (ASA^T)^{-1}\{P\}$$

The several load conditions for a rigid frame or truss in an industrial building might include:

Load combination	NLC
Dead + live	1
Dead + live in alternate bays	2
Dead + live + snow	3
Dead + live + $\frac{1}{2}$ snow	4
Dead + live + snow one side + wind	5
Dead + live + wind	6

The several load conditions for a bridge would include

> Dead
> Dead + live, including impact
> Dead + live + wind

The live load is actually an influence-line problem, since the stresses in various bridge elements are load-position-dependent. A very practical method is to develop a live-load P matrix by positioning the H or HS truck or the Cooper railroad loading at intervals of, say, 2 or 5 ft along the bridge span. A computer program can be used for this, with the output punched on cards in a format suitable for using the computer program of the previous section.

Impact and combining of dead plus live loads may be automated via computer programming; however, impact may not be a constant for all the elements in a bridge, and since impact is applied only to the live loads, the designer must make a judgment on how to most efficiently combine the dead plus live load and impact effects to obtain the design force for the given bridge element. This is a consideration that warrants very careful initial analysis, since there may be 40 to 60 conditions on long bridges due to live loading and the output must be scanned to obtain the maximum force in any element from all the several load conditions.

Computer programs are included in the Appendix to develop the load matrix via punched cards for analysis of a highway or railroad truss.

2-10 CHECKING COMPUTER OUTPUT

It is essential that computer output be checked for correctness. This checking consists of two basic steps:

1. Always check the member and load data input for a decimal error. If the P–X coding is mispunched, the program may or may not work, depending on whether NP is so large that it exceeds the stiffness matrix DIMENSION or whether the superposition effect is contributed to the incorrect global matrix location. Note that a correct analysis can be made for incorrect member data such as A or I. A "correct" analysis may be made for incorrect member length data. This type of "correct" analysis will usually satisfy the statics checks.
2. Always perform a statics check at selected nodes and/or ΣF_h and $\Sigma F_v = 0$ for the entire structure. Generally, if a node near the beginning and one near the end of the coding sequence is checked and both satisfy statics, the problem has been correctly solved *for that input data*.

Problem checks can be performed in several ways, as illustrated in the examples displaying computer output. Generally, a corner where the end moments of the two contributing members should be equal and opposite (within computer roundoff error) should be made first, as these can be made with a minimum of computational effort—often by inspection. Summing forces above

72 STRUCTURAL STEEL DESIGN

a story are also often convenient, as the column axial forces are directly given. Advantage can often be taken of symmetry of structure and loadings, as in the two office building examples that follow (without wind). Sometimes the *X* matrix can be used to advantage, as illustrated in the column check of the office building.

Truss checking is similar to that for rigid frames. Always take advantage of truss geometry by checking members in which the internal force is known to be zero. If the computer output does not give zero and the load matrix is correct, there is something surely wrong with the input data, such as mispunching an NP or a mismatch of *H* and *V* (either length or signs). Perform any additional statics checks near each end of the truss at joints where a minimum of bars connect. If these joints do not satisfy statics, the more complicated joints do not either and the problem needs to be reprogrammed.

2-11 DESIGN EXAMPLES

The following several design examples will further illustrate frame coding and computer input/output for the analysis computer program in the Appendix. The computer output from these examples will be used in many of the design examples in later chapters to illustrate the problems in structural (steel) design in a manner somewhat like that which would take place in actual design. In the interest of saving text space and maintaining reader interest, the examples are considerably edited from actual design problems.

> **Example 2-3** A small three-story office building is as shown in plan and elevation on Fig. E2-3*a*. Wind bracing will be used in the E-W direction (at bay and locations marked wb) together with a simple framing scheme. A rigid frame (with resulting member end moments) will be used in the N-S direction. A brick veneer exterior will be carried by lintel beams to the exterior columns and at each floor. The framing method allows a corridor between the interior columns with a clear space for the office areas. Rental space will be in blocks: a, b, c, and so on. This space will contain miscellaneous office furniture, partitions, and so on, based on rentor needs and desires. With three stories, a flight of stairs at each end plus a service elevator on the west end for large equipment will be the total entrance/exit allowance. Air conditioning and heating will utilize a heat pump in an auxiliary shed (not shown). The basement will contain the remainder of the environmental equipment and provide extra storage. Use the NBC for general design.
>
> We will use a flat roof which may be used for worker exercising during the day. For this additional activity, the roof will be designed for a live load of 80 psf (as opposed to 20 to 30 psf for usual live and/or snow). The reader should note that building codes and material specifications stipulate *minimum* requirements—the designer can always use larger values.

ELEMENTS OF FRAME, TRUSS, AND BRIDGE DESIGN **73**

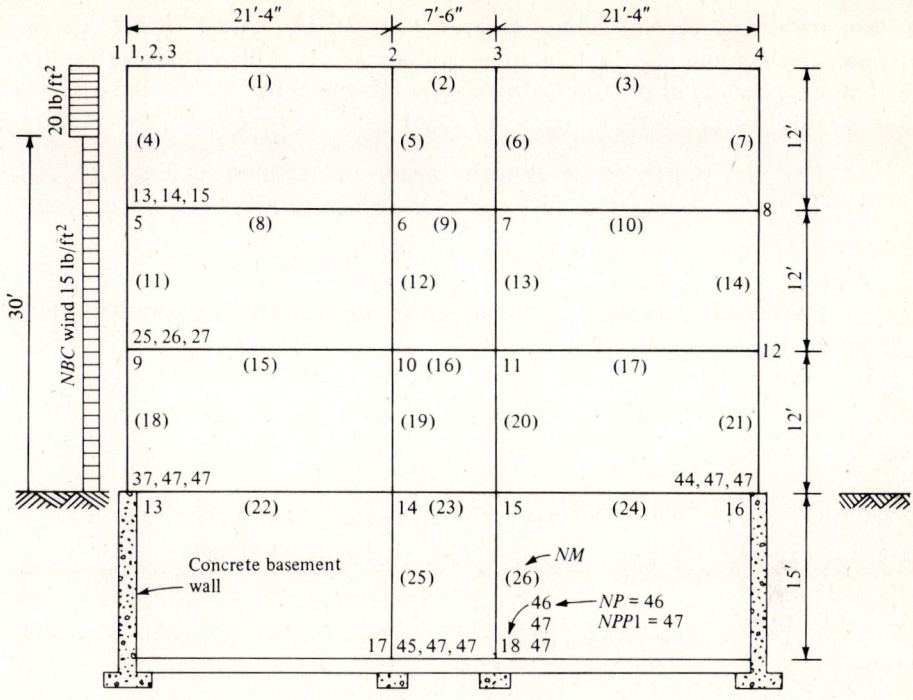

Elevation of typical interior bent

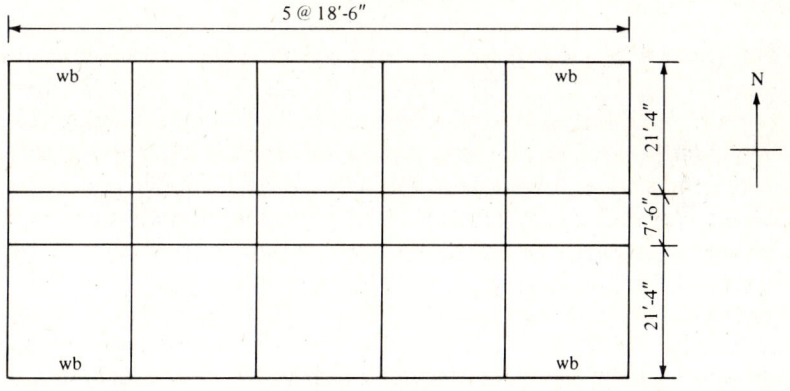

Plan view

Figure E2-3a

SOLUTION The general computer coding is partially shown in Fig. E2-3a and also on Fig. E2-3d (computer output sheet). This coding gives the following computer program control data information:

$$NM = 26 \text{(members)} \quad NP = 46 \quad NBAND = 11$$

The next step is to develop the beam and column loadings for this typical interior bent (refer to Fig. E2-3b, which displays loads after computations).

Roof:

Dead load: estimate 5 in of concrete on a metal deck supported by steel bar joists.

Concrete: (5/12)144	= 60 psf
Metal deck and joists (estimated)	= 5 psf
Ceiling, ductwork, electrical, etc.	= 5 psf
Total	= 70 psf

Live load: check reduction in 21.3 × 18.5 spaces. Normally, the roof live load is not reduced; however, we are not using a standard roof live loading.

$$R = 21.3(18.5)(0.08) = 31.5 \text{ percent}$$

or

$$R = \frac{D + L}{4.33 L}(100) = \frac{70 + 80}{4.33(80)}(100) = 43.3 \text{ percent}$$

Note that the $L = 80$ psf value is from Table IV-4. Use a 30 percent reduction.

$$\text{Live load} = 80(1.00 - 0.30) = 56 \text{ psf}$$

In a 7.5-ft span, $R = 7.5(18.5) = 138.8 \text{ ft}^2 < 150$ (no reduction).

The equivalent beam loads for the roof are:

$$\text{Dead load} = 0.070(18.5) = 1.295 \text{ kips/ft}$$
$$\text{Live load in 18.5-ft span} = 0.056(18.5) = 1.036 \text{ kips/ft}$$
$$\text{Live load in 7.5-ft span} = 0.080(18.5) = 1.48 \text{ kips/ft}$$

Other floors:

Estimate dead load with increase for floor surfacing and ceiling finishing	= 80 psf
Corridors (Table IV-4 and conservative use)	= 100 psf
Office space	= 80 psf

The office space will have a 30 percent reduction in live load, but no

ELEMENTS OF FRAME, TRUSS, AND BRIDGE DESIGN **75**

live-load reduction in corridor. This gives the following beam loads:

$$\text{Dead load} = 0.080(18.5) \qquad = 1.48 \text{ kips/ft}$$
$$\text{Live load office} = 0.080(0.70)(18.5) = 1.036 \text{ kips/ft}$$
$$\text{Live load corridor} = 0.10(18.5) \qquad = 1.85 \text{ kips/ft}$$

These loads are shown in Fig. E2-3b.

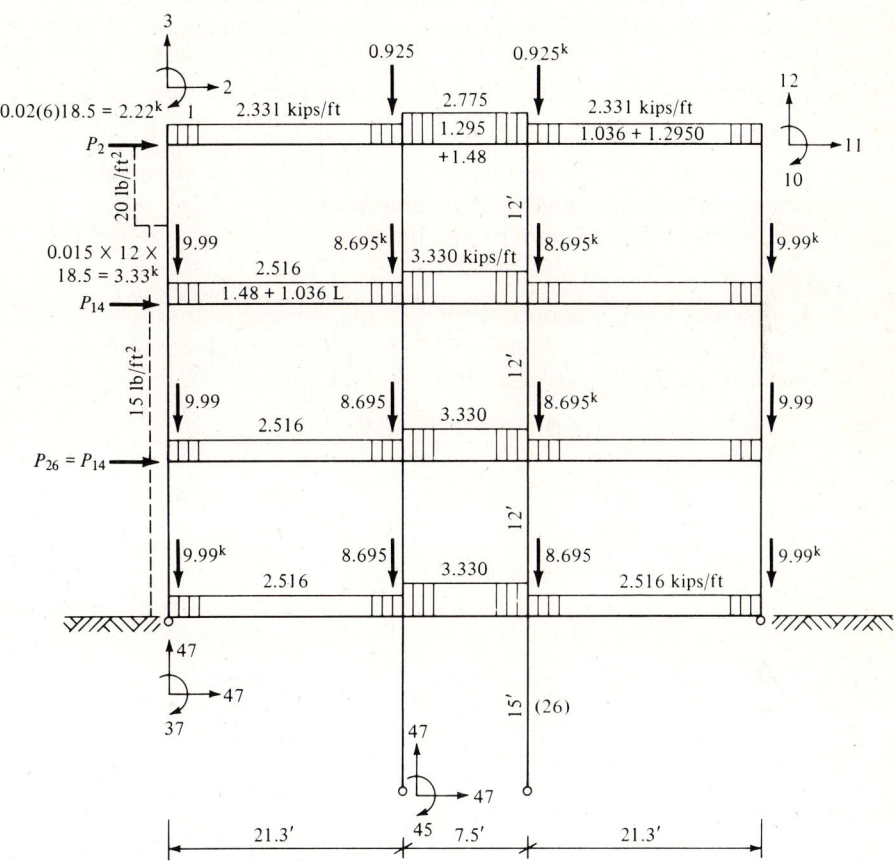

Figure E2-3b General frame loading.

The exterior brick veneer wall will contribute column loads for two upper stories as shown. The values are obtained as:

Obtain weight of 4-in brick veneer (Table IV-3) = 80/2 = 40 psf plus interior wall finish, fixtures, etc., of 5 psf gives a total = 45 psf.

$$\text{Column load} = 12(18.5)(0.045) = 9.99 \text{ kips}$$

The longitudinal walls defining the corridor and longitudinal beam contribute interior column loads:

> Tile wall at 8 in = 35 psf
> Beam at 50 lb/ft (est.) = 50 plf
> Column shear = 12(18.5)(0.035) + 18.5(0.050) = 8.695 kips
> For roof = 0.050(18.5) = 0.925 kip (interior columns)

Neglect any beam weight contributing shear to the roof line of exterior columns, since the values are two small to be either reliable or to affect the design.

Now make some preliminary member size estimates with the following practical considerations:

1. Use continuous column (no splices) for full 36 ft of height. [Splice interior columns (maybe) at first floor level.]
2. Use constant-depth beams across the bent.

For beams:

If the roof beam is simply supported, the moment is $M = wL^2/8$, and if fully fixed, the moment is $M = wL^2/12$. The actual value is somewhere in between, so taking an average:

$$M_s = \frac{(1.036 + 1.48)(21.3)^2}{8} = 142.7 \text{ ft} \cdot \text{kips}$$

$$M_f = (142.7)\frac{8}{12} = 95.1 \text{ ft} \cdot \text{kips}$$

$$M_{av} = 118.9, \text{ say } 120 \text{ ft} \cdot \text{kips}$$

The required section modulus based on an allowable bending stress of 24 ksi (Chap. 7) is

$$S_x = \frac{120(12)}{24} = 60 \text{ in}^3$$

Use W16 × 40:

$$S_x = 64.7 \qquad A = 11.80 \text{ in}^2 \qquad I_x = 518.0 \text{ in}^4$$

For columns:

Estimate a length coefficient (Chap. 7) $K = 1.2 \rightarrow KL = 12 \times 1.2 = 14.4$ ft. Estimate $F_a = 16$ ksi (something less than 22 ksi).

$$P_{approx} = 0.925 + (8.695)(3) + 12.5(3) + (2.775)\frac{7.5}{2} + (2.331)\frac{21.3}{2}$$

$$+ 26.79(3) = 180.1 \text{ kips}$$

This computation accumulates the beam shears from longitudinal contributions + the uniform beam loads from roof to the first floor of an interior column.

STIFFNESS METHOD OF RIGID FRAME AND TRUSS ANALYSIS

```
NO OF NP =        46      NO OF MEMBERS =   26      NO OF LOAD CONDIT =   2
NO NLC TC WIND NLC =   2  NO NON-ZERO P =    3      METRIC (1 = METRIC) = 0
NO ISTIFF ZEROED =    800 MOD ELASTICITY = 29000.0

   NLC FOR D + L =  0     NLC FOR WIND ON BEAM =  0
```

MEMNO	W	P1	P2	X1	X2	V1	V2
1	2.331	0.000	0.000	0.000	0.000	0.000	0.925
2	2.331	0.000	0.000	0.000	0.000	0.000	0.925
3	2.775	0.000	0.000	0.000	0.000	0.000	0.000
4	2.775	0.000	0.000	0.000	0.000	0.000	0.000
5	2.331	0.000	0.000	0.000	0.000	0.925	0.000
6	2.331	0.000	0.000	0.000	0.000	0.925	0.000
7	2.516	0.000	0.000	0.000	0.000	8.695	8.695
8	2.516	0.000	0.000	0.000	0.000	8.695	8.695
9	3.330	0.000	0.000	0.000	0.000	0.000	0.000
10	2.516	0.000	0.000	0.000	0.000	9.990	9.990
11	2.516	0.000	0.000	0.000	0.000	8.695	8.695
12	2.516	0.000	0.000	0.000	0.000	0.000	0.000
13	3.330	0.000	0.000	0.000	0.000	9.990	9.990
14	3.330	0.000	0.000	0.000	0.000	9.990	9.990
15	2.516	0.000	0.000	0.000	0.000	8.695	8.695
16	2.516	0.000	0.000	0.000	0.000	0.000	0.000
17	2.516	0.000	0.000	0.000	0.000	9.990	9.990
18	2.516	0.000	0.000	0.000	0.000	0.000	0.000
19	3.330	0.000	0.000	0.000	0.000	0.000	0.000
20	2.516	0.000	0.000	0.000	0.000	8.695	8.695
21	2.516	0.000	0.000	0.000	0.000	9.990	9.990

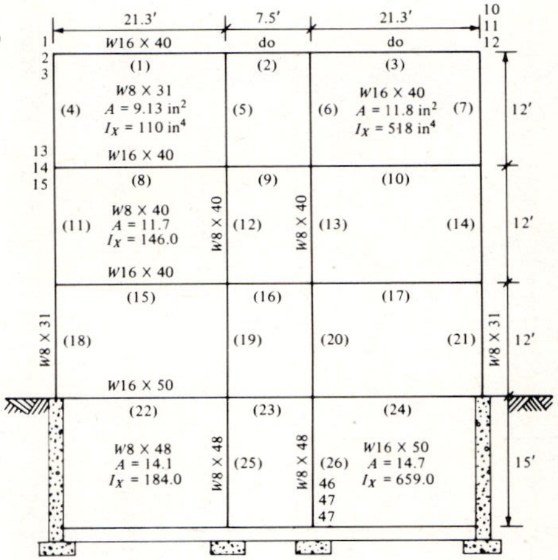

MEMBER	NP1	NP2	NP3	NP4	NP5	NP6	H	V	A	I	L	COS	SIN	FEM1	FEM2
1	1	2	3	4	5	6	255.60	0.00	11.8	518.	255.60	1.00000	0.00000	1057.55	-1057.55
2	4	5	6	7	8	9	90.00	0.00	11.8	518.	90.00	1.00000	0.00000	156.09	-156.09
3	7	8	9	10	11	12	255.60	0.00	11.8	518.	255.60	1.00000	0.00000	1057.55	-1057.55
4	13	14	15	16	17	18	0.00	144.00	9.13	110.	144.00	0.00000	1.00000	0.00	0.00
5	16	17	18	19	20	21	0.00	144.00	11.7	146.	144.00	0.00000	1.00000	0.00	0.00
6	19	20	21	22	23	24	0.00	144.00	11.8	518.	144.00	0.00000	1.00000	0.00	0.00
7	22	23	24	1	2	3	255.60	0.00	11.8	518.	255.60	1.00000	0.00000	1141.48	-1141.48
8	13	14	15	25	26	27	0.00	144.00	11.7	146.	144.00	0.00000	1.00000	187.31	-187.31
9	25	26	27	4	5	6	255.60	0.00	11.8	518.	255.60	1.00000	0.00000	1141.48	-1141.48
10	25	26	27	28	29	30	0.00	144.00	9.13	110.	144.00	0.00000	1.00000	0.00	0.00
11	28	29	30	31	32	33	0.00	144.00	11.7	146.	144.00	0.00000	1.00000	0.00	0.00
12	31	32	33	34	35	36	0.00	144.00	11.8	518.	144.00	0.00000	1.00000	0.00	0.00
13	34	35	36	7	8	9	255.60	0.00	11.8	518.	255.60	1.00000	0.00000	1141.48	-1141.48
14	28	29	30	37	38	39	0.00	144.00	11.7	146.	144.00	0.00000	1.00000	187.31	-187.31
15	31	32	33	40	41	42	0.00	144.00	11.9	146.	144.00	0.00000	1.00000	0.00	0.00
16	37	38	39	10	11	12	255.60	0.00	11.7	146.	255.60	1.00000	0.00000	0.00	0.00
17	37	38	39	40	41	42	0.00	144.00	9.13	110.	144.00	0.00000	1.00000	0.00	0.00
18	40	41	42	43	44	45	0.00	90.00	14.7	659.	90.00	0.00000	1.00000	0.00	0.00
19	43	44	45	46	47	48	0.00	180.00	14.1	184.	180.00	0.00000	1.00000	0.00	0.00
20	44	45	46	47	48	49	0.00	180.00	14.1	184.	180.00	0.00000	1.00000	0.00	0.00

```
(Frame diagram shows a 4-bay, 3-story rigid frame with member designations and sections:
W8 × 31, W16 × 40, W8 × 40, W8 × 48, W16 × 50, W16 × 40
Bay widths: 21.3', 7.5', 12', 12', 12', 15'
Column heights: 21.3', 12', 10'
Members labeled (1)-(26), nodes 1-47)
A = 9.13 in², I_x = 110 in⁴
A = 11.8 in², I_x = 518 in⁴
A = 11.7, I_x = 146.0
A = 14.7, I_x = 659.0
A = 14.1, I_x = 184.0
```

Figure E2-3c

```
NO STIFF(I) ENTRIES =     690          BAND WIDTH =   15

THE P-MATRIX, K AND K-IN                THE X-MATRIX, IN OR RADIANS

NP =  1      1057.55      1057.55       NX =  1      0.00394      0.00402
NP =  2         0.00         2.22       NX =  2      0.00487      0.40556
NP =  3       -24.83       -24.83       NX =  3     -0.09541     -0.09344
NP =  4      -901.46      -901.46       NX =  4     -0.00169     -0.00157
NP =  5         0.00         0.00       NX =  5      0.00033      0.39968
NP =  6       -36.16       -36.16       NX =  6     -0.19343     -0.19873
NP =  7       901.46       901.46       NX =  7      0.00169      0.00180
NP =  8         0.00         0.00       NX =  8     -0.00033      0.39872
NP =  9       -36.16       -36.16       NX =  9     -0.19343     -0.19813
NP = 10     -1057.55     -1057.55       NX = 10     -0.00394     -0.00386
NP = 11         0.00         0.00       NX = 11     -0.04487     -0.39386
NP = 12       -24.83       -24.83       NX = 12     -0.09541     -0.09738
NP = 13      1141.48      1141.48       NX = 13      0.00272      0.00303
NP = 14         0.00         3.33       NX = 14     -0.00081      0.33816
NP = 15       -36.79       -36.79       NX = 15     -0.08087     -0.08122
NP = 16      -954.17      -954.17       NX = 16     -0.00121     -0.00101
NP = 17         0.00         0.00       NX = 17     -0.00006      0.33683
NP = 18       -47.98       -47.98       NX = 18     -0.17249     -0.17249
NP = 19       954.17       954.17       NX = 19      0.00121      0.00140
NP = 20         0.00         0.00       NX = 20      0.00006     -0.33652
NP = 21       -47.98       -47.98       NX = 21     -0.17721     -0.18194
NP = 22     -1141.48     -1141.48       NX = 22     -0.00272     -0.00240
NP = 23         0.00         0.00       NX = 23      0.00082      0.33687
NP = 24       -36.79       -36.79       NX = 24     -0.08302     -0.08483
NP = 25      1141.48      1141.48       NX = 25      0.00275      0.00335
NP = 26         0.00         3.33       NX = 26      0.00061      0.19844
NP = 27       -36.79       -36.79       NX = 27     -0.17721     -0.04997
NP = 28      -954.17      -954.17       NX = 28     -0.00125     -0.00097
NP = 29         0.00         0.00       NX = 29      0.00007      0.19594
NP = 30       -47.98       -47.98       NX = 30     -0.14018     -0.13600
NP = 31       954.17       954.17       NX = 31      0.00125      0.00153
NP = 32         0.00         0.00       NX = 32     -0.00007      0.19537
NP = 33       -47.98       -47.98       NX = 33     -0.14018     -0.14437
NP = 34     -1141.48     -1141.48       NX = 34     -0.00275     -0.00216
NP = 35         0.00         0.00       NX = 35      0.00061      0.19433
NP = 36       -36.79       -36.79       NX = 36     -0.05121     -0.05245
NP = 37      1141.48      1141.48       NX = 37     -0.00353     -0.00387
NP = 38      -954.17      -954.17       NX = 38      0.00136      0.00123
NP = 39         0.00         0.00       NX = 39      0.00023      0.00194
NP = 40       -47.98       -47.98       NX = 40     -0.08232     -0.07989
NP = 41       954.17       954.17       NX = 41      0.00136      0.00149
NP = 42         0.00         0.00       NX = 42     -0.00023     -0.00149
NP = 43       -47.98       -47.98       NX = 43     -0.08232     -0.08475
NP = 44     -1141.48     -1141.48       NX = 44     -0.00353     -0.00319
NP = 45         0.00         0.00       NX = 45     -0.00068      0.00063
NP = 46         0.00         0.00       NX = 46     -0.00068     -0.00073
```

Figure E2-3d

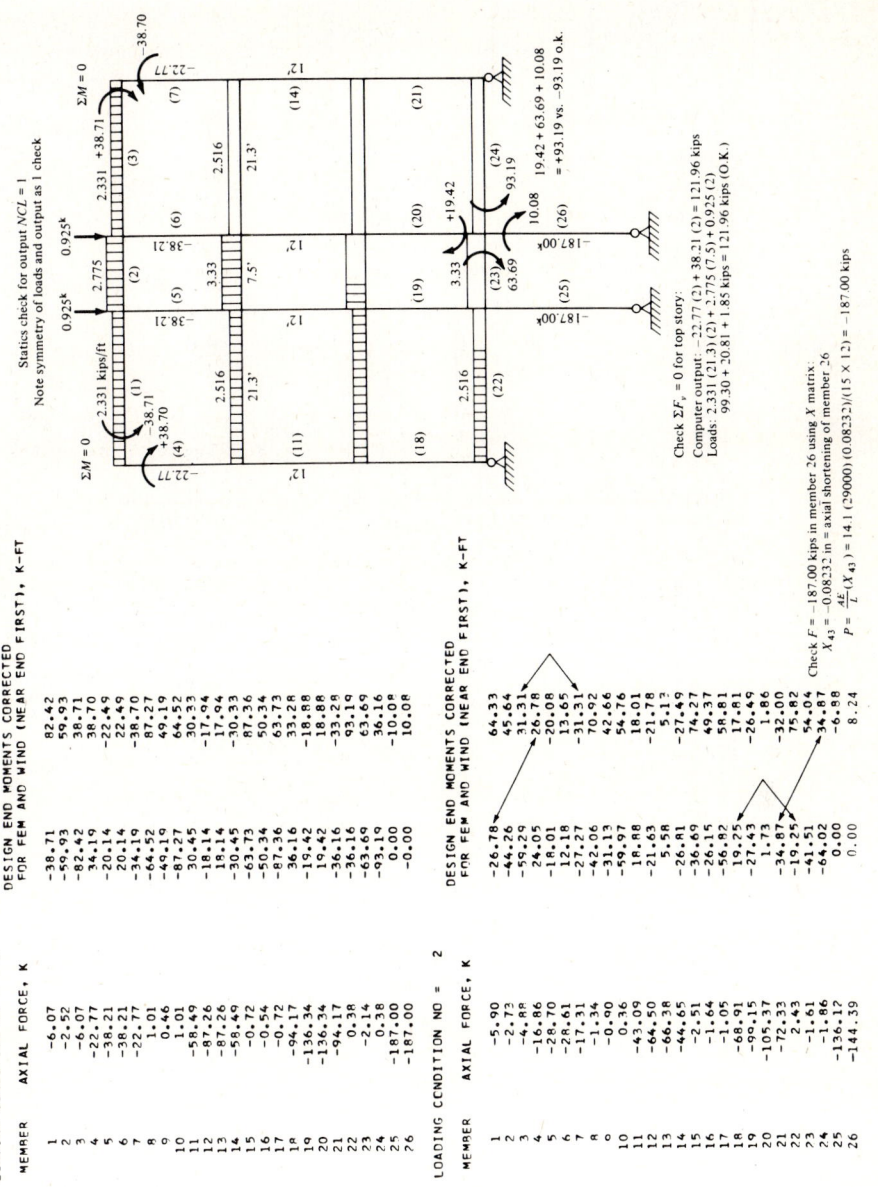

Figure E2-3e

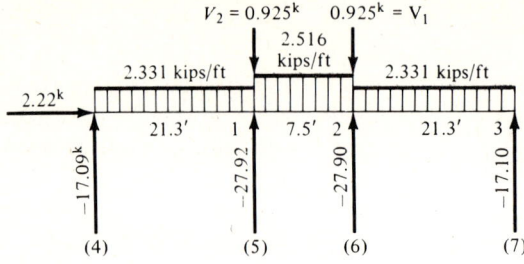

Check ΣF_v for top story:

$\Sigma F_{v(\text{loads})} = 2.331(21.3)(2) + 2.516(7.5) + 0.925(2) = 120.02$ kips

From computer output for values shown which have been reduced 25 percent for wind so that output is directly comparable to $D + L$ output:

$\Sigma F_v = (17.09 + 27.92 + 27.90 + 17.10)/0.75$

$\Sigma F_v = 120.01$ kips O.K.

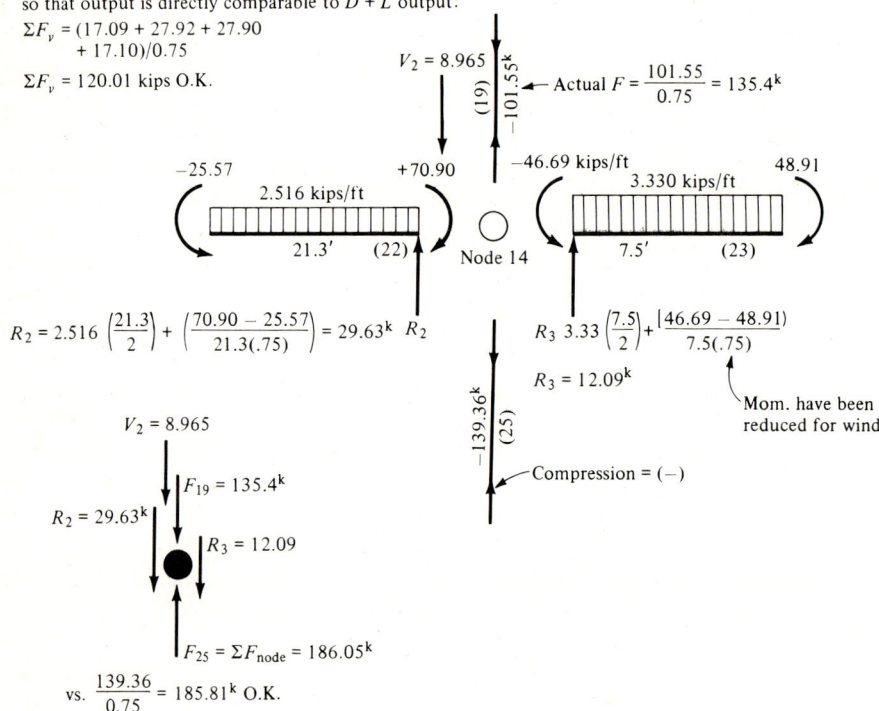

vs. $\dfrac{139.36}{0.75} = 185.81^k$ O.K.

Checking computer output for column (member 25) using computer output and the statics of node 14.

Figure E2-3f Output checking for NLC = 2 (with wind).

$$A_{\text{col}} = \frac{P}{F_a} = \frac{180.1}{16} = 11.2 \text{ in}^2$$

Use W8 × 40:

$$A = 11.70 \text{ in}^2 \qquad I_x = 146 \text{ in}^4$$

Spot check if O.K.: $KL/r_x = 14.4(12)/3.53 = 49$. The allowable column stress from Table II-5 = 18.5 ksi.

We will try this value for the first design iteration, since there will be end moments on the columns of an unknown amount that will effectively increase P. We will use a smaller section for exterior columns. Use W8 × 31:

$$A = 9.13 \text{ in}^2 \qquad I_x = 110 \text{ in}^4$$

For the two basement columns, use W8 × 48:

$$A = 14.10 \text{ in}^2 \qquad I_x = 184.0 \text{ in}^4$$

These data are used to make up a set of member data cards, beam loading cards (interspersed as appropriate in member data), and a P-matrix data set. Note the P-matrix entries read (PR(i, j)) are zero for NLC = 1 and take the values shown in Fig. E2-3b for NLC = 2. The remainder of the P matrix is built by the computer program using beam loading data and the column shears separately computed and also shown on Fig. E2-3b. The input is written back as a part of the output for designer checking (Fig. E2-3c). The remainder of the output is shown on Figs. E2-3d and E2-3e together with some output checks. Figure E2-3f further illustrates output check of NLC = 2.   ///

The reader should note that when wind is considered, the allowable stresses can be increased by one-third. So that the design is all on the same basis, the program multiplies all member forces by 0.75 for NLC that have wind forces as input. With this adjustment, the designer merely has to scan all the output for the maximum moment or axial force. The largest value in any load condition is then used with the allowable stress to determine if the member is adequate. Note that in checking the output (as illustrated in Fig. E2-1f) the fact that this 0.75 factor has been used must be considered so that statics is satisfied.

Example 2-4 We will design the small office building of Example 2-3 using SI units. Note that the dimensions are slightly different than when using a soft conversion. The same general design considerations/parameters will apply as in that example except that the brick veneer wall will be taken at 200 mm (approximately 2 × thickness of Example 2-3). Figure E2-4a displays the general building layout and SI dimensions.

SOLUTION The computer program will solve either SI or fps problems. It is necessary to include a "UNITS" card with fps or SI identification for use in the FORMAT statements to identify the units of the output.

Since the same building design criteria are being used, we may proceed directly to coding the joints (refer back to Fig. E2-3d since frame is similar) and developing the frame loads.

82 STRUCTURAL STEEL DESIGN

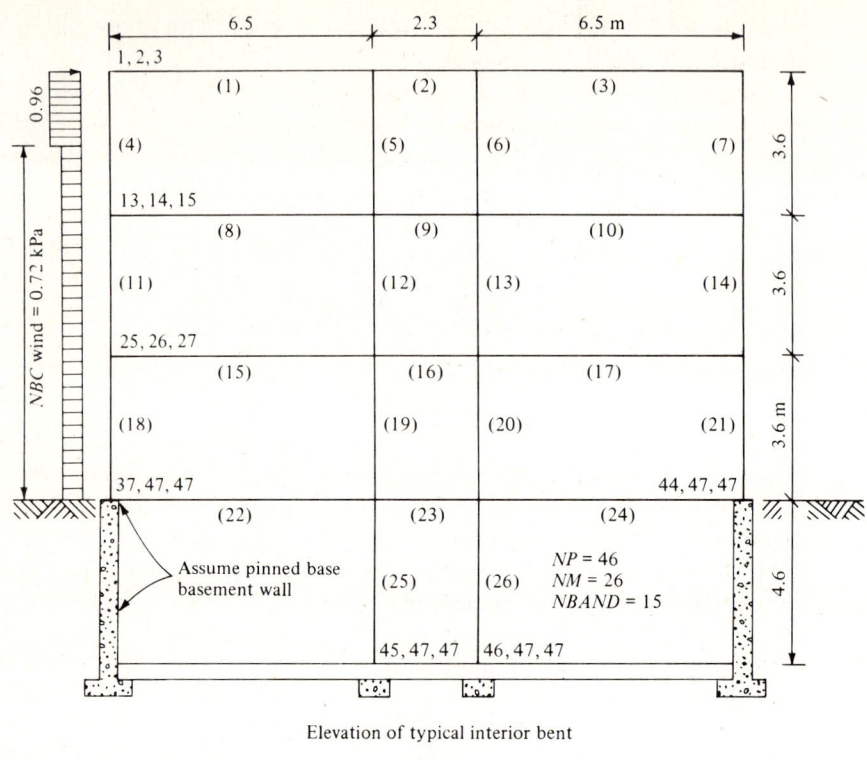

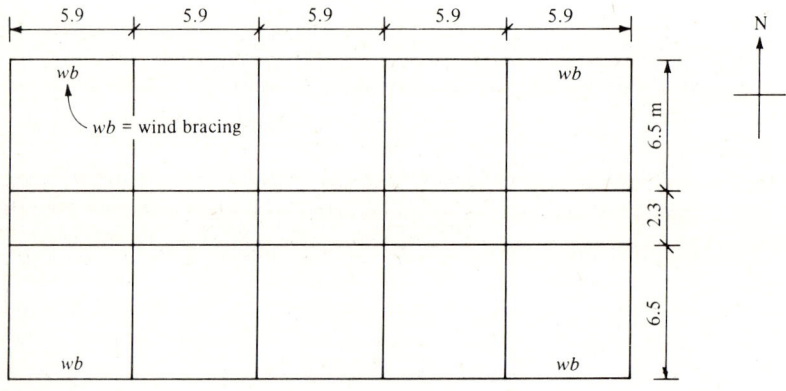

Figure E2-4a

Roof loads:

Dead load: estimate 130 mm of concrete on a metal deck supported by steel bar joists.

$$\begin{array}{lr}
\text{Concrete: } 0.13(23.5 \text{ kN/m}^3) & = 3.055 \text{ kPa} \\
\text{Metal deck and joists (estimated)} & = 0.263 \text{ kPa} \\
\text{Ceiling, finishing, electrical, etc.} & = 0.263 \text{ kPa} \\
\text{Total} & \overline{3.581 \text{ kPa}}
\end{array}$$

Live load: a live-load reduction can be used in the 6.5 × 5.9 m area. Basic live load = 4.0 kPa (Table IV-4).

$$R = 6.5(5.9)(0.0086) = 0.329 < 0.60$$

or

$$R = \frac{D + L}{4.33 L} = \frac{3.581 + 4.0}{4.33(4)} = 0.438 > 0.329$$

Use a 30 percent live-load reduction. The reduced live load = 4.0(1.0 − 0.30) = 2.8 kPa. Do not use a live-load reduction across the 2.3-m span.

The roof beam loads are:

Dead load = 3.581(5.9)	= 21.1 kN/m
Live load in 6.5-m span: 2.8(5.9)	= 16.5 kN/m
Live load in 2.3-m span: 4(5.9)	= 23.6 kN/m

Frame loads for other floors:

Use 30 percent reduction for office area live loads but none in corridor.

Dead load, including floor, finish, ceiling, ducting, etc.	= 3.9 kPa
Corridor live load (Table IV-4)	= 5.0 kPa
Office area	= 4.0 kPa

These load intensities give the following beam loads:

Dead load: 3.9(5.9)	= 23.01 kN/m
Office live load: 4.0(0.70)(5.9)	= 16.5 kN/m
Corridor live load: 5.0(5.9)	= 29.5 kN/m

The exterior brick veneer at 200 mm will contribute column loads (use Table IV-3):

3.77 kPa (brick) + finish: use 4.0 kPa

$$P = 4.0(5.9)(3.6) \qquad = 85.0 \text{ kN}$$

Longitudinal walls defining corridor + longitudinal beam:

Wall board + framing estimated	= 1.08 kPa
Beam at	= 0.75 kN/m

84 STRUCTURAL STEEL DESIGN

Interior column shear:

$$P = 1.08(5.9)(3.6) + 5.9(0.75) = 27.2 \text{ kN}$$

For the interior roof columns, use shear from longitudinal beam only (omit for exterior columns as excessive refinement):

$$P = 5.9(0.75) = 4.4 \text{ kN}$$

Using Example 2-3 as a guide, make preliminary beam and columns selections as follows:

All beams: W410 × 59.9 $A = 7.61 \times 10^{-3}$ $I_x = 215.6 \times 10^{-6}$

Exterior columns: W200 × 46.1 $A = 5.89 \times 10^{-3}$ $I_x = 45.8 \times 10^{-3}$

Interior columns: W200 × 59.5 $A = 7.55 \times 10^{-3}$ $I_x = 60.8 \times 10^{-3}$

Basement columns: W200 × 71.4 $A = 9.10 \times 10^{-3}$ $I_x = 76.6 \times 10^{-3}$

Figures E2-4b to E2-4d illustrate the input and computer output. Selected statics checks that can be made by inspection of force matrix are shown on Fig. E2-4d for NLC = 2. We shall make a statics check for node 1 and for ΣF_v of the entire structure for both NLC = 1 and 2. Note that the force matrix has been reduced by the factor of 0.75 (NLC = 2 with wind), so that the output from the several NLC can be readily compared for the maximum force to use in design and using the basic allowable column and bending stresses. We will have to put this factor back into the force values for any statics checks as illustrated on page 85, top.

Now check ΣF_v and $\Sigma F_h = 0$ for the entire structure. Compute the total weight for both vertical force checks:

$$37.6(6.5)(2) + 44.7(2.3) + 39.5(6.5)(6) + 52.5(2.3)(3) + 4.4(2)$$
$$+ 27.2(6) + 85(4)\dagger = 3006.36 \text{ kN}$$

For NLC = 1 (see diagram, page 85, bottom):
Vertical resisting forces: $R_1 + R_2$ + forces in basement columns.

$$F_r = 642.28 + 642.28 + 861.35 + 861.35 = 3007.28 \text{ kN (vs. 3006.36 kN)}$$

Difference is due to computer roundoff error. By inspection, $F_h = 0$, since loads are symmetrical.

For NLC = 2: Do not show distributed beam loads.
The vertical statics check is made by referring to Fig. E2-4d, where part of

† Note: two 85-kN forces go directly into reactions and are not considered.

ELEMENTS OF FRAME, TRUSS, AND BRIDGE DESIGN 85

At Node 1

For *NLC* = 1: Not shown on the beam is the distributed load but it is used to compute the end shear as shown. Note direction of arrows on members to illustrate compressive axial stresses.

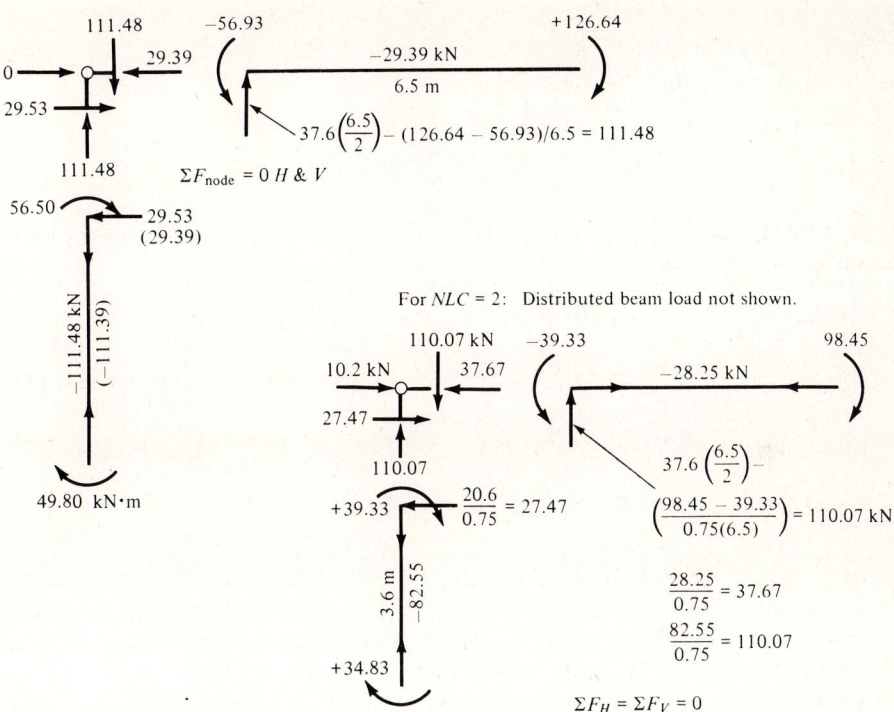

For *NLC* = 2: Distributed beam load not shown.

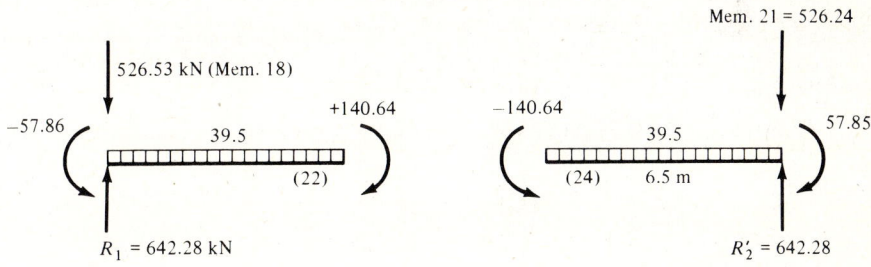

STIFFNESS METHOD OF RIGID FRAME AND TRUSS ANALYSIS

NO OF NP = 46 NO OF MEMBERS = 26 NO OF LOAD CONDIT = 2
NO NLC TO WIND NLC = 2 NO NON-ZERO P = 3 METRIC (1 = METRIC) = 1
NO ISTIFF ZEROED = 800 MOD ELASTICITY = 200000.0

NLC FOR D + L = 0 NLC FOR WIND ON BEAM = 0

MEMNO	W	P1	P2	X1	X2	V1	V2
1	37.600	0.000	0.000	0.000	0.000	0.000	4.400
1	37.600	0.000	0.000	0.000	0.000	0.000	4.400
2	44.700	0.000	0.000	0.000	0.000	0.000	0.000
2	44.700	0.000	0.000	0.000	0.000	0.000	0.000
3	37.600	0.000	0.000	0.000	0.000	4.400	0.000
3	37.600	0.000	0.000	0.000	0.000	4.400	0.000
8	39.500	0.000	0.000	0.000	0.000	85.000	27.200
8	39.500	0.000	0.000	0.000	0.000	85.000	27.200
9	52.500	0.000	0.000	0.000	0.000	0.000	0.000
9	52.500	0.000	0.000	0.000	0.000	0.000	0.000
10	39.550	0.000	0.000	0.000	0.000	27.200	85.000
10	39.550	0.000	0.000	0.000	0.000	27.200	85.000
15	39.550	0.000	0.000	0.000	0.000	85.000	27.200
15	39.550	0.000	0.000	0.000	0.000	85.000	27.200
16	52.500	0.000	0.000	0.000	0.000	0.000	0.000
16	52.500	0.000	0.000	0.000	0.000	0.000	0.000
17	39.500	0.000	0.000	0.000	0.000	27.200	85.000
17	39.500	0.000	0.000	0.000	0.000	27.200	85.000
22	39.500	0.000	0.000	0.000	0.000	85.000	27.200
22	39.500	0.000	0.000	0.000	0.000	85.000	27.200
23	52.500	0.000	0.000	0.000	0.000	0.000	0.000
23	52.500	0.000	0.000	0.000	0.000	0.000	0.000
24	39.500	0.000	0.000	0.000	0.000	27.200	85.000
24	39.500	0.000	0.000	0.000	0.000	27.200	85.000

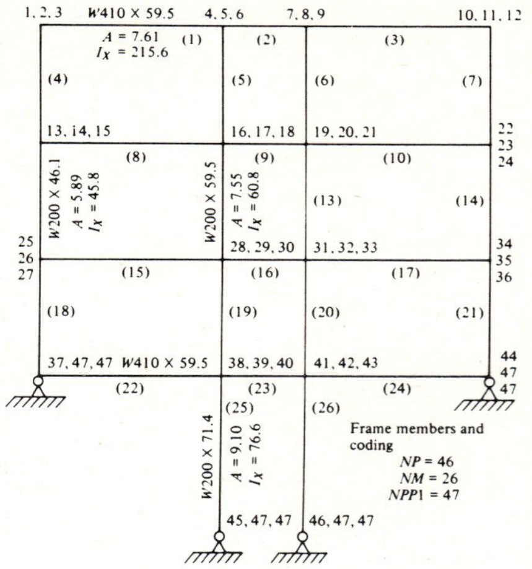

MEMBER	NP1	NP2	NP3	NP4	NP5	NP6	H	V	A	I	L	COS	SIN	FEM1	FEM2
1	1	2	3	4	5	6	6500.00	0.00	0.761E-02	0.216E-03	6500.00	1.00000	0.00000	132383.30	-132383.30
2	4	5	6	7	8	9	2300.00	0.00	0.761E-02	0.216E-03	2300.00	1.00000	0.00000	19705.23	-19705.23
3	7	8	9	10	11	12	6500.00	0.00	0.761E-02	0.216E-03	6500.00	1.00000	0.00000	132383.30	-132383.30
4	13	14	15	1	2	3	0.00	3600.00	0.589E-02	0.458E-04	3600.00	0.00000	1.00000	0.00	0.00
5	16	17	18	4	5	6	0.00	3600.00	0.755E-02	0.608E-04	3600.00	0.00000	1.00000	0.00	0.00
6	19	20	21	7	8	9	0.00	3600.00	0.755E-02	0.609E-04	3600.00	0.00000	1.00000	0.00	0.00
7	22	23	24	10	11	12	0.00	3600.00	0.589E-02	0.458E-04	3600.00	0.00000	1.00000	0.00	0.00
8	13	14	15	16	17	18	6500.00	0.00	0.761E-02	0.216E-03	6500.00	1.00000	0.00000	139072.80	-139072.80
9	16	17	18	19	20	21	2300.00	0.00	0.761E-02	0.216E-03	2300.00	1.00000	0.00000	23143.71	-23143.71
10	19	20	21	22	23	24	6500.00	0.00	0.761E-02	0.216E-03	6500.00	1.00000	0.00000	139248.90	-139248.90
11	25	26	27	13	14	15	0.00	3600.00	0.589E-02	0.458E-04	3600.00	0.00000	1.00000	0.00	0.00
12	28	29	30	16	17	18	0.00	3600.00	0.755E-02	0.608E-04	3600.00	0.00000	1.00000	0.00	0.00
13	31	32	33	19	20	21	0.00	3600.00	0.755E-02	0.608E-04	3600.00	0.00000	1.00000	0.00	0.00
14	34	35	36	22	23	24	0.00	3600.00	0.589E-02	0.458E-04	3600.00	0.00000	1.00000	0.00	0.00
15	25	26	27	28	29	30	6500.00	0.00	0.761E-02	0.216E-03	6500.00	1.00000	0.00000	139248.90	-139248.90
16	28	29	30	31	32	33	2300.00	0.00	0.761E-02	0.216E-03	2300.00	1.00000	0.00000	23143.71	-23143.71
17	31	32	33	34	35	36	6500.00	0.00	0.761E-02	0.216E-03	6500.00	1.00000	0.00000	139072.80	-139072.80
18	37	47	47	25	26	27	0.00	3600.00	0.589E-02	0.458E-04	3600.00	0.00000	1.00000	0.00	0.00
19	38	39	40	28	29	30	0.00	3600.00	0.755E-02	0.608E-04	3600.00	0.00000	1.00000	0.00	0.00
20	41	42	43	31	32	33	0.00	3600.00	0.755E-02	0.608E-04	3600.00	0.00000	1.00000	0.00	0.00
21	44	47	47	34	35	36	0.00	3600.00	0.589E-02	0.458E-04	3600.00	0.00000	1.00000	0.00	0.00
22	37	47	47	38	39	40	6500.00	0.00	0.761E-02	0.216E-03	6500.00	1.00000	0.00000	139072.80	-139072.80
23	38	39	40	41	42	43	2300.00	0.00	0.761E-02	0.216E-03	2300.00	1.00000	0.00000	23143.71	-23143.71
24	41	42	43	44	47	47	6500.00	0.00	0.761E-02	0.216E-03	6500.00	1.00000	0.00000	139072.80	-139072.80
25	45	47	47	38	39	40	0.00	3600.00	0.910E-02	0.766E-04	3600.00	0.00000	1.00000	0.00	0.00
26	46	47	47	41	42	43	0.00	3600.00	0.910E-02	0.766E-04	3600.00	0.00000	1.00000	0.00	0.00

Figure E2-4b

```
NO STIFF(I) ENTRIES =    690           BAND WIDTH =   15

THE P-MATRIX, KN AND K-MM               THE X-MATRIX, MM OR RADIANS

NP =  1     132383.30     132383.30     NX =  1     0.00422      0.00430
NP =  2          0.00         10.20     NX =  2     0.13429     10.25240
NP =  3       -122.20       -122.20     NX =  3    -2.92370     -2.87463
NP =  4     -11267.00    -11267.00     NX =  4    -0.00192     -0.00181
NP =  5          0.00          0.00     NX =  5     0.00877     10.09155
NP =  6       -178.00       -178.00     NX =  6    -4.64149     -4.52804
NP =  7     112678.00     112678.00     NX =  7     0.00192     -0.00203
NP =  8          0.00          0.00     NX =  8    -0.00793     10.06711
NP =  9       -178.00       -178.00     NX =  9    -4.64194     -4.75541
NP = 10    -132383.30    -132383.30     NX = 10    -0.00422     -0.00413
NP = 11          0.00          0.00     NX = 11    -0.13351      9.93314
NP = 12       -122.20       -122.20     NX = 12    -2.92418     -2.97336
NP = 13     139072.80     139072.80     NX = 13     0.00280      0.00312
NP = 14          0.00         15.30     NX = 14    -0.02431      8.55236
NP = 15       -213.38       -213.38     NX = 15    -2.58330     -2.53825
NP = 16    -115929.10    -115929.10     NX = 16    -0.00133     -0.00114
NP = 17          0.00          0.00     NX = 17    -0.00056      8.52131
NP = 18       -215.95       -215.95     NX = 18    -4.19129     -4.07779
NP = 19     116105.10     116105.10     NX = 19     0.00133      0.00153
NP = 20          0.00          0.00     NX = 20     0.00289      8.51323
NP = 21       -216.11       -216.11     NX = 21    -4.19181     -4.30535
NP = 22    -139248.90    -139248.90     NX = 22    -0.00281     -0.00248
NP = 23          0.00          0.00     NX = 23     0.02671      8.52660
NP = 24       -213.54       -213.54     NX = 24    -2.58378     -2.62890
NP = 25     139248.90     139248.90     NX = 25     0.00279      0.00339
NP = 26          0.00         15.30     NX = 26     0.03047      5.07696
NP = 27       -213.54       -213.54     NX = 27    -1.60909     -1.57808
NP = 28    -116105.10    -116105.10     NX = 28    -0.00135     -0.00107
NP = 29          0.00          0.00     NX = 29    -0.00412      4.99844
NP = 30       -216.11       -216.11     NX = 30    -3.21213     -3.11130
NP = 31     115929.10     115929.10     NX = 31     0.00135      0.00163
NP = 32          0.00          0.00     NX = 32    -0.00206      4.98083
NP = 33       -215.95       -215.95     NX = 33    -3.21199     -3.31094
NP = 34    -139072.80    -139072.80     NX = 34    -0.00278     -0.00218
NP = 35          0.00          0.00     NX = 35    -0.02834      4.94196
NP = 36       -213.38       -213.38     NX = 36    -1.60911     -1.64010
NP = 37     139072.80     139072.80     NX = 37     0.00430      0.00474
NP = 38    -115929.10    -115929.10     NX = 38    -0.00170     -0.00155
NP = 39          0.00          0.00     NX = 39     0.00716      0.06471
NP = 40       -215.95       -215.95     NX = 40    -1.70377     -1.65167
NP = 41     115929.10     115929.10     NX = 41     0.00170      0.00185
NP = 42          0.00          0.00     NX = 42    -0.00710      0.05042
NP = 43       -215.95       -215.95     NX = 43    -1.70373     -1.75594
NP = 44    -139072.80    -139072.80     NX = 44    -0.00430     -0.00388
NP = 45          0.00          0.00     NX = 45    -0.00085     -0.00090
NP = 46          0.00          0.00     NX = 46    -0.00085     -0.00090
```

Figure E2-4c

```
LOADING CONDITION NO =   1
                          DESIGN END MOMENTS CORRECTED
                          FOR FEM AND WIND (NEAR END FIRST), KN-M
MEMBER  AXIAL FORCE, KN
  1       -29.39          -56.50      126.81
  2       -11.05          -91.80       91.76
  3       -29.41         -126.80       56.52
  4      -111.38           49.30       56.50
  5      -188.84          -31.02      -35.01
  6      -188.80           31.05       35.04
  7      -111.30          -49.34      -56.52
  8         5.56          -92.23      131.08
  9         2.29          -73.00       73.15
 10         5.58         -131.28       92.30
 11      -318.78           42.86       42.83
 12      -410.70          -27.18      -77.06
 13      -410.08           27.17       27.08
 14      -318.93          -42.83      -42.96
 15        -6.17           93.00      130.63
 16        -4.09          -73.76       73.60
 17        -6.15         -130.43       92.93
 18      -526.53           57.86       50.14
 19      -632.67          -32.06      -29.68
 20      -632.63           32.04       29.65
 21      -526.54          -57.84      -50.10
 22         1.68          -86.86      140.64
 23        -0.44          -57.86       86.88
 24         1.66         -140.64       57.85
 25      -861.35            0.00      -21.72
 26      -861.33            0.00       21.72

LOADING CONDITION NO =   2
                          DESIGN END MOMENTS CORRECTED
                          FOR FEM AND WIND (NEAR END FIRST), KN-M
MEMBER  AXIAL FORCE, KN
  1       -28.25          -39.33*      98.45
  2       -12.13          -67.73       69.94
  3       -23.53          -91.73       45.49*
  4       -92.55          -34.83      -39.33*
  5      -141.64          -27.81      -30.73
  6      -141.58           19.23       21.79
  7       -84.54          -30.92      -45.44*
  8        -5.65          -44.84      105.70
  9        -4.01          -66.55       63.01
 10         2.35          -91.05       77.90
 11      -235.64          -26.76       25.72
 12      -303.41          -31.45*     -31.79
 13      -312.82            9.32        8.81
 14      -242.67          -37.55      -38.70
 15       -13.79          -54.60      109.88*
 16        -8.74          -13.79       71.59
 17        -8.83          -85.97       84.73
 18      -387.29           38.93*      27.84
 19      -459.80          -41.93      -39.50*
 20      -489.18            6.19        5.06
 21      -402.51          -53.64      -47.18
 22        11.36          -32.98*     113.07
 23      -242.67          -56.15       74.17
 24        -8.85          -97.96      -14.99*
 25      -626.26           -0.00       53.64
 26      -665.79            0.00       17.60
```

Figure E2-4d

*Increase 25% due to use of 0.75 wind factor.
Part of computations to check $\Sigma F_V = 0$

Selected output shown for wind load condition ($NLC = 2$)

$R'_L = \dfrac{(32.98 - 113.07)}{0.75}/6.5 + 39.5\,(6.5)/2$
$= 111.95$ kN*

$R'_R = \dfrac{(53.64 - 97.96)}{0.75}/6.5 + 39.5\,(6.5)/2$
$= 119.28$ kN*

the reactions at the basement wall are computed:

$$F_v = R'_L + \text{axial load}_{18} + R'_R + \text{axial load}_{21} + \text{axial load}_{25} + \text{axial load}_{26}$$

$$= 111.95 + 119.28 + \frac{387.29 + 402.51 + 626.26 + 665.79}{0.75}$$

$$= 231.23 + \frac{2081.85}{0.75} = 3007.03 \text{ (vs. 3006.36 kN)} \quad \text{O.K.}$$

Now consider $\Sigma F_h = 0$:

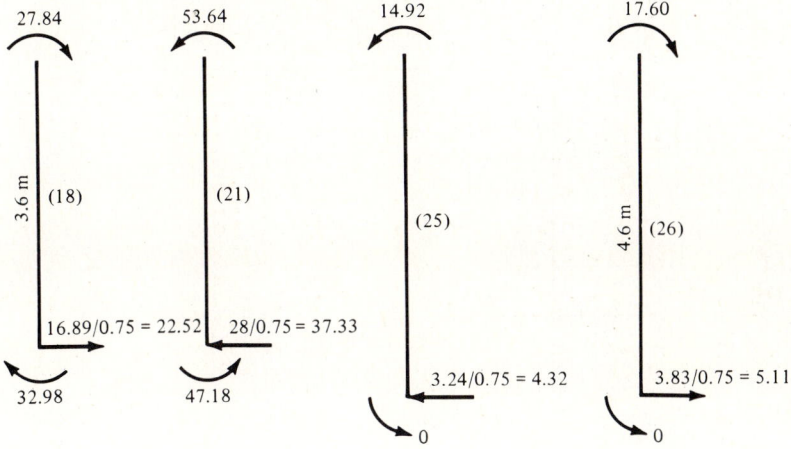

Member forces

Forces on two basement reactions, including effects from horizontal beam members 22 and 24:

$$\Sigma F_{h(\text{applied externally})} = 10.2 + 15.3(2) = 40.8 \text{ kN}$$

$$\Sigma F_{h(\text{nodes})} = -22.52 + 15.15 + 37.33 + 11.80 + 4.32 - 5.11$$

$$= 40.93 \text{ kN} \quad \text{O.K.} \qquad ///$$

Example 2-5 This is a partial design of an industrial building (for a manufacturing process) with a general elevation view shown in Fig. E2-5a, together with a reduced size plan view. With the large unsupported clear span and height required, roof beams are not deemed practical, so roof trusses with purlins will be used. These trusses will both reduce roof deflections and increase the overall frame rigidity, particularly by making the columns continuous to the top of the main roof truss. Making the column continuous will reduce column rotation and translation at the roof

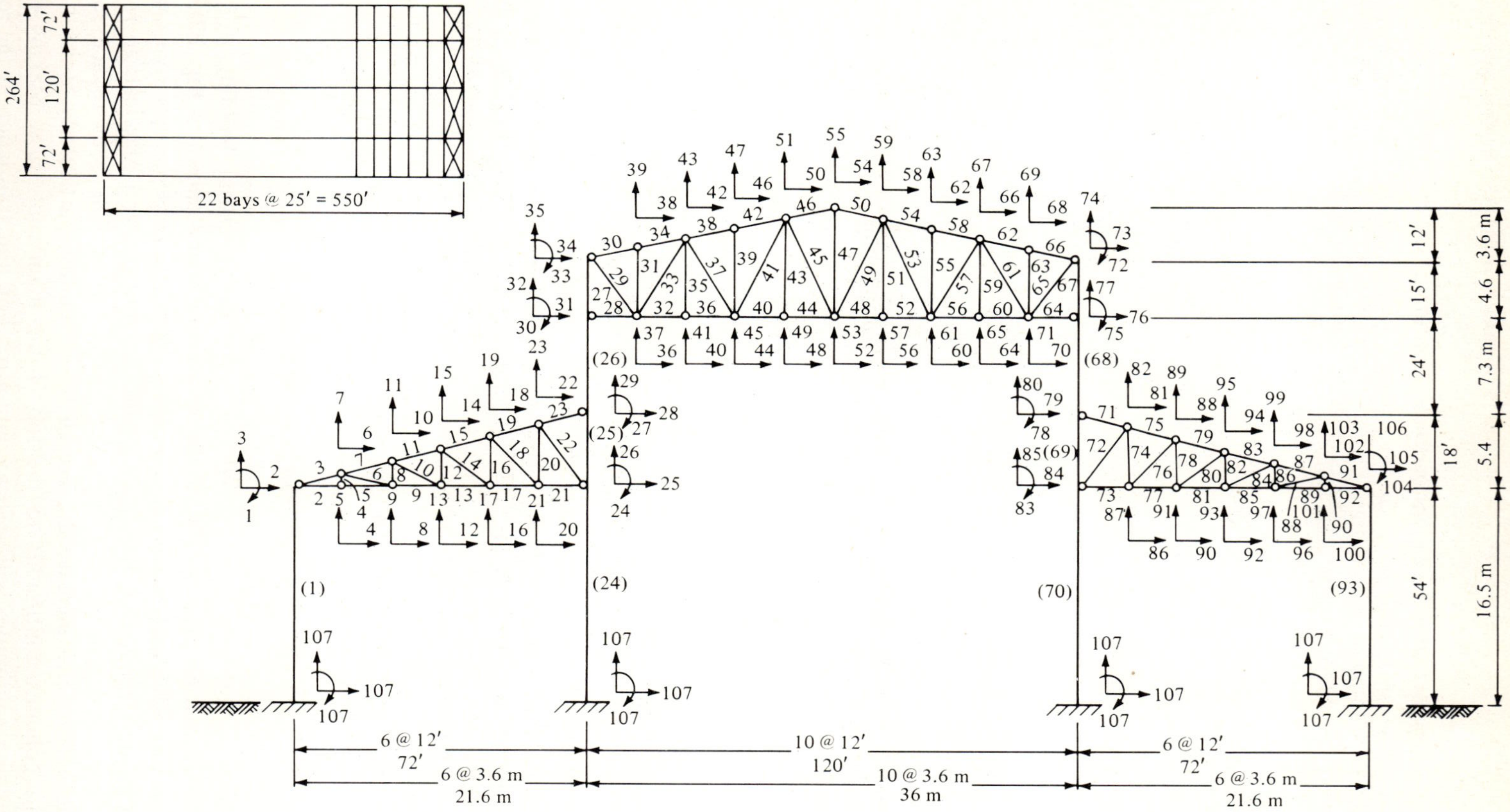

Figure E2-5a

line. The purlins (beams across bays to support the roofing) will likely be spaced at about 6-ft centers; thus one will fall between panel points. This will produce bending in the top chord members in addition to the axial force from the truss analysis.

It will be necessary to make an initial estimate of member sizes, program these data, inspect the computer output, and iterate as required. A first iteration (not shown) used W14 × 68 columns. The most severe loading condition (with wind) produced lateral displacements on the order of 36 in, which would never be acceptable. The columns were resized using W27 × 178 for the second iteration shown here.

The building plan shows 22 bays at 25-ft spacing. Selected bays will carry wind bracing (not shown here or designed in this example). There will be longitudinal beams between columns at approximately 8-ft vertical spacings to carry the exterior walls. These members will carry additional load to the column and also provide lateral bracing for the column about the weak axis. The wall loads are not considered in the analysis at this point. The wall contribution can be readily accounted for by adding the wall beam shears directly to the axial force on the computer output when it is determined what the exterior siding will be (sheet metal, metal and wood, etc.).

General design parameters for a typical *interior* bent (or frame) include:

$$\text{Roof + purlins, estimated at} = 15 \text{ psf}$$
$$\text{Miscellaneous additional details} = 5 \text{ psf}$$
$$\text{Total} = \overline{20 \text{ psi}}$$

Load/ft of roof = 25(0.02) = 0.50 kip/ft

SOLUTION The dead load of the truss is obtained after an approximate truss analysis (not indeterminate truss) has been made to estimate possible bar forces. This analysis is very approximate and not shown here, as there are very crude assumptions and there are several which various persons may make to obtain similar results. From this initial estimate the following truss member sizes are selected:

Top and bottom chords of shed and main roof truss: two L6 × 4 × $\frac{5}{16}$:

$$\text{wt/ft} = 20.6 \text{ lb/ft} \quad A = 6.06 \text{ in}^2$$

Vertical and diagonal members: two L4 × 3 × $\frac{1}{4}$:

$$\text{wt} = 11.6 \text{ lb/ft} \quad A = 3.38 \text{ in}^2$$

The approximate truss weight is computed as follows:

Length of top + bottom chord (approx.) main truss = 2 × 120   = 240 ft

Verticals: av. height × 11 = (15 + 27)(11)/2   = 231 ft

Diagonals: av. length × 10 = (27.3 + 19.2)(10)/2   = 232 ft

The average truss weight:
$$W(120) = 240(0.0206) + 0.0116(231 + 232) = 10.31 \text{ kips}$$
$$W = \frac{10.31}{120} = 0.09 \text{ kip/ft} \quad \text{use } 0.10 \text{ kip/ft}$$

Truss weight for side sheds is as follows:

Length of top + bottom chord (approx.) = 2 × 72	= 144 ft
Verticals: (3 + 15)5/2	= 45 ft
Diagonals: (12.4 + 19.2)(5)/2	= 80 ft

The average truss weight:
$$W(72) = 144(0.0206) + 0.0116(80 + 45) = 4.42 \text{ kips}$$
$$W = \frac{4.42}{72} = 0.06 \text{ kip/ft} \quad \text{use } 0.07 \text{ kip/ft}$$

The weights of the trusses have been rounded up to account for connections. The truss will be analyzed for two load conditions.

NLC = 1: dead load + snow
NLC = 2: wind + snow + dead on left shed + wind (suction) + dead on main truss + 1.5 snow + dead on right shed

The truss will be symmetrically designed, since wind can blow from either direction. The rationale for this combination of loads is as follows:

1. Wind from the right will not blow snow off the left shed. Also, the vertical wall will create some stagnation, so the direction of the wind is normal downward.
2. On the main truss, the slope is such that aerodynamic action will result (in addition to blowing snow off) in a suction.
3. The right shed should be protected from the wind, but the snow from the roof can accumulate; to account for this, the snow is increased to 1.5 × S.

We can now begin to compute the node forces.
For dead load: Apply truss weight (computed as kips/ft of span) to the top node along with the roof dead load since the small total weight causes too much additional work for the increase in computation precision.
Shed dead loads: note that both shed and main roof truss have 12-ft panels.

Interior nodes: (0.07 + 0.50)(12)	= 6.84 kips
Exterior nodes: 6.84/2	= 3.42 kips

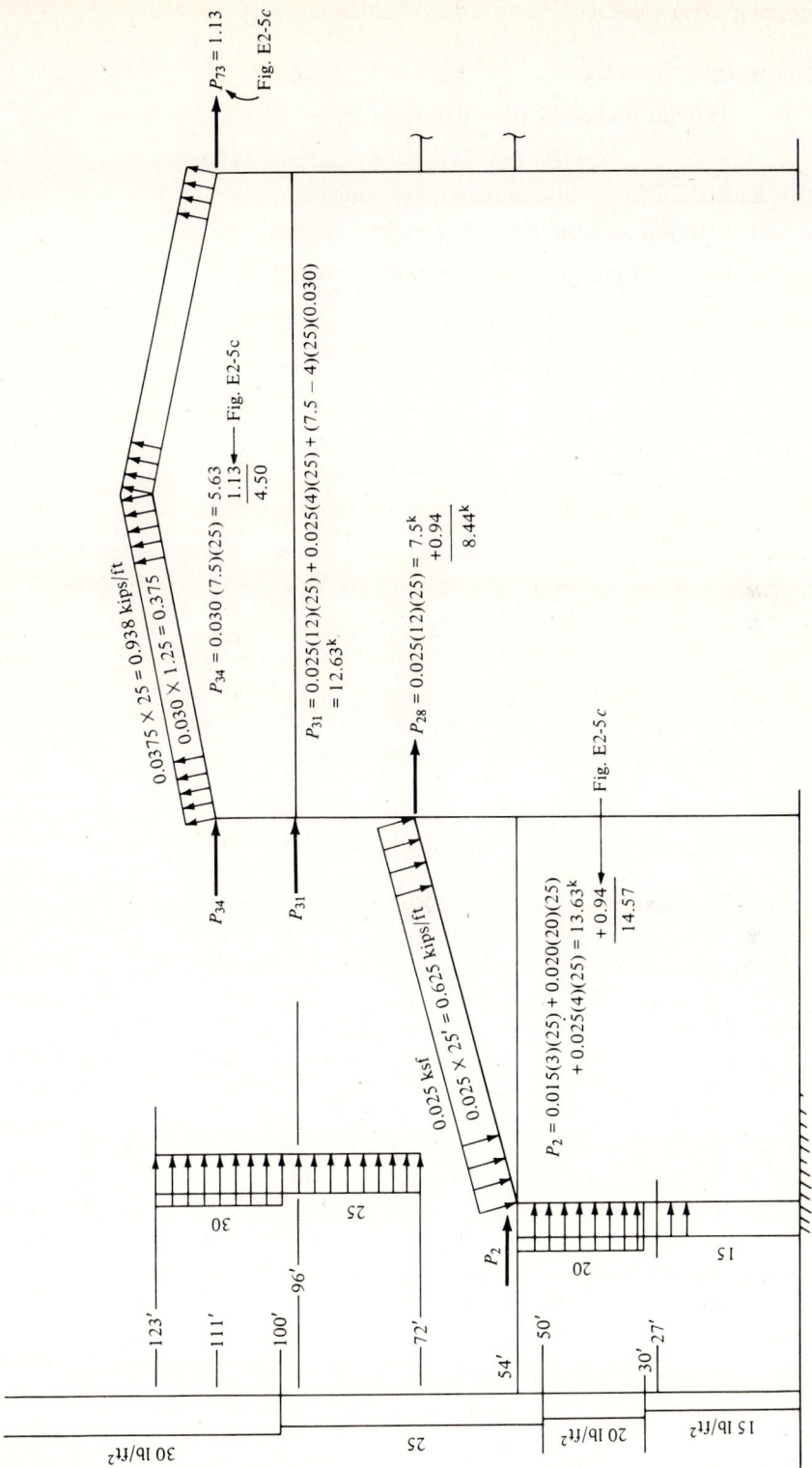

Figure E2-5b

Main truss:

$$\text{Interior nodes: } (0.10 + 0.50)(12) = 7.20 \text{ kips}$$
$$\text{Exterior nodes: } 7.20/2 = 3.60 \text{ kips}$$

Snow load at 25 psf of horizontal (span) projection:

$$\text{Interior nodes: } 0.025(25)(12) = 7.50 \text{ kips}$$
$$\text{Exterior nodes: } 7.50/2 = 3.75 \text{ kips}$$

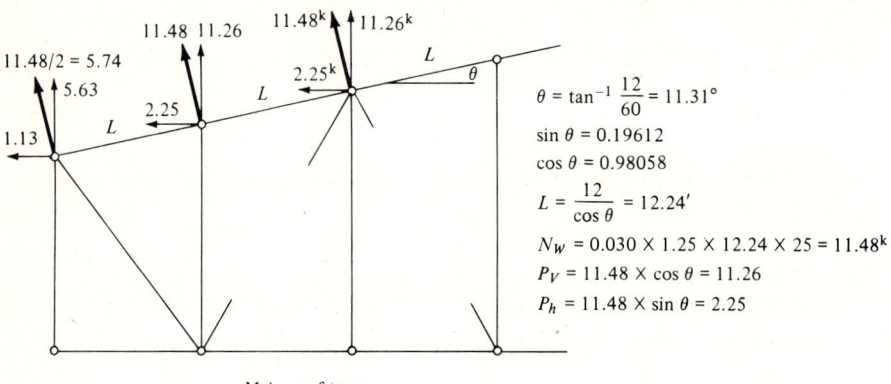

$\theta = \tan^{-1} \dfrac{12}{60} = 11.31°$

$\sin \theta = 0.19612$

$\cos \theta = 0.98058$

$L = \dfrac{12}{\cos \theta} = 12.24'$

$N_W = 0.030 \times 1.25 \times 12.24 \times 25 = 11.48^k$

$P_V = 11.48 \times \cos \theta = 11.26$

$P_h = 11.48 \times \sin \theta = 2.25$

Main roof truss

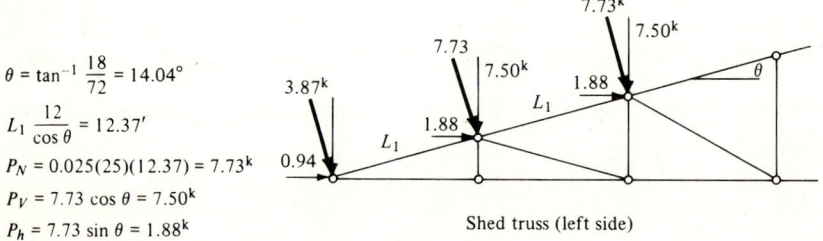

$\theta = \tan^{-1} \dfrac{18}{72} = 14.04°$

$L_1 \dfrac{12}{\cos \theta} = 12.37'$

$P_N = 0.025(25)(12.37) = 7.73^k$

$P_V = 7.73 \cos \theta = 7.50^k$

$P_h = 7.73 \sin \theta = 1.88^k$

Shed truss (left side)

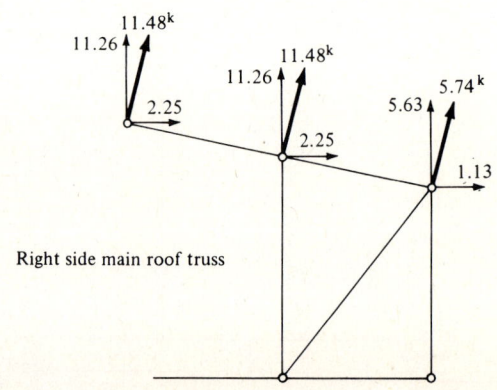

Right side main roof truss

Figure E2-5c

ELEMENTS OF FRAME, TRUSS, AND BRIDGE DESIGN 95

The resulting $D + S$ for NLC = 1 is:

Sheds: exterior nodes: $D + S = 3.42 + 3.75 = 7.17$ kips (in P matrix)

interior nodes: $\quad 6.84 + 7.50 = 14.34$ kips

Main truss: interior nodes: $\quad 7.2 + 7.5 = 14.70$ kips

exterior nodes: $\quad 14.70/2 = 7.35$ kips

These values are shown in the P matrix on the computer output sheets with a $(-)$ sign for the sign convention; for the node locations, refer to Fig. E2-5a.

It will be necessary to refer to Figs. E2-5b and E2-5c for the developing of the node forces for NLC = 2. The NBC wind profile against the frame is shown in Fig. E2-5b. Selected P-matrix entries are also directly developed on this diagram as a convenience for the reader. Strict attention is necessary so that the correct signs are placed with the node forces. Note that forces in the same direction as for those shown in Fig. 2-6 are $(+)$, and when the direction is reversed the sign is $(-)$.

Figures E2-5d to E2-5f show computer outputs for the two load conditions. Figure 2-5g displays selected computer output for design (using the largest value from either of the 2 NLCs). These data will be used in later

```
STIFFNESS METHOD OF RIGID FRAME AND TRUSS ANALYSIS

 NO OF NP  =   106         NO OF MEMBERS =   93       NO OF LOAD CONDIT =    2
 NO NLC TO WIND NLC =    2         NO NON-ZERO P =   43       METRIC (1 = METRIC) =  0
 NO ISTIFF ZEROED =   999      MOD ELASTICITY =  29000.0

       NLC FOR D + L =    0       NLC FOR WIND ON BEAM =    0

MEMBER NP1 NP2 NP3 NP4 NP5 NP6      H          V         A         I            L         COS       SIN         FEM1      FEM2
   1   107 107 107   1   2   3    0.00     648.00    47.7    0.517E 04       648.00   0.00000  1.00000       0.00      0.00
   2     2   3   4   5   0   0  144.00       0.00     6.06   0.000           144.00   1.00000  0.00000       0.00      0.00
   3     2   3   6   7   0   0  144.00      36.00     6.06   0.000           148.43   0.97014  0.24254       0.00      0.00
   4     4   5   6   7   0   0    0.00      36.00     3.38   0.000            36.00   0.00000  1.00000       0.00      0.00
   5     4   5   8   9   0   0  144.00       0.00     6.06   0.000           144.00   1.00000  0.00000       0.00      0.00
   6     6   7   8   9   0   0  144.00     -36.00     3.38   0.000           148.43   0.97014 -0.24254       0.00      0.00
   7     6   7  10  11   0   0  144.00      36.00     6.06   0.000           148.43   0.97014  0.24254       0.00      0.00
   8     8   9  10  11   0   0    0.00      72.00     3.38   0.000            72.00   0.00000  1.00000       0.00      0.00
   9     8   9  12  13   0   0  144.00       0.00     6.06   0.000           144.00   1.00000  0.00000       0.00      0.00
  10    10  11  12  13   0   0  144.00     -72.00     3.38   0.000           161.00   0.89443 -0.44721       0.00      0.00
  11    10  11  14  15   0   0  144.00      36.00     6.06   0.000           148.43   0.97014  0.24254       0.00      0.00
  12    12  13  14  15   0   0    0.00     108.00     3.38   0.000           108.00   0.00000  1.00000       0.00      0.00
  13    12  13  16  17   0   0  144.00       0.00     6.06   0.000           144.00   1.00000  0.00000       0.00      0.00
  14    14  15  16  17   0   0  144.00    -108.00     3.38   0.000           180.00   0.80000 -0.60000       0.00      0.00
  15    14  15  18  19   0   0  144.00      36.00     6.06   0.000           148.43   0.97014  0.24254       0.00      0.00
  16    16  17  18  19   0   0    0.00     144.00     3.38   0.000           144.00   0.00000  1.00000       0.00      0.00
  17    16  17  20  21   0   0  144.00       0.00     6.06   0.000           144.00   1.00000  0.00000       0.00      0.00
  18    18  19  20  21   0   0  144.00    -144.00     3.38   0.000           203.65   0.70711 -0.70711       0.00      0.00
  19    18  19  22  23   0   0  144.00      36.00     6.06   0.000           148.43   0.97014  0.24254       0.00      0.00
  20    20  21  22  23   0   0    0.00     180.00     6.06   0.000           180.00   0.00000  1.00000       0.00      0.00
  21    20  21  25  26   0   0  144.00       0.00     6.06   0.000           144.00   1.00000  0.00000       0.00      0.00
  22    22  23  25  26   0   0  144.00    -180.00     3.38   0.000           230.51   0.62470 -0.78087       0.00      0.00
  23    22  23  28  29   0   0  144.00      36.00     6.06   0.000           148.43   0.97014  0.24254       0.00      0.00
  24   107 107 107  24  25  26    0.00     648.00    52.3    0.699E 04       648.00   0.00000  1.00000       0.00      0.00
  25    24  25  26  27  28  29    0.00     216.00    52.3    0.699E 04       216.00   0.00000  1.00000       0.00      0.00
  26    27  28  29  30  31  32    0.00     288.00    52.3    0.699E 04       288.00   0.00000  1.00000       0.00      0.00
  27    30  31  32  33  34  35    0.00     180.00    52.3    0.699E 04       180.00   0.00000  1.00000       0.00      0.00
  28    31  32  36  37   0   0  144.00       0.00     6.06   0.000           144.00   1.00000  0.00000       0.00      0.00
  29    34  35  36  37   0   0  144.00    -180.00     3.38   0.000           230.51   0.62470 -0.78087       0.00      0.00
  30    34  35  38  39   0   0  144.00      28.80     6.06   0.000           146.85   0.98058  0.19612       0.00      0.00
  31    36  37  38  39   0   0    0.00     208.80     3.38   0.000           208.80   0.00000  1.00000       0.00      0.00
  32    36  37  40  41   0   0  144.00       0.00     6.06   0.000           144.00   1.00000  0.00000       0.00      0.00
  33    36  37  42  43   0   0  144.00     237.60     3.38   0.000           277.83   0.51830  0.85520       0.00      0.00
  34    38  39  42  43   0   0  144.00      28.80     6.06   0.000           146.85   0.98058  0.19612       0.00      0.00
  35    40  41  42  43   0   0    0.00     237.60     3.38   0.000           237.60   0.00000  1.00000       0.00      0.00
  36    40  41  44  45   0   0  144.00       0.00     6.06   0.000           144.00   1.00000  0.00000       0.00      0.00
  37    40  41  44  45   0   0  144.00    -237.60     3.38   0.000           277.83   0.51830 -0.85520       0.00      0.00
  38    42  43  46  47   0   0  144.00      28.80     6.06   0.000           146.85   0.98058  0.19612       0.00      0.00
  39    44  45  46  47   0   0    0.00     266.40     3.38   0.000           266.40   0.00000  1.00000       0.00      0.00
  40    44  45  48  49   0   0  144.00       0.00     6.06   0.000           144.00   1.00000  0.00000       0.00      0.00
  41    44  45  50  51   0   0  144.00     295.20     3.38   0.000           328.65   0.43842  0.89877       0.00      0.00
  42    46  47  50  51   0   0  144.00      28.80     6.06   0.000           146.85   0.98058  0.19612       0.00      0.00
  43    48  49  50  51   0   0    0.00     295.20     3.38   0.000           295.20   0.00000  1.00000       0.00      0.00
  44    48  49  52  53   0   0  144.00       0.00     6.06   0.000           144.00   1.00000  0.00000       0.00      0.00
  45    50  51  52  53   0   0  144.00    -295.20     3.38   0.000           328.65   0.43842 -0.89877       0.00      0.00
  46    50  51  54  55   0   0  144.00      28.80     6.06   0.000           146.85   0.98058  0.19612       0.00      0.00
  47    52  53  54  55   0   0    0.00     324.00     3.38   0.000           324.00   0.00000  1.00000       0.00      0.00
```

Figure E2-5d

96 STRUCTURAL STEEL DESIGN

NO STIFF(I) ENTRIES =   954   BAND WIDTH =   9

THE P-MATRIX, K AND IN-K   THE X-MATRIX, IN OR RADIANS

NP =	1	0.00	0.00		NX =	1	-0.00023	0.01043
NP =	2	0.00	14.57		NX =	2	-0.09800	4.50431
NP =	3	-7.17	-10.92		NX =	3	-0.01613	-0.01999
NP =	4	0.00	0.00		NX =	4	-0.00880	4.60253
NP =	5	0.00	0.00		NX =	5	-1.42992	-1.62598
NP =	6	0.00	1.88		NX =	6	0.15761	4.79184
NP =	7	-14.34	-21.84		NX =	7	-1.42992	-1.62598
NP =	8	0.00	0.00		NX =	8	0.08040	4.70075
NP =	9	0.00	0.00		NX =	9	-1.55415	-1.70321
NP =	10	0.00	1.88		NX =	10	0.11525	4.73349
NP =	11	-14.34	-21.84		NX =	11	-1.54888	-1.69502
NP =	12	0.00	0.00		NX =	12	0.14610	4.76242
NP =	13	0.00	0.00		NX =	13	-1.36939	-1.45396
NP =	14	0.00	1.88		NX =	14	0.02006	4.62983
NP =	15	-14.34	-21.84		NX =	15	-1.35359	-1.42938
NP =	16	0.00	0.00		NX =	16	0.18830	4.78753
NP =	17	0.00	0.00		NX =	17	-1.01955	-1.04842
NP =	18	0.00	1.88		NX =	18	-0.09197	4.52340
NP =	19	-14.34	-21.84		NX =	19	-0.98795	-0.99926
NP =	20	0.00	0.00		NX =	20	0.20700	4.77608
NP =	21	0.00	0.00		NX =	21	-0.56980	-0.56118
NP =	22	0.00	1.88		NX =	22	-0.20456	4.43285
NP =	23	-14.34	-21.84		NX =	23	-0.51713	-0.47925
NP =	24	0.00	0.00		NX =	24	-0.00180	0.00193
NP =	25	0.00	0.00		NX =	25	0.20220	4.72807* At shed truss
NP =	26	0.00	0.00		NX =	26	-0.05346	-0.02798 and int. col.
NP =	27	0.00	0.00		NX =	27	-0.00167	-0.00243
NP =	28	0.00	8.44		NX =	28	-0.28645	4.39819* Top of shed
NP =	29	-7.17	-10.92		NX =	29	-0.06617	-0.02936 truss
NP =	30	0.00	0.00		NX =	30	0.00133	-0.00044
NP =	31	0.00	12.63		NX =	31	-0.26690	4.10000
NP =	32	0.00	0.00		NX =	32	-0.08013	-0.02502
NP =	33	0.00	0.00		NX =	33	0.00183	-0.00057
NP =	34	0.00	4.50		NX =	34	0.03280	4.00589* Roof line
NP =	35	-7.35	2.03		NX =	35	-0.08895	-0.02230
NP =	36	0.00	0.00		NX =	36	-0.26637	4.10056
NP =	37	0.00	0.00		NX =	37	-0.55227	0.12680
NP =	38	0.00	-2.25		NX =	38	0.09677	3.98208
NP =	39	-14.70	4.06		NX =	39	-0.58359	0.13641
NP =	40	0.00	0.00		NX =	40	-0.20621	4.08144
NP =	41	0.00	0.00		NX =	41	-0.96568	0.26435
NP =	42	0.00	-2.25		NX =	42	0.13821	3.96639
NP =	43	-14.70	4.06		NX =	43	-0.96568	0.26435
NP =	44	0.00	0.00		NX =	44	-0.14604	4.06233
NP =	45	0.00	0.00		NX =	45	-1.21898	0.35097
NP =	46	0.00	-2.25		NX =	46	0.12801	3.96998
NP =	47	-14.70	4.06		NX =	47	-1.25893	0.36323
NP =	48	0.00	0.00		NX =	48	-0.07301	4.03843
NP =	49	0.00	0.00		NX =	49	-1.35411	0.39891
NP =	50	0.00	-2.25		NX =	50	0.07819	3.98816
NP =	51	-14.70	4.06		NX =	51	-1.35411	0.39891
NP =	52	0.00	0.00		NX =	52	0.00001	4.01453
NP =	53	0.00	0.00		NX =	53	-1.35785	0.40439
NP =	54	0.00	0.00		NX =	54	0.00001	4.02054
NP =	55	-14.70	4.06		NX =	55	-1.30304*	0.37563* Ridge line
NP =	56	0.00	0.00		NX =	56	0.07304	3.98858 w/wind
NP =	57	0.00	0.00		NX =	57	-1.35411	0.41300
NP =	58	0.00	2.25					

(W/o wind bracket spans NX = 49 through 54)

Figure E2-5e

ELEMENTS OF FRAME, TRUSS, AND BRIDGE DESIGN

LOADING CONDITION NO = 1		DESIGN END MOMENTS CORRECTED FOR FEM AND WIND (NEAR END FIRST), FT-K		LOADING CONDITION NO = 2		DESIGN END MOMENTS CORRECTED FOR FEM AND WIND (NEAR END FIRST), FT-K	
MEMBER	AXIAL FORCE, K			MEMBER	AXIAL FORCE, K		
1	-34.43	8.75	0.00	1	-32.00	-301.56	-0.00
2	108.86	----	----	2	119.87	----	----
3	-112.38	----	----	3	-130.90	----	----
4	0.00	----	----	4	0.00	----	----
5	108.86	----	----	5	119.87	----	----
6	-29.56	----	----	6	-45.99	----	----
7	-82.82	----	----	7	-86.85	----	----
8	7.17	----	----	8	11.15	----	----
9	80.18	----	----	9	75.26	----	----
10	-32.07	----	----	10	-49.88	----	----
11	-53.25	----	----	11	-42.79	----	----
12	14.34	----	----	12	22.31	----	----
13	51.50	----	----	13	30.64	----	----
14	-35.85	----	----	14	-55.77	----	----
15	-23.69	----	----	15	1.27	----	----
16	21.51	----	----	16	33.46	----	----
17	22.82	----	----	17	-13.97	----	----
18	-40.56	----	----	18	-63.10	----	----
19	5.88	----	----	19	45.32	----	----
20	28.68	----	----	20	44.62	----	----
21	-5.86	----	----	21	-58.59	----	----
22	-45.91	----	----	22	-71.42	----	----
23	35.44	----	----	23	89.37	----	----
24	-125.12	-142.83	-236.85	24	-49.12	-780.46	-704.98
25	-89.27	236.85	258.35	25	-7.29	704.98	193.25
26	-73.50	-258.35	93.45	26	17.16	-193.25	-17.83
27	-73.50	-93.45	-0.00	27	17.16	17.83	0.00
28	0.64	----	----	28	0.68	----	----
29	74.41	----	----	29	-24.36	----	----
30	-41.05	----	----	30	9.32	----	----
31	-14.70	----	----	31	4.51	----	----
32	73.43	----	----	32	-23.33	----	----
33	-50.75	----	----	33	16.97	----	----
34	-41.05	----	----	34	11.61	----	----
35	0.00	----	----	35	0.00	----	----
36	73.43	----	----	36	-23.33	----	----
37	24.45	----	----	37	-8.59	----	----
38	-80.79	----	----	38	27.42	----	----
39	-14.70	----	----	39	4.51	----	----
40	89.12	----	----	40	-29.16	----	----
41	-6.91	----	----	41	3.16	----	----
42	-80.79	----	----	42	29.71	----	----
43	0.00	----	----	43	0.00	----	----
44	89.12	----	----	44	-29.17	----	----
45	-9.22	----	----	45	1.98	----	----
46	-79.76	----	----	46	32.53	----	----
47	16.58	----	----	47	-8.70	----	----
48	89.12	----	----	48	-31.67	----	----
49	-9.22	----	----	49	7.70	----	----
50	-79.76	----	----	50	32.53	----	----
51	0.00	----	----	51	0.00	----	----
52	89.12	----	----	52	-31.67	----	----
53	-6.90	----	----	53	-3.81	----	----
54	-80.79	----	----	54	35.38	----	----
55	-14.70	----	----	55	4.51	----	----
56	73.43	----	----	56	-32.69	----	----
57	24.45	----	----	57	-1.27	----	----
58	-80.79	----	----	58	33.09	----	----
59	0.00	----	----	59	0.00	----	----
60	73.43	----	----	60	-32.69	----	----
61	-50.75	----	----	61	7.62	----	----
62	-41.05	----	----	62	26.10	----	----
63	-14.70	----	----	63	4.06	----	----
64	0.64	----	----	64	-22.53	----	----
65	74.41	----	----	65	-13.54	----	----
66	-41.05	----	----	66	26.10	----	----
67	-73.50	93.46	0.00	67	13.29	180.00	-0.01
68	-73.50	258.35	-93.46	68	13.29	62.48	-180.01
69	-89.27	-236.85	-258.35	69	6.28	-42.31	-62.48
70	-125.12	142.83	236.85	70	-27.64	-145.39	42.32
71	35.44	----	----	71	1.27	----	----
72	-45.91	----	----	72	-57.92	----	----
73	-5.86	----	----	73	30.97	----	----
74	28.68	----	----	74	36.18	----	----
75	5.88	----	----	75	-36.02	----	----
76	-40.56	----	----	76	-51.17	----	----
77	22.82	----	----	77	67.15	----	----
78	21.51	----	----	78	27.14	----	----
79	-23.69	----	----	79	-73.32	----	----
80	-35.85	----	----	80	-45.23	----	----
81	51.50	----	----	81	103.34	----	----
82	14.34	----	----	82	18.09	----	----
83	-53.25	----	----	83	-110.62	----	----
84	-32.07	----	----	84	-40.46	----	----
85	80.19	----	----	85	139.52	----	----
86	7.17	----	----	86	9.04	----	----
87	-82.82	----	----	87	-147.92	----	----
88	-29.56	----	----	88	-37.30	----	----
89	108.87	----	----				
90	0.00	----	----				
91	-112.38	----	----				
92	108.87	----	----				
93	-34.43	-8.75	-0.00				

Figure E2-5f

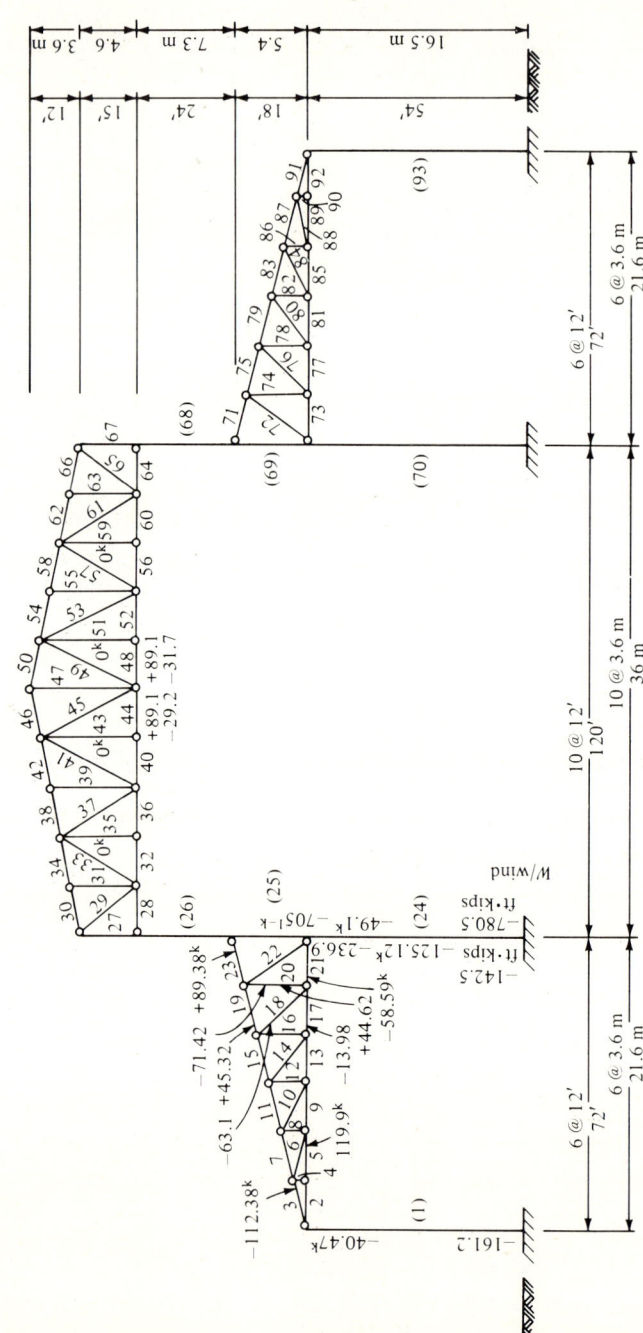

Figure E2-5g

chapters to design (or redesign the members). Note that the maximum lateral displacement is now on the order of 4.7 in at the base of the shed truss and only 4.0 in at the top of the main truss, indicating some additional bending in the interior columns. Note that this displacement is occurring in the wind NLC, as one would reasonably expect. ///

Example 2-6 Design the industrial warehouse building in SI units with the general dimensions shown in Fig. E2-6a. Use the same assumptions and problem parameters as used for Example 2-5.

SOLUTION The initial member sizes are selected as follows (and being considerably guided by the computer output of Example 2-5, which in professional terminology is using "experience"):

Top and bottom chord members of both main and shed truss: two L152 × 102 × 7.9 mm:

$$\text{wt} = 0.301 \text{ kN/m} \qquad A = 3.91 \times 10^{-3} \text{ m}^2$$
$$I_x = \text{not needed for truss-type members}$$

Intermediate truss members (verticals and diagonals): two L102 × 76 × 6.3 mm:

$$\text{wt} = 0.169 \text{ kN/m} \qquad A = 2.18 \times 10^{-3} \text{ m}^2$$

Shed columns (member numbers 1 and 93): W610 × 241.1:

$$\text{wt} = 0.72 \text{ kN/m} \qquad A = 30.77 \times 10^{-3} \text{ m}^2 \qquad I_x = 2151.9 \times 10^{-6} \text{ m}^4$$

Two interior columns (members 24, 25, 27, 28, 67, 68, 69, 70): W690 × 264.9:

$$\text{wt} = 0.99 \text{ kN/m} \qquad A = 33.74 \times 10^{-3} \text{ m}^2 \qquad I_x = 2909.5 \times 10^{-6} \text{ m}^4$$

The computation of frame loads now follows:

Roof†	= 0.72 kPa
Miscellaneous, including purlins	= 0.24 kPa
Total	= 0.96 kPa

$$\text{Load/m on bent} = 0.96(7.6) = 7.3 \text{ kN/m}$$

Dead load of truss: note that this is computed directly as a horizontal projection.

Main truss:

Length of top and bottom chord = 2(36)		= 72 m
Approx. length of diagonals = 10(5.84 + 8.3)/2		= 70.7 m
Approx. length of verticals = 11(4.6 + 8.2)/2		= 70.4 m

$$36W = 72(0.301) + 0.169(70.4 + 70.7) = 45.52 \text{ kN}$$

$$W = \frac{45.52}{36} = 1.26 \text{ kN/m}$$

† Take as horizontal projection, as truss slope is very small and values are estimates.

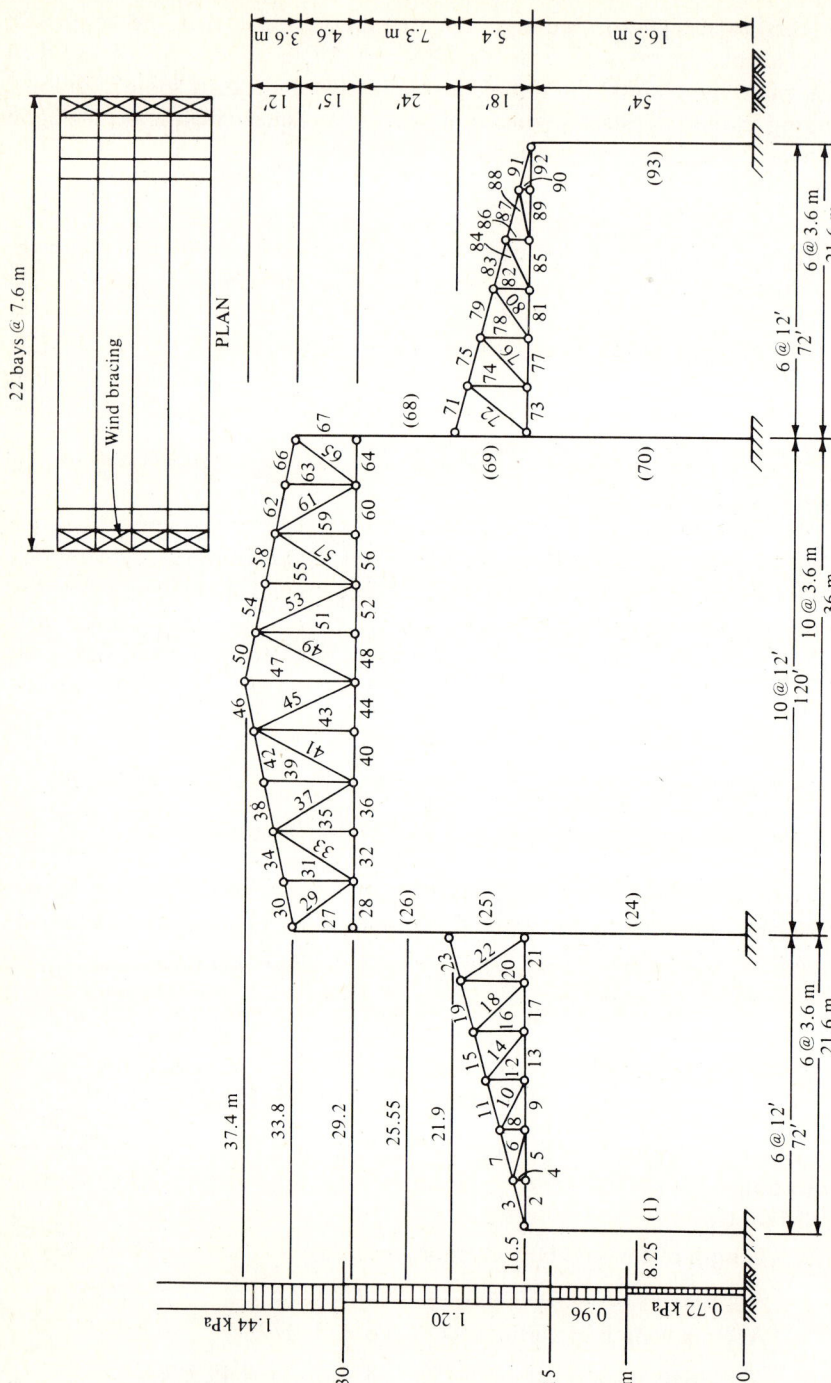

Figure E2-6a General plan and typical bent elevation. Wind pressure profile (NBC) shown on the left.

Shed truss dead weight:

Length of top + bottom chord = 2(21.6) = 43.2 m
Approx. length of verticals = 5(0.9 + 4.5)/2 = 13.5 m
Approx. length of diagonals = 5(3.71 + 5.26)/2 = 22.4 m

$$21.6W = 43.2(0.301) + 0.169(13.5 + 22.4) = 19.1 \text{ kN}$$

$$W = \frac{19.1}{21.6} = 0.88 \text{ kN/m} \quad \text{use } 0.90 \text{ kN/m}$$

Snow load is taken as 1.20 kPa (horizontal projection):

$$\frac{S}{m} = 1.2(7.6) = 9.12 \text{ kN/m}$$

Figure E2-6b shows typical shed and main truss node forces for dead plus snow loading as just computed.

Compute the wind loadings as follows. Assume that the wind effects are as shown in Fig. E2-6c and use the NBC wind profile on the windward side of the frame as shown in Fig. E2-6a. Also take the wind normal and downward on the windward shed, normal and outward on the main roof truss plus dead load, and 1.55 snow plus dead load on the leeward shed truss

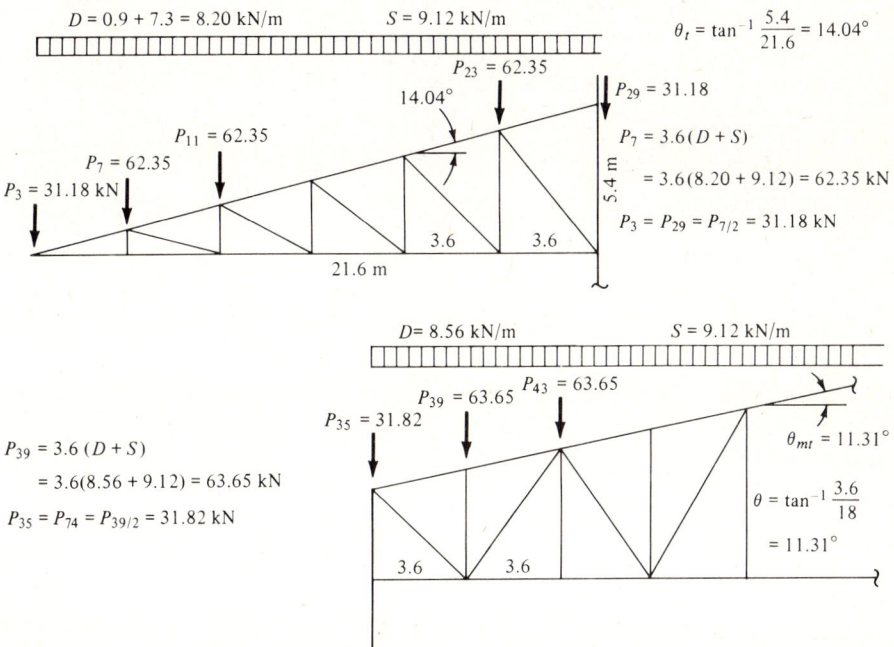

Figure E2-6b

102 STRUCTURAL STEEL DESIGN

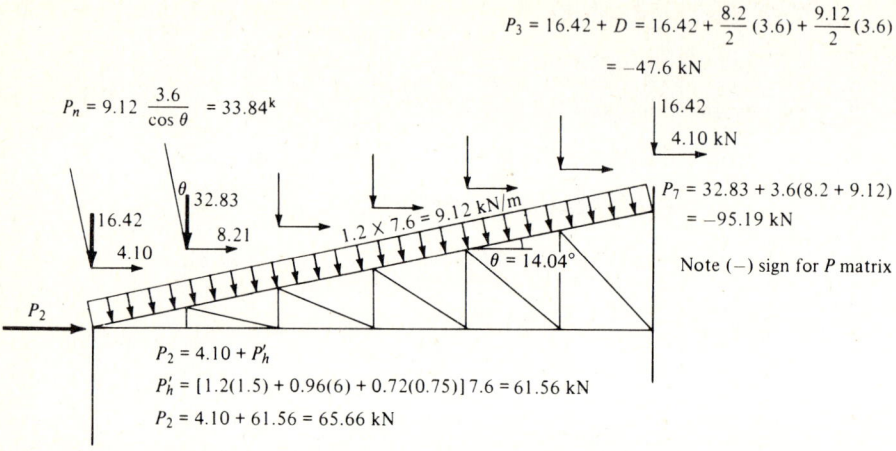

$P_3 = 16.42 + D = 16.42 + \dfrac{8.2}{2}(3.6) + \dfrac{9.12}{2}(3.6)$

$= -47.6$ kN

$P_n = 9.12 \dfrac{3.6}{\cos \theta} = 33.84^k$

$P_7 = 32.83 + 3.6(8.2 + 9.12)$
$= -95.19$ kN

Note $(-)$ sign for P matrix

$P_2 = 4.10 + P'_h$
$P'_h = [1.2(1.5) + 0.96(6) + 0.72(0.75)]7.6 = 61.56$ kN
$P_2 = 4.10 + 61.56 = 65.66$ kN

$P_{35} = 24.62 - D = 24.62 - 8.56 \dfrac{3.6}{2} = 9.21$ kN

$P_{39} = 49.24 - 8.56(3.6)$

$P_{39} = 18.42$ kN

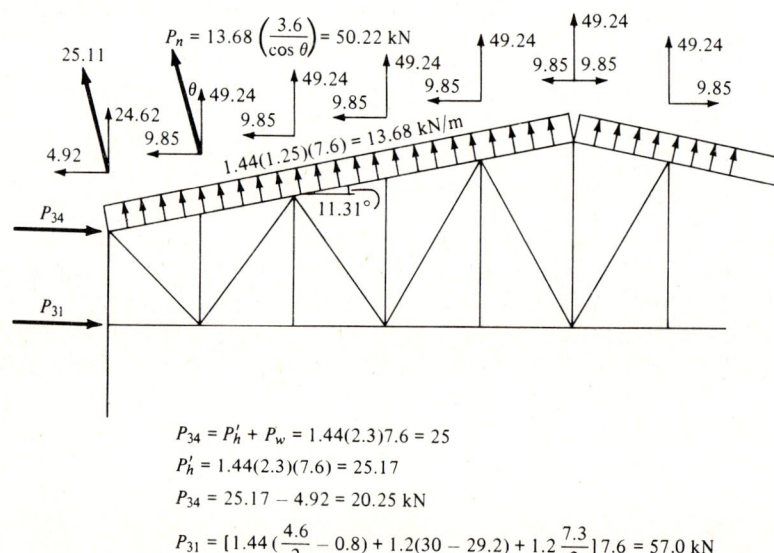

$P_{34} = P'_h + P_w = 1.44(2.3)7.6 = 25$
$P'_h = 1.44(2.3)(7.6) = 25.17$
$P_{34} = 25.17 - 4.92 = 20.25$ kN

$P_{31} = [1.44(\dfrac{4.6}{2} - 0.8) + 1.2(30 - 29.2) + 1.2\dfrac{7.3}{2}]7.6 = 57.0$ kN

Figure E2-6c

```
STIFFNESS METHOD OF RIGID FRAME AND TRUSS ANALYSIS

   NO OF NP =   106       NO OF MEMBERS =    93     NO OF LOAD CONDIT =    2
   NO NLC TO WIND NLC =    2       NO NON-ZERO P =   43     METRIC (1 = METRIC) =  1
   NO ISTIFF ZEROED =    999       MOD ELASTICITY = 200000.0

        NLC FOR D + L =   0     NLC FOR WIND ON BEAM =    0

MEMBER NP1 NP2 NP3 NP4 NP5 NP6      F           V         A          I         L        COS       SIN        FEM1      FEM2
  1    107 107 107   1   2   3    0.00    16500.00  0.308E-01  0.215E-02  16499.99   0.00000  1.00000     0.00      0.00
  2      2   3   4   5   0   0  3600.00       0.00  0.391E-02  0.000       3600.00   1.00000  0.00000     0.00      0.00
  3      2   3   6   7   0   0  3600.00     900.00  0.391E-02  0.000       3710.79   0.97014  0.24254     0.00      0.00
  4      4   5   6   7   0   0     0.00     900.00  0.218E-02  0.000        900.00   0.00000  1.00000     0.00      0.00
  5      4   5   8   9   0   0  3600.00       0.00  0.391E-02  0.000       3600.00   1.00000  0.00000     0.00      0.00
  6      6   7   8   9   0   0  3600.00    -900.00  0.218E-02  0.000       3710.79   0.97014 -0.24254     0.00      0.00
  7      6   7  10  11   0   0  3600.00     900.00  0.391E-02  0.000       3710.79   0.97014  0.24254     0.00      0.00
  8      8   9  10  11   0   0     0.00    1800.00  0.218E-02  0.000       1800.00   0.00000  1.00000     0.00      0.00
  9      8   9  12  13   0   0  3600.00       0.00  0.391E-02  0.000       3600.00   1.00000  0.00000     0.00      0.00
 10     10  11  12  13   0   0  3600.00   -1800.00  0.218E-02  0.000       4024.92   0.89443 -0.44721     0.00      0.00
 11     10  11  14  15   0   0  3600.00     900.00  0.391E-02  0.000       3710.79   0.97014  0.24254     0.00      0.00
 12     12  13  14  15   0   0     0.00    2700.00  0.218E-02  0.000       2700.00   0.00000  1.00000     0.00      0.00
 13     12  13  16  17   0   0  3600.00       0.00  0.391E-02  0.000       3600.00   1.00000  0.00000     0.00      0.00
 14     14  15  16  17   0   0  3600.00   -2700.00  0.218E-02  0.000       4500.00   0.80000 -0.60000     0.00      0.00
 15     14  15  18  19   0   0  3600.00     900.00  0.391E-02  0.000       3710.79   0.97014  0.24254     0.00      0.00
 16     16  17  18  19   0   0     0.00    3600.00  0.218E-02  0.000       3600.00   0.00000  1.00000     0.00      0.00
 17     16  17  20  21   0   0  3600.00       0.00  0.391E-02  0.000       3600.00   1.00000  0.00000     0.00      0.00
 18     18  19  20  21   0   0  3600.00   -3600.00  0.218E-02  0.000       5091.17   0.70711 -0.70711     0.00      0.00
 19     18  19  22  23   0   0  3600.00     900.00  0.391E-02  0.000       3710.79   0.97014  0.24254     0.00      0.00
 20     20  21  22  23   0   0     0.00    4500.00  0.218E-02  0.000       4500.00   0.00000  1.00000     0.00      0.00
 21     20  21  25  26   0   0  3600.00       0.00  0.391E-02  0.000       3600.00   1.00000  0.00000     0.00      0.00
 22     22  23  25  26   0   0  3600.00   -4500.00  0.218E-02  0.000       5762.81   0.62470 -0.78087     0.00      0.00
 23     22  23  28  29   0   0  3600.00     900.00  0.391E-02  0.000       3710.79   0.97014  0.24254     0.00      0.00
 24    107 107 107  24  25  26     0.00   16500.00  0.337E-01  0.291E-02  16499.99   0.00000  1.00000     0.00      0.00
 25     24  25  26  27  28  29     0.00    5400.00  0.337E-01  0.291E-02   5399.99   0.00000  1.00000     0.00      0.00
 26     27  28  29  30  31  32     0.00    7300.00  0.337E-01  0.291E-02   7300.00   0.00000  1.00000     0.00      0.00
 27     30  31  32  33  34  35     0.00    4600.00  0.337E-01  0.291E-02   4600.00   0.00000  1.00000     0.00      0.00
 28     31  32  36  37   0   0  3600.00       0.00  0.391E-02  0.000       3600.00   1.00000  0.00000     0.00      0.00
 29     34  35  36  37   0   0  3600.00   -4600.00  0.218E-02  0.000       5841.23   0.61631 -0.78751     0.00      0.00
 30     34  35  38  39   0   0  3600.00     720.00  0.391E-02  0.000       3671.29   0.98058  0.19612     0.00      0.00
 31     36  37  38  39   0   0     0.00    5320.00  0.218E-02  0.000       5319.99   0.00000  1.00000     0.00      0.00
 32     36  37  40  41   0   0  3600.00       0.00  0.391E-02  0.000       3600.00   1.00000  0.00000     0.00      0.00
 33     36  37  42  43   0   0  3600.00    6040.00  0.218E-02  0.000       7031.46   0.51198  0.85900     0.00      0.00
 34     38  39  42  43   0   0  3600.00     720.00  0.391E-02  0.000       3671.29   0.98058  0.19612     0.00      0.00
 35     40  41  42  43   0   0     0.00    6040.00  0.218E-02  0.000       6039.99   0.00000  1.00000     0.00      0.00
 36     40  41  44  45   0   0  3600.00       0.00  0.391E-02  0.000       3600.00   1.00000  0.00000     0.00      0.00
 37     42  43  44  45   0   0  3600.00   -6040.00  0.218E-02  0.000       7031.46   0.51198 -0.85900     0.00      0.00
 38     42  43  46  47   0   0  3600.00     720.00  0.391E-02  0.000       3671.29   0.98058  0.19612     0.00      0.00
 39     44  45  46  47   0   0     0.00    6760.00  0.218E-02  0.000       6760.00   0.00000  1.00000     0.00      0.00
 40     44  45  48  49   0   0  3600.00       0.00  0.391E-02  0.000       3600.00   1.00000  0.00000     0.00      0.00
 41     44  45  50  51   0   0  3600.00    7480.00  0.218E-02  0.000       8301.22   0.43367  0.90107     0.00      0.00
 42     46  47  50  51   0   0  3600.00     720.00  0.391E-02  0.000       3671.29   0.98058  0.19612     0.00      0.00
 43     48  49  50  51   0   0     0.00    7480.00  0.218E-02  0.000       7479.99   0.00000  1.00000     0.00      0.00
 44     48  49  52  53   0   0  3600.00       0.00  0.391E-02  0.000       3600.00   1.00000  0.00000     0.00      0.00
 45     50  51  52  53   0   0  3600.00   -7480.00  0.218E-02  0.000       8301.22   0.43367 -0.90107     0.00      0.00
 46     50  51  54  55   0   0  3600.00     720.00  0.391E-02  0.000       3671.29   0.98058  0.19612     0.00      0.00
 47     52  53  54  55   0   0     0.00    8200.00  0.218E-02  0.000       8199.99   0.00000  1.00000     0.00      0.00
```

Figure E2-6d

to account for the aerodynamic effects on the main truss and the accumulation of additional snow on the lee shed truss from the main roof truss.

From these data the edited input/output is shown in Figs. E2-6d to E2-6f. Note that the more critical wind effects are produced on the portion of the output shown except for the windward column, which is listed as follows:

Member	F_1 (axial), kN	F_2, kN	F_3, kN·m
24	−206.2	−941.98	−873.24
25	−23.86	873.24	296.57
26	78.75	−296.56	−53.11
27	78.75	53.11	0.01

The building members must be designed for the maximum member force from any of the loading conditions, and is, of course, symmetrical. The

```
THE P-MATRIX, KN AND K-MM          THE X-MATRIX, MM OR RADIANS

NP =   1      0.00      0.00       NX =   1    -0.00019      0.00901
NP =   2      0.00     65.66       NX =   2    -2.11252     99.13850
NP =   3    -31.18    -47.60       NX =   3    -0.40057     -0.51275
NP =   4      0.00      0.00       NX =   4     0.06161    101.61240
NP =   5      0.00      0.00       NX =   5   -34.86252    -41.08029
NP =   6      0.00      8.21       NX =   6     4.11880    106.38340
NP =   7    -62.36    -95.19       NX =   7   -34.86252    -41.08029
NP =   8      0.00      0.00       NX =   8     2.23574    104.08630
NP =   9      0.00      0.00       NX =   9   -37.88237    -43.23219
NP =  10      0.00      8.21       NX =  10     3.08638    104.91360
NP =  11    -62.36    -95.19       NX =  11   -37.75365    -43.03146
NP =  12      0.00      0.00       NX =  12     3.83557    105.66430
NP =  13      0.00      0.00       NX =  13   -33.37656    -37.04097
NP =  14      0.00      8.21       NX =  14     0.76925    102.24680
NP =  15    -62.36    -95.19       NX =  15   -32.99039    -36.43877
NP =  16      0.00      0.00       NX =  16     4.86116    106.34650
NP =  17      0.00      0.00       NX =  17   -24.85272    -26.79018
NP =  18      0.00      8.21       NX =  18    -1.95573     99.45392
NP =  19    -62.36    -95.19       NX =  19   -24.08038    -25.58569
NP =  20      0.00      0.00       NX =  20     5.31259    106.13320
NP =  21      0.00      0.00       NX =  21   -13.89939    -14.36411
NP =  22      0.00      8.21       NX =  22    -4.69145     97.00620
NP =  23    -62.36    -95.19       NX =  23   -12.61215    -12.35670
NP =  24      0.00      0.00       NX =  24    -0.00175      0.00130
NP =  25      0.00      0.00       NX =  25     5.18986    105.02440
NP =  26      0.00      0.00       NX =  26    -1.32775     -0.67222
NP =  27      0.00      0.00       NX =  27    -0.00163     -0.00227
NP =  28      0.00      4.10       NX =  28    -6.67495     95.89003
NP =  29    -31.18    -47.60       NX =  29    -1.63753     -0.69767
NP =  30      0.00      0.00       NX =  30     0.00125     -0.00023
NP =  31      0.00     57.00       NX =  31    -6.41213     90.31894
NP =  32      0.00      0.00       NX =  32    -1.98181     -0.58411
NP =  33      0.00      0.00       NX =  33     0.00173     -0.00051
NP =  34      0.00     20.25       NX =  34     0.79493     88.39206
NP =  35    -31.82      9.21       NX =  35    -2.19875     -0.51255
NP =  36      0.00      0.00       NX =  36    -6.39507     90.42130
NP =  37      0.00      0.00       NX =  37   -13.27372      2.99750
NP =  38      0.00     -9.85       NX =  38     2.32960     87.80621
NP =  39    -63.65     18.42       NX =  39   -14.05036      3.24629
NP =  40      0.00      0.00       NX =  40    -4.95209     90.01476
NP =  41      0.00      0.00       NX =  41   -23.18555      6.29029
NP =  42      0.00     -9.85       NX =  42     3.32098     87.41142
NP =  43    -63.65     18.42       NX =  43   -23.18555      6.29029
NP =  44      0.00      0.00       NX =  44    -3.50910     89.60822
NP =  45      0.00      0.00       NX =  45   -29.25648      8.38120
NP =  46      0.00     -9.85       NX =  46     3.08191     87.49551
NP =  47    -63.65     18.42       NX =  47   -30.24336      8.69736
NP =  48      0.00      0.00       NX =  48    -1.75330     89.06926
NP =  49      0.00      0.00       NX =  49   -32.50523      9.54463
NP =  50      0.00     -9.85       NX =  50     1.88365     87.93967
NP =  51    -63.65     18.42       NX =  51   -32.50523      9.54463
NP =  52      0.00      0.00       NX =  52     0.00250     88.53032
NP =  53      0.00      0.00       NX =  53   -32.58763      9.68993
NP =  54      0.00      0.00       NX =  54     0.00251     88.73569
NP =  55    -63.65     18.42       NX =  55   -31.26768      8.98293
NP =  56      0.00      0.00       NX =  56     1.75830     87.93411
NP =  57      0.00      0.00       NX =  57   -32.50520      9.91492
NP =  58      0.00      9.85       NX =  58    -1.87863     89.60577
NP =  59    -63.65     18.42       NX =  59   -32.50520      9.91492
NP =  60      0.00      0.00       NX =  60     3.51409     87.33791
NP =  61      0.00      0.00       NX =  61   -29.25644      9.19504
NP =  62      0.00      9.85       NX =  62    -3.07688     90.27214
NP =  63    -63.65     18.42       NX =  63   -30.24335      9.51114
NP =  64      0.00      0.00       NX =  64     4.95706     86.71877
NP =  65      0.00      0.00       NX =  65   -23.18552      7.20518
NP =  66      0.00      9.85       NX =  66    -3.31594     90.50999
NP =  67    -63.65     18.42       NX =  67   -23.18552      7.20518
NP =  68      0.00      9.85       NX =  68    -2.32453     90.46103
NP =  69    -63.65     18.42       NX =  69   -14.05033      4.19319
NP =  70      0.00      0.00       NX =  70     6.40001     86.09964
NP =  71      0.00      0.00       NX =  71   -13.27369      3.94440
NP =  72      0.00      0.00       NX =  72    -0.00173      0.00058
NP =  73      0.00      4.92       NX =  73    -0.78986     90.08044
NP =  74    -31.82      9.21       NX =  74    -2.19877     -0.23628
NP =  75      0.00      0.00       NX =  75    -0.00125      0.00167
NP =  76      0.00      0.00       NX =  76     6.41703     85.73734
NP =  77      0.00      0.00       NX =  77    -1.98183     -0.29029
NP =  78      0.00      0.00       NX =  78     0.00163      0.00427
NP =  79      0.00      0.00       NX =  79     6.67801     62.99757
NP =  80    -31.18    -39.39       NX =  80    -1.63755     -0.37599
```

Figure E2-6e

ELEMENTS OF FRAME, TRUSS, AND BRIDGE DESIGN 105

LOADING CONDITION NO = 1				LOADING CONDITION NO = 2			
MEMBER	AXIAL FORCE, KN	DESIGN END MOMENTS CORRECTED FOR FEM AND WIND (NEAR END FIRST), KN-M		MEMBER	AXIAL FORCE, KN	DESIGN END MOMENTS CORRECTED FOR FEM AND WIND (NEAR END FIRST), KN-M	
1	-149.40	10.02	0.00	1	-143.43	-352.62	-0.00
2	472.27	----	----	2	537.40	----	----
3	-487.43	----	----	3	-592.25	----	----
4	0.00	----	----	4	0.00	----	----
5	472.27	----	----	5	537.38	----	----
6	-128.59	----	----	6	-200.52	----	----
7	-358.84	----	----	7	-400.24	----	----
8	31.18	----	----	8	48.62	----	----
9	347.52	----	----	9	342.77	----	----
10	-139.46	----	----	10	-217.47	----	----
11	-230.27	----	----	11	-208.25	----	----
12	62.36	----	----	12	97.24	----	----
13	222.78	----	----	13	148.20	----	----
14	-155.90	----	----	14	-243.13	----	----
15	-101.71	----	----	15	-16.28	----	----
16	93.54	----	----	16	145.88	----	----
17	98.06	----	----	17	-46.34	----	----
18	-176.38	----	----	18	-275.06	----	----
19	26.85	----	----	19	175.72	----	----
20	124.72	----	----	20	194.50	----	----
21	-26.66	----	----	21	-240.86	----	----
22	-199.65	----	----	22	-311.34	----	----
23	155.41	----	----	23	367.71	----	----
24	-543.01	-190.09	-313.62	24	-206.19	-941.52	-872.91
25	-387.11	313.62	338.99	25	-23.85	872.91	296.55
26	-318.24	-338.99	120.58	26	78.73	-296.54	-53.07
27	-318.24	-120.58	0.00	27	78.73	53.08	-0.00
28	3.71	----	----	28	22.23	----	----
29	320.24	----	----	29	-112.97	----	----
30	-174.54	----	----	30	34.66	----	----
31	-63.65	----	----	31	20.39	----	----
32	313.45	----	----	32	-88.31	----	----
33	-219.49	----	----	33	79.83	----	----
34	-174.54	----	----	34	44.70	----	----
35	0.00	----	----	35	0.00	----	----
36	313.45	----	----	36	-88.31	----	----
37	106.53	----	----	37	-41.63	----	----
38	-344.76	----	----	38	118.12	----	----
39	-63.65	----	----	39	20.39	----	----
40	381.40	----	----	40	-117.07	----	----
41	-30.92	----	----	41	17.06	----	----
42	-344.76	----	----	42	128.16	----	----
43	0.00	----	----	43	0.00	----	----
44	381.40	----	----	44	-117.07	----	----
45	-38.95	----	----	45	6.58	----	----
46	-341.21	----	----	46	142.80	----	----
47	70.18	----	----	47	-37.59	----	----
48	381.40	----	----	48	-129.51	----	----
49	-38.95	----	----	49	35.14	----	----
50	-341.21	----	----	50	142.80	----	----
51	0.00	----	----	51	0.00	----	----
52	381.40	----	----	52	-129.51	----	----
53	-30.92	----	----	53	-17.59	----	----
54	-344.76	----	----	54	156.05	----	----
55	-63.65	----	----	55	20.39	----	----
56	313.44	----	----	56	-134.49	----	----
57	106.53	----	----	57	-5.28	----	----
58	-344.76	----	----	58	146.01	----	----
59	0.00	----	----	59	0.00	----	----
60	313.44	----	----	60	-134.49	----	----
61	-219.49	----	----	61	33.67	----	----
62	-174.54	----	----	62	115.59	----	----
63	-63.65	----	----	63	20.39	----	----
64	3.70	----	----	64	-78.70	----	----
65	320.24	----	----	65	-62.62	----	----
66	-174.54	----	----	66	105.54	----	----
67	-318.24	120.61	0.00	67	59.42	206.89	-0.00
68	-318.24	338.99	-120.61	68	59.41	104.18	-206.90
69	-387.11	-313.63	-338.99	69	24.40	-90.00	-104.18
70	-543.02	190.09	313.63	70	-123.31	-140.66	90.00
71	155.41	----	----	71	30.08	----	----
72	-199.66	----	----	72	-252.22	----	----
73	-26.66	----	----	73	113.70	----	----
74	124.73	----	----	74	157.56	----	----
75	26.85	----	----	75	-132.34	----	----
76	-176.39	----	----	76	-222.83	----	----
77	98.07	----	----	77	271.24	----	----
78	93.55	----	----	78	118.18	----	----
79	-101.72	----	----	79	-294.79	----	----
80	-155.92	----	----	80	-196.96	----	----
81	222.81	----	----	81	428.79	----	----
82	62.37	----	----	82	78.79	----	----
83	-230.30	----	----	83	-457.23	----	----
84	-139.47	----	----	84	-176.20	----	----
85	347.56	----	----	85	586.36	----	----
86	31.18	----	----	86	39.40	----	----
87	-358.88	----	----	87	-619.71	----	----
88	-128.61	----	----	88	-162.48	----	----
89	472.33	----	----	89	743.96	----	----
90	0.00	----	----	90	0.00	----	----
91	-487.49	----	----	91	-782.21	----	----
92	472.33	----	----	92	743.95	----	----
93	-149.41	-10.03	-0.00	93	-171.83	-184.61	-0.00

Figure E2-6f

present maximum horizontal deflections are:

$$X_4 = 101.7 \text{ mm} \qquad X_{25} = 105.1 \text{ mm}$$
$$X_6 = 106.4 \text{ mm} \qquad X_{31} = 90.4 \text{ mm}$$
$$X_{16} = 106.4 \text{ mm} \qquad X_{34} = 88.5 \text{ mm}$$

The difference

$$X_{31} - X_{76} = 90.4 - 85.8 = 4.6 \text{ mm}$$

is the net axial shortening of the bottom chord. In later chapters selected members of this frame will be designed to illustrate the method of design of tension and compression members and connections. ///

PROBLEMS

The following problems are of two types. The first four problems are for the student to obtain familiarity with using the computer program given in the Appendix to solve structural problems (other computer programs, such as STRESS or STRUDL, may also be used, but the program in the Appendix is likely to be considerably faster to run). Use $E = 200\,000$ MPa or $29\,000$ ksi. Obtain the solution in either fps or SI units, as assigned by the instructor.

One or more of the last three problems should be used for the design projects to be carried along with other later chapter problems.

2-1 Given the following beam. Obtain a W section that limits the deflection at point A to 1 in or 25 mm. Note that all NPE(2) can be made NPP1.
Answer: $M_A = 171.6$ ft · kips or 236.2 kN · m.

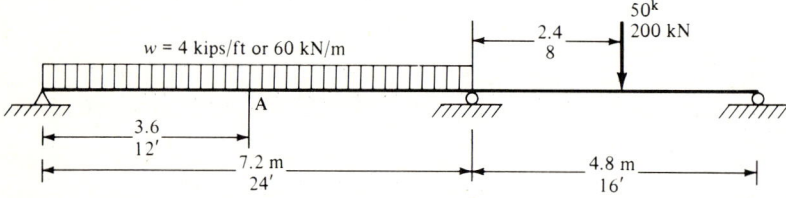

Figure P2-1

2-2 Given the following academic beam/frame. Find a W section that limits the deflection at point A to 0.5 in or 12.5 mm.

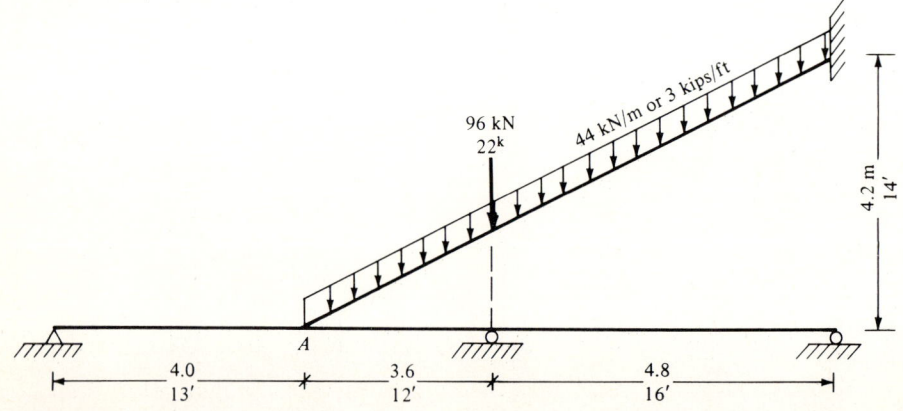

Figure P2-2

2-3 For the given truss and dead load shown applied to top chord and using a constant-size pair of angles with long legs back to back for both top and bottom chords, find the angle size such that either the largest pair is used or the deflection at point C is limited to 2 in or 50 mm for the moving live load shown. Apply the live load only to the points B, C, and D and not to any intermediate points. Use $E = 29\ 000$ ksi or 200 000 MPa. The live load is 20 kips or 90 kN applied to only one node at a time plus dead load.

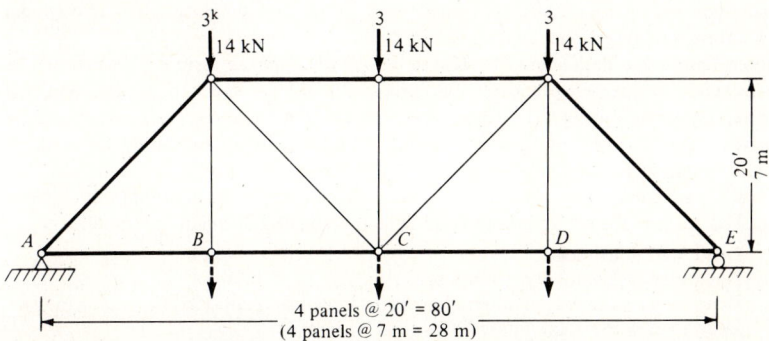

Figure P2-3

2-4 Iterate the frame shown until the lateral deflection and vertical deflection at point C is not more than 0.5 in or 12.7 mm. The use of the most massive W section is not acceptable until you show that this is all that works. Other data:

Roof dead load= 22 psf (or metric equivalent from table)
  Snow = 30 psf
  Wind = NBC value (take height as 9 m to points B and E for wind in either direction).
Vertical load from a hoist at point $D = 20$ kips or 90 kN
Bay spacing = 25 ft or 8 m
Try to limit iterations to five.

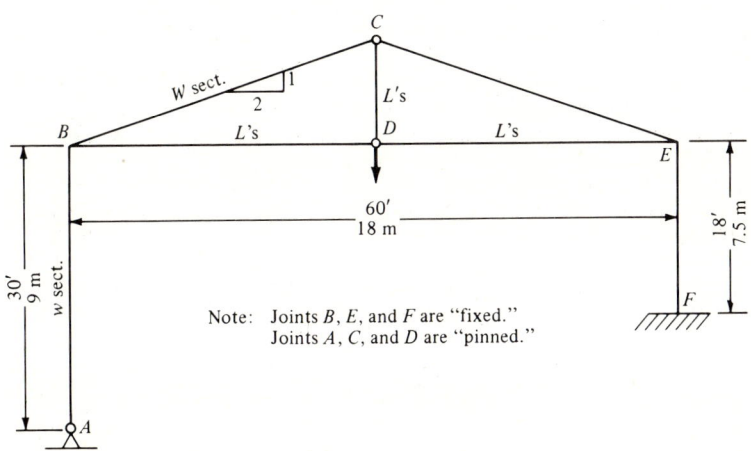

Note: Joints B, E, and F are "fixed."
   Joints A, C, and D are "pinned."

Figure P2-4

Design Problems

General comments It is suggested that one or more of the following problems be solved using some initial estimates of A and I as appropriate to the problem. These solutions should be neatly made up into line drawings which clearly show general dimensions, coding, and any member sizes used or other assumptions. Design computations for live, dead, wind, and snow loads or truck loads should be neatly tabulated so that gross errors can be readily identified. Be sure to state all assumptions and where design loads such as live floor loads, snow loads, and so on, were obtained (such as National Building Code, Uniform Building Code, etc.).

These solutions and design data should be kept in a notebook and the various elements will be designed (or redesigned) in later chapters, as appropriate. At or near the end of the term, the problem is to be reprogrammed using the "design" values for a new iteration. Elements should be checked for adequacy and if sufficient time is available, additional iterations should be made to converge to a better/working solution.

This set of work is academic; therefore, considerable detail work must be omitted in the interest of available time. The student should limit load conditions investigated in building structures to

    Dead load + live load
    Dead load + live load + snow load
    Dead load + live load in alternate bays + snow load (optional)
    Dead load + live load + wind load (last NLC = 4)

This number of NLC for the building structures can be solved with one computer run (if data are correctly input). Note that there will be four cards (one for each NLC) following each beam member data card and with NPE(7) = 0 on this (all beam) card.

2-5 Given the office building frame shown in Fig. P2-5. Assume that all stairs and elevator wells are located on the west end. Make a reasonable floor plan layout (same for all floors) for corridors and

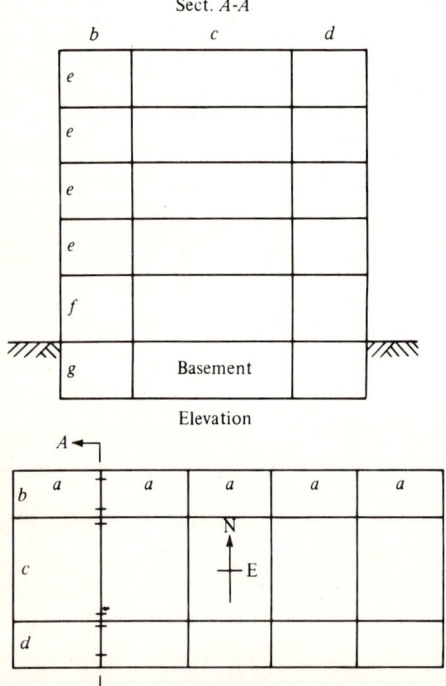

Figure P2-5

office space, and using appropriate building code data, compute floor loads, live load reductions, and so on, and estimating relative A and I (limit columns to W12 or W310) on the order of 1.2 to 1.5 (i.e., beam $I = 1.2I_{col}$). Obtain and set up dimensions based on the following table. Table data are such that several different problems are developed. The computer output will be somewhat different, depending on the input (although selected members may be the same size as the design proceeds).

Dimensions	fps	SI, m
Part a	26 ft ± 2 ft	8.0 ± 0.6
b	13 ft 8 in ± 16 in	4.20 ± 0.40
c	30 ft ± 2 ft	9.1 ± 0.60
d	Same as b	Same as b
e	12 ft ± 6 in	3.75 ± 0.15
f	14 ft ± 1 ft	4.20 ± 0.30

2-6 Given the framing shown in Fig. P2-6 for an industrial building, assume initially that the crane loading produces a lateral thrust at the column of 5 kips or 22 kN, as appropriate. Note that this analysis will not include all the NLC of the building frame of Prob. 2-5. Use the following

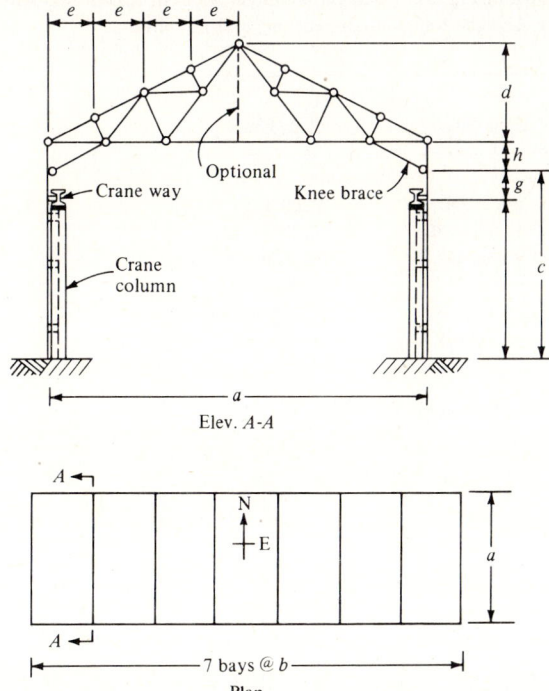

Figure P2-6

110 STRUCTURAL STEEL DESIGN

dimensions for your analysis:

Dimensions	fps, ft	SI, m
Part a[a]	60 ± 1.5	18.4 ± 0.40
b	30 ± 3.0	9.1 ± 1.0
c	29 ± 0.5	8.8 ± 0.20
d	13.75 ± 0.5	4.20 ± 0.15
e	7.5 ± 0.5	2.3 ± 0.2
f	20.0 ± 2.0	6.1 ± 0.6
g	3.25	1
h	6.0 ± 1.5	2.0 ± 0.5

[a] Make this dimension consistent with dimension e.

Notes:

1. The knee brace is pinned to the column, but the column is continuous to the truss.
2. The lateral crane impact load will be applied to the column at the attachment of the runway girder to the column, as shown in the figure.
3. An optional vertical member is used to reduce the KL/r of the bottom chord. If you use this member, it should be pinned to the bottom chord, but the bottom chord is continuous across this connection (thus will have bending).
4. Sketch where you would place wind bracing, but do not design.

2-7 Given the highway bridge truss shown in Fig. P2-7, make a computer analysis for the standard truck loading assigned by the instructor. Use the following dimensions and initial data:

Dimensions/section, s[a]	fps: 25 ± 3 ft	SI: 7.5 ± 1 m
Members		
Bottom chord (1, 4, ... , 21, 25)	W8 × 48	W200 × 71.4
Top chord (6, 10, 15, 19, 23)	2 C12 × 30	2 C305 × 44.6
Verticals (3, 7, ... , 20, 24)	W8 × 40	W200 × 59.5
Diagonals except 2, 13, 14, 26	W8 × 40	W200 × 50.5
Diagonals 2, 26	2 C12 × 30	2 C305 × 44.6
Diagonals 13, 14	2 L$3\frac{1}{2}$ × 3 × $\frac{7}{16}$	2 L89 × 76 × 11.1

[a] Use increments of 0.5 ft or 0.15 m to give s = 25, 25.5–28 ft.

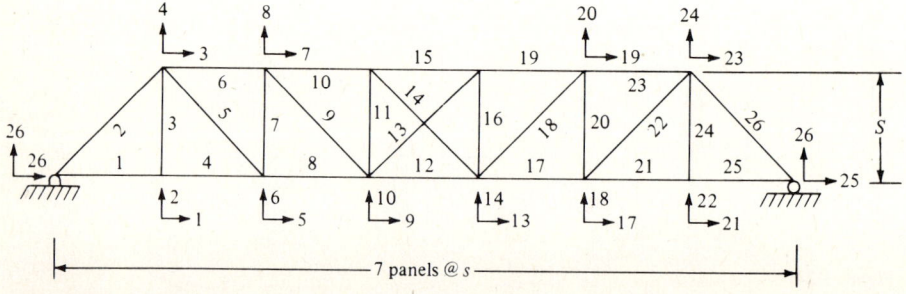

Figure P2-7

2-8 Given an open-deck railroad bridge truss with geometry shown in Fig. P2-7, make a computer analysis for either Cooper's E-80 or E-110 loading, as assigned by the instructor. Take the truss spacing at 17 ft or 5.2 m on centers. It is suggested that the tentative section areas shown in Prob. 2-7 (for a highway bridge truss) be approximately doubled for the initial computer run.

Notes (for both Probs. 2-7 and 2-8):
1. The given section data may be applicable for the basic dimensions and HS 20 loading.
2. Increasing or decreasing panel lengths may necessitate changing the initial section sizes up or down one step.
3. It is necessary to obtain influence lines (or their equivalent) to determine maximum (and minimum) member forces. Impact and dead load must also be considered.
4. The program for influence lines in the Appendix can be used to punch a set of load matrix cards for the analysis program for the wheel loads at selected positions along the span. It is suggested to use wheel distance increments of 5 ft or 1.8 m.
5. Do not design any lateral bracing (not enough time).

Figure III-1 The Chicago "Picasso". A massive sculpture using corrosion-resistant steel designed by the famous artist Picasso. Note the background building using a steel frame.

CHAPTER
THREE

ELASTIC, PLASTIC, AND BUCKLING BEHAVIOR OF STRUCTURAL STEEL

3-1 INTRODUCTION

Two modes of structural behavior under stress are of particular importance and have considerable influence on the design of steel members. One of these modes is the behavior of steel in the plastic region of the stress–strain curve (see Fig. 1-3b). Prior to the early 1970s, this particular behavior was often termed "plastic behavior." Presently, the term inelastic behavior is often used; however, the reader must be cautious in using the term "inelastic behavior," since many persons still confine inelastic behavior to the strain region following the onset of strain hardening.

The tendency of unsupported structural elements to buckle under compression stresses is a second behavioral characteristic of particular interest. We may define "buckling" as the sudden bending, warping, curling, or crumpling of the element under compressive stresses. Plate elements, as well as column members, are designed using equations that have been developed as a combination of theory and laboratory testing and in recognition of buckling in the presence of compressive stresses. These two behavioral modes will be briefly considered in this chapter to provide the necessary theoretical background material for the later design-oriented chapters.

The stress–strain diagram for steel shown in Fig. 1-3b illustrates that material behavior is nearly linear to the proportional limit ($E = \sigma/\varepsilon$), is elastic (strains are recoverable) to the elastic limit (ε_y), and exhibits a plastic flow (inelastic) type of behavior to the onset of strain hardening ε_{st}. Plastic behavior can be described as that strain due to the ductility of the steel and occurs at a constant stress above the elastic limit. After some amount of plastic strain, the steel tends to strain-harden, and an increase in load accompanied by additional strains is possible. This region of the stress–strain curve represents an additional reserve strength capacity of steel beyond the elastic limit. The slope of the curve after the onset of strain hardening gives the *strain hardening modulus* (sometimes called a tangent modulus; however, a tangent modulus may also be developed for the material from the onset of strain beyond the proportional limit).

Structural designers commonly design steel members to function in the elastic portion of the stress–strain curve. Some designers use the plastic region of the curve as the basis for design, although the actual loads produce stresses that are in the elastic region of the stress–strain curve. Many of the design equations used for "elastic" design, particularly those for relatively short columns and plate girders, are based on steel behavior in the inelastic region.

3-2 ELASTIC VERSUS PLASTIC DESIGN THEORY

In the mid-1950s and early 1960s, a considerable effort was expended at several leading university structural laboratories to develop a method of steel design based on material behavior in the plastic region of the stress–strain curve. This effort made considerable headway as a new design tool until the widespread availability and use of digital computers. Widespread use of digital computers with the great ease of solving structural frames using the elastic (classical) methods, and particularly the stiffness method of problem formulation, brought these methods into favor, so that at present much less use is made of the plastic† design concept. As we shall see, however, plastic design is a very rapid method for many beams and for many one-story rigid frames and often results in somewhat more economical (lighter-weight) members. There is not even a particular advantage in using the digital computer for design of these several structures.

Elastic design is based on the premise that stress is proportional to strain, so that plane sections before stressing remain plane after stress appplication such that for axial load the stress is directly computed as

$$f_a = \frac{P}{A}$$

and the corresponding deformation is $e = PL/AE = f_a L/E = \varepsilon L$. The familiar mechanics-of-materials equation for bending stresses,

$$f_b = \frac{Mc}{I} = \frac{M}{S}$$

---

† The author will use "plastic design" to refer to a design method which makes use of the fact that the stress–strain curve is nearly flat from ε_y to ε_{st}.

is used. We also recall that as long as the stresses are in the elastic region, the principle of superposition of effects is valid and the order of load application is not important. The principle of superposition is not, in general, valid for deformations beyond ε_y. The deflections are not so directly computed for bending as for axial loads; however, use can be made of the differential equations for a beam given in Sec. 2-2 to compute beam slope and deflection. Closed-form solutions can be obtained using these equations, but they are valid only for the given beam, geometry, and loads.

The following example is used to illustrate the concept of "elastic" analysis.

Example 3-1 Given a three-bar structure of A-36 steel with $F_y = 36$ ksi connected by a rigid bar ABC such that there is vertical translation but no rotation (Fig. E3-1). It is necessary to find the axial force in each member.

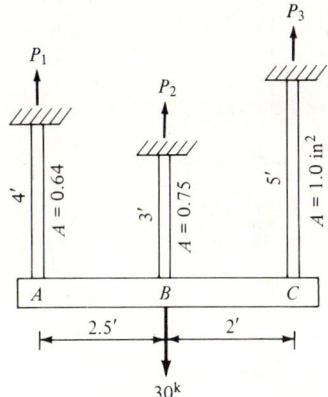

Figure E3-1

SOLUTION It should be noted that for member ABC not to rotate, there must be displacement compatibility between the member lengths and cross-sectional areas.

$$P_1 + P_2 + P_3 = 30 \text{ kips} \qquad (a)$$

Since member ABC is rigid and does not rotate,

$$e_1 = e_2 = e_3 = e \qquad (b)$$

From mechanics of materials,

$$e_1 = \frac{P_1(48)}{0.64 \times 29 \times 10^3} \qquad e_2 = \frac{P_2(36)}{0.75 \times 29 \times 10^3} \qquad e_3 = \frac{P_3(60)}{1.0 \times 29 \times 10^3} \qquad (c)$$

The additional equation needed to solve for the bar forces (three equations with three unknowns) is obtained by taking moments at a convenient location (say point B) to obtain

$$2.5P_1 - 2P_3 = 0$$

$$P_1 = \frac{2}{2.5}P_3 = 0.8P_3$$

116 STRUCTURAL STEEL DESIGN

From Eqs. (c) and (b),

$$e_1 = e_3 = \frac{0.8P_3(48)}{0.64(29\,000)} = \frac{P_3(60)}{1.0(29\,000)}$$

from which we obtain

$$60P_3 = 60P_3 \quad \text{which checks displacements} = \text{constant}$$

Also, from Eq. (c):

$$\frac{48P_1}{0.64(29\,000)} = \frac{36P_2}{0.75(29\,000)}$$

$$P_2 = 1.5635P_1$$

$$P_2 = 1.5625(0.8P_3) = 1.25P_3$$

Now substituting into Eq. (a), we obtain

$$0.8P_3 + 1.25P_3 + P_3 = 30 \text{ kips}$$

$$P_3 = 30/3.05 \quad = 9.836 \text{ kips}$$

$$P_2 = 1.25(9.836) = 12.295 \text{ kips}$$

$$\underline{P_1 = 0.8(9.836) = 7.869 \text{ kips}}$$

Total $\qquad\qquad\qquad = \overline{30.000 \text{ kips}} \qquad$ (checks)

Back substitution for e in each of Eq. (c) gives $e = 0.02035$ in (which the reader should verify). ///

Now let us reconsider Example 3-1 using "plastic analysis" in the following example.

Example 3-2 For the sketch shown in Fig. E3-2 (same as Example 3-1), what are the bar forces when all three bars have yielded?

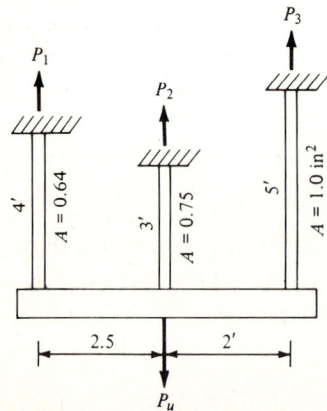

Figure E3-2

SOLUTION We must apply a factored load of sufficient magnitude to develop $f_a = F_y$ in the three bars. At this time, the bar forces are simply

$$P_1 = A_1 F_y \qquad P_2 = A_2 F_y \qquad P_3 = A_3 F_y$$

Also,

$$P_1 + P_2 + P_3 = P_{\text{ultimate}} \qquad \text{or simply } P_u$$
$$P_u = 36(0.64) + 36(0.75) + 36(1.0) = 84.04 \text{ kips}$$

Since the actual load is only 30 kips, the load factor is P_u/P:

$$LF = \frac{86.04}{30} = 2.87$$

Several comments are in order:

1. Plastic analysis is much simpler.
2. The rigid bar ABC will rotate under the applied load $P_u = 84.04$ kips, which was not the case in the elastic analysis. Why?
3. The elastic analysis (Example 3-1) indicates bar 2 yields first. Why? When $F_y = 36$ ksi in bar 2, it carries no additional load but merely elongates with any additional load being carried by adjacent bars until they reach F_y in turn. When F_y is reached in bar 2, the load in the bar is $P_2 = 0.75(36) = 27$ kips.

By proportion from Example 3-1 the load at this point is

$$P = \frac{27.0}{12.295}(30) = 65.88 \text{ kips} \qquad ///$$

Beam behavior based on a plastic analysis is similar to the bar problem. Consider the beam shown in Fig. 3-1. If we apply a bending moment to the section, the moment–rotation (M–ϕ) curve is linear† to M_y. From the point at which the most stressed beam fiber is at F_y (producing the yield moment M_y) to the point at which all of the beam fibers are at F_y (either tension or compression depending on which side of the neutral axis we are inspecting) and producing the plastic moment M_p, the curve is nonlinear. When M_p is reached, the beam simply rotates at this point, with no further increase in moment capacity (or stress) and we say that a plastic "hinge" has formed. There is some small additional increase in moment capacity when some of the beam fibers most distant from the neutral axis reach strains into the strain-hardening region. This effect depends on beam cross-sectional geometry of the flanges and web and, of course, the beam span and boundary conditions. If the beam is loaded with a moment greater than M_y (but not M_p) and unloaded, the curve branch BE is obtained with a permanent amount of residual beam rotation OE.

---

† Within material homogeneity and rolling tolerances, as well as practical measuring limitations.

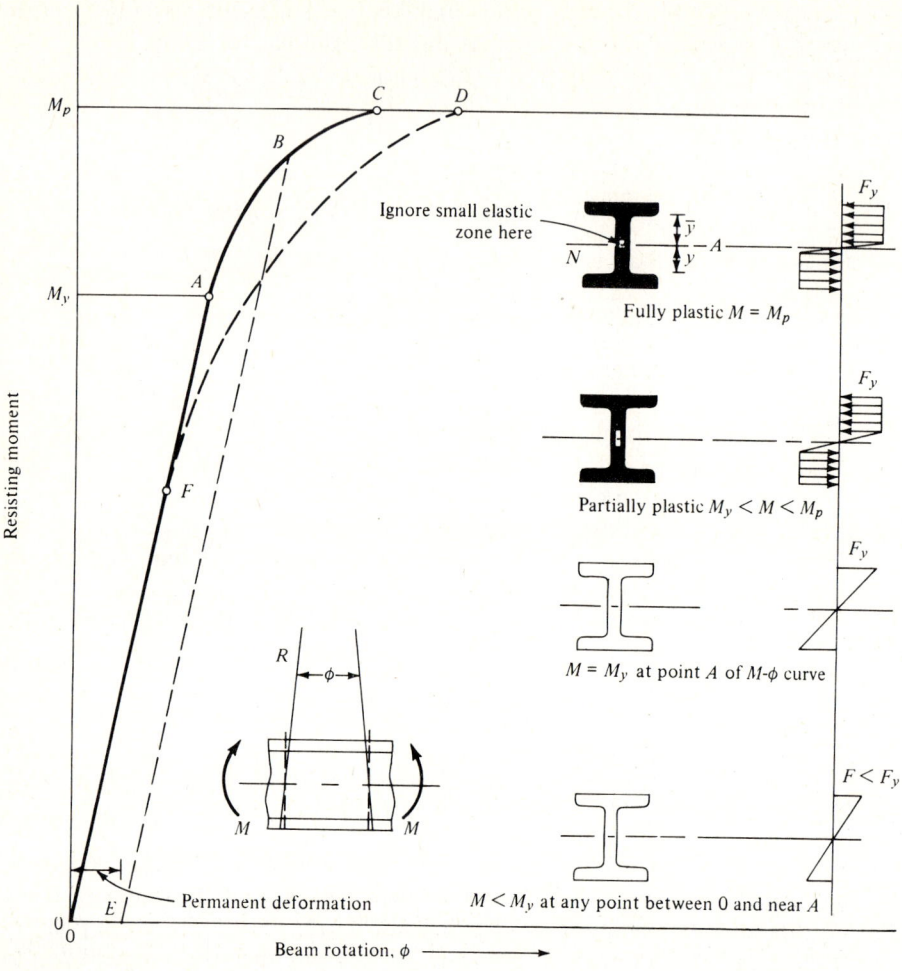

Figure 3-1 Moment versus rotation for a W shape.

The rolled structural shapes (W, M, C, etc.) nearly always contain residual stresses caused by differential cooling. The flange tips and interior web parts, being thinner and more exposed, always cool more quickly than the other parts of the flange. The junction of the flange and web is the thickest and most protected part of the section and always cools last. Residual *tension* stresses develop in those parts that cool last as the metal tends to contract but is restrained by the colder metal. These residual tension stresses produce residual *compressive* stresses in the adjacent metal, which has cooled earlier. Welding also produces residual stresses as the hot metal and flame cutting in the vicinity is restrained from contraction by the cooler-to-cold surrounding metal. Residual stresses can be produced by straightening of bent members and in the process of

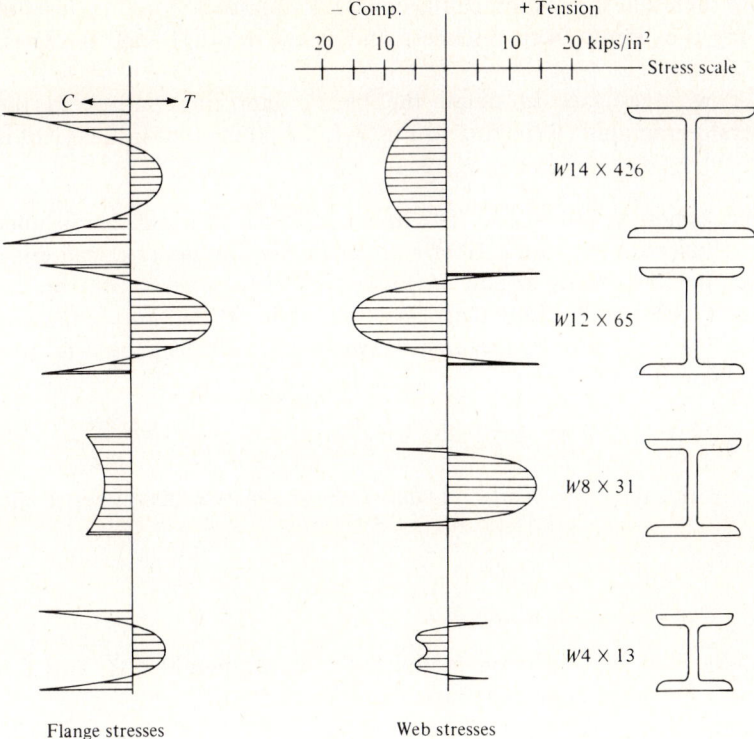

Figure 3-2 Residual stresses in W-shapes (after Johnston, *Guide to Stability Design Criteria for Metal Structures*, 3rd ed. (New York: John Wiley & Sons, Inc.).

cold-forming structural shapes by bending cold metal plates to the desired section geometry. In any case, nearly all steel members contain both tension and compression residual stresses, which tends to produce a net eccentricity on the cross section. Figure 3-2 illustrates in a qualitative (and somewhat quantitative) manner the distribution of residual stresses from thermal causes on several rolled shapes.

Residual stresses from differential cooling tend to be heavily dependent on section geometry, with thicker sections having lesser stresses. For example, a W12 × 65 can have compression stresses in the web and flange tips on the order of 15 ksi and tension stresses at the flange-to-web junction of about 12 ksi. For a W14 × 426, the corresponding stresses on the flange tips is about 20 ksi, but for the flange-to-web the tension stresses are only about 4 or 5 ksi. Interestingly, for a W8 × 31, the entire flange seems to be in compression with maximum values of about 8 ksi, while the maximum web stress is about 12 ksi (tension).

The plastic moment capacity of a rolled section is hardly affected by the presence of residual stresses (based on a number of full-scale laboratory tests), as indicated by the dashed line *FD* of Fig. 3-1. This is true only if the section proportions are such that the section can become fully plastic before the onset of strain hardening (i.e., depth/web thickness and flange width/flange thickness not too large).

We will now investigate in detail the plastic moment concept in the following several paragraphs. Referring to Fig. 3-1, the moment at initial yield is

$$M_y = S_x F_y$$

where S_x is the section modulus, I/c. The moment of inertia I and the distance from the neutral axis to the extreme fiber c are as in any mechanics-of-materials textbook. The plastic moment M_p, by inspection of the stresses shown on the cross section in Fig. 3-1 with a fully plastic section, $M = M_p$ and noting that the neutral axis at this point divides the area in two parts with distance $\bar{y}$ to area centroid from neutral axis, is

$$M_p = 2F_y \frac{A}{2} \bar{y} = A\bar{y} F_y$$

The value of $A\bar{y}$ is called the *plastic modulus*, Z, so that we may rewrite the moment as

$$M_p = ZF_y$$

The ratio of Z/S is termed the *shape factor, f*.

Example 3-3 What is the section modulus S_x, plastic modulus Z, and the shape factor f for the rectangular shape shown in Fig. E3-3?

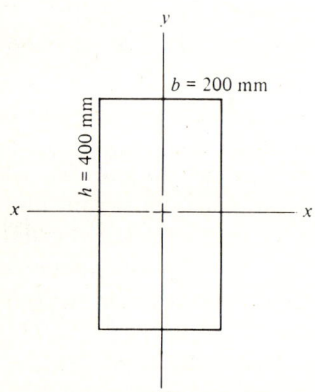

Figure E3-3

SOLUTION The elastic section modulus is computed using mechanics-of-materials equations:

$$S = \frac{I}{c} = \frac{bh^3}{12(h/2)} = \frac{bh^2}{6}$$

$$S = \frac{200(0.4)^2}{6} = 5.333 \times 10^{-3} \text{ m}^3$$

ELASTIC, PLASTIC, AND BUCKLING BEHAVIOR OF STRUCTURAL STEEL **121**

The plastic section modulus is obtained as the statical moment of area about the neutral axis, which divides the area equally. Note that this is necessary in order to satisfy statics on the section of $\Sigma F_h = 0$.

$$\frac{A}{2} = \frac{bh}{2} \qquad 2\bar{y} = h - \frac{h}{4} - \frac{h}{4} = \frac{h}{2}$$

$$Z = \frac{A\bar{y}}{2} = \frac{bh}{2}\frac{h}{2} = \frac{bh^2}{4}$$

$$= \frac{200(0.4)^2}{4} = 8.00 \times 10^{-3} \text{ m}^3$$

The shape factor is computed as

$$f = \frac{Z}{S} = \frac{8}{5.333} = 1.50 \qquad\qquad ///$$

The plastic modulus and shape factor for a W shape can be computed in a manner similar to the rectangular shape of Example 3-3. Here convenient use is made of the tables for T shapes, as illustrated in the following example.

Example 3-4 Compute the plastic section modulus and shape factor for a W610 × 241.1 rolled shape.

SOLUTION The value of $A/2$ is readily obtained from the WT table (WT305 × 120.5), since this T shape is made from splitting a W610 shape. The $\bar{y}$ value in the table also locates the center of the area of the T but is with respect to the flange.

From Table V-18 of SSDD, obtain

$$\frac{A}{2} = 15.39 \times 10^{-3} \text{ m}^2 \qquad \bar{y} = 68.6 \text{ mm}$$

From Table V-3, the depth of a W610 × 241.1 is 635 mm. The total area = 30.77×10^{-3} m².

$$\bar{y}_{\text{sect}} = d - 2\bar{y}_T = 635 - 2(68.6) = 497.8 \text{ mm} = 0.498 \text{ m}$$

$$Z = \frac{A\bar{y}}{2} = 15.39 \times 10^{-3} \times 0.498 = 7.664 \times 10^{-3} \text{ m}^3$$

The value given in Table V-3 for $Z_x = 7.659 \times 10^{-3}$ m³ and the discrepancy is due to the extra digits used by the computer in computing Z_x directly as opposed to rounding for Table V-18 and the use of 0.4978 vs. 0.498 above.

From Table V-3, the section modulus of a W610 × 241.1 is

$$S_x = 6.78 \times 10^{-3} \text{ m}^3$$

and the shape factor f can be directly computed as

$$f = \frac{7.659}{6.78} = 1.13 \qquad\qquad ///$$

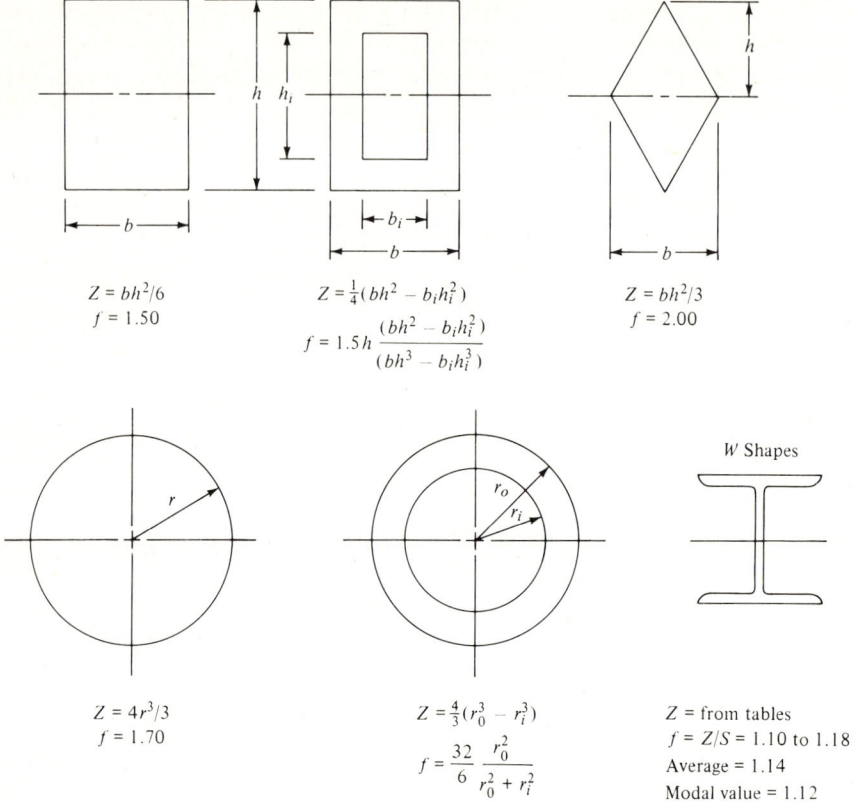

Figure 3-3 Plastic section modulus and shape factor for selected cross sections.

The shape factor is a measure of the increase in plastic moment capacity M_p over the value of yield moment M_y, and we have

$$M_p = ZF_y = f(SF_y)$$

The modal value of f for all of the W shapes is approximately 1.12, with a range of about 1.10 to 1.18 when all the rolled sections are considered.

The plastic section modulus and resulting shape factor for several cross sections is shown in Fig. 3-3.

3-3 SAFETY FACTORS IN ELASTIC AND PLASTIC DESIGN

Steel design may be based on the yield strength (termed plastic, or limit states, design) or on an elastic design. In limit states design the analysis proceeds based on assumed plastic behavior—the member continues to strain from ε_y to ε_{st} (see Fig. 1-3b) with no increase in load. Elastic design allows for this unique behavior of steel but limits the working stresses to the elastic region of the stress–strain curve, where stress is proportional to strain as defined by the modulus of elasticity being a constant.

The present method of design of steel structures is based on the elastic design method. The current elastic design procedures as found in the several design specifications is based on the linear stress–strain response to the elastic limit, but there is implicit recognition of the steel behavior beyond the elastic limit. Elastic design as commonly used places the limiting steel stress as the yield stress F_y. A design stress that used the limiting stress of F_y in the elastic design procedure would have a safety factor $F = 1$. A safety factor of 1 is unacceptable, as it allows for no future changes in structural use/occupancy, or for corrosion and fatigue effects. A value of $F = 1$ would not allow any load position changes. For these reasons and including some uncertainty in the material properties (flaws, under dimensions of sections, and minor metallurgical differences within steel grades), load uncertainties, and stress differences resulting from use of an idealized mathematical model for the real structure, a value of $F > 1$ is required.

Ideally, every element of a steel structure should have the same factor of safety. In practice, this is not the case. Flexural member response tends to be the most reliable to predict and those members have a minimum value of F. Columns are more sensitive to structure and section geometry and have a F that has a lower limit the same as for flexural members. Connections, whose failure results in a structure collapse, rightfully have the largest values of F.

The basic safety factor for the steel members in building construction is obtained as follows. Let the computed strength of the member be defined as the value S and let the computed service load be defined as R. The safety factor can be defined as

$$F = \frac{\text{computed strength of member}}{\text{computed service load}} = \frac{S}{R}$$

Uncertainties exist in both the service loads and in the actual member strength, and we may look at that combination which results in the poorest S/R ratio. This combination may be taken as $S - \Delta S$ and for service load as $R + \Delta R$.

For $F = 1$, this new ratio becomes

$$1 = \frac{S - \Delta S}{R + \Delta R} = \frac{S}{R} \frac{1 - \Delta S/S}{1 + \Delta R/R}$$

Now if we take $\Delta S/S = \Delta R/R = 0.25$ and noting that $S/R = F$, we obtain

$$1 = F\frac{0.75}{1.25}$$

Solving for F, we obtain

$$F = \frac{1}{0.6} = 1.667 \quad \left(\text{or } \tfrac{5}{3}\right)$$

This value of F is taken as the basic value of F for use in the elastic design method for structural steel for structures other than bridges for highways and railroads. Railroad and highway bridges are generally subjected to a more hostile environment and a greater possibility of overloading, so the uncertainty factor $\Delta S = \Delta R$ is taken as 0.29, which gives $F = 1/0.55 = 1.82$.

The value of $F = 1/0.6$ is modified to $1/0.66$ when the cross-sectional geometry is such that a plastic hinge can fully develop at the most highly stressed point. Rolled shapes whose section geometry is such that the plastic hinge can fully develop so that the basic value of $F = 1/0.66$ can be used are termed *compact shapes*. The geometry criteria for these shapes will be considered in Chap. 4.

For A-36 steel the basic allowable stresses using the previously defined safety factor becomes

$$F_a = 0.6 F_y = 0.6(36) = 21.6 \text{ ksi}$$

(the AISC specification allows use of 22 ksi for this case)

$$F_a = 0.6(250) = 150 \text{ MPa} \quad \text{(in SI units)}$$

For the AREA and AASHTO specifications, we have

$$F_a = 0.55(36) = 19.8 \text{ ksi}$$

(these specifications allow use of 20 ksi for this single case)

$$F_a = 0.55(250) = 137.5 \text{ MPa}$$

We note that the optional rounding of 21.6 ksi to 22 ksi (as should be done using desk calculations) can create a slight computational discrepancy if a digital computer is used, unless a rounding procedure for this grade of steel is set in the computer program. The author suggests that the rounding to 22 ksi be done, since it is allowed and A-36 steel is the most common grade used. It is not recommended (at this time) to round 137.5 MPa to 140 MPa, since some rounding up has already taken place to obtain 250 MPa from 36 ksi.

3-3.1 Factor of Safety for Current Plastic Design

The factor of safety used (called *load factor*) for plastic design according to the present AISC procedure is obtained by using the average shape factor f defined in Sec. 3-2 and illustrated in the computations for a typical rectangular shape in Example 3-3. In elastic design with compact sections, the value of $F = 1/0.66 = 1.52$. The value of plastic moment $M_p = f M_y$, where the shape factor = 1.12 as the modal value for all the rolled W shapes. Now using the same value of working stress f_b for either design method, we have

$$\frac{M_y}{1.52 S} = \frac{M_p}{F_1 S} = \frac{f M_y}{F_1 S}$$

Canceling the section modulus S, we obtain

$$F_1 = 1.52 f = 1.52(1.12) = 1.70 \quad \text{(as used in Part 2 of the AISC specifications)}$$

This value of F is used in plastic (or limit states) design as a load factor by which the working or design loads are multiplied to obtain the "ultimate" loads.

Stresses obtained based on section properties from these ultimate loads are compared to the yield stress F_y and adjustments made until computed $f_i \leq F_y$.

3-4 ELASTIC VERSUS PLASTIC DESIGN DEFLECTIONS

Should a plastic hinge develop at a point along a beam or column, a very large deflection would result. This deflection would, however, have no meaning, since it would result in a structure collapse. No structure is designed for this event, so the deflection under the actual working load is that deflection of interest. The working load for plastic design is obtained by applying the load factor of 1.7 to the "ultimate" loads to w_u, P_u, M_u, and so on. This factor ensures that the deflections under actual working load conditions will be elastic values.

Since plastic deflections result in a structure collapse, only elastic deflections have any significance, so for this reason deflections are always computed using elastic analytical procedures for both general methods of design.

3-5 LENGTH OF PLASTIC HINGE

The length of the plastic hinge for a W shape can be derived for a uniformly loaded fixed-end beam (refer to Fig. 3-4) as follows. With the slope of the moment curve horizontal at midspan, the offset to the parabolic-shaped moment curve is

$$y_0 = M_p - (-M_p) = 2M_p \text{ at } x = \frac{L}{2}$$

The equation of a parabola with the origin of x as shown is

$$y = kx^2$$

Inserting y_0 and $x = L/2$, we obtain

$$y = 8M_p\left(\frac{x}{L}\right)^2 \tag{3-1}$$

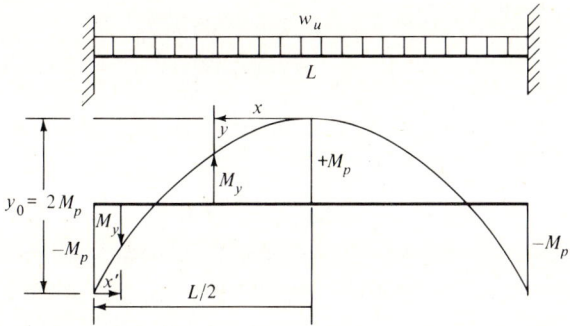

Figure 3-4 Length of plastic hinge for uniformly loaded fixed-end beam.

Now we need to find x such that $M_x = M_y = S_x F_y$. Since $y = M_p - M_y$, we have

$$y = F_y(Z_x - S_x) = F_y(fS_x - S_x)$$

Substituting for y and M_p in Eq. (3-1), we obtain

$$8(fF_y S_x)\left(\frac{x}{L}\right)^2 = fF_y\left(1 - \frac{1}{f}\right)S_x$$

Canceling S_x and F_y and rearranging, we obtain for the hinge length defined by the length x (which is half the midspan hinge length),

$$x = \frac{L}{2.83}\sqrt{1 - \frac{1}{f}} \qquad (3\text{-}2)$$

When $f = 1.12$, the length $2x$ of the plastic hinge in the center of the beam is

$$L_{\text{hinge}} = 2\frac{L}{8.64} \qquad \text{(nearly 20 percent of span length)}$$

The hinge length for other beam loadings, such as cantilevered beams, beams with concentrated loads, and load combinations, can be obtained in a manner similar to that just presented and taking into account the moment gradient (slope of the moment curve).

3-6 ELASTIC VERSUS PLASTIC DESIGN

There are several advantages in using plastic design for continuous beams and small one- or two-story structures:

1. The rapidity of obtaining design moments.
2. There is some economy of steel (lighter sections can often be used).
3. Gives some idea of the collapse mode and strength of the structure.

Offsetting these advantages are several disadvantages:

1. Widespread availability of computer programs, which can rapidly solve both simple and complicated structures using elastic methods.
2. Most designers have more familiarity with elastic design methods.
3. Difficulty of obtaining the collapse mode if the structure is reasonably complicated.
4. There is little savings in column design (and sometimes for other members depending on the fabrication methods).
5. Difficult to design for fatigue.
6. Lateral bracing requirements are more stringent than for elastic design.

In plastic design it is necessary to determine the location of the plastic hinges that form at locations where M_p develops. It is necessary that enough

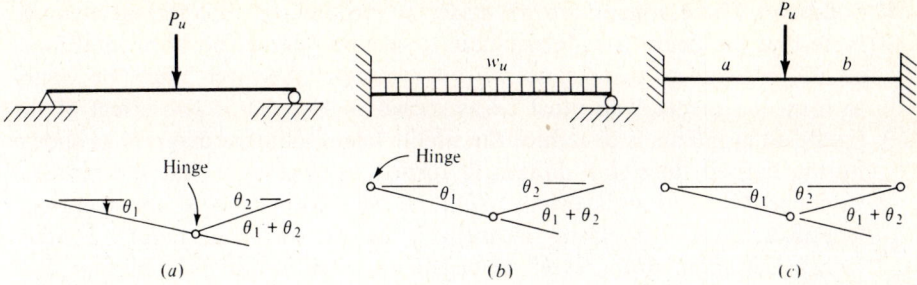

Figure 3-5 Plastic hinge formation for several beams loaded as shown. (*a*) Simply supported one hinge for failure. (*b*) Propped cantilever two hinges for failure. (*c*) Fixed-end, three hinges for failure.

hinges form to develop a collapse (structure is incipiently unstable) mechanism. Thus a simple beam requires one hinge, a propped cantilever two hinges, a beam fixed on both ends three hinges, and so on. Figure 3-5 illustrates the beam and collapse mechanisms for the three beams just considered. The reader should note that a continuous beam is similar to the fixed-end beam and the exterior span of a continuous beam is similar to a propped cantilever.

There are two methods of analysis commonly used to determine the value of M_p when the collapse mechanism has been determined. These two methods are the *equilibrium* (also called *statics method*) *method* and the *virtual work method*. Only the equilibrium method will be considered further. This method will be illustrated by the following examples. The reader should appreciate the fact that the beams considered in the following examples will produce hinges (and M_p) at clearly defined locations. A rigid frame consisting in several spans (and stories) must develop sufficient hinges to produce a collapse mechanism for making a plastic analysis. These hinge locations must generally be determined by trial (i.e., assume hinge locations and compute M_p). That set of hinges producing the minimum M_p is considered the critical set and is used for design. Again this process of iteration is not necessary for the beams in the following examples.

Example 3-5 Derive an expression for M_p for the fixed-end beam shown and select a W shape with adequate Z for $P_w = 120$ kN.

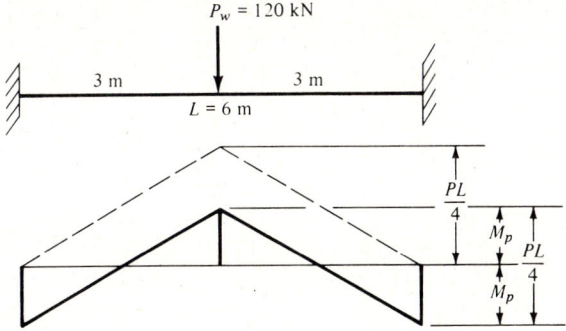

Figure E3-5

SOLUTION Three hinges are necessary to produce a collapse mechanism. Note that the beam is indeterminate to second degree (no horizontal load, so $\Sigma F_h = 0$ has no significance). From symmetry the three hinges necessary to form the mechanism must be as shown in Fig. E3-5. The effect of the fixed-end moments is to reduce the simple beam moment diagram as shown by the dashed lines. For hinges to form it is necessary that the moment value be M_p, and it is evident that M_p will form first at the fixed-end locations, since the elastic moment is largest at those points. Further increases in moment increases the elastic moment into the plastic range. It is also evident that the only other possible location for M_p is under the concentrated load, since the moment at this point will be the next location where the elastic moment is large enough that increases in P_w to P_u will force the moment into the plastic range. When this hinge forms, the structure collapses (theoretically) and no further increase in load is possible.

With this consideration, we have (again referring to Fig. E3-5)

$$2M_p = \frac{P_u L}{4}$$

from which

$$M_p = \frac{P_u L}{8}$$

For A-36 steel, $F_y = 250$ MPa.

$$P_u = P_w \times \text{load factor} = 120(1.7) = 204 \text{ kN}$$

$$M_u = M_p = \frac{P_u L}{8} = \frac{204(6)}{8} = 153 \text{ kN} \cdot \text{m}$$

The required plastic section modulus is

$$Z = \frac{153}{250} = 0.612 \times 10^{-3} \text{ m}^3$$

From Table VI-2 of SSDD, select

$$\text{W360} \times 38.7 \qquad Z_x = 0.6566 \times 10^{-3} \text{ m}^3$$

The beam must carry its own weight, so for self-weight the simple beam moment is $M = wL^2/8$. For plastic analysis use the same concept as for the concentrated load, which gives

$$2M_p = \frac{w_u L^2}{8}$$

$$M_p = \frac{w_u L^2}{16}$$

For the W360 × 38.7, the weight/m = 0.38 kN/m (Table V-3).

$$\Delta M_p = \frac{0.38(1.7)(6)^2}{16} = 1.454 \text{ kN} \cdot \text{m}$$

By proportion,

$$\Delta Z_x = \frac{1.454}{153}(0.612) = 0.0058$$

The total Z_x required is

$Z_x = 0.612 + 0.0058 = 0.6178 < 0.6566 \times 10^{-3}$ m³ furnished   O.K.

Use a W360 × 38.7 beam.

It is still necessary to check bracing requirements. For an elastic design using $F_a = 0.6F_y$ (commonly used allowable stress), the beam would be

$$M = \frac{PL}{8} = \frac{120(6)}{8} = 90 \text{ kN} \cdot \text{m}$$

The required section modulus S is

$$S = \frac{90}{0.6(250)} = 0.60 \times 10^{-3} \text{ m}^3$$

Use a W410 × 38.7 section.

$$\Delta S_x = \frac{0.38(6)^2}{12(150)} = 0.0076$$

$S_{x(\text{reqd})} = 0.60 + 0.0076 = 0.6076 < 0.629 \times 10^{-3}$ furnished   O.K.

By coincidence we have found a section that has exactly the same mass per meter; in most cases sections obtained by plastic design methods are somewhat lighter than those obtained using elastic design, at least when the beam is indeterminate.   ///

Example 3-6 Given the propped cantilever beam shown in Fig. E3-6, it is required to obtain a general expression for M_p and design the beam if $w_w = 5$ kN/m and $F_y = 250$ MPa. Also derive a general expression for the location of M_p in the span.

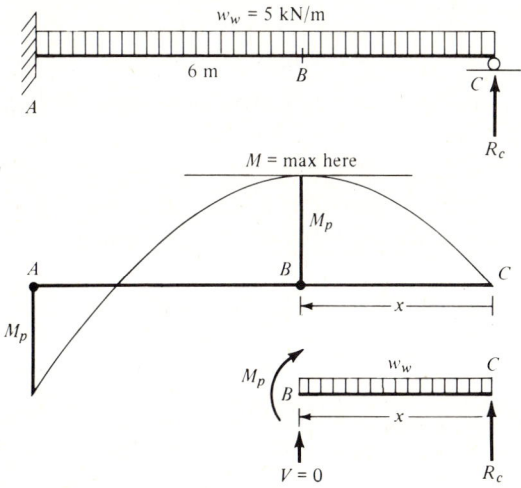

Figure E3-6

130 STRUCTURAL STEEL DESIGN

SOLUTION The collapse mechanism will consist of two hinges located as shown. From statics the moment M_p at B is a maximum. From mechanics of materials, $V = 0$, where M = maximum. This gives

$$R_c = w_u x \qquad (a)$$

Also, using statics $\Sigma M_B = 0$ for segment BC, which gives

$$M_p + \frac{w_u x^2}{2} - R_c x = 0 \qquad (b)$$

Taking moments of beam segment AC about A ($\Sigma M_A = 0$) gives

$$M_p + R_c L - \frac{w_u L^2}{2} = 0 \qquad (c)$$

Substituting Eq. (a) for R_c in Eq. (c), then substituting Eq. (c) into Eq. (b) for M_p, we obtain

$$x^2 + 2xL - L^2 = 0$$

Solve this by completing the square, we obtain

$$x = 1.414L - L = 0.414L \qquad (3\text{-}3)$$

Now a general expression for M_p can be obtained from Eqs. (a) and (b):

$$M_p = 0.08579 w_u L^2 \qquad (3\text{-}4)$$

Using Eq. (3-4) with the given beam length and loading, the value of M_p is

$$M_p = 0.08579(45 \times 1.7)(6)^2 = 236.26 \text{ kN} \cdot \text{m}$$

$$Z_x = \frac{236.26}{250} = 0.9451 \times 10^{-3} \text{ m}^3$$

Try a W460 × 52.1 section. $Z_x = 1.0861 \times 10^{-3}$ m³. The increase in Z_x for beam weight is

$$\Delta Z_x = \frac{0.51}{45}(0.9451) = 0.0107$$

$$Z_{x(\text{reqd})} = 0.9451 + 0.0107 = 0.9558 < 1.0861 \times 10^{-3} \text{ m}^3 \quad \text{O.K.}$$

Use a W460 × 52.1 section (note that the section was not checked for lateral bracing requirements). ///

Example 3-7 Given the two-span continuous beam shown in Fig. E3-7, select an economical W section using plastic design and A-36 steel.

SOLUTION Two hinges are necessary to collapse at least one span. The values of M_p to accomplish this are:
  Left span:

$$M_p + \frac{M_p}{3} = \frac{2(90)L}{9}$$

$$M_p = 15L$$

ELASTIC, PLASTIC, AND BUCKLING BEHAVIOR OF STRUCTURAL STEEL

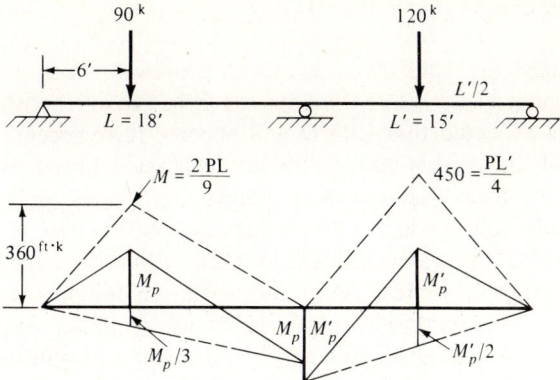

Figure E3-7

Right span:

$$M'_p + \frac{M'_p}{2} = \frac{120 L'}{4}$$

$$M'_p = 20 L'$$

The maximum value of M_p from either span is used for design (since beam runs across both spans using a constant section).

$$M_p = 15(1.7)(18) = 459 \text{ ft} \cdot \text{kips}$$

$$M'_p = 20(1.7)(15) = 510 \text{ ft} \cdot \text{kips} \quad \text{use this}$$

$$Z = \frac{510(12)}{36} = 170 \text{ in}^3$$

From Table II-2, select a W24 × 68 with $Z_x = 176.4$ in³. Check the beam weight effect as *approximately* (if a borderline case is found, one may be justified in the additional work for an exact analysis):

$$\Delta M'_p + \frac{\Delta M'_p}{2} = \frac{w_u L^2}{8}$$

$$\Delta M'_p = \frac{2}{3} \frac{0.068 \times 1.7 \times 15^2}{8} = 2.17 \text{ ft} \cdot \text{kips}$$

and the required ΔZ_x is

$$\Delta Z_x = \frac{2.17(12)}{36} = 0.72 \text{ in}^3$$

Total $Z_{x(\text{reqd})} = 170.0 + 0.72 = 170.72 < 176.4$ furnished O.K.

Use a W24 × 68 section with $Z_x = 176.4$ in³.   ///

3-7 LOAD RESISTANCE FACTOR DESIGN

Load resistance factor design (or LRFD) is a recent proposal which is still undergoing some development as an alternative approach to the currently used elastic design method. It is expected that LRFD will become fully accepted by the AISC within the useful life of this textbook. This forecast is based on the facts that this procedure (at least the essentials, called *limit states design*) is already accepted in Canada and several other countries outside the United States. The current AASHTO bridge specifications (12th edition) provide an alternative design method in steel termed load factor design for simple and continuous beams and girders of moderate length which use compact sections. All the LRFD designs are very similar to each other and to the strength design procedure used in reinforced concrete design. In LRFD, as in reinforced concrete, ϕ factors are used to reflect uncertainties in the material (in this case the specified steel strength, F_y). These factors are under current study with the current suggestions as in Table 3-1.

LRFD uses an equation of the general form

$$\phi R = \psi(F_d D + F_L L)$$

where ψ = analysis factor (also termed importance factor); value currently suggested, 1.1

F_d = uncertainty factor for dead load with a value of 1.1 suggested

F_L = uncertainty factor for live load with a value of 1.4 suggested for

Table 3-1 Current recommendations for ϕ factors

Stress condition	Suggested[a]	Canada	AASHTO
Tension members			
Yielding (F_y)	0.88	0.90	1.0
Fracture (F_u)	0.74	0.90	1.0
Bending			
Rolled sections and plate girders	0.86	0.90	1.0
Columns[b]		0.90	1.0
$\eta \leq 0.16$	0.86		
$0.16 < \eta \leq 1.0$	$0.90 - 0.25\eta$		
$\eta > 1.0$	0.65		
Shear			
Webs of beams and girders	0.86	0.90	1.0
Connections			
Bolts[c]	0.70–1.00	0.90	—
Welds	0.80	0.90	—

[a] See *Journal of Structural Division*, ASCE ST9, September 1978 (contains eight papers on LRFD).

[b] $\eta = (KL/\pi r)\sqrt{F_y/E}$ (K = length factor as given in Chap. 6).

[c] See Sec. 8-10.

building occupancy; other values are also used (e.g., 1.5 for maximum snow, 1.6 for maximum wind, etc.)

The equations for several loadings including wind and snow might read as follows:

$$\phi R \geqslant 1.1(1.1D + 1.4L + 1.6W_{max})$$
$$\phi R \geqslant 1.1(1.1D + 1.5S_{max})$$
$$\phi R \geqslant 1.1(1.1D + 1.4L)$$

The value of resistance will generally be $R = F_y$ (the yield stress of the steel).

The general objective with LRFD is to assess each item that influences the design of a structure rather than "lumping" several effects together, as, for example, simply adding the dead and live loads to obtain the composite load. Larger factors are used with those items that carry more uncertainty, such as snow and wind loads (live-load factors of 1.5 and 1.6 versus the dead-load factor of 1.4) and lesser factors for material properties (ϕ factor) and dead loads, the latter being reasonably well identified, at least after the structure is designed. The factors ϕ, F_d, F_L, and ψ are based on extensive probability studies (which are still continuing) and may eventually be rounded up or down for conservative and/or practical considerations.

3-8 LOCAL BUCKLING OF PLATES

It is readily observed that most structural members are assemblages of flat plates. The W sections consist of three flat plates, two flat plates make an angle, and so on. When a plate is subjected to direct compression, bending, or shear stresses or to combinations of these stresses, the plate may buckle locally before the entire member fails.

Consider a rectangular plate with dimensions of $a \times b$ (see Fig. 3-6) free of residual stresses, perfectly plane, homogeneous, and isotropic that is subjected to a uniform compressive load along opposite edges. Under this stress the plate will compress uniformly until the buckling stress is reached. When the buckling stress is reached, the plate will deflect in a single wave or a series of waves, depending on the edge (boundary) conditions and length to width (a/b) ratio with a resultant redistribution of the compressive stresses until, with the addition of load, the entire plate is buckled.

From the theory of plates as proposed by several authorities,† the critical elastic buckling stress F_{cr} is

$$F_{cr} = \frac{k_c \pi^2 E}{12(1 - \mu^2)(b/t)^2} \qquad (3\text{-}5)$$

---

† For example, Timoshenko and Goodier, *Theory of Elasticity*, or Bleich, *Buckling Strength of Metal Structures* (New York: McGraw-Hill Book Company), or Johnston, *Guide to Stability Design Criteria for Metal Structures*, 3rd ed. (New York: John Wiley & Sons, Inc.).

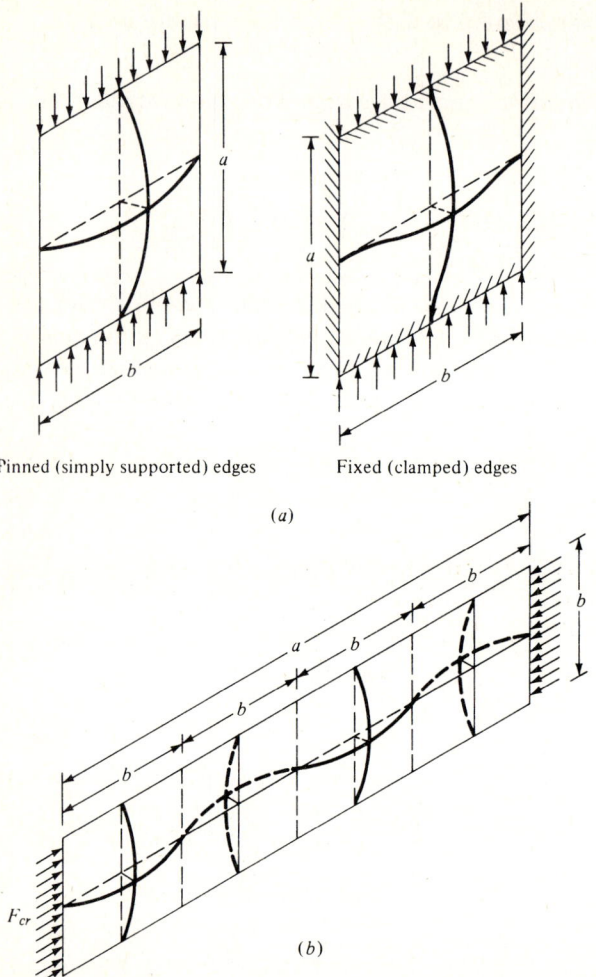

Figure 3-6 Buckling of thin plates under edge compression stresses. (*a*) Buckling of rectangular plate. (*b*) Buckling of a long, rectangular plate.

Inelastic buckling, here defined as the critical stress F_{cr} above the proportional limit, is also possible and can be described by use of a λ term in Eq. (3-5):

$$F_{cr} = \frac{k_c \pi^2 \lambda E}{12(1 - \mu^2)(b/t)^2} \quad (3\text{-}6)$$

where E = modulus of elasticity, 29 000 ksi or 200 000 MPa
 μ = Poisson's ratio (may use 0.33 or 0.3 for steel)
 a, b = plate length and width, respectively; note that F_{cr} is applied across the width b

t = plate thickness

k_c = nondimensional coefficient that depends on edge supports; the a/b ratio and is given in the general form

$$k = \left(m\frac{b}{a} + \frac{a}{b}\frac{n^2}{m}\right)^2$$

When $a/b = m$ and $n = 1$ and the edges parallel to the compressive stress are simply supported, $k = 4.00$. The terms m and n are from use of a series solution to develop the plate equation. Other values of k are not so directly obtained but have been computed and are tabulated for several boundary conditions in Table 3-2.

λ = factor used so that the tangent modulus is obtained as $E_t = \lambda E$. When $\lambda = 1$, we have $E_t = E$. In general, λ is

$$\lambda = \frac{F_y - F_{cr}F_{cr}}{F_y - F_{pl}F_{pl}} \qquad (3\text{-}7)$$

Table 3-2 Values of buckling coefficient for several plate edge conditions shown

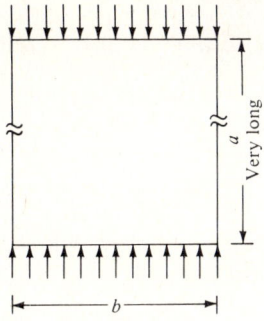

	k'_c	k_s	
	0.38	For k_s all four plate edges have same fixity	
	1.15	$\alpha = a/b$	
	3.615	$4.00 + 5.34/\alpha^2$	$\alpha \leq 1$
		$5.34 + 4.00/\alpha^2$	$\alpha \geq 1$
	4.90		
	6.30	$5.60 + 8.98/\alpha^2$	$\alpha \leq 1$
		$8.98 + 5.60/\alpha^2$	$\alpha \geq 1$

136 STRUCTURAL STEEL DESIGN

where F_{pl} = steel stress at the proportional limit (a value of F_{pl} = 0.70 to $0.75 F_y$ may be used)
F_y = yield stress of steel
F_{cr} = critical buckling stress of Eq. (3-6).

If we attempt to solve Eq. (3-6) for the critical buckling stress, several problems develop, particularly if $\lambda < 1$. First, we must determine k_c. While the general expression for k_c has been given, it is necessary to adjust this for the various boundary conditions that are possible. This has been done by several authorities, but as a convenience the author has further combined the effect of $\pi^2/[12(1 - \mu^2)] = 0.9038$ to give values shown in Table 3-2 for k'_c [i.e., $4.00(0.9038) = 3.615$]. If the λ term is less than 1, it is necessary to iterate to F_{cr}. This is illustrated as follows.

Rewrite Eq. (3-6) in terms of k'_c, to obtain

$$F_{cr} = \lambda E k'_c \left(\frac{t}{b}\right)^2 \tag{3-8}$$

Now divide through by

$$\frac{F_{cr}}{\lambda} = E k'_c \left(\frac{t}{b}\right)^2 \tag{3-8a}$$

We will temporarily hold this equation. Now using Eq. (3-7) for λ with $F_{pl} = 0.75 F_y$, we can, with some rearranging, obtain

$$\frac{F_{cr}}{\lambda} = \frac{0.1875 F_y^2}{F_y - F_{cr}} \tag{3-9}$$

Since F_{cr} is on both sides of Eq. (3-9), we must solve for F_{cr}/λ by trial. Once the value is obtained, this can be used in Eq. (3-8a) for the ratio of b/t, which is usually the item of interest. From Eq. (3-8a) the b/t ratio is

$$\frac{b}{t} = \sqrt{\frac{E k_c}{F_{cr}/\lambda}}$$

For $k_c = 3.615$, $\lambda = 1$, and $F_{cr} = 0.75 F_y$, the limiting b/t ratio for A-36 steel is

$$\frac{b}{t} = \sqrt{\frac{(29\,000)(3.615)}{0.75 \times 36}} = 62.3$$

It is often useful in using Eq. (3-9) to set up a table of λ vs. F_{cr} with values from $F_{cr} = 0.75 F_y$ to F_y. For A-36 steel, typical values are as follows:

F_{cr}, ksi	λ	E_t, ksi
27.0	1.00	29 000
28.0	0.922	26 738
30.0	0.741	21 489
33.0	0.407	11 803
36	0.0	0

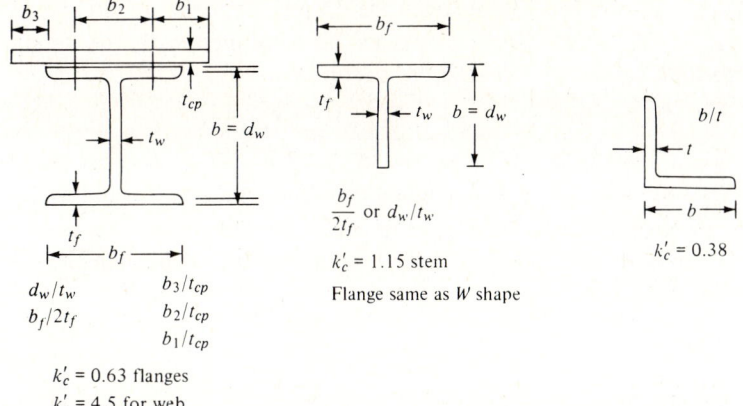

Figure 3-7 Compression characteristics for rolled shapes shown. Note that it is generally necessary to investigate the critical b/t ratio, which may be as shown for a W shape with a cover plate that is welded or bolted.

Compression characteristics (k_c and width b in compression) of three rolled shapes are given in Fig. 3-7. The k_c values shown have been found to agree reasonably with tests. Adjustments in k_c are necessary because very few plates are free of imperfections and residual stresses.

If we use the value of $k_c' = 0.63$ shown in Fig. 3-7 for the flange of a W section and a SF = 2.00 and $F_{cr} = F_y$, we obtain

$$\frac{b}{t} = 0.5\sqrt{\frac{(29\ 000)(0.63)}{36}} = 11.3 \quad \text{say } 11$$

or, in general,

$$\frac{b}{t} = 0.5\sqrt{\frac{(29\ 000)(0.63)}{F_y}} = \frac{67.5}{\sqrt{F_y}} \quad \text{say } \frac{65}{\sqrt{F_y}}$$

The current AISC specification allows a b/t (uses $b_f/2t_f$) ratio of $65/\sqrt{F_y}$. Note also that if we consider the web, $k_c' = 4.9$, we obtain

$$\frac{d}{t_w} = 0.5\sqrt{\frac{(29\ 000)(4.9)}{F_y}} = \frac{188.5}{\sqrt{F_y}} \quad \text{say } \frac{190}{\sqrt{F_y}}$$

which is also in AISC.

Values of $k_i = k_s$ are also shown in Table 3-2 for the critical stress to produce shear buckling. The critical buckling stress for shear can be derived in a similar manner to that for compression, with the substitution of an appropriate buckling coefficient k_s to obtain, from Eq. (3-6),

$$F_{crs} = \frac{k_s \pi^2 \lambda E}{12(1 - \mu^2)(b/t)^2}$$

138 STRUCTURAL STEEL DESIGN

It is usual to assume that the four plate edges are simply supported in shear and the shear stress $F_{cfs} = F_y/\sqrt{3}$. This value must be combined with the safety factor of $1/0.6$ so that the design shear stress becomes

$$F_s = \frac{F_{crs}}{SF \times \sqrt{3}} = \frac{F_y}{1.67 \times \sqrt{3}} = \frac{F_y}{2.89}$$

Most practical steel design problems consider either buckling in compression or buckling in shear. Where both stresses act simultaneously, the reader should consult books such as those by Bleich, Johnston, and Timoshenko and Goodier cited in an earlier footnote.

3-9 POST-BUCKLING STRENGTH OF PLATES

Experimental evidence shows that a buckled plate does not result in immediate failure. Rather, there is a considerable strength reserve attributed to the effect of the adjacent plate material, which restrains the buckling and allows transfer of any post-buckling load increase to the unbuckled zones. This situation is idealized in Fig. 3-8, which illustrates the central buckled zone in the loaded width b. On either side are strips that confine the buckling and are loaded to a lesser effective load f_e. The concept of effective width b_e is applied as the sum of the two strip widths on each side of the buckled zone. When the effective stress on these two edge strips (stress on width of b_e) reaches a value such that deformation is constant with no further increase in load, the full load capacity of the plate has been reached. The difference between the initial buckling load and this new value is the post-buckling strength of the plate.

The stress f_e may be evaluated using Eq. (3-8), to obtain

$$f_e = f'_c \lambda E \left(\frac{t}{b_e}\right)^2$$

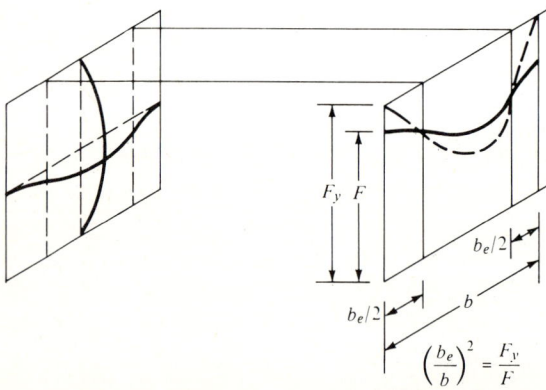

Figure 3-8 Effective width for post-buckling plate capacity.

where all terms have been previously identified except b_e, which is shown in Fig. 3-8. For a very long, thin, simply supported plate, it appears that the theoretical value of k'_c is 3.615 for this equation. The use of post-buckling strength of plates is not often directly evaluated. It is used more often in a more indirect manner; for example, AISC allows use indirectly via Appendix C3, which states: "When the width-thickness ratio of a uniformly compressed stiffened element exceeds the applicable limit given in Sec. 1-9.2.2, a reduced effective width, b_e, shall be used in computing the . . . ."

PROBLEMS

3-1 What is the allowable $b_f/2t_f$ for any W section using a steel with $F_y = 50$ ksi?
Answer: 9.2.

3-2 What is the allowable $b_f/2t_f$ for any W section with $F_y = 250$ MPa?

3-3 What is the plastic moment capacity (kN · m) of a W920 × 200.9 section for $F_y = 345$ MPa?
Answer: 2865.2 kN · m.

3-4 Verify Z_x in Table V-3 for a W610 × 241.1 rolled shape.

3-5 What is Z_x for the geometrical shape shown in Fig. P3-5 if it is used for a beam?

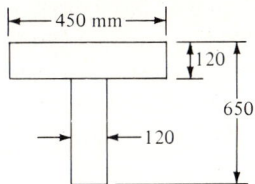

Figure P3-5

3-6 Select the lightest W section to *satisfy bending* for the span and loading shown in Fig. P3-6. Use $F_y = 36$ ksi and plastic design.

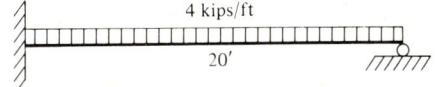

Figure P3-6

3-7 Select the lightest W section *to satisfy bending* for the span and loading shown in Fig. P3-7. Use $F_y = 250$ MPa and plastic design.
Answer: W610 × 101.2.

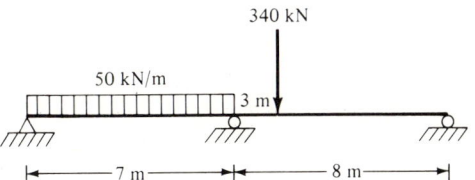

Figure P3-7

3-8 Select the lightest W section for bending for the beam shown in Fig. P3-8. Use $F_y = 345$ MPa and plastic design.

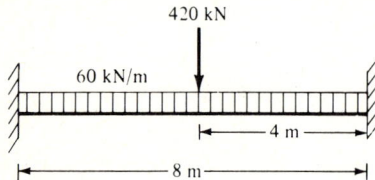

Figure P3-8

3-9 What is the uncertainty factor on both S and R to produce $F_a = 0.5F_y$? What are two reasonable alternate nonequal values to produce the same effect?
   Answer: $S = R = \frac{1}{3}$.

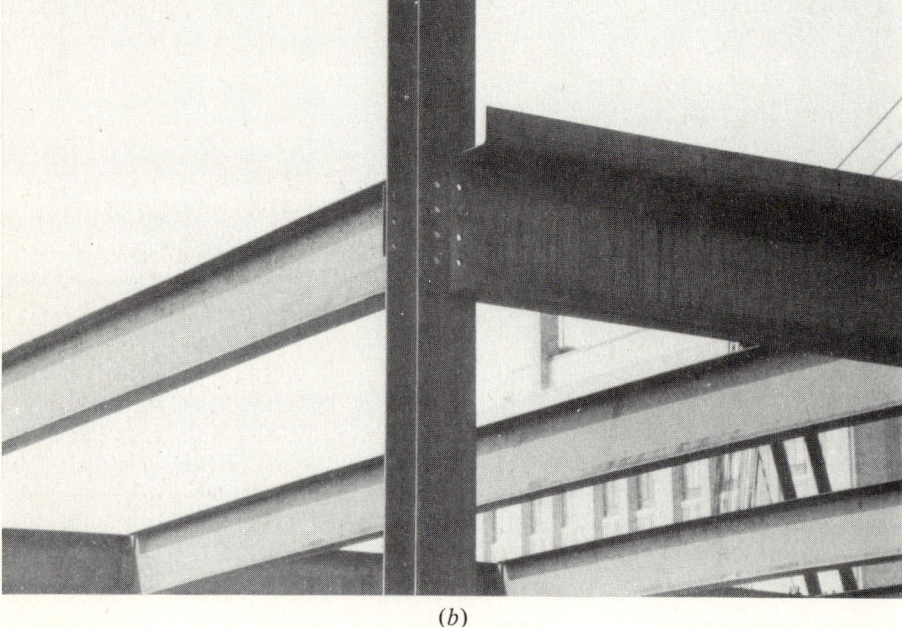

Figure IV-1 Steel beams. (*a*) Massive pair of beams used for crane and crane runway. Crane hoist can be seen between two beams (or girders). Note that crane runway columns are stepped with a smaller column section above the base of the crane runway girder. (*b*) Beams being erected for medium-size office complex. Note the erection bolts used to temporarily hold parts. Beam in foreground is a spandrel built using a channel and angle.

CHAPTER
FOUR

DESIGN OF BEAMS FOR BENDING

4-1 GENERAL CONSIDERATIONS

Beams are structural members that carry transverse loads which produce bending moments and shear with bending resistance the design parameter of particular significance. Beams may be horizontal (most common), sloping (as roof beams), or vertical. Sloping and vertical beams may carry both axial and transverse loadings. The latter members, called beam–columns, will be considered in Chap. 7. These need to be considered separately, since bending and buckling behavior is considerably influenced by a combination of bending and compressive axial stresses.

Beams may be termed *simple beams* where the end connections either do not, or are assumed not to, carry any end moments from any continuity developed by the connection. A beam is continuous when it extends continuously across more than two supports; is a fixed beam if the ends are rigidly attached to other members so that a moment can be carried across the connection. In a steel frame the term "fixed end" is somewhat of a misnomer, since the ends of rigid connections are not fixed in the sense that a fixed-end beam is analyzed in mechanics-of-materials textbooks. There is usually some joint rotation in building frames so that the actual end moment is not directly obtained from fixed-end-moment equations but is computed considering the general frame rigidity. We shall see later that in selected circumstances we may design the end connection for a portion of the moment capacity of the member. Most of our attention will be directed to cases where the beam

1. Has no end moments (simple beam), or
2. Has moments at the ends of each span (continuous or fixed end).

Beam loadings will consist of both dead and live loads. The effects of these in terms of whether a series of concentrated loads or a uniform loading is produced depends on the general framing plan and the beam location. A part of the beam dead load is its own weight. Where long, heavy members are used, the self-weight may be a significant part of the total beam load. Where the span is small and/or the external loads are also small, the beam weight may be quite small, and here the strength/weight ratio of steel is a particular advantage. In any case, the section should always be checked for adequacy for the applied loads, including the beam weight.

Beams may be classified as:

1. *Girders:* main load-carrying members into which floor beams or joists frame, as shown in Fig. 4-1. Chapter 10 considers built-up members used as girders for bridges.
2. *Joists:* members used to carry roofing and floors of buildings.
3. *Lintels:* beam members used to carry wall loads over wall openings.
4. *Spandrels:* exterior beams at the floor level in building construction used to carry part of the floor load and the exterior wall. When the wall surfacing is brick or tile, the spandrel may carry the wall surfacing for a story to reduce compressive stresses in the masonry.
5. *Stringers:* members used in bridges parallel to the traffic to carry the deck slab and commonly frames into transverse (floor beams) members.
6. *Floor beams:* secondary members of a floor system (Fig. 4-1) and main members in bridge construction into which stringers frame (see Figs. 4-13 and 4-14).

In most cases, particularly for maximizing economy, a rolled steel shape is loaded so that the bending is about the strong axis (X–X axis, as shown in the section property tables in SSDD). Occasionally, the bending takes place about the weak (Y–Y) axis and in some instances there is simultaneous bending about both the X and Y axes. In nearly all of the applications involving a single axis of bending, the load is considered to be applied through the shear center of the W or S shapes. The shear center for these shapes is in the center of area, and this load position produces simple bending about either axis.

When the load does not pass through the shear center as may happen for channels, angles, and some built-up sections, unless a special load device is used, a torsional moment is produced along with the bending moment and should be taken into account to avoid overstressing the member.

The design of beams requires an analytical iterative analysis to determine the shear and moment diagrams based on the several possible load combinations

Figure 4-1 Several illustrations of beams in building construction. (*a*) Use of beams in an industrial building. Crane runway girder is a fabricated beam. Shown on top are purlins resting on roof trusses. (*b*) Rigid frame building using tapered columns and girder to maintain a large interior space free of columns. (*c*) Use of bar joists for both girders and roof joists. (*d*) Simple one-story building with roof beams simple framed to columns. Wind is resisted in both directions using cross bracing at four corners. (*e*) Girder (left to right) and floor beam system in an office building. (*f*) Closeup view showing framing of floor beams and joists. (*g*) Roof beams and spandrel for flat roof building. (*h*) Roof joists (partly in place); also floor beams, girders, and exterior spandrel beams.

$(D + L, D + L + S, D + L + W$, etc.). This is a relatively simple process for simple beams even where a moving live load is involved. Where continuous beams and moving live loads are involved, it may be necessary to use influence lines. The use of a computer program such as that given in the Appendix (used in Chap. 2) is of considerable advantage, since one may move a live load across the beam with the output being directly the influence ordinates. Inspection of the output directly gives the critical design moment. If shear is required, the moment together with the loads can be used to back-compute the critical shear.

The stresses f_i (use lowercase letters for computed values and uppercase letters for allowable values) produced by shear and moment are compared to the allowable stresses F_i and a section is selected. When the design is based on material behavior below the proportional limit, the procedure is termed *elastic design* or *working stress design* and the equations used are found in any textbook on mechanics of materials:

For bending:

$$f_b = \frac{My}{I}$$

For extreme fiber stresses:

$$f_b = \frac{Mc}{I} = \frac{M}{S} \leq F_b \quad (4\text{-}1)$$

For shear:

$$f_v = \frac{VQ}{It} \leq F_v \quad (4\text{-}2)$$

where $M =$ bending moment from the moment diagram or direct computations
$y =$ distance from neutral axis of beam section to point where stress is desired (see Fig. 4-2)
$c =$ distance from neutral axis to extreme fiber
$I =$ moment of inertia of cross section
$S = I/c =$ section modulus of section (both I and S are tabulated in section property tables such as Tables I-4 and V-4 of SSDD)
$Q =$ statical moment of area above point where shear stress is being determined $= A\bar{y}$ (refer also to Fig. 4-2)
$t =$ thickness of beam at point of shear stress investigation
$V =$ critical shear force from shear diagram or direct computations

The section is checked to be sure that it is adequate to carry its own weight, and finally the working load deflections are checked. Sometimes deflections rather than stresses will control the section size, especially where deflections are rigidly limited. Values of deflections commonly range from $L/360$ to as small as $L/1000$.

DESIGN OF BEAMS FOR BENDING **147**

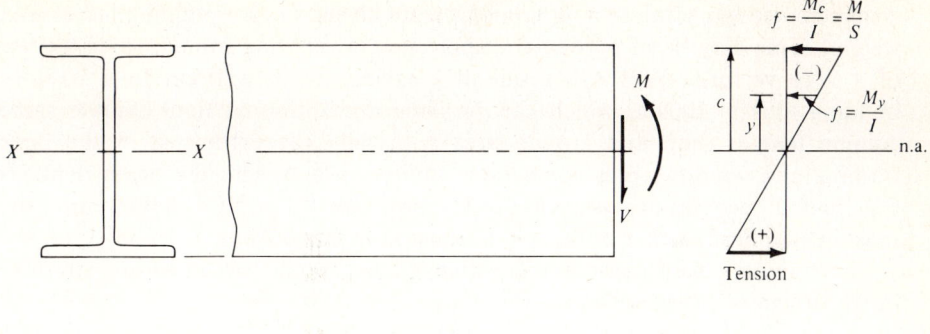

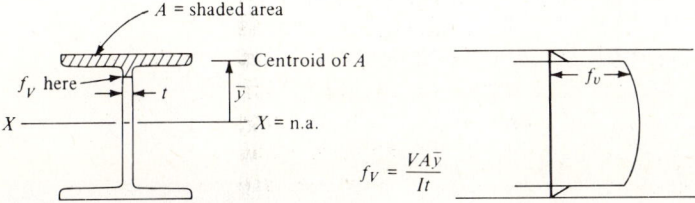

Figure 4-2 Bending and shear stresses for beams.

4-2 DESIGN OF BEAMS BY THE ELASTIC METHOD

The W shapes will usually be used for beams. On occasion, S shapes, M shapes, and even channels are used, depending on location, mode of connection to rest of frame, and loads to be carried. The W shapes, having relatively wide flanges, are generally more stable relative to lateral buckling and are generally more economical in terms of weight than any of the other shapes produced.

Inspection of Eq. (4-1) indicates that for all factors equal the preferred section will have a large section modulus S. Also note that the larger section modulus is nearly always accompanied by an increase in beam depth. Since several of the rolled sections have almost the same value of S, it is necessary to find both the lightest section and at the same time one with an adequate S. Tables II-1 and VI-1 in SSDD (and in AISC) have the sections commonly used for beams ranked in descending order of S and with respect to the X–X (strong) axis. The ranking is also with respect to weight, so that considering a group as defined by extra line spacings, we have:

1. The largest S in any group is at the top of that group.
2. The lightest section in the group is at the top and the heaviest is at the bottom.

This ranking allows rapid determination of the most economical rolled section to use as a beam based on stress and weight considerations. On occasion it may be necessary to use a section of less depth (and corresponding increase in weight). This may be of particular importance in building frames, since the use of a W16 versus a W21 can result in a savings of 5 in/floor. In a 20-story building, this is 100 in, which can produce an additional floor for the same column length, roof, and ground area with only the extra cost of the floor framing and windows plus some extra utility costs. Where the beam depth is specified, it becomes necessary to investigate Table I-4 or VI-4 (SSDD) until the most economical section in the given nominal depth is found.

Two factors exert considerable influence on the allowable bending stress F_b in the design of rolled sections. These are:

1. Section geometry (based on flange width/thickness and depth/web thickness).
2. Lateral support (or lateral bracing spacing).

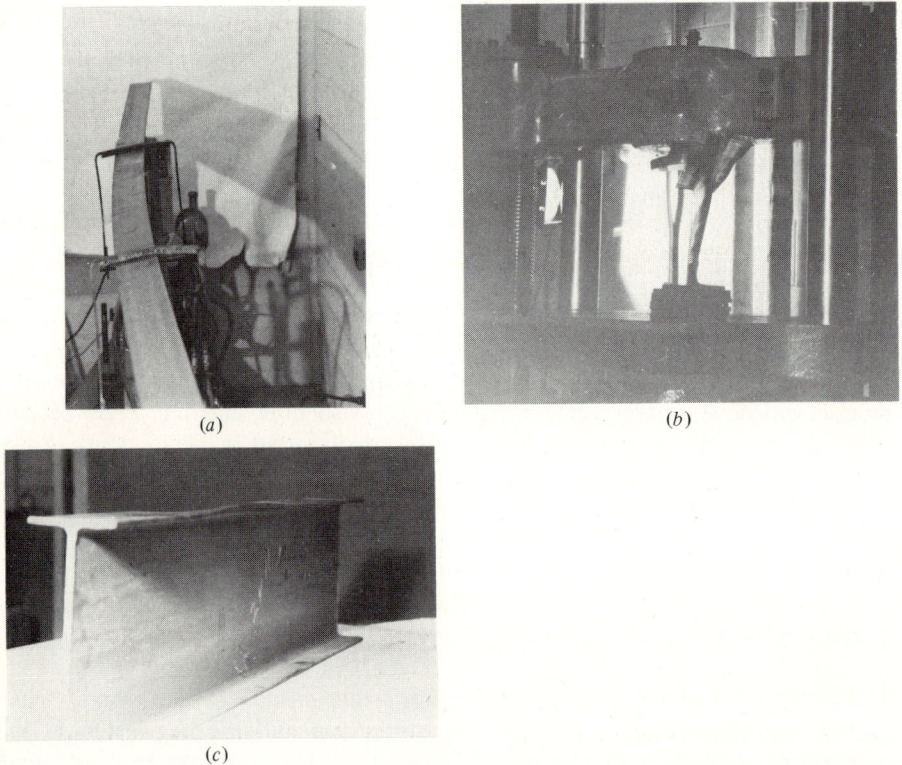

Figure 4-3 Photographs of beam failures caused by laboratory loading. (*a*) Lateral buckling of a W section. (*b*) Web buckling of a short beam loaded with concentrated load at midspan. (*c*) Flange buckling accompanying the web buckling of (*b*).

Both of these factors are direct consequences of the plate buckling problem posed in Chap. 3. If the flange width/thickness ($b_f/2t_f$) and depth/web thickness (d/t_w) ratios are not adequate, the section elements will tend to buckle at low compression stresses (compression due to bending + any due to axial loads). If the compression flange is not supported at intervals along the compression zone, it will either buckle in plane or out-of-plane coupled with twisting. These modes of failure are illustrated in Fig. 4-3 (laboratory test) and Fig. 4-4 (line detail).

The designer has little control over section geometry; however, steel producers, by means of a combination of refined analysis and spacing of the rolls, produce sections with $b_f/2t_f$ and d/t_w ratios such that this factor is not as important as the lateral bracing requirements, which are designer-controlled.

The geometry of a rolled shape may be such to produce a *compact* or *noncompact* section according to AISC specifications. AASHTO and AREA specifications do not make this distinction when using elastic design methods. The latest AASHTO specification does now use compact section criteria when using the alternative method of load factor design (AASHTO Sec. 1-7.52). There is some advantage in having a compact section when using AISC design specifications (for most buildings), since the basic bending stress is:

$$F_b = 0.6 F_y \quad \text{(for noncompact sections)}$$

$$F_b = 0.66 F_y \quad \text{(for compact sections)}$$

A compact section is one (see also Sec. 1-5.1.4 in Part III of SSDD) that meets the following requirements:

1. The compression flange must be continuously connected to the beam web.

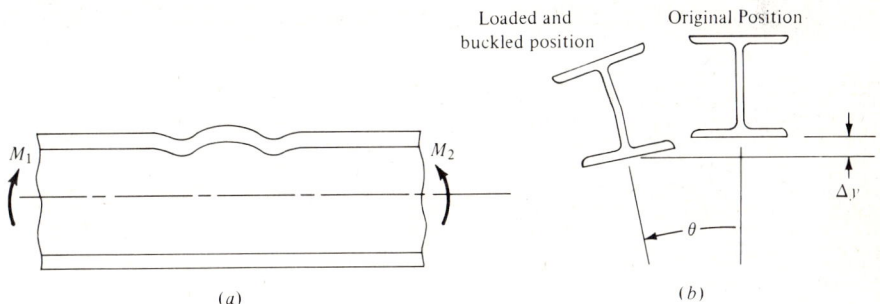

Figure 4-4 Line details of beam stability considerations caused by stresses in compression flange due to bending. (a) Flange buckling when compression is very large and is laterally supported so that lateral bending does not occur (illustrated in Fig. 4-3c). Note that vertical web buckling may also occur with this failure. (b) Lateral buckling and torsion. Note that section has moved laterally, vertically, and has twisted. Angle θ referenced to original beam position. This mode of failure is also illustrated in Fig. 4-3a.

2. It has the following flange width/thickness ratios:

$$\text{fps:} \quad \frac{b_f}{2t_f} \leq \frac{65}{\sqrt{F_y}} \qquad SI: \quad \frac{b_f}{2t_f} \leq \frac{170}{\sqrt{F_y}}$$

Use F_y = ksi or MPa in these and all other equations following. Also note that $b_f/2t_f$ and d/t_w ratios are dimensionless, so consistent units of in or mm must be used.

We may solve for $F_y' = F_y$ in the foregoing expressions to obtain the upper limit of yield stress for any given $b_f/2t_f$ ratio of a section, as in Tables II-1 and VI-1. These values are presented so that as the section is selected, one checks to see if the F_y of steel being used is less than or equal to the value tabulated. One readily notes that there are no shapes that do not meet this part of the "compact" criteria for A-36 steel, and the W610 × 154.8 will be "noncompact" only for F_y > 403 MPa (W24 × 104 for F_y > 58 ksi).

3. The d/t_w ratio is partially controlled by the presence of an axial load. At this point we are considering only a beam where the axial load is zero; thus the d/t_w ratio is (referring to SSDD as previously cited)

$$\text{fps:} \quad \frac{d}{t_w} \leq \frac{640}{\sqrt{F_y}} \qquad SI: \quad \frac{d}{t_w} \leq \frac{1690}{\sqrt{F_y}}$$

We can solve for $F_y'' = F_y$ in these equations for the limiting d/t_w for any rolled shape. This value of F_y'' is given in Tables II-1 and VI-1, where we note that every shape in every grade of F_y currently produced for beams (up to 65 ksi or 450 MPa) is compact (as indicated by ---).

4. The compression flange must be adequately braced to provide the last "compact" criteria. The bracing distance L_b must be the lesser of the following distances (points of inflection *may* be considered bracing):

$$L_1 = \frac{76 b_f}{12\sqrt{F_y}} \quad \text{(ft)} \qquad L_1 = \frac{0.20 b_f}{\sqrt{F_y}} \quad \text{(m)}$$

or

$$L_2 = \frac{20\,000}{12 \times F_y \times d/A_f} \quad \text{(ft)} \qquad L_2 = \frac{139\,000}{F_y \times d/A_f} \quad \text{(m)}$$

where b_f = width of compression flange, in or mm, from Table I-4 or V-4
d = depth of section, in or mm
A_f = area of compression flange = $b_f \times t_f$; note that d/A_f is already computed in Tables I-4 and V-4
F_y = yield stress, in kips/in² or MPa

The lesser of the values of L_1 and L_2 above is called L_c (brace length for a "compact" section) and the larger value is L_u. Both of these values are given in Tables II-1 and VI-1 as a convenience.

When the unbraced length of compression flange $L_b \leq L_c$ and the section geometry is "compact," we may use

$$F_b = 0.66 F_y \qquad (F_b = \tfrac{2}{3} F_y)$$

When the unbraced length $L_b \geq L_u$ the largest possible value of F_b is

$$F_b = 0.6 F_y$$

When either

$$L_c < L_b < L_u \qquad \text{or} \qquad \frac{65}{\sqrt{F_y}} < \frac{b_f}{2 t_f} \leq \frac{95}{\sqrt{F_y}}$$

we may use a linear transition between $0.66 F_y$ and $0.6 F_y$ as

$$F_b = F_y \left[0.79 - A \left(\frac{b_f}{2 t_f} \right) \sqrt{F_y} \right] \qquad (4\text{-}3)$$

where $A = 0.002$ fps and F_y in ksi
       $= 0.00076$ SI and F_y in MPa

The use of Eq. (4-3) is not recommended by the author for most situations as the computational effort is considerable for at most a 10 percent (most values will be only 2 or 3 percent) increase in allowable stress over simply using $F_b = 0.6 F_y$.

The AASHTO and AREA basic bending stress for beams laterally supported is

$$F_b = 0.55 F_y \qquad (4\text{-}4)$$

For A-36 steel, this equation gives a value that is slightly rounded of 20 ksi or 138 MPa.

Example 4-1 Given a beam with the span and loading shown in Fig. E4-1, select the lightest section using A-36 steel if any beam depth may be used.

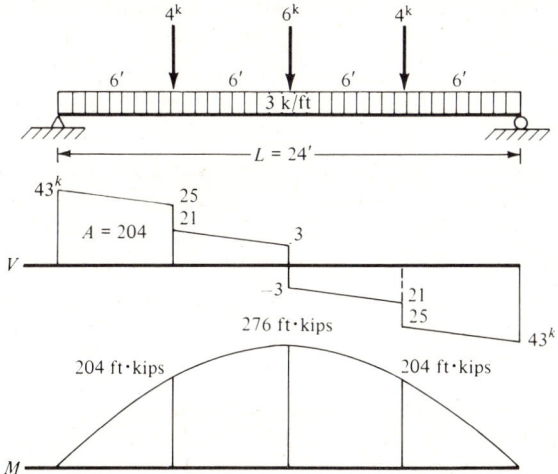

Figure E4-1

152 STRUCTURAL STEEL DESIGN

SOLUTION Draw shear and moment diagrams as shown in Fig. E4-1. Next, obtain $M_{max} = 276$ ft · kips.

With a uniform loading on the beam, the top flange is in contact with something; thus we will assume full lateral support, so that

$$L_b = 0 < L_c$$

and use $F_b = 0.66F_y = \frac{2}{3} \times 36 = 24$ ksi.

Rearranging Eq. (4-1), the required section modulus with strong axis bending is

$$S_x = \frac{M(12)}{F_b} = \frac{276(12)}{24} = 138 \text{ in}^3$$

We must find a section with a somewhat greater value of S so that the beam weight can be carried. From Table II-1, select

$$\text{W21} \times 68: \quad S_x = 140 \text{ in}^3$$

Checking the beam weight, the additional S required is

$$\Delta S_x = \frac{wL^2(12)}{8(F_b)} = \frac{0.068(24)^2(12)}{8(24)} = 2.45 \text{ in}^3$$

$$S_{reqd} = 138 + 2.45 = 140.45 > 140 \quad \text{furnished}$$

If we use this beam, the actual $f_b = 24(140.45)/140 = 24.077$ ksi, or this beam will be overstressed 77 psi based on the given load conditions. We may:

1. Use engineering judgment that this small overstress is acceptable, or
2. Use the next lightest section, which in this case is a

$$\text{W24} \times 68: \quad S_x = 154.0 > 140.45$$

With S_x so much larger and beam weight the same as in the earlier section, it is not necessary to again check the beam weight.

The reader should observe that from a deflection standpoint the W24 is the better selection. ///

Example 4-2 What size of beam should be used in Example 4-1 if the beam depth is limited to 16 in?

SOLUTION S_{reqd} is approximately 145 in³, since the weight will have to be larger than 68 lb/ft of the most economical section selected in the example. Always use the maximum depth possible when the beam is laterally supported. Therefore, by inspection of the W16's in Table I-3, select

W16 × 89: $S = 155.0$ in³ $d = 16.75$ in > 16 N.G.

Try   W14 × 90: $S = 143.0$ in³ $d = 14.02$ in   (next try)

W12 × 106: $S = 145.0$ in³ $d = 12.89$ in

Checking the beam weight, we obtain

$$\Delta S = \frac{0.090(24)^2(12)}{8(24)} = 3.24 \text{ in}^3$$

$$S_{\text{reqd}} = 138 + 3.24 = 141.2 < 143 \text{ in}^3 \quad \text{furnished}$$

Use a W14 × 90 beam. ///

Example 4-3 Given the beam span and loading shown in Fig. E4-3, which is approximately the beam system of Example 4-1. Most of the SI problems will be approximately the same as the fps problems (but not a direct conversion of units), so the reader may be assisted in developing a feeling for the SI units and selection of section sizes. Use $F_y = 250$ MPa.

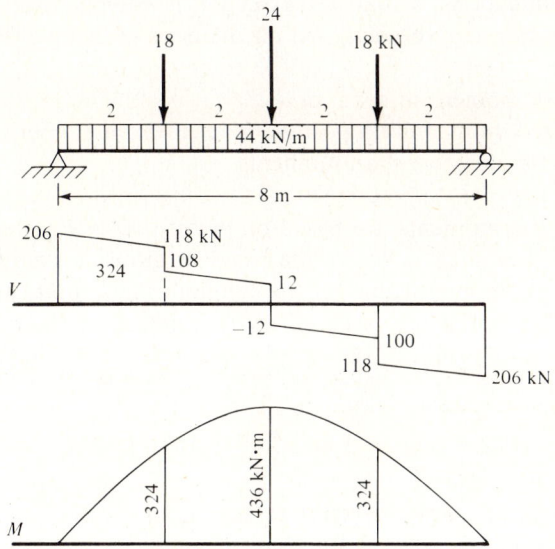

Figure E4-3

SOLUTION First, draw shear and moment diagrams.
Second, obtain $M_{\text{max}} = 436$ kN · m. Assuming full lateral support, we obtain

$$F_b = 0.67 \times 250 = 167 \text{ MPa}$$

The required section modulus is

$$S_x = \frac{M}{F_b} = \frac{436}{167} = 2.611 \times 10^{-3} \text{ m}^3$$

Note that 167 MPa × 10^3 = 167 × 10^3 kPa, which is consistent with 436 kN · m. Check Table VI-1 for S_x somewhat larger. Find:

$$\text{W610} \times 113.1: \quad S_x = 2.8841 \times 10^{-3} \text{ m}^3$$

Check ΔS_x for beam weight = 1.11 kN/m:

$$\Delta S_x = \frac{1.11(8)^2}{8(167)} = 0.0532 \times 10^{-3} \text{ m}^3$$

$S_{\text{reqd}} = S_x + \Delta S_x = 2.611 + 0.0532 = 2.6642 < 2.8841 \times 10^{-3}$ m^3 O.K.

Use a W610 × 113.1 beam.   ///

4-3 DESIGN OF CONTINUOUS BEAMS

Continuous beams are designed similar to simple beams. A major difference when using AISC specifications is that if the section is compact and is not a cantilever beam, the section may be designed on the basis of using either:

1. 0.9 × largest negative moment in span, or
2. Positive moment based on maximum positive moment from moment diagram + 0.1 × average of the negative span moments,

whichever is larger. These moments are based on gravity ($D + L$, $D + L + S$, etc., and not wind) load moments. Where the beam or girder is rigidly framed into a column, the design moment value for the column at this point can also be reduced 10 percent. This procedure is based on recognition of the method of plastic hinge formation and resultant transfer of moment from the negative zone to the positive zone until a hinge finally forms at that point. This procedure will now be illustrated.

Example 4-4 Use the computer output of Example 2-2 and check and/or redesign members 22, 23, and 24 (refer to Fig. E4-4 for the gravity load moment (which inspection of Fig. E2-4d indicates are the critical values for design). Use $F_y = 250$ MPa.

SOLUTION From the computer, output obtain beam end moment values of

− 57.86   + 140.64   − 86.86   + 86.88   − 140.64   + 57.86 kN · m

These values are needed to complete the shear and moment diagrams shown in Fig. E4-4.

Since we are using the same section for all spans, the largest design moment is in span 1 or 3, by inspection. Note the moment in span 2 is all negative, requiring possible lateral bracing on the bottom (compression) flange if $L_b = 2.3 > L_c$. We must check this possibility when the section is selected.

The $-M$ for design is

$$M = -140.6(0.9) = -126.5 \text{ kN} \cdot \text{m}$$

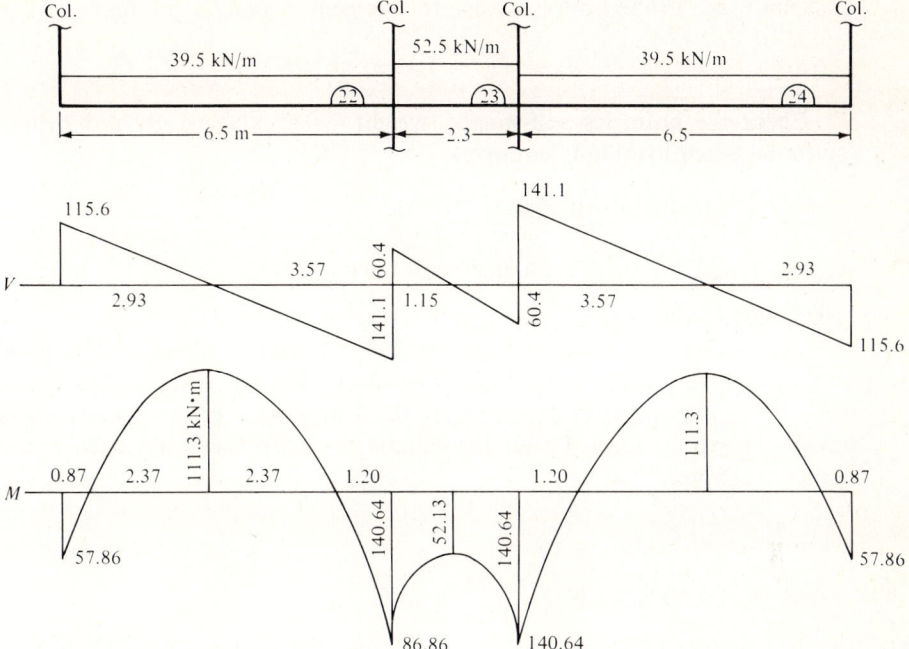

Figure E4-4

The $+M$ for design is

$$+M = 111.3 + \frac{57.86 + 140.6}{2}(0.1) = 121.2 < 126.5 \text{ kN} \cdot \text{m}$$

Since the largest moment is -126.5 kN · m, this is used to select the section. Note that the compression flange is part on the top and part on the bottom, with inflection points as shown on the moment diagram of Fig. E4-4. Tentatively assume that L_c will be larger than the values for the bottom flange; if so, no lateral bracing will be required and we can use $F_b = 0.66F_y$.

$$S_{\text{reqd}} > \frac{126.5}{167} = 0.7575 \times 10^{-3} \text{ m}^3$$

From Table VI-1, select a

W410 × 46.1: $S = 0.7735 \times 10^{-3}$ m^3

$L_c = 1.78$ m > 1.20 O.K. for no bracing

$L_u = 2.16$ m

The point of inflection in either outside span produces 0.87 m of unsupported compression flange on bottom of beam. This will be deemed adequate compression flange bracing, since the remainder of the compression flange (on the top) is braced by the floor. We will have to use a midspan brace to the bottom flange for the center span, to produce

$$L_b = \frac{2.3}{2} = 1.13 \text{ m} < 1.78$$

Check the beam for self-weight (weight = 0.45 kN/m). By proportion (since the beam loading is uniform)

$$\Delta S = \frac{0.45}{39.5}(0.7575) = 0.0086 \times 10^{-3} \text{ m}^3$$

$$S_{\text{reqd}} = S + \Delta S = 0.7575 + 0.0086 = 0.7611 \times 10^{-3} \text{ m}^3 < 0.7735 \text{ furnished}$$

Use a W410 × 46.1 beam.

The reader should verify that the method used to obtain ΔS is both correct and the most practical means available. Since a W410 × 59.5 beam was used in the initial computer analysis, it appears that the problem will have to be reprogrammed after the column design/revision has been made in a later chapter. ///

4-4 WEB BUCKLING AND CRIPPLING

Web buckling is an out-of-plane web distortion resulting from a combination of large d/t_w ratio and bending stress. The unbraced length of compression flange may also contribute to web buckling. Web buckling is controlled by limiting either the d/t_w ratio or the stress that can be used for the given d/t_w ratio. This is allowed for in the several specifications. Web buckling is illustrated in Fig. 4-5b.

Web crippling can occur if the web in-plane compressive stresses are sufficiently large. This can occur if reaction distances or load-bearing plates used to deliver column loads to the beam flange are too narrow. Web crippling can also occur if a uniform load on the flange is too large for the web thickness. Web crippling control will be obtained by determining the required reaction

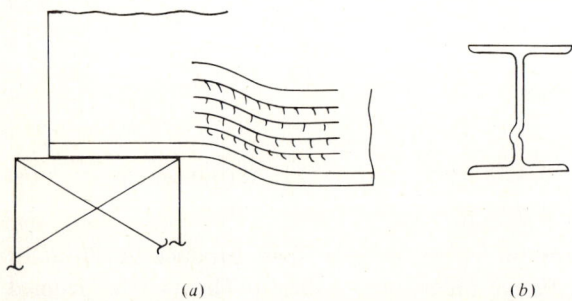

(a)   (b)

Figure 4-5 Web failures to avoid in design. (a) Web crippling. (b) Web buckling.

distance or column base plate width in the following way. The needed reaction distance is obtained by considering an area in web compression defined by the reaction length + an additional distance using a 1 : 1 (45°) slope through the k distance of the section. The section property tables tabulate k for the several rolled sections. The k distance is measured from the outer flange face to the top of the fillet transitioning the web-to-flange interface. At this location the resulting web area in compression is nearly (if not exactly) a minimum. At a reaction the area in web compression is

$$A_c = (N + k)t_w$$

The allowable stress at this location is taken by AISC (see SSDD Sec. 1-10.10.1) to be

$$F_a = 0.75 F_y$$

At a reaction with $f = R/A_c$, we obtain

$$\frac{R}{(N + k)t_w} \leq 0.75 F_y \qquad (4\text{-}5)$$

At a concentrated load in the span, the distance k can develop on both sides of the load as illustrated in Fig. 4-6. For this condition, we obtain

$$\frac{R}{(N + 2k)t_w} \leq 0.75 F_y \qquad (4\text{-}6)$$

where N = reaction width; a basic value of $3\frac{1}{2}$ in or 89 mm (width of standard brick) is often assumed; N = width of column or load-delivering element for interior loads
R = reaction or other concentrated load, kips or kN

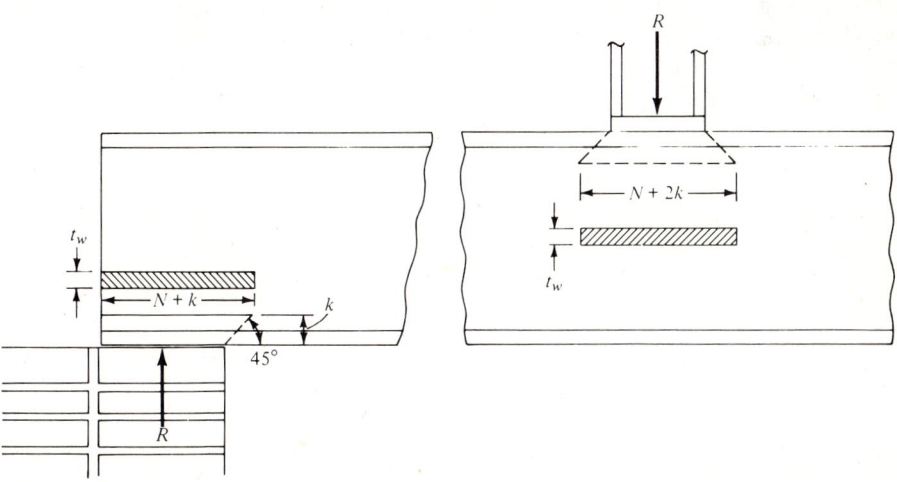

Figure 4-6 Bearing length for concentrated loads on beams according to AISC specifications.

t_w = web thickness of section, in or mm
k = distance from exterior flange face to fillet toe, in or mm

When large uniform loads are carried through the flange to the web, it may be necessary to check the compression stress f_c and limit the value to

$$f_c \leq 0.75 F_y$$

Example 4-5 What is the allowable reaction for a W16 × 40 using the basic value of $N = 3\frac{1}{2}$ in, with A-36 steel? What column load can be transmitted using a W8 × 31?

SOLUTION From Table I-3, obtain for a W16 × 40:

$k = 1.03$ in
$t_w = 0.305$ in
$R = (N + k)t_w(0.75 F_y) = (3.5 + 1.03)(0.305)(0.75 \times 36) = 37.3$ kips

The column load is limited (assuming that the column base is exactly the same size as column depth as in Fig. 4-6) to: for a W8 × 31: $d = 8.00$ in $= N$.

$$P = (N + 2k)t(0.75 F_y)$$
$$= [8.00 + 2(1.03)](0.305)(0.75 \times 36) = 82.8 \text{ kips} \qquad ///$$

Example 4-6 What is the allowable reaction for a W460 × 74.4 section using the basic value of $N = 89$ mm and $F_y = 345$ MPa? What column load can be transmitted using a W200 × 46.1?

SOLUTION From Table V-3 for a W460 × 74.4:

$k = 27.8$ mm
$t_w = 9.0$ mm
$R = (N + k)t_w(0.75 F_y) = (89 + 27.8)(0.009)(0.75 \times 345) = 272$ kN

Note that MPa × mm² × 10^{-3} = kN (10^{-3} not shown). The column load (assuming base plate same size as column depth) is: for a W200 × 46.1, the depth, $d = N = 203$ mm.

$$P = [203 + 2(27.8)](0.009)(0.75 \times 345) = 602.2 \text{ kN} \qquad ///$$

Bridge reactions are generally supplied by special fabricated bearings, and AASHTO specifications require that web stiffeners will be used when the web shear at the bearing is $f_v > 0.75 F_v$. Column loads are usually not carried by bridge beams. The AREA specifications are more stringent than those of the AASHTO and require pairs of web stiffeners at the end bearings and at points of concentrated loads of girders and beams. Design of girder (built-up beams)-type members will be considered in Chap. 10.

4-5 SHEAR CRITERIA

The shear stress distribution across any section subjected to bending can be computed using the equation presented earlier:

$$f_v = \frac{VQ}{It}$$

A plot of shear stress using this equation is illustrated in Fig. 4-7. We note that the average shear stress based on

$$f_v = \frac{V}{dt_w} \qquad (4\text{-}7)$$

differs somewhat from the maximum value shown in Fig. 4-7 (in this case about 23 percent) but is considerably easier to compute. AISC allows use of Eq. (4-7) for either rolled or fabricated (plate girders) sections.

The AASHTO and AREA specifications simply allow computation of f_v based on the area of the "gross" section. This can be interpreted as in the AISC specifications (i.e., $f_v = V/dt_w$).

The allowable shear stress F_v for rolled sections is computed as:

$$F_v = 0.40 F_y \qquad \text{(AISC)}$$

$$F_v = 0.33 F_y \qquad \text{(AASHTO)}$$

$$F_v = 0.35 F_y \qquad \text{(AREA)}$$

Shear stresses seldom govern in building construction unless the section is both very short and heavily loaded, as illustrated in the following example.

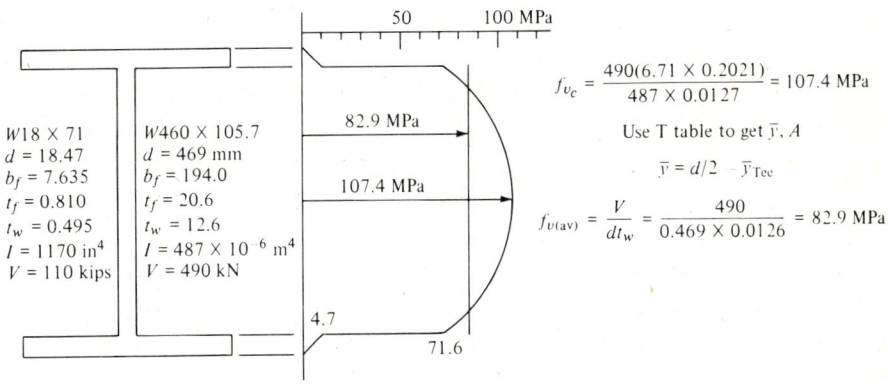

Figure 4-7 Theoretical and average shear stress distribution on section for conditions shown.

Example 4-7 What is the span length and load/ft for a beam of A-36 steel shown in Fig. E4-7 such that either F_v or F_b will control? What length N is required for the reaction?

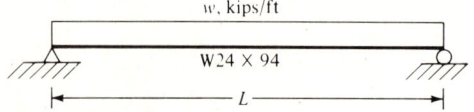

Figure E4-7

SOLUTION Data for a W24 × 94:

$$d = 24.31 \text{ in} \qquad k = 1.53 \text{ in}$$
$$t_w = 0.515 \text{ in} \qquad S_x = 222.0 \text{ in}^3$$

$$f_v = \frac{wL}{2t_w d} = 0.4 f_y \tag{a}$$

$$f_b = \frac{wL^2}{8 S_x} = 0.66 F_y \tag{b}$$

From Eq. (a),

$$w = \frac{0.4(36)(2)(24.31)(0.515)}{L} = \frac{360.6}{L}$$

Substituting this w into Eq. (b), we obtain

$$\frac{360.6 L}{8 S_x} = 0.66(36)$$

$$L = \frac{24(8)(222)}{360.6 \times 12} = 9.85 \text{ ft}$$

Back substitution gives $w = 360.6/9.85 = 36.61$ kips/ft. Checking, we obtain

$$f_b = \frac{36.61(9.85)^2(12)}{8(222.0)} = 24.0 \text{ ksi} \qquad \text{O.K.}$$

$$f_v = \frac{36.61(9.85)}{2(24.31)(0.515)} = 14.4 \text{ ksi} \qquad \text{O.K.}$$

The reaction length N based on $R = wL/2 = 180.4$ kips and $F_a = 0.75 F_y = 27$ ksi is

$$(N + k) t_w F_a = R$$

$$N = \frac{R}{t_w F_a} - k$$

$$= \frac{180.3}{0.515(27)} - 1.53 = 11.44 \text{ in} \qquad ///$$

4-6 STRONG- VERSUS WEAK-AXIS BENDING

Sections should be oriented whenever possible so that bending is developed about the strong (X–X) axis. This should produce both the minimum weight section and least deflection due to load.

Depth limitations or other considerations may require orientation so that bending is produced about the Y–Y axis. The allowable bending stress in the AISC specification for compact rolled sections bent about the Y–Y axis and for solid round, square, or rectangular bars may be taken as

$$F_b = F_y \left[1.075 - B\left(\frac{b}{2t_f}\right)\sqrt{F_y} \right] < F_y \tag{4-8}$$

where $B =$ 0.005 for fps units
$\phantom{\text{where } B}=$ 0.0019 for SI units

For most rolled sections this gives approximately $F_b = 0.75 F_y$, which may be used in lieu of Eq (4.8). Where a solid bar is used, $b/2t_f$ becomes width/depth, and for a round bar take $b/2t_f = 1$. The use of Eq. (4-8) will be illustrated later in the design of laterally-unsupported beams, where bending about both the X and Y axis takes place.

4-7 DEFLECTIONS

Deflection estimates under working load are often required to ensure that floors are reasonably plane and level and that frame warping does not cause doors and windows to fail to function properly. Cracks can be produced in masonry, plastered, and tiled walls if later application of live load causes excessive deflections.

Large deflections are generally indicative of lack of structural rigidity and possible vibrations under live-load movement (including people walking) or elevator starts and stops, and lateral movements from less-than-design wind forces.

Deflections of flat roofs result in local accumulation of water (from either rain or snow), which in turn increases the deflection and more accumulation of water, possibly terminating in a roof collapse. It might be noted in passing that this type of failure is a combination of water accumulation from local deflection plus a failure of the roof drainage system by plugging from accumulated windblown debris or freezing of debris-retained water.

Deflections for simple beams can be computed using superposition of the effects of the beam loading and the tabulated equations for selected loading given in handbooks (also in Table IV-5 of SSDD) and several in the next paragraph. For other cases of loading, including continuous beams, the deflection can be computed using the general differential equation

$$EIy^{iv} = -w$$

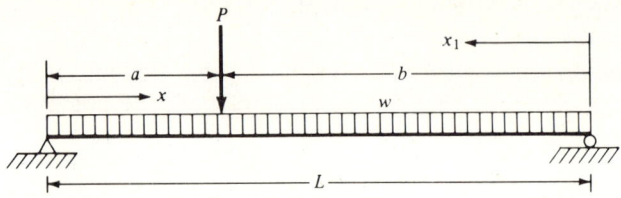

	Δ_x	Δ_{load}	Δ_{cen}
Concentrated load:	$\dfrac{Pbx}{6EIL}(L^2 - b^2 - x^2)$	$\dfrac{Pb^2a^2}{3EIL}$	$\dfrac{PL^3}{48EI}$ when $a = b$
	$\dfrac{Pax_i}{6EIL}(L^2 - a^2 - x_1^2)$		
Uniform load:	$\dfrac{wx}{24EI}(L^3 - 2Lx^2 + x^3)$		$\dfrac{5wL^4}{384EI}$

Figure 4-8 Simple beam loading and resulting deflections.

and integrating and incorporating known boundary conditions to evaluate the constants of integration. This method has been outlined in Sec. 2-2 and will be illustrated in Example 4-9.

It is usually not necessary to find the "exact" value of deflection (particularly for live load) since the loads are not very precisely known. Figure 4-8 tabulates several deflection equations for a uniform beam load and a concentrated load.

Inspection of these equations indicates that deflection is inversely proportional to the moment of inertia I. The use of smaller sections, such as from the use of high-strength steel, will tend to increase the deflections and may even control whether high-strength steel can be used.

Example 4-8 What is the approximate deflection of the beam selected in Example 4-1?

SOLUTION The beam selected was a W21 × 68 with $I_x = 1480$ in^4. The maximum Δ is at the center and will be due to

$$\Delta_t = \Delta_w + \Delta_{p1} + \Delta_{p2} + \Delta_{p3}$$

Note that a unit conversion factor of 1728 is needed for both w and P in fps units when span length is in feet and Δ is inches. From tabulated equations:

$$\Delta_w = \frac{5wL^4}{384EI} = \frac{5(3)(24)^4(1728)}{384(29\ 000)(1480)} = 0.522 \text{ in}$$

$$\Delta_{p2} = \frac{PL^3}{48EI} = \frac{6(24)^3(1728)}{48(29\ 000)(1480)} = 0.070 \text{ in}$$

$$\Delta_{p2} = \Delta_{p3} \to 2\Delta_{p2} = \frac{2Pax}{6EIL}(L^2 - a^2 - x_1^2)$$

$$= \frac{2(4)(6)(1728)}{6(29\ 000)(1480)(24)}(24^2 - 6^2 - 12^2) = 0.005 \text{ in}$$

$$\Delta_t = \overline{0.597 \text{ in}}$$

The maximum deflection based on $L/360$ is

$$\frac{L}{360} = \frac{24(12)}{360} = 0.80 > 0.597 \text{ in}$$

Thus this beam would satisfy the $L/360$ deflection criteria. ///

Example 4-9 What is the estimated deflection of the continuous beam (member 22) of Example 4-4? Data are shown on Fig. E4-9.
  Section data: W410 × 46.1:

$$I = 156.1 \times 10^{-6} \text{ m}^4$$
$$E = 200\ 000 \text{ MPa}$$

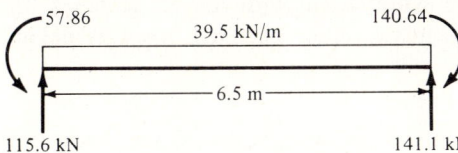

Figure E4-9

SOLUTION It will be necessary to use the differential equation for deflection

$$EIy^{iv} = -w = -39.5$$
$$EIy''' = -39.5x + C_1 \quad \text{but } c_1 = \text{reaction at } x = 0$$
$$EIy''' = -39.5x + 115.6$$
$$EIy'' = -\frac{39.5x^2}{2} + 115.6x + C_2 \quad \text{but } c_2 = M \text{ at } x = 0$$
$$EIy'' = -\frac{39.5x^2}{2} + 115.6x - 57.86$$
$$EIy' = -\frac{39.5x^3}{6} + \frac{115.6x^2}{2} - 57.86x + C_3 \quad C_3 = \text{slope at } x = 0$$
$$EIy = -\frac{39.5x^4}{24} + \frac{115.6x^3}{6} - \frac{58.86x^2}{2} + C_3 x + C_4$$

Deflection = 0 at $x = 0$ and is also 0 at $x = L$, from which we obtain $C_4 = 0$. Upon substitution of $x = L$, we obtain

$$C_3 = \frac{39.5L^3}{24} - \frac{115.6L^2}{6} + \frac{57.86L}{2}$$

For $L = 6.5$ m, we obtain $C_3 = -173.98$ (computer output $= 0.0043$ rad) and

$$y = \frac{1}{EI}\left(-\frac{39.5x^4}{24} + \frac{115.6x^3}{6} - \frac{57.86x^2}{2} - 173.98x\right)$$

It appears that Δ_x occurs at $+M = M_{max}$ at $x = 0.87 + 2.37 = 3.24$ m from the left end. For this value of x in the preceding equation we compute

$$\Delta_x = \frac{-393.5 \times 10^{-3} \text{ MN} \cdot \text{m}}{(200\ 000)(0.0001561 \text{ m}^4)} = 0.0126 \text{ m}$$

$$= 12.6 \text{ mm (approximately } \tfrac{1}{2} \text{ in)} \qquad ///$$

4-8 BIAXIAL BENDING AND BENDING ON UNSYMMETRICAL SECTIONS

The preceding sections have considered bending about either the X or the Y axis of W, M, or S shapes. In all these cases the moment produced by the load was applied perpendicular to the X or Y axis with a line of action through the origin of axes. This produced the type of bending stress that can be computed using Eq. (4-1). The W, M, and S sections are symmetrical with respect to the X–Y plane containing the X and Y axes and thus produce principal axes. Principal axes are such that

$$I_x = \int^A y^2 \, dA$$

$$I_y = \int^A x^2 \, dA$$

$$I_{xy} = \int^A xy \, dA = 0$$

The latter equation for I_{xy} is a product of the inertia term produced when at least one of the X and Y axes is not an axis of symmetry. Principal axes are produced as mutually perpendicular axes through the centroid of area such that the moments of inertia are a maximum with respect to one of the axes and a minimum with respect to the other. The principal axes are axes of symmetry for symmetrical sections but can be found for unsymmetrical sections. If the product of inertia I_{xy} is not zero, the axes of interest are not principal axes.

4-8.1 Symmetrical Sections with Biaxial Bending

When the X and Y axes are principal axes but the load is not applied perpendicular to the principal axes (but the resultant does pass through the origin of axes), biaxial bending is produced. For this loading condition it is necessary to determine the components of load perpendicular to each axis and use superposition to find the stresses at critical locations, using

$$f_b = \pm \frac{M_x}{S_x} \pm \frac{M_y}{S_y} \qquad (4\text{-}9)$$

and for shear

$$f_v = \pm \frac{V_x Q_x}{I_x t_x} \pm \frac{V_y Q_y}{I_y t_y} \qquad (4\text{-}9a)$$

Equation (4-9) is commonly used for spandrel beams and roof purlins, where there are both vertical and horizontal force components. When the load is applied to the top flange of the member (common for purlins on sloping roofs) and is separated into components perpendicular and parallel to the X and Y axes, the force parallel to the Y axis does not pass through the origin of axes. This is a very complex stress state that involves biaxial bending and torsion. An approximate solution is obtained as

$$f_b = \frac{M_x}{S_x} + \frac{2M_y}{S_y}$$

that is, using one-half the section modulus of the section for the tangential bending component and the (+) sign used to obtain the stress.

An alternative form of Eq. (4-9) is widely used for biaxial bending and in similar form for combining bending and axial stresses. To obtain this form, we set $f_b = F_b$ and then divide both sides of the equation by the allowable stress F_b (noting that this is done even if $F_{bx} \neq F_{by}$), delete the (−), and reorder to obtain

$$\frac{M_x}{F_{bx} S_x} + \frac{M_y}{F_{by} S_y} = 1.0$$

When this is done, a section with any combination of moments and section modulus producing 1.0, or less, is satisfactory; one should try to obtain a value as nearly 1.0 as possible. Any value larger than 1.0 represents an overstress.

Example 4-10 Design a roof purlin using a channel section for the side sheds and for the most critical loading condition of Example 2-5. Space the purlins 6 ft horizontally as shown in Fig. E4-10 and use sag rods at the midbay spacing, giving an unsupported length of 12.5 ft for bending moments about the Y axis of the purlin.

SOLUTION Additional data from Example 2-5:

$$\text{Dead load} = 20.0 \text{ psf}$$

Live load = snow load = $25 \cos \theta$

$$= 25 \cos 14.04 = 24.25 \text{ psf}$$
$$\text{Total} = \overline{44.25 \text{ psf}}$$

Vertical load/ft $= 6.18(0.04425)$

$= 0.273$ kip/ft

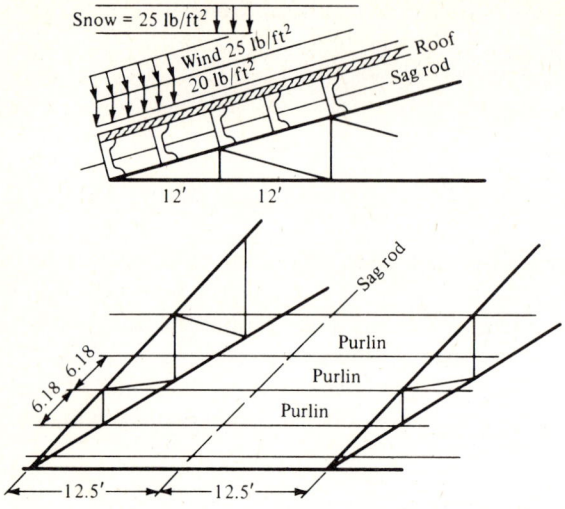

Figure E4-10

Normal load due to wind at 25 psf: $6.18(0.025) = 0.1545$ kip/ft

With wind, the allowable stress can be increased one-third so check if the normal component of $D + S + W > 1.33(D + S)$:

$$D + S + W = 0.273 \cos 14.04 + 0.1545$$
$$= 0.4195 \text{ kip/ft}$$
$$1.33(D + S) = 1.33(0.273 \cos 14.04)$$
$$= 0.352 < 0.4195 \text{ kip/ft}$$

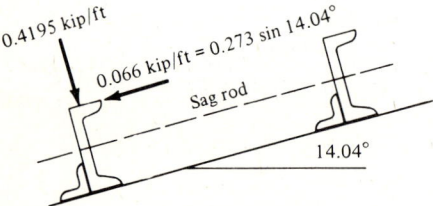

The wind case controls.

With sag rods shown and loads applied to the channel as indicated:

$$M_x = \frac{wL^2}{8} = \frac{0.4195(25)^2(12)}{8} = 393 \text{ in} \cdot \text{kips}$$

$$M_y = -2\left(\frac{wL^2}{16}\right) \quad \text{where } L = 12.5 \text{ ft} \quad \text{(Table IV-5 and both spans loaded)}$$

$$= -2\frac{(0.066)(12.5)^2(12)}{16} = 15.5 \text{ in} \cdot \text{kips}$$

The allowable bending stress is $F_b = 0.6F_y$ for an unsymmetrical section.

Increase by one-third: $F_b = 0.6(36)(1.33) = 28.7$ ksi. The critical stress will be tension in the flange at the bottom of the purlin, since M_y causes compression along the web side due to negative moment at the sag rod. Note that S_y of a channel is based on the largest c distance (from centroid to flange tip).

$$\frac{M_x}{S_x} + \frac{2M_y}{S_y} = F_b$$

$$\frac{393}{S_x} + \frac{2(15.5)}{S_y} = 28.7$$

We must simply try sections to find one that satisfies this equality.
First, try C12 × 30:

$$S_x = 27.00 \text{ in}^3 \qquad S_y = 2.06 \text{ in}^3$$

$$\frac{393}{27.00} + \frac{31}{2.06} = 29.6 > 28.7 \text{ ksi} \qquad \text{N.G.}$$

Second, try C15 × 33.9:

$$S_x = 42.00 \text{ in}^3 \qquad S_y = 3.11 \text{ in}^3$$

$$\frac{393}{42.00} + \frac{31.0}{3.11} = 19.3 > 28.7 \text{ ksi} \qquad \text{O.K.}$$

Use a C15 × 33.9 section, because no other section is lighter (a W section may be preferable). ///

4-8.2 Bending about Axes Other than Principal Axes

We now consider the case where the X and Y axes are not principal axes and the load is at some orientation. This may be parallel and perpendicular to the principal axes or otherwise but through the origin of axes. If either of the given X or Y axes is an axis of symmetry, use case A or the concept of shear center given in the next section. For this case the standard solutions are given as

$$f_b = \frac{M_y I_x - M_x I_{xy}}{I_x I_y - I_{xy}^2}(x) + \frac{M_x I_y - M_y I_{xy}}{I_x I_y - I_{xy}^2}(y) \qquad (4\text{-}10)$$

Strict attention to signs is required when using this equation, and additionally a + sign is obtained for compression when the sign convention as shown in Fig. 4-9 is used. The term I_{xy} is the product of inertia defined along with I_x and I_y as

$$I_x = \int^A x^2 \, dA \qquad I_y = \int^A y^2 \, dA \qquad I_{xy} = \int^A xy \, dA$$

The moments of inertia I_x and I_y are always positive; however, I_{xy} may be either (+) or (−). The orientation of the principal axes with respect to the given X and Y axes (refer also to Fig. 4-9) is

$$\tan 2\theta = \frac{2I_{xy}}{I_y - I_x} \qquad (4\text{-}11)$$

168 STRUCTURAL STEEL DESIGN

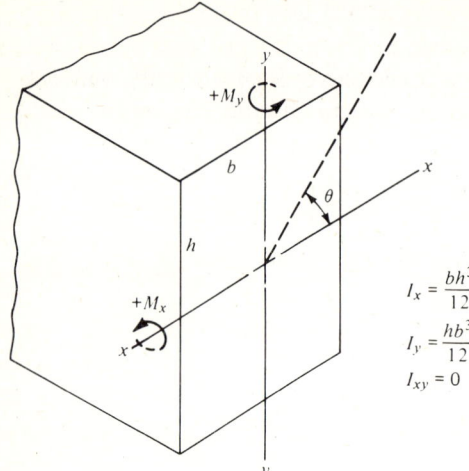

Figure 4-9 Sign convention and identification of terms for Eq. (4-10) and Eq. (4-11).

The moments of inertia about the principal axes I_{xp} and I_{yp} can be computed once θ is known, and we can use Eq. (4-9) in lieu of Eq. (4-10) if the principal moments are used. The principal moments of inertia are:

$$I_{xp} = I_x \cos^2 \theta + I_y \sin^2 \theta - 2I_{xy} \sin \theta \cos \theta$$
$$I_{yp} = I_y \cos^2 \theta + I_x \sin^2 \theta + 2I_{xy} \sin \theta \cos \theta \qquad (4\text{-}12)$$

Note that if we add Eqs. (4-12), we obtain the following useful equality:

$$I_{xp} + I_{yp} = I_x + I_y$$

Example 4-11 A 203 × 152 × 25.4 angle is used as a lintel over a sliding door. Investigate the bending stresses and estimate the deflection produced in this unequal leg angle for the loading shown in Fig. E4-11.

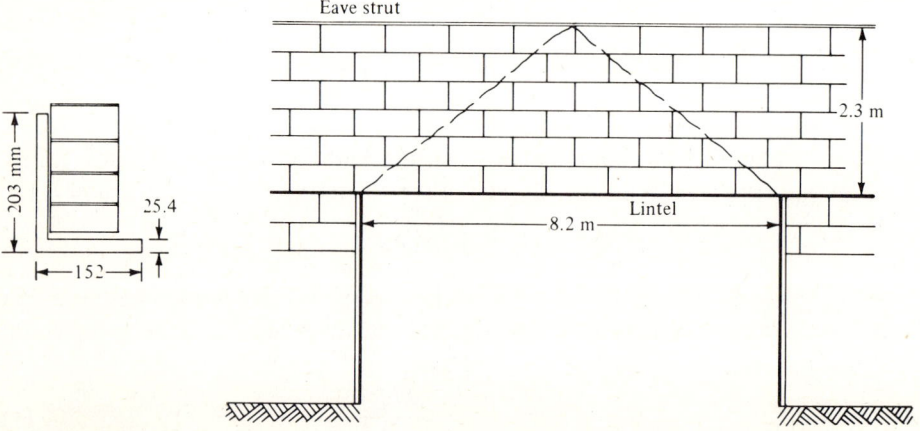

Figure E4-11

DESIGN OF BEAMS FOR BENDING **169**

SOLUTION Assume:

1. Masonry cannot take any appreciable tension stress, so any deflection produces the triangular load diagram shown via crackling.
2. The masonry + vertical leg of angle resists any out-of-plane deflections.
3. Unsymmetrical bending occurs due to brick not being placed so the resultant coincides with center of area.

Using weights given in Table IV-3 of SSDD and using veneer one brick wide:

$$\text{Weight}/m^2 = \frac{\text{wt of 200-mm brick wall}}{2} \times \text{height of wall over lintel}$$

$$w = \frac{3.77}{2} \times 2.3 \text{ m} = 1.89 \text{ kN/m} \quad \text{(see Table IV-3, SSDD)}$$

Wt of angle $\quad = 0.645$ kN/m
Total $\quad = \overline{2.535}$ kN/m

$M = WL/6$ from Table IV-5, where $W = wL/2$.

$$M_x = \frac{10.394(8.2)}{6} = 14.21 \text{ kN} \cdot \text{m}$$

Note that we are neglecting any twisting due to the bricks not being placed so that the weight acts through the center of area of the angle.

The section properties of L203 × 152 × 25.4 are:

$I_x = 33.63 \times 10^{-6} \text{ m}^4 \quad I_y = 16.1 \times 10^{-6} \text{ m}^4 \quad A = 8.387 \times 10^{-3} \text{ m}^2$

$r_z = 32.5 \text{ mm} \quad \tan \alpha = \tan \theta = 0.543$

$\theta = \tan^{-1} 0.543 = 28.5°$

From Eq. (4-11),

$$I_{xy} = (I_y - I_x) \tan 2\theta$$
$$= (16.1 - 33.63) \tan 2(28.5) = -27.0 \times 10^{-6} \text{ m}^4$$

The Z–Z axis of Table V-8 is the I_{xp} axis as used here.

$$I_{xp} = Ar_z^2 = 8.387 \times 10^{-3} \text{ m}(0.0325 \text{ m})^2 = 8.859 \times 10^{-6} \text{ m}^4$$

Since $I_{xp} + I_{yp} = I_x + I_y$,

$$I_{yp} = 33.63 + 16.1 - 8.859 = 40.87 \times 10^{-6} \text{ m}^4$$

Using Eq. (4-1) and noting that $M_y = 0$, we obtain

$$f_b = -\frac{M_x I_{xy}}{D} x + \frac{M_x I_y}{D} y$$

where $D = (I_x I_y - I_{xy}^2) \times 10^3$ (the 10^3 is to convert kN to MN). Substituting

values, we obtain
$$f_b = -2046x - 1220y \quad \text{(if set to 0, gives location of neutral axis)}$$
At point A, $x = -0.0419$, $y = +0.1357$ m.
$$f_b = -2046(-0.0419) - 1220(0.1357)$$
$$= -79.8 \text{ MPa} \quad \text{(compression)}$$
At point B, $x = -0.0419$, $y = -0.0673$.
$$f_b = 167.8 \text{ MPa} \quad (+ = \text{tension})$$

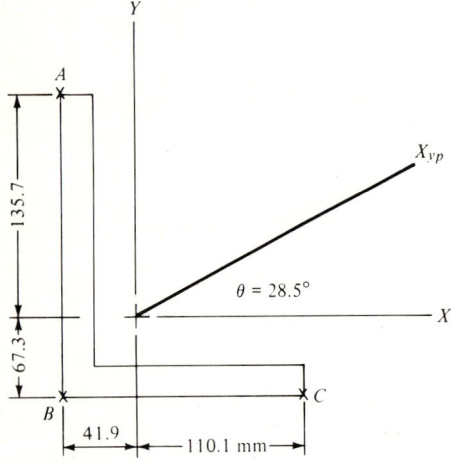

At point C, $x = 0.110$, $y = -0.0673$.
$$f_b = -142.9 \text{ MPa} \quad (- = \text{compression})$$

If we use $f_b = Mc/I$:
At point A:
$$f_b = \frac{14.21(0.1357)}{33.63}(10^3) = 57.3 \text{ MPa} \quad \text{(compression)}$$
At points B and C:
$$f_b = \frac{14.21(0.0673)}{33.63}(10^3) = 28.4 \text{ MPa} \quad \text{(tension)}$$

These values are considerably different from those computed using unsymmetrical bending.

Check the deflections approximately using Table IV-5 for a simply supported beam:
$$\Delta = \frac{wL^3}{60EI} = \frac{10.394(8.2)^3}{60(200\,000)(33.63 \times 10^{-6})} = 0.0142 \text{ m}$$

This deflection is computed at 14.2 mm (over $\frac{1}{2}$ in), which could cause the veneer to crack if it actually occurred. Engineering judgment would have to be applied to decide if this angle is satisfactory with a deflection that could be as large as 14 mm but has a good probability of being much less than this. ///

4-9 SHEAR CENTER OF OPEN SECTIONS

The shear center locates the point with respect to a cross section to apply the flexural load so that no twisting (or torsion) occurs when shear stresses due to bending act on the plane through the point. Thus, if the loading passes through the shear center, the section may be analyzed for simple bending and shear using Eqs. (4-9) and (4-9a). If the beam loading does not pass through the shear center, a torsion moment is developed that produces torsional shearing stresses of

$$f_s = \frac{Ve't}{J} \tag{4-13}$$

where Ve' = shear and shear eccentricity with respect to the shear center
t = thickness of element where shear stress is desired
J = torsional constant of section, for a thin rectangle

$J = \dfrac{bt^3}{3}$ $b/t \to 10$ (section webs and some flanges)

$J = \dfrac{bt^3}{3} - 0.21t^4$ $b/t \to 4$ (stubby flanges as for channels)

The computation of the shear center is complicated for all but the simplest shapes. Fortunately, most sections have the shear center at a convenient location (see Fig. 4-10), for example:

1. If a section contains an axis of symmetry, the shear center is on the axis.
2. From (1) it follows that the shear center of all sections with symmetry about both axes is on the intersection of the two axes (all W, M, and S shapes).
3. For all sections consisting of two intersecting plate elements (angles, tees, etc.), the shear center is at the plate intersection.

The shear center E_0 (using symbols as in Table I-6) for channels is readily derived to be

$$E_0 = \frac{b_1 t_f}{2b_1 t_f + h' t_w/3} \tag{4-14}$$

where $b_1 = b_f - t_w/2$
$h' = d - t_f$ = average section depth
t_f, t_w = flange and web thickness, respectively

Using these values for a C10 × 30 with d = 10.00 in, b_f = 3.033 in, t_f = 0.436 in, and t_w = 0.673 in, we obtain E_0 = 0.705 in, as in AISC and Table I-6 of SSDD.

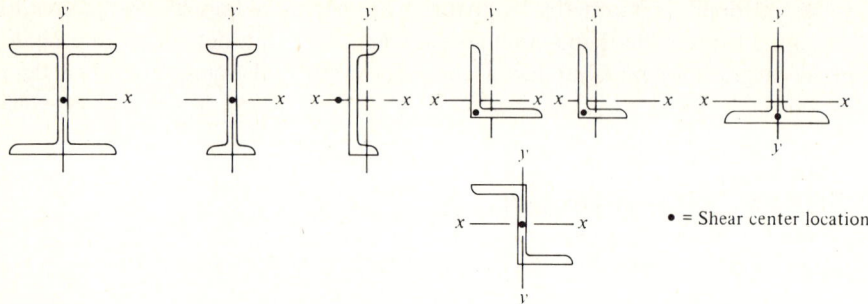

• = Shear center location

Figure 4-10 Shear center location for several rolled shapes.

The reader should note that there are several equations used in various design tables which give E_0 values for channels slightly different from those of Eq. (4-14), depending on the assumptions made for shear flow in the flanges. The computation for shear center of more complicated shapes is beyond the scope of this text, and the reader is referred to any text on advanced mechanics of materials.

The bending stress for conditions where torsion is developed by the applied load not passing through the shear center is approximately

$$f_b = \frac{Mc}{I} + \frac{Vxb_f}{2I_f} \qquad (4\text{-}15)$$

where $Mc/I =$ usual bending stress computation
$V =$ shear at a distance x along beam from origin
$b_f =$ width of flange being stressed
$x =$ distance from origin of axis to V and where f_b is desired
$I_f =$ moment of inertia of flange being stressed; approximately $I_y/2$ for a channel

Example 4-12 What is the approximate bending and torsion stress in a C10 × 30 channel loaded as shown in Fig. E4-12?

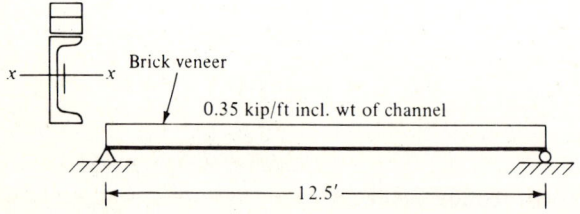

Brick veneer

0.35 kip/ft incl. wt of channel

12.5'

Figure E4-12

SOLUTION From Table I-6, the section properties of a C10 × 30 are:

$$S_x = 20.70 \text{ in}^3 \qquad E_0 = 0.705 \text{ in}$$
$$I_y = 3.94 \simeq 2I_f \qquad b_f = 3.033 \text{ in}$$

At midspan:

$$x = \frac{L}{2} \quad V = 0$$

$$M = \frac{wL^2}{8} = \frac{0.35(12.5)^2(12)}{8} = 82.0 \text{ in} \cdot \text{kips}$$

$$f_b = \frac{82}{20.7} = 3.96 \text{ ksi} \ll 0.6F_y(22 \text{ ksi})$$

At $x = L/4 = 3.13$ ft:

$$M_x = \frac{w(L-x^2)}{2} = \frac{0.35(12.5 - 3.13^2)(12)}{2} = 5.7 \text{ in} \cdot \text{kips}$$

$$V_x = \frac{wL}{2} - wx = 0.35\left(\frac{12.5}{2} - 3.13\right) = 1.092 \text{ kips}$$

$$f_b = \frac{5.7}{20.7} + \frac{1.092(3.13)(3.03)(12)}{3.94}$$

$$= 0.3 + 31.6 = 31.9 \text{ ksi} \gg 0.6F_y$$

Note that if the channel beam were designed without considering torsion and shear center location, the apparent bending stress at midspan indicates that the section is considerably overdesigned. When considering the shear center effects, the section is not adequate unless it can be assumed that the analysis is too approximate and that the end connections are such as to restrain rotation so that the larger bending stresses do not develop. The author would assume that the section is considerably underdesigned and use a different section, as there is too much design risk involved for the small amount of savings obtainable from using a lighter section.

The torsion shear stress can be evaluated as

$$f_s = \frac{Ve't_f}{J}$$

J = contribution of web + two flanges

$$J = \frac{bt^3}{3} + 2\left(\frac{bt^3}{3} - 0.21t^4\right)$$

$$= 10 - \frac{2(0.436)(0.673)^3}{3} + \frac{2(3.033)(0.436)^3}{3} - 0.21(0.436)^4$$

$$= 0.927 + 0.152 = 1.08 \text{ in}^4 \text{ (vs. 1.22 in SSDD)}$$

The discrepancy between J as computed above and in the SSDD tables is due to using a more refined computation in the tables, which accounts for the increased stiffness at the flange-to-web intersection. We will use the table value of $J = 1.22$.

$$e' = E_0 + \bar{x} = 0.751 + 0.649 = 1.40 \text{ in}$$

$$f_s = \frac{1.092(1.40)(0.436)}{1.22} = 0.546 \text{ ksi which appears satisfactory} \quad ///$$

4-10 DESIGN OF LATERALLY UNSUPPORTED BEAMS

In a very large number of cases, flexural members have the compression flange adequately restrained to satisfy lateral support, or lateral bracing, requirements. This is particularly true of beams supporting floors and roofs, where the beam is in continuous direct contact or has the top flange embedded somewhat into the floor or roof. Cases where the floor system consists of metal decking placed onto the beams may not provide adequate lateral flange support if the resistance is solely due to friction. In these situations, carefully placed welds to take, say, 2 to 3 percent of the flange force (0.02 to $0.03 A_f f_b$) will provide adequate lateral bracing. A very large lateral restraint is not required, as evidenced by the application of only hand force to restrain reasonably sized beams against lateral buckling in laboratory tests.

The lateral bending and warping of laterally unsupported beams was illustrated in Figs. 4-2 and 4-3. There are several situations, particularly involving vertical flexural members and beams used to carry column loads across large open areas, crane runway girders, and continuous beams in large spans, where the compression flange is not in contact with decking and others, where it is not practical to provide lateral bracing except at the ends and/or at only a few interior span points. For conditions where the compression flange is laterally unsupported for some distance, the resulting column-type action may result in flange buckling, a warping (partial-to-full section rotation), and lateral bending. Just the superposition effect of vertical compressive bending stresses adding to the compressive stresses due to lateral bending as the section deflects out of plane may produce yield or buckling stresses in one side of the compression flange. In any case, where this situation is possible, the allowable bending stresses are reduced. This reduction is critical when the section is light and deep, since with lateral bending the warping resistance of the flanges and web is small. It is also true for long unsupported lengths, since, similar to buckling of a long column, it takes a smaller load (or stress) to cause the compression flange to bend laterally out of plane and producing a tendency for warping.

A theoretical combination of torsion and lateral bending resistance can be made to obtain the critical compression flange buckling stress as

$$F_{cr} = \frac{\pi^2 E \sqrt{C_w I_y}}{L^2 S_x} \sqrt{1 + \frac{L^2 GJ}{\pi^2 E C_w}} \qquad (4\text{-}16)$$

or, alternatively,

$$F_{cr} = \frac{\pi \sqrt{E I_y GJ}}{L S_x} \sqrt{\frac{\pi^2 E C_w}{L^2 GJ} + 1} \qquad (4\text{-}17)$$

where J = torsion constant previously defined and given in the table of section properties in SSDD
C_w = warping constant $\simeq \frac{1}{24} t_f b_f^3 (d - t_f)^2$

DESIGN OF BEAMS FOR BENDING

G = shear modulus = $\dfrac{E}{2(1+\mu)}$ = 11 150 ksi for $\mu = 0.3$
I_y = moment of inertia about the Y axis
S_x = section modulus about the X axis
L = unsupported length

If we make the following approximations:

$$I_y = A r_y^2 \qquad r_x = 0.41 d$$

$$S_x = \dfrac{2 A r_x^2}{d} \qquad h = 0.95 d$$

and substitute into Eq. (4-16), we obtain

$$F_{cr} = \dfrac{14E}{(L/r_y)^2} \sqrt{1 + 0.0476 \left(\dfrac{L/r_y}{d/t_f} \right)^2} \qquad (4\text{-}18)$$

Substituting similarly into Eq. (4-17), we obtain

$$F_{cr} = \dfrac{3E}{Ld/r_y t_f} \sqrt{1 + 21 \left(\dfrac{d/t_f}{L/r_y} \right)^2} \qquad (4\text{-}19)$$

For beams with low torsional stiffness, Eq. (4-18) is very nearly

$$F_{cr} = \dfrac{14E}{(L/r_y)^2} \qquad (4\text{-}20)$$

and for beams with high torsional stiffness, Eq. (4-19) is approximately

$$F_{cr} = \dfrac{3E}{Ld/r_y t_f} \qquad (4\text{-}21)$$

One method of obtaining F_{cr} is to compute values from both Eqs. (4-20) and (4-21) and use the largest value. Note, however, that the original expression of Eq. (4-17) is the following form when using the approximate values from Eqs. (4-20) and (4-21):

$$F_{cr} = \sqrt{\left[\dfrac{14E}{(L/r_y)^2} \right]^2 + \left[\dfrac{3E}{(L/r_y)^2} \right]^2} \qquad (4\text{-}22)$$

This equation may be interpreted as shown in Fig. 4-11. If we do not use the resultant for F_{cr}, the next best solution is to use the largest leg of the right triangle, which produces a maximum error of $(\sqrt{2} - 1.0)100 = 41$ percent. Most cases will not be this extreme.

The AISC specifications can now be developed using the preceding several equations and some additional simplification and rounding of numbers. Since about 1946, the AISC specifications have used the following formula based on substitution of $r_y = 0.22b$, applying a safety factor of 1.67 and a factor C_b to

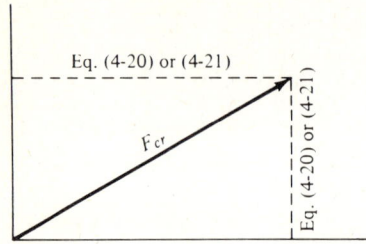

Figure 4-11 Graphical interpretation of Eq. (4-22).

account for moment gradient (the latest modification) into Eq. (4-21), to obtain

$$F_b = \frac{12\,000 C_b}{Ld/A_f} \leq 0.6 F_y \quad [\text{AISC Eq. (1.5-7)}] \qquad (4\text{-}23)$$

$$F_b = \frac{82\,700 C_b}{Ld/A_f} \quad (\text{SI units}) \qquad (4\text{-}23m)$$

where A_f = area of compression flange = $b_f t_f$
  d = depth of section; note that the ratio d/A_f is computed and tabulated in tables of section properties
  $C_b = 1.75 + 1.05(M_1/M_2) + 0.3(M_1/M_2)^2 \leq 2.3$. In this equation M_1 is always the *smaller* and M_2 the larger moment at the end of the unbraced length. $M_1/M_2 = (+)$ when moments are of same sign (producing reversed curvature) and $M_1/M_2 = (-)$ when of opposite sign. Use $C_b = 1.0$ when the moment in the interior of the unbraced length is larger than either of the end moments and regardless of sign.

The Column Research Council proposed as an alternative to Eq. (4-20) the basic column formula

$$F_{cr} = \left[1.0 - \frac{(L/r)^2}{2C_c^2 C_b}\right] F_y \qquad (4\text{-}24)$$

Taking SF = 1.67, this equation becomes

$$F_b = \left[1.0 - \frac{(L/r)^2}{2C_c^2 C_b}\right] 0.60 F_y \qquad (4\text{-}25)$$

where the term C_c = slenderness ratio for which residual stresses cause inelastic buckling or the transition L/r from inelastic to Euler (or elastic) buckling:

$$C_c = \sqrt{\frac{2\pi^2 E}{F_y}}$$

Substituting this value and using 0.667 instead of 0.6 for the first term (1.0) and approximately 0.75 for the second term in Eq. (4-24), we obtain

$$F_b = \left[\frac{2}{3} - \frac{F_y(L/r_T)^2}{1530 \times 10^3 C_b}\right] F_y \quad [\text{AISC Eq. (1.5-6}a\text{)}] \qquad (4\text{-}26)$$

$$F_b = \left[\frac{2}{3} - \frac{F_y(L/r_T)^2}{10.6 \times 10^6 C_b}\right]F_y \quad \text{(SI units)} \quad (4\text{-}26m)$$

where r_T = radius of gyration of the compression flange + 1/3 of the compression web area taken about the Y axis; these values are also given in the AISC and SSDD tables.

Equation (4-26) accounts for the possibility of inelastic buckling; however, elastic buckling is also a possibility depending on section proportions and the unbraced length. For this possibility we may use Eq. (4-18) with $r_x = 0.38d$ instead of $0.41d$, to obtain

$$F_{cr} = \frac{16E}{(L/r_y)^2}$$

Now dividing by SF = 1.92 and taking $r_T \simeq 1.2r_y$, we obtain an allowable bending stress of

$$F_b = \frac{170 \times 10^3 C_b}{(L/r_T)^2} \quad \text{(fps units)} \quad [\text{AISC Eq. (1.5-6b)}] \quad (4\text{-}27)$$

$$F_b = \frac{1.17 \times 10^6 C_b}{(L/r_T)^2} \quad \text{(SI units)} \quad (4\text{-}27m)$$

The minimum L/r_T ratio for using Eq. (4-26) is found by equating $F_b = 0.6F_y$ to obtain (with slight rounding):

$$\frac{L}{r_T} = \sqrt{\frac{102 \times 10^3 C_b}{F_y}} \quad \text{(fps units)}$$

$$\frac{L}{r_T} = \sqrt{\frac{706 \times 10^3 C_b}{F_y}} \quad \text{(SI)}$$

Criteria other than L/r_T will determine F_b (such as compact section or AISC Eq. 1.5-7) when (L/r_T is less than that obtained from the equation above. The minimum L/r_T ratio at which Eq. (4-27) applies is found by equating Eqs. (4-26) and (4-27) to obtain (again slightly rounded):

$$\frac{L}{r_T} = \sqrt{\frac{510 \times 10^3 C_b}{F_y}} \quad \text{(fps)}$$

$$\frac{L}{r_T} = \sqrt{\frac{3.54 \times 10^6 C_b}{F_y}} \quad \text{(SI)}$$

In all cases it is necessary to compute the allowable bending stress by AISC Eq. (1.5-6a) or (1.5-6b), depending on L/r_T and by AISC Eq. (1.5-7). The largest value of F_b from the two equations is used for design. A "more refined analysis"

178 STRUCTURAL STEEL DESIGN

may be used according to AISC, which means that we may use

$$F_b = \sqrt{\left[\begin{array}{c}\text{Eq. (1.5-6}a)\\ \text{or}\\ \text{Eq. (1.5-6}b)\end{array}\right]^2 + \left[\text{Eq. (1.5-7)}\right]^2}$$

according to Eq. (4.22) and as illustrated in Fig. 4-11.

The design of laterally unsupported beams may be summarized as follows:

1. Initially assume that $F_b = 0.6F_y$ and make a tentative section selection using a table such as Table II-1 of SSDD, which also gives L_c and L_u. If the actual unbraced length $L_b \leq L_c$ or L_u, a direct solution can be obtained since the unbraced length will not be a factor.
2. If the tentative section indicates $L_u < L_b$, the unbraced length may be a design factor. Using the tentative section or one somewhat larger, compute F_b using Eq. (4-23). If this equation gives a value of F_b that satisfies bending, a solution (but possibly not the best) is obtained.
3. If Eq. (4-23) does not supply a (satisfactory) solution, the designer must use either Eq. (4-26) or (4-27), depending on the L/r_T ratio. Use the largest F_b from either Eq. (4-23) or from the controlling equation (4-26) or (4-27) as determined by L/r_T.

A table of unbraced lengths versus allowable bending moments such as Table II-3 or VI-3 may be used to obtain a direct design or to give an indication of sections that may possibly prove to be adequate.

The following examples illustrate the use of the equations for laterally unsupported beams.

Example 4-13 Given a girder using a W36 × 300 supporting two columns as shown in Fig. E4-13, what is the maximum column load using the AISC specifications and A-36 steel? Assume that the girder is restrained against rotation only at the ends. The columns may provide some lateral restraint

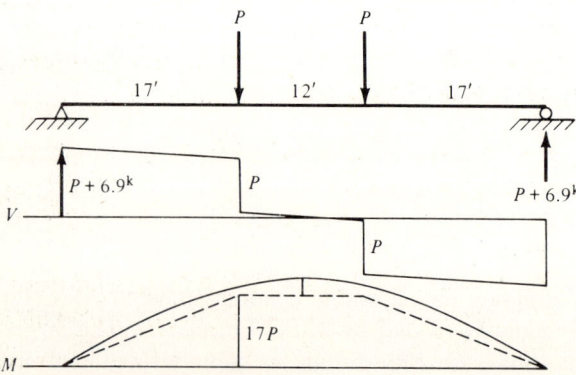

Figure E4-13

but will be neglected.

SOLUTION Using the given load geometry, complete Fig. E4-13 by drawing the shear and moment diagrams also shown. Next, from Table II-1 of SSDD for a W36 × 300, obtain $L_c = 17.6$ ft and $L_u = 35.3$ ft. Since $L_b > L_u$ (46 > 35.3), we immediately note that $F_b < 0.6 F_y$.

$$M = 17P(12) + \frac{0.30(46)^2(12)}{8}$$

$$M = 204P + 952.2 \text{ in} \cdot \text{kips}$$

From Table I-3, obtain the section properties for a W36 × 300:

$$S_x = 1110 \text{ in}^3 \qquad \frac{d}{A_f} = 1.31 \text{ in}^{-1} \qquad r_T = 4.422 \text{ in}$$

Alternative computation of d/A_f:

$$d = 36.74 \text{ in} \qquad b_f = 16.655 \text{ in} \qquad t_f = 1.680 \text{ in}$$

$$\frac{d}{A_f} = \frac{d}{b_f t_f} = \frac{36.74}{16.655(1.680)} = 1.31 \text{ in}^{-1}$$

Alternative computation of r_T:

$$I_f = \frac{t_f b_f^3}{12} = \frac{1.690(16.655)^3}{12} = 646.8 \text{ in}^4$$

$$A_f + \frac{A_w}{6} = 16.655(1.680) + \frac{36.74 - 2(1.680)0.945}{6}$$

$$= 33.24 \text{ in}^2$$

$$r_T = \sqrt{\frac{I_f}{A}} = \sqrt{\frac{646.8}{33.24}} = 4.411 \text{ in}$$

The slight discrepancy is due to the somewhat approximate computations used; however, this method is satisfactory where r_T must be computed.

$C_b = 1.0$, since the moment diagram shows that the end moments are 0 and the interior span moment is larger.

$$\frac{L}{r_T} = \frac{46(12)}{4.42} = 124.89$$

Check to see if AISC Eq. 1.5-6b [Eq. (4-27)] applies:

$$\frac{L}{r_T} = \sqrt{\frac{510\ 000(1)}{36}} = 119 < 124.9$$

Use AISC Eq. 1.5-6b [Eq. (4-27)]:

$$F_b = \frac{170\ 000(1)}{124.9^2} = 10.9 \text{ ksi}$$

180 STRUCTURAL STEEL DESIGN

Also use AISC Eq. 1.5-7 [Eq. (4-23)]:

$$F_b = \frac{12\ 000(1)}{Ld/A_f} = \frac{12\ 000}{46(12)(1.31)} = 16.6 \text{ ksi}$$

Use $F_b = 16.6$ ksi (largest value):

$$f_b = \frac{M}{S} = F_b \rightarrow M = SF_b$$

$$204P + 952.2 = 16.6(1110)$$

$$P = \frac{17\ 473.8}{204} = 85.6 \text{ kips} \qquad ///$$

The preceding example was easy to check, since the beam size has been selected and it is only necessary to determine the allowable bending stress. In most design situations the problem is more of an iterative process, in that while the loads are given, we do not know what section will be the most economical. On occasion one may use charts such as those in AISC, which give the allowable moment for several unbraced lengths or, alternatively, computer-generated tables such as Table II-3 or VI-3 of SSDD, which give the allowable moment for selected shapes for several unbraced lengths L_b. We should also observe that if Eq. (4-23) controls the design, the use of A-36 steel is the most economical solution (the reader should verify why this is true).

Example 4-14 Given the laterally unsupported girder shown in Fig. E4-14 for a crane runway in an industrial warehouse. Select the lightest W section that also limits deflection to $L/360$. Use A-36 steel. Note that the crane trolley travels on a 90-lb railroad rail fastened to the top of the flange (90 lb = 90 lb/yd = 30 lb/ft).

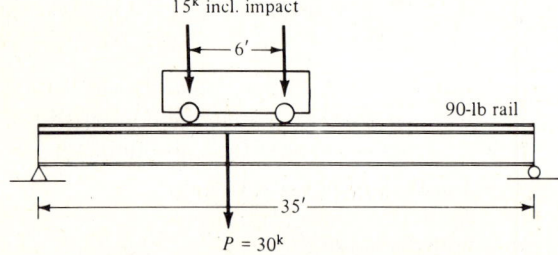

Figure E4-14

SOLUTION Find the maximum moment. Write an equation for M in terms of x and take $dM/dx = 0$.

$$M = \frac{[15x + 15(x + 6)][L - (x + 6)]}{L}$$

$$\frac{dM}{dx} = 0 = -60x - 270 + 30L \qquad x = 13 \text{ ft from left end}$$

$$M = \frac{(30 \times 13 + 90)(35 - 13 - 6)}{35} = 219.4 \text{ ft} \cdot \text{kips}$$

Estimate the beam weight as approximately $0.1 \times$ load:

$$\frac{(0.03 \times 35 + 30)(0.1)}{35} = 0.09 \text{ kips/ft}$$

$$M \text{ due to beam} = \frac{wL^2}{8} = \frac{(0.09 + 0.03)(35)^2}{8} = 18.4 \text{ ft} \cdot \text{kips}$$

$$M_{\text{total}} = 237 \text{ ft} \cdot \text{kips}$$

With a large unsupported compression flange length L_b, we must somehow estimate the beam size. Note that in SSDD, Table II-3, we have selected sections with moments given up to 34 ft unbraced length. Since 35 ft is only slightly more, use this table as a guide and select the following tentative sections:

W12 × 96: $S_x = 131.0$ $L_u = 39.9$ (Table II-4 for W12)

W14 × 90: $S_x = 143.0$ $L_u = 34$ $\dfrac{d}{A_f} = 1.36$ (Table I-3)

W16 × 100: $S_x = 175.0$ $L_u = 28$ $\dfrac{d}{A_f} = 1.65$

W18 × 97: $S_x = 188.0$ $L_u = 24.1$ $\dfrac{d}{A_f} = 1.92$

Try a W14 × 90 and check AISC Eq. (1.5-7). Take $C_b = 1.0$.

$$F_b = \frac{12\,000}{(35)(12)(1.36)} = 21 \text{ ksi}$$

$$M_a = F_b S_x = \frac{21(143)}{12} = 250 \text{ ft} \cdot \text{kips} > 237 \qquad \text{O.K.}$$

Check the deflections:

$$I_x = 999.0 \text{ in}^4$$

From Fig. 4-8, obtain the needed deflections at forward load:

$$\Delta_{\text{load}} = \frac{15(19)^2(16)^2}{3EIL} + \frac{15(13)(16)}{6EIL}(L^2 - 13^2 - 16^2)$$

$$= \frac{1}{3IL}(47.80 + 43.03)(1728) = 1.5 \text{ in}$$

$$\frac{L}{360} = \frac{35(12)}{360} = 1.2 < 1.5 \text{ in}$$

Use a larger section; a side computation indicates that a W18 × 97 does not work. Try a W16 × 100:

$$I = 1490 \text{ in}^4$$

$$F_b = \frac{12\,000}{(35)(12)(1.65)} = 17.3 \text{ ksi}$$

$$M_a = \frac{17.3(175.0)}{12} = 252 \text{ ft} \cdot \text{kips} \quad \text{O.K.}$$

$$\Delta = 1.5\left(\frac{999.0}{1490.0}\right) + \frac{5(0.10 + 0.03)(35)^4 1728}{384E(1490)} = 1.10 < 1.2 \text{ in} \quad \text{O.K.}$$

Use a W16 × 100 section. ///

These two examples illustrate the method of applying reduced allowable bending stresses when the compression flange is laterally unsupported using the AISC specifications. The reader should note (see also Part III of SSDD for AASHTO and AREA) that the AASHTO specifications are somewhat easier to apply in that the bending stress is reduced from the basic value based on an L_b/b_f ratio.

4-11 BEAMS WITH NONPARALLEL FLANGES

Structural economy or appearance may dictate use of a tapered beam as illustrated in Fig. 4-12. As the beam depth increases, it is evident that the moment capacity increases rapidly. Theoretical analyses of these types of beams is rather complicated, and since the loads are usually not precisely known, simplified analyses are commonly made. Figure 4-12a illustrates one method of analysis for a tapered beam. The neutral axis is taken so that the angle $\phi/2$ is as shown. By careful inspection of the figure it is evident that the bottom flange (in this particular geometry) is less effective or, conversely, is the more highly

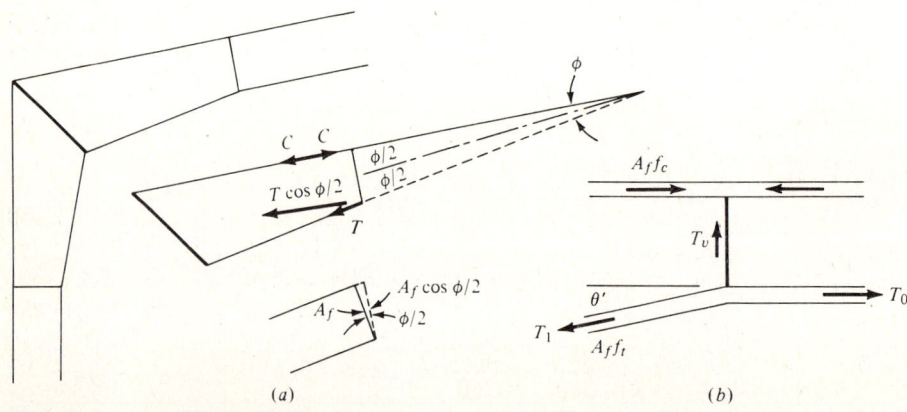

Figure 4-12 Tapered beam analysis.

stressed flange, since $\Sigma F_H = 0$ gives

$$C = T \cos \frac{\phi}{2}$$

$$A_f f_c = A_f f_t \cos \frac{\phi}{2} = A_f \cos \frac{\phi}{2} f_t \cos \frac{\phi}{2}$$

or $\quad\quad f_t = \dfrac{f_c}{\cos^2 \phi/2} \quad$ for equal flange areas (commonly used)

Since a value of $\phi = 20°$ produces a stress increase of

$$f_t = \frac{f_c}{\cos^2 10°} = 1.03 f_c$$

consideration of the slope is generally not necessary in computing I, S, and so on, as the loads are more uncertain than this discrepancy of approximately 3 percent. It is preferable to use a small amount of overdesign (f_b somewhat less than F_b). The AISC allowable bending stresses for tapered beam members is given in Appendix D of the AISC manual and will not be presented here.

Provision should be made to account for the vertical force developed when the flange changes direction as illustrated in Fig. 4-12b. This force increases the web shear, and commonly a stiffener is placed at this point to carry the vertical force (compression). The stiffener force is directly computed from statics as

$$T_v = T_1 \sin \phi'$$
$$T_1 \cos \phi' = T_0 \quad (\Sigma F_h = 0)$$

The stiffener can be readily designed as a column, taking into consideration the minimum b/t ratio. AISC does not have specific design criteria for this member; thus the designer has considerable control.

4-12 DESIGN OF BRIDGE STRINGERS AND FLOOR BEAMS

Figures 4-13 and 4-14 illustrate details of the location of bridge stringers and floor beams (also "transverse" beams). The dead loads carried by these members are determined by treating as simple beams and dead-load geometry. The live-load distribution is specified by AASHTO and AREA specifications.

4-12.1 AASHTO Criteria

The distribution of wheel loads for multilane bridges is given by AASHTO and is based on the standard truck or lane loading occupying a 10-ft (3.05-m) lane width. The roadway width is taken as the curb-to-curb distance. Roadway widths from 20 to 24 ft (6.1 to 7.3 m) inclusive are assumed to have two loaded lanes. For greater roadway widths, the number of loaded lanes can be obtained as

$$N = \frac{W}{L_w}$$

184 STRUCTURAL STEEL DESIGN

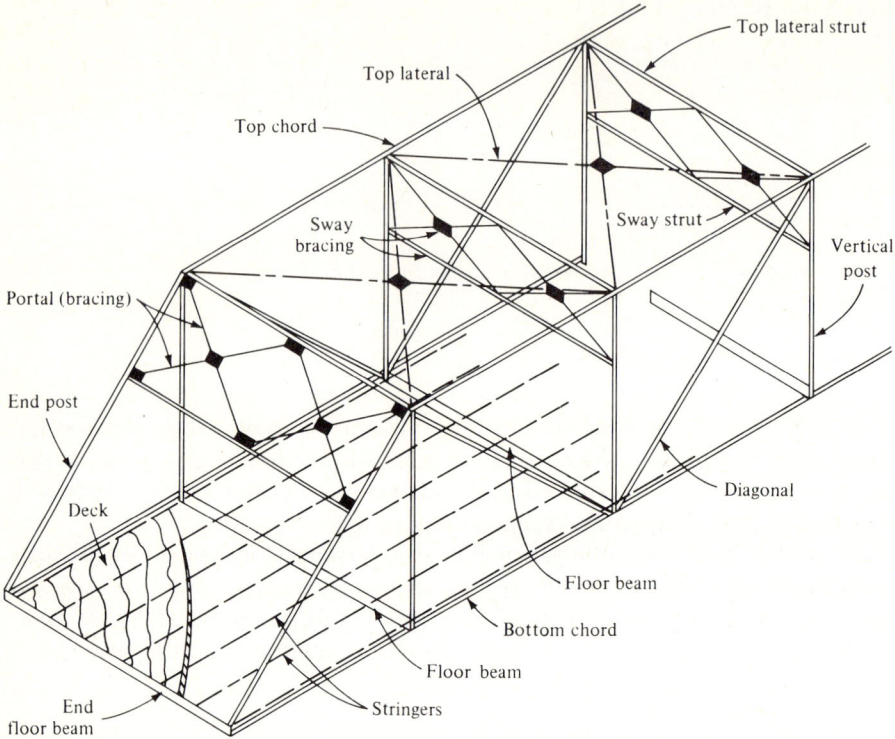

Figure 4-13 Framing of a typical through truss highway bridge for multiple lanes. Only principal members shown. Refer also to Fig. VI-1 for as-built highway bridge truss.

Figure 4-14 Bridge components of in-service bridges. (*a*) View of a through highway bridge truss illustrating stringers, floor beams and other bracing/framing details. (*b*) Rolled beam stringers for an open-deck railroad bridge.

where $L_w = $ 10 ft or 3.05 m
  $W = $ roadway width (including any shoulder allowance)

The number of lanes N is an integer; thus a lane fraction should be rounded up to the next integer. A reduction in load intensity is allowed when the number of lanes $N > 2$ as follows:

N	Percent of live load
2	100
3	90
4 or more	75

The distribution of wheel loads to stringers (or longitudinal beams) as well as transverse floor beams is based on a paper by Newmark ("Design of I-Beam Bridges," *Transactions, ASCE*, Vol. 114, 1949) and is given in terms of span S divided by a coefficient as given in Table 4-1.

The computation of shear and bending moments for deck stringers is based on a simple beam analysis for the critical load (either truck or lane). After these values are computed and adjusted for impact, the distribution factors from Table 4-1 are used to obtain the design effect on any of the interior stringers.

Table 4-1 Wheel load distribution coefficients for AASHTO bridge design used as S/coefficient where $S = $ stringer (beam) spacing

Bending moment and shear (lateral distribution on steel interior[a] beams)	1 lane	2 + lanes
		( ) = SI value
For: concrete floor	7.0 (2.134)	5.5 (1.676)
steel grid deck $\leq$ 4 in (102 mm)	4.5 (1.372)	4.0 (1.219)
steel grid deck $>$ 4 in (102 mm)	6.0 (1.828)	5.0 (1.524)

Note: For exterior stringers use statics to obtain effective wheel loads causing bending and shear (refer to Fig. 4-15).

Bending moment in transverse floor beams without stringers:
  With concrete deck   6.0 (1.828)
  With steel grid $\leq$ 4 in (102 mm)   4.5 (1.372)
       $>$ 4 in (102 mm)   6.0 (1.828)

Bending in transverse floor beams with stringers: see Fig. 4-15.

Bending moments in concrete deck slabs (main reinforcement transverse):
  $S = $ effective span length = clear flange edge-to-flange edge distance + $b_f/2$
  $P_i = $ rear wheel load of H 20, H 15, H 10, etc. (H 20 or HS 20 = 32/2 = 16 kips = 72 kN)

$$M = \frac{S+2}{32} P_i \text{ ft} \cdot \text{kips/ft} \qquad M = \frac{S+0.61}{9.74} P_i \text{ kN} \cdot \text{m/m}$$

Use $M' = 0.8M$ if three or more stringers.

[a] Exterior stringers to be at least as large as interior stringers, to allow for future bridge widening.

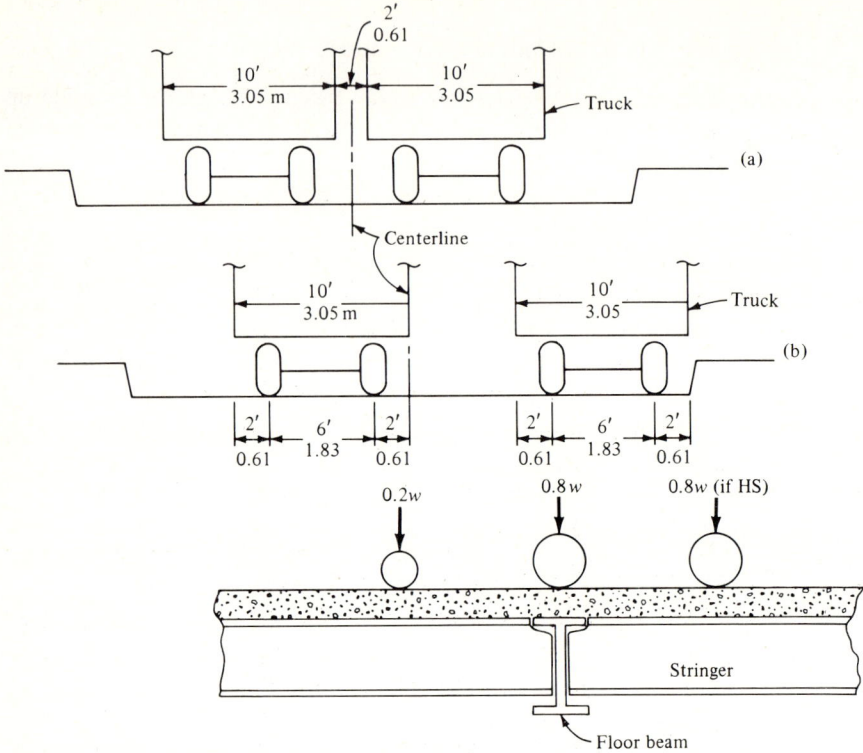

Figure 4-15 Truck wheel placement for maximum shear and moments in floor beams. Note that lane loading would seldom (if ever) control design of these members.

These factors allow for deck continuity across several stringers via plate action, causing the load to distribute—even if a wheel is directly over a stringer—as the deck tends to deflect. The computation of shears and bending moments for the exterior stringers is based entirely on statics and simple beam analysis; however, AASHTO currently requires the exterior stringers to be at least as large as the interior stringers in case of future bridge widening.

The computation of bending moment in the deck slab is empirical and depends on the direction of the main reinforcement (usually transverse), with the applicable AASHTO equations given in Table 4-1.

The computation of shear and bending moments in the transverse or floor beams into which stringers frame (generally used only for truss bridges at present) is based on Fig. 4-15, with one truck in each lane placed so as to produce either maximum shear at one end of the beam or maximum moment. The analysis is based on statics and no distribution factors are applied.

4-12.2 AREA Criteria

AREA wheel load distribution to cross beams in open-deck bridges (see Figs. 10-1c and 10-4) is based on the full value of the critical axle load applied.

DESIGN OF BEAMS FOR BENDING **187**

Cross-tie design considers an equal distribution onto three ties but not more than 4 ft (1.22 m) of a wheel load.

The portion of track load on transverse floor beams of ballasted-deck bridges (see Figs. 10-4d and 10-5) may be taken as

$$P = \frac{1.15 P_w D}{s} \tag{4-28}$$

where P_w = critical axle load
D = effective beam spacing ($D < s$) and can be reduced
s = axle spacing at critical load P_w

One-half the effective load P is placed on each rail at the gage distance apart. The gage distance may be taken as 4.71 ft (1.436 m). AREA specifications give a complicated expression for D which allows use of a value less than the actual beam spacing. The designer may, however, use D = actual spacing to produce a conservative design moment. The actual spacing is required to compute the floor beam shear. If the beam spacing is $D > s$, one should use $P = P_w$ directly.

Example 4-15 Design the floor beams for the through-ballasted plate girder bridge of Example 10-5. Use A-36 steel and the AREA specifications. Use Cooper's E-110 loading. The general bridge geometry is shown in Fig. E4-15. Refer to Fig. 10-4 for location and illustration of actual floor beams in service on a railroad bridge.

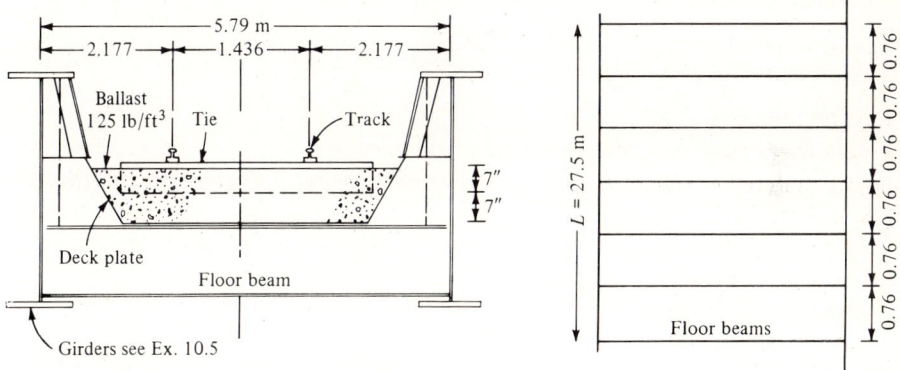

Figure E4-15

SOLUTION
   Step 1. Determine the dead load on the floor beam.
   The track at 0.200 kip/ft × 14.59 × 0.76 × 2 (AREA allowance) = 4.44 kN. Ballast and ties: ties are 8 ft 6 in × 6 in or 8 in (or 7 in × 9 in) and are embedded in 6 to 7 in of ballast with 7 in of ballast below the tie to the ballast plate. Assume no displacement and 125 pcf for unit weight of ballast, including ice, water, etc. In SI:

125(9.807/62.4)[(7 + 7)/12](0.3048)(0.76)   = 5.31 kN/m

Ballast plate: include 10 percent for corrosion protection
0.015(77 kN/m³)(0.76)(1.10) $\qquad$ =0.97 kN/m

Estimate cross-beam weight $\qquad$ =1.50 kN/m

Miscellaneous maintenance, storage of ties, material $\qquad$ =1.23 kN/m

$\qquad$ Total =9.01 kN/m

The dead-load moment is

$$M_d = \frac{wL^2}{8} + \text{track}$$

$$= \frac{9.01(5.79)^2}{8} + \frac{4.44/2(5.79 - 1.44)}{2} = 37.8 + 4.8 = 42.6 \text{ kN} \cdot \text{m}$$

The dead-load shear is

$$V_d = \frac{9.01(5.79)}{2} + \frac{4.44}{2} = 28.3 \text{ kN}$$

Step 2. Find the live load on the beam. ($D = 0.76$ m, $s = 5$ ft $= 1.524$ m)

$$P = \frac{1.15 P_w D}{s}$$

$$= \frac{1.15(110/80)(80 \times 4.448)(0.76)}{1.524} = 280.6 \text{ kN}$$

with $P/2$ placed on each rail. The live-load shear is

$$\frac{P_w}{2} = \frac{280.6}{2} = 140.5 \text{ kN}$$

The impact factor is

$$I = \frac{30.5}{S} + 40 - \frac{3L^2}{150} \qquad \text{where } S = \text{beam spacing, } L = \text{length}$$

$$= \frac{30.5}{0.76} + 40 - \frac{3(5.79)^2}{150} = 79.5 \text{ percent}$$

The design live-load shear is

$$140.5 \times 1.795 = 252.2 \text{ kN}$$

The design live-load moment is

$$M_L = \frac{280.6(2.177 \times 1.795)}{2} = 548 \text{ kN} \cdot \text{m}$$

The total design moment is

$$M_{\text{design}} = M_d + M_L = 42.6 + 548 = 591 \text{ kN} \cdot \text{m}$$

The total design shear is

$$V_{\text{design}} = V_d + V_L$$

$$= 28.3 + 252.2 = 280.5 \text{ kN}$$

Step 3. Find the beam section.

The required section modulus is based on $F_b = 0.55F_y$, since the compression flange is laterally supported.

$$S_x = \frac{M}{F_b} = \frac{591}{137.5} = 4.30 \times 10^{-3} \text{ m}^3$$

We could use a W760 × 147.3 section; however, we will arbitrarily go to the next most economical section, W760 × 160.7/1.58:

$$S_x = 4.8997 \times 10^{-3} \text{ m}^3 \qquad t_w = 13.8 > 8.50 \text{ mm}$$
$$t_f = 19.3 \text{ mm} \qquad d = 758 \text{ mm}$$

We note that the weight is 0.08 kN/m larger than assumed, but the section modulus is more than adequate for this small difference. Check the shear:

$$f_v = \frac{V}{dt_w} = \frac{280.5}{0.758(13.8)} = 26.8 \ll 0.35F_y \qquad \text{O.K.}$$

(The end connections to the plate girder will be designed in Example 8-9.) Use a W760 × 160.7/1.58 rolled section. ///

4-13 COMPOSITE BEAMS

A *composite beam* is one whose strength depends upon the mechanical interaction between two or more materials. Reinforced concrete beams are actually composite members but are not generally designated as such. Most often the term "composite beam" in building and bridge construction is applied to a steel section on which a concrete floor or bridge deck has been cast. The concrete is securely bonded to the steel section via carefully designed shear connectors so that the concrete and steel act together as a tee beam. Figures 4-16 and 4-17 illustrate shear connectors in position to bond the concrete to the beams so that composite action is obtained. The shear connectors shown are called *shear studs* and are welded to the beam (and through the deck pan in Fig. 4-16). Other types of shear connectors can be used, but shear studs are most common.

When there is no particular effort to bond the steel beam and concrete floor or deck, relative slip occurs at the interface of the two materials and the result is a noncomposite section. In some cases we may design for a specified amount of composite action. Actually, there will always be some small slip due to the unequal deformations in the shear studs, concrete, and steel beam, but for practical purposes it can be neglected in composite design. The brief introduction to composite design made here will only consider full composite action.

The method of construction and code specifications are significant design parameters in design of composite sections. Figure 4-18 illustrates the effective width of concrete to use in a transformed section b' using the modular ratio $n = E_s/E_c$ as in several mechanics-of-materials texts.

The modular ratio depends on the 28-day concrete design strength f'_c, as shown in Table 4-2.

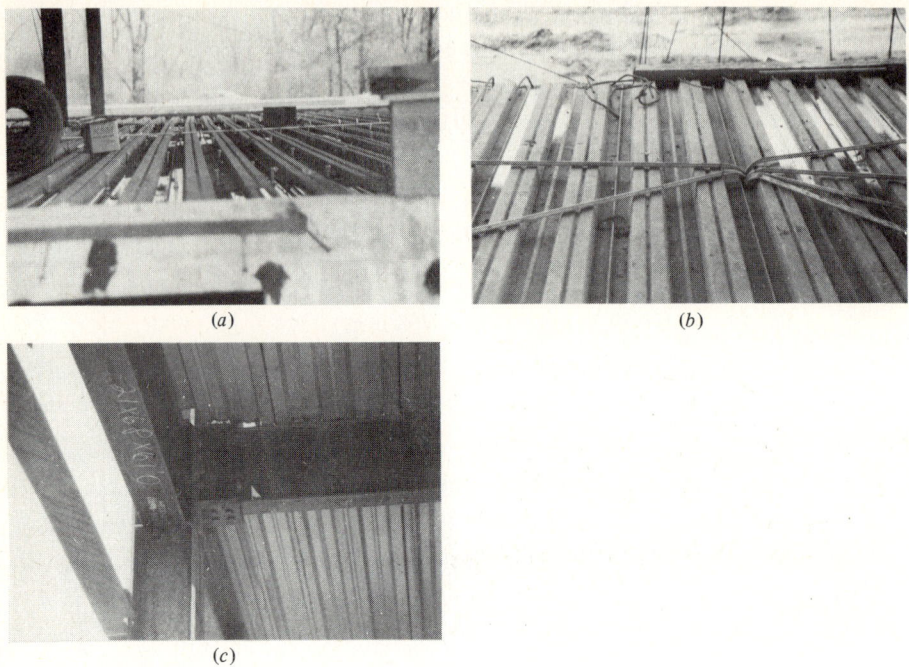

Figure 4-16 Composite building construction. (*a*) Shear studs in place for building using unshored construction and metal deck for form. (*b*) View of metal deck showing electrical conduit to be encased by concrete floor. (*c*) Bottom view of metal deck resting on floor beams and girders. Metal deck is welded to floor beams with shear studs.

The two basic construction methods for producing composite beams are:

1. *Shored construction.* The steel beams are put in place and the formwork for the concrete slab is added. This assembly is then shored (braced or propped) so that no (or relative small amounts of) deflection can occur, and the concrete is poured. After the concrete has hardened for about 7 days (about 70 to 75 percent of f_c' obtained), the shoring is removed. At this point the stresses in the composite beam are due to the dead weight of the steel beam plus a proportionate share of the concrete deck.
2. *Unshored construction.* The steel beams are placed and formwork (metal decking may be the necessary formwork as in Fig. 4-16) supplied for the concrete deck (refer to Fig. 4-17). The concrete is poured and at this time the steel beam carries the dead load of steel, formwork (as used), and the concrete. After the concrete hardens, any formwork is removed. It is evident at this stage of the construction that the steel beam has been permanently stressed with the weight of the steel beam plus a proportionate share of the weight of the concrete deck.

DESIGN OF BEAMS FOR BENDING

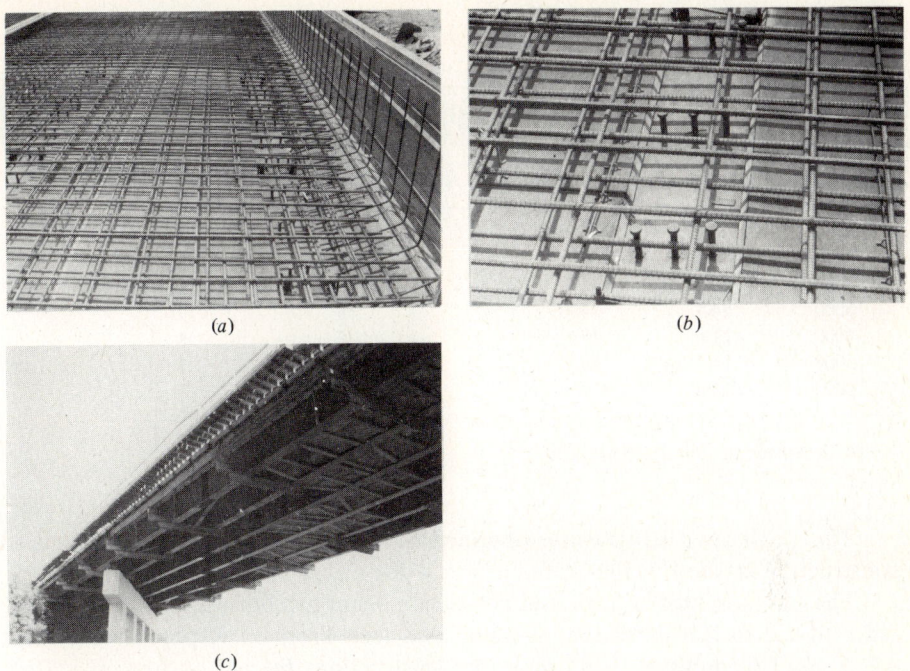

Figure 4-17 Composite bridge deck (two lanes of interstate highway using six stringers. (*a*) Composite bridge deck nearly ready for concrete with reinforcing and shear studs in place. Side vertical steel is for parapet. (*b*) Closeup view of shear studs. (*c*) Underview showing stringers and unshored formwork for deck.

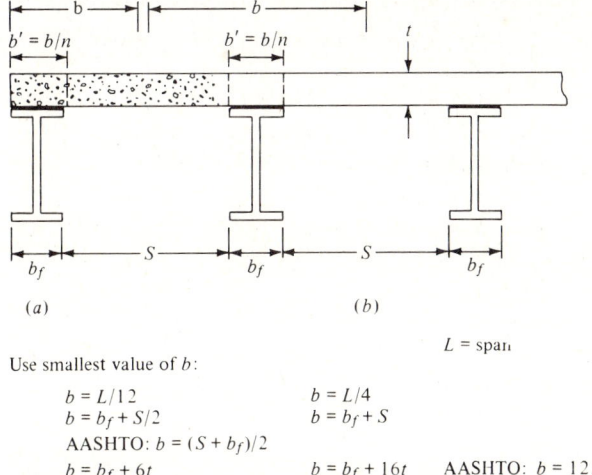

Use smallest value of b:

$b = L/12$ $\qquad$ $b = L/4$
$b = b_f + S/2$ $\qquad$ $b = b_f + S$
AASHTO: $b = (S + b_f)/2$
$b = b_f + 6t$ $\qquad$ $b = b_f + 16t$ $\qquad$ AASHTO: $b = 12t$

Figure 4-18 Effective flange width of composite sections by both AISC and AASHTO specifications with any differences identified. (*a*) Edge. (*b*) Interior.

Table 4-2 Modular ratio for AISC and AASHTO composite design for various concrete strengths

f'_c		$E_c = w^{1.5}33(f'_c)^{\frac{1}{2}}$	Modular ratio, n	
psi	MPa	$= 57(f'_c)^{\frac{1}{2}}$	ACI[a]	AASHTO
2000	13.8	2550 (psi)	11	11
2500	17.2	2850	10	10
3000	20.7	3100	9	9
3500	24.1	3400	9	9
4000	27.6	3600	8	8
5000	34.5	4000	7	7
6000	41.4	4400	7	6

[a] ACI, American Concrete Institute.

The qualitative stress state obtained at the end of shored and unshored construction methods is illustrated in Fig. 4-19.

The ultimate load of a shored composite beam has been found to be on the order of 2.2 to 2.5 times the working load (see Progress Report of the Joint ASCE–ACI Committee on Composite Construction, *Journal of Structural Division, ASCE*, ST12, December 1960) and even higher for unshored composite beams. Since this compares to a load ratio of about 1.7 to 1.8 for rolled steel beams, a modified method of analysis seems practical and is used by AISC. Define

S_s = section modulus of steel beam referred to tension flange

S_{tr} = section modulus of composite beam referred to tension flange

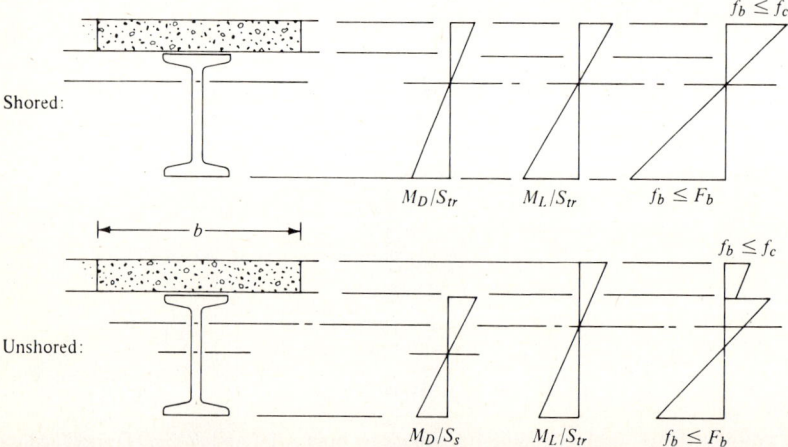

Figure 4-19 Qualitative stress profiles for composite beams for construction procedure indicated.

If we use an overstress factor of 135 percent (1.35 to account for the large ultimate capacity), the stress on the bottom flange of a steel composite beam under final load conditions in terms of moments is

$$f_{b(\text{final})} = \frac{M_D}{S_s} + \frac{M_L}{S_{tr}} = \frac{1.35(M_D + M_L)}{S_{tr}}$$

Multiplying through by S_{tr} and solving, we obtain the AISC equation, which limits the section modulus of the bottom flange to not more than the following:

$$S_{tr} = \left(1.35 + \frac{0.35 M_L}{M_D}\right) S_s \qquad (4\text{-}29)$$

When the computed value of S_{tr} based on the actual cross section is greater than given in this equation, the value from Eq. (4-29) should be used to compute the actual bending stress in the composite-section tension flange.

The actual bending stresses in a composite beam are limited by the material. In general, the steel stress is limited to $F_b = 0.66 F_y$ for compact sections and $F_b = 0.6 F_y$ for other sections. The concrete stress is limited to $0.45 f'_c$ using working stress methods and to $0.85 f'_c$ using strength design methods.

From consideration of Fig. 4-20, the shear developed at the neutral axis is either compression or tension, neglecting any concrete in the tension zone. Since it is difficult to find the neutral axis and to account for working stresses, AISC and AASHTO allow computation of the horizontal shear to be resisted at the interface of steel and concrete as the smaller of

$$V_h = \frac{A_s F_y}{2} \quad \text{or} \quad V_h = \frac{0.85 f'_c (bt)}{2} \qquad (4\text{-}30)$$

where the factor 2 is used to reduce the ultimate shear to a working value. When steel reinforcement is in the concrete flange, Eq. (4-30), using concrete f'_c, should be adjusted, but the reader is referred to the AISC manual or the AASHTO specifications.

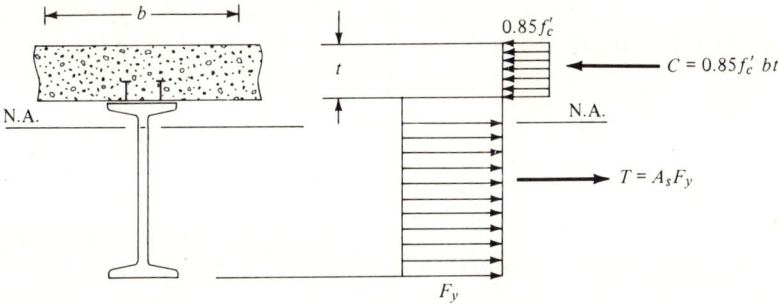

Figure 4-20 Stress distribution at ultimate load in composite beam.

The number of shear studs depends on the stud capacity which for AISC specifications is as follows:

Stud size (diameter × length), in (mm)	Horizontal stud resistance, q, kips (kN) f'_c, ksi (MPa)	
	3.0 (20.7)	4.0 (27.6)
$\frac{1}{2} \times 2$ (12 × 50)	5.1 (22.7)	5.9 (26.2)
$\frac{5}{8} \times 2\frac{1}{2}$ (16 × 65)	8.0 (35.6)	9.2 (40.9)
$\frac{3}{4} \times 3$ (20 × 75)	11.5 (51.2)	13.3 (59.2)
$\frac{7}{8} \times 3\frac{1}{2}$ (22 × 90)	15.6 (69.4)	18.0 (80.1)

The number of shear connectors by AISC specifications is not based on VQ/I, as for welding or bolting, but is based on

$$N_1 = \frac{V_h}{q} \tag{4-31}$$

where V_h = minimum value of horizontal shear force from Eq. (4-30)
q = shear stud capacity from foregoing table

This computation of number of required shear studs assumes a uniform spacing of shear studs (usually in pairs) from the point of zero to maximum moment. When a concentrated load is in the span, the number of shear stud connectors between the load and the nearest point of zero moment is computed as

$$N_2 = \frac{N_1((S_{tr}M)/(S_s M_{max}) - 1)}{(S_{tr}/S_s - 1)} \tag{4-32}$$

where M = moment at concentrated load (M_{max} = maximum moment in span)
N_1 = number of shear connectors previously defined by Eq. (4-31)

The appropriate specifications (AASHTO or AISC) should be consulted for shear connector design for continuous spans.

The AISC handbook contains a number of composite sections for which the section properties are already provided for selected steel sections and thickness of concrete. Similar tables can be readily developed by the reader for selected steel sections, slab thickness, and f'_c using a digital computer.

4-13.1 AASHTO Shear Stud Requirements

AASHTO requirements are very similar to AISC specifications; the differences are given in the following paragraphs. The allowance for fatigue is the major difference and depends on the number of fatigue cycles expected. AASHTO requires computation of the horizontal shear range based on the range of shear

V_r due to live loads from the live-load shear envelope, where

$$V_r = V_{max} - V_{min}$$

to obtain

$$S_r = \frac{V_r Q}{I}$$

where Q and I are as used in Eq. (4-2) and I is the moment of inertia based on the composite section.

The allowable horizontal shear force developed by a shear stud F_r depends on the stud diameter and the number of fatigue cycles expected (and for a stud H/d ratio ≥ 4) as:

$$Z_r = G(d)^2 \quad \text{(kips or kN)} \tag{4-33}$$

where

Number of cycles =	0.1×10^6	0.5×10^6	2×10^6	over 2×10^6
fps: $G =$	13.0	10.6	7.85	5.5
SI: $G =$	0.0895	0.0731	0.0541	0.0379

Use the stud diameter $d =$ in or mm as in Eq. (4-33).

The required pitch of connectors at any section based on the number of connectors n used across the beam (steel) flange is

$$s_i = \frac{nZ_r}{S_r} \tag{4-34}$$

The total number of shear connectors N_1 must be at least the same as that given by the AISC equations [Eqs. (4-30) and (4-31)].

Example 4-16 Design a composite interior beam section for a 30-ft span in an office building (Fig. E4-16a) using a 4-in concrete slab and a metal deck for the form, with steel beams spaced 10 ft on centers. No depth restrictions are assumed for the total depth of the section. No shoring will be required. Use $f'_c = 3$ ksi, A-36 steel, and the AISC specifications. Limit dead-load deflection to 1.5 in and live-load deflection to $L/360$. Other data:

$$\text{live load} = 100 \text{ psf}$$

$$\text{Partition weight} = 20 \text{ psf}$$

$$\text{Ceiling weight} = 10 \text{ psf}$$

196 STRUCTURAL STEEL DESIGN

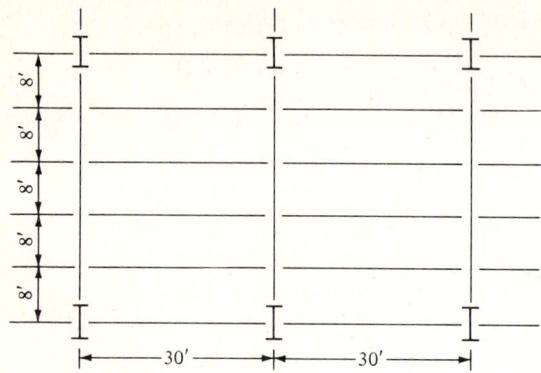

Figure E4-16a

SOLUTION
  Step 1. Find the bending moments.
    (a) Dead loads carried by steel beam:

Steel, including reinforcing, studs, metal decking   = 0.080 ksf

Concrete: $4(\frac{1}{12})(0.144)$ = 0.048 ksf

Total = 0.128 ksf

$$M_D = \frac{wL^2}{8} = \frac{0.128(30)^2(8)}{8} = 115.2 \text{ ft} \cdot \text{kips}$$

  (b) Live loads carried by composite section:

Live load = 0.100 ksf

Partitions = 0.020 ksf

Ceiling = 0.010 ksf

Total = 0.130 ksf

$$M_L = \frac{wL^2}{8} = \frac{0.130(30)^2(8)}{8} = 117.0 \text{ ft} \cdot \text{kips}$$

$$V_{max} = \frac{(0.128 + 0.130)(8 \times 30)}{2} = 30.96 \text{ kips}$$

Step 2. Find the required section modulus of S_s of the steel beam (and $F_b = 0.66 F_y$).

$$S_s = \frac{M_D}{F_b} = \frac{115.2(12)}{24} = 57.6 \text{ in}^3$$

$$\text{Minimum } S_{tr} = \frac{M_D + M_L}{F_b} = \frac{(115.2 + 117.0)12}{24} = 116.1 \text{ in}^3$$

For a dead-load deflection of less than 1.5 in, the moment of inertia is at least

$$I = \frac{5wL^4}{384(E)\Delta} = \frac{5(0.128 \times 8)30^4 12^3}{384(29\ 000)(1.5)} = 429 \text{ in}^3$$

After several preliminary computations (not shown to save space), try:

W18 × 50: $I_x = 800$ in^4 $S_x = 88.9$ in^3

$d = 17.99(18.0)$ in $A = 14.70$ in^2

$b_f = 7.495(7.5)$ in $t_w = 0.355$ in

Step 3. Find the width of the composite beam flange b.

$$b = \frac{L}{4} = \frac{30(12)}{4} = 90 \text{ in}$$

$S = 8(12) - 7.5 = 88.5$ in

$b = 7.5 + 88.5 = 96$ in

$b = 7.5 + 16(4) = 71.5$ in   controls

Step 4. Compute the properties of the composite section (refer to Fig. E4-16*b*).

$$n = 9 \quad b' = \frac{b}{n} = \frac{71.5}{9} = 7.9 \text{ in} \quad M_{xx} = Ay$$

$$(7.9 \times 4 + 14.70)y = \left(\frac{18}{2} + 2\right)(7.9 \times 4)$$

$$y = \frac{347.6}{46.3} = 7.51 \text{ in}$$

With y computed, the neutral axis is located as shown in Fig. E4-16*b*. We can now compute I, S_{top}, and $S_{bot} = S_{tr}$.

$$I_{xx} = 800 + 14.7(7.51)^2 + \frac{7.9(4)^3}{12} + 7.9 \times 4[2 + (9 - 7.51)]^2$$

$$= 800 + 829.1 + 42.1 + 384.9$$

$$= 2056.1 \text{ in}^4$$

$$S_{top} = \frac{2056.1}{5.49} = 347.5 \text{ in}^3$$

$$S_{bot} = S_{tr} = \frac{2056.1}{16.51} = 124.5 \text{ in}^3 > 116.1 \quad \text{O.K.}$$

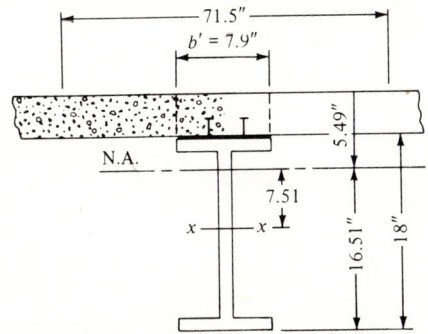

Figure E4-16*b*

Check the AISC equation for S_{tr} maximum:

$$S_{tr} = \left[1.35 + 0.35\left(\frac{M_L}{M_D}\right)\right] S_s$$

$$S_{tr} = \left[1.35 + 0.35\left(\frac{115.2}{117}\right)\right] 88.9$$

$$= 150.6 > 124.5 \quad \text{O.K. to use } 124.5 \text{ in}^3$$

Step 5. Check the final stresses.

Initial: $\quad f_b = \dfrac{M_D}{S_s} = \dfrac{115.2(12)}{88.9} = 15.6 \text{ ksi}$

Final: top:

$$f_b = \frac{117.0(12)}{347.5} = 4.04 \text{ ksi} \quad \text{(transformed)}$$

$$f_c = \frac{4.04}{9} = 0.45 \text{ ksi} \ll 0.45 f'_c \quad \text{O.K.}$$

Bottom:

$$f_b = \frac{117.0(12)}{124.5} = 11.3 \text{ ksi}$$

Based on AISC,

$$f_b = \frac{(117 + 115.2)(12)}{124.5} = 22.4 \text{ ksi} < 24 \quad \text{O.K.}$$

The stress profile for the section based on mechanics considerations is plotted in Fig. E4-16c for comparison with the AISC composite section stress of 22.4 ksi.

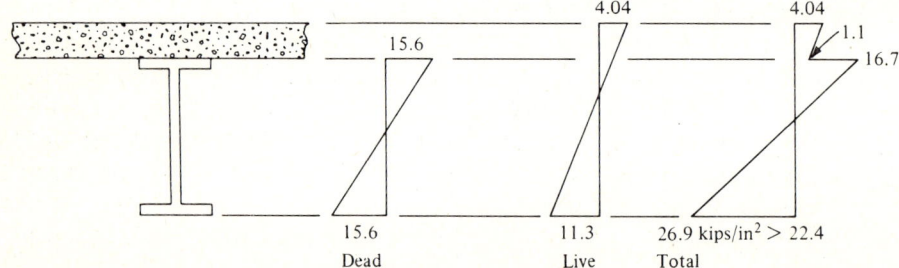

```
       4.04              4.04
15.6                        1.1
                            16.7

15.6        11.3    26.9 kips/in² > 22.4
Dead        Live    Total
```

Figure E4-16c

Step 6. Find the number of shear connectors required:

$$V_h = \frac{0.85 f'_c bt}{2} = \frac{0.85(3)(71.5)(4)}{2} = 364.6 \text{ kips}$$

or $\quad V_h = \dfrac{A_s F_y}{2} = \dfrac{14.7(36)}{2} = 264.6 \text{ kips} \quad \text{controls}$

Use $\tfrac{5}{8}$ connectors with a capacity of 8 kips/connector; the number required is

$$N_1 = \frac{V_h}{q} = \frac{264.6}{8} = 33.07 \qquad \text{use 34 connectors}$$

Use two connectors at each section at a spacing $w = 4d = 4(\tfrac{5}{8}) = 2.5$ in center to center. The longitudinal connector spacing is

$$s = \frac{30(12)}{2(34)/2} = 10.6 \text{ in} > 6D \text{ (minimum)}$$

$$< 8t \text{ (maximum)}$$

Step 7. Check the shear with both dead and live loads acting.

$V = 30.96$ kips

$$f_v = \frac{V}{dt_w} = \frac{30.96}{18(0.355)} = 4.9 \text{ ksi} \ll 0.4F_y \qquad \text{O.K.}$$

Step 8. Check the deflection under live load. Note that the dead-load deflection has been built out of the floor via flow of the wet concrete during placing.

$$\Delta = \frac{5wL^4}{384EI} = \frac{5(0.13 \times 8)(30^4)(12^3)}{384(29\,000)(2056.1)} = 0.32 \text{ in} < \frac{L}{360} \qquad \text{O.K.} \quad ///$$

Example 4-17 Design a composite bridge section for a two-lane bridge (two 12-ft lanes + two 8-ft shoulders and pedestrian walkways) as shown in Fig. E4-17a. Other data: HS 20 loading; $f_c' = 27.6$ MPa (class A concrete at 4 ksi); $f_y = 415$ MPa (reinforcing steel); A-36 steel for the stringers; AASHTO specifications; 2×10^6 loading cycles.

SOLUTION The two parapets and railings will be placed after the deck slab is placed. We will assume that this loading is carried by the two outer stringers, which are not designed here (nor is the interior stringer checked to see if it is adequate as an exterior stringer, since the exterior stringer must be at least as large as the interior stringers). We will also assume that a future 40 mm of wearing surface will be added as the initial surface deteriorates.

Step 1. Design the concrete deck slab.

The steel beam spacing is selected at 2033 mm, as shown in Fig. E4-17a. Estimate the stringer flange width $b_f = 420$ mm (a W920 section), and compute the effective stringer spacing (AASHTO Sec. 1-3.2) of the interior beams as

$$s = s_{\text{act}} - b_f + \frac{b_f}{2} = 2033 - 210 = 1823 \text{ mm}$$

The slab dead load, including future wearing surface based on $D = 230$ mm

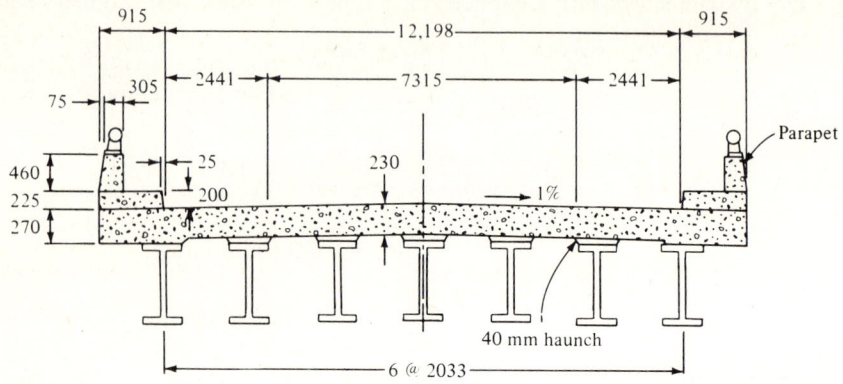

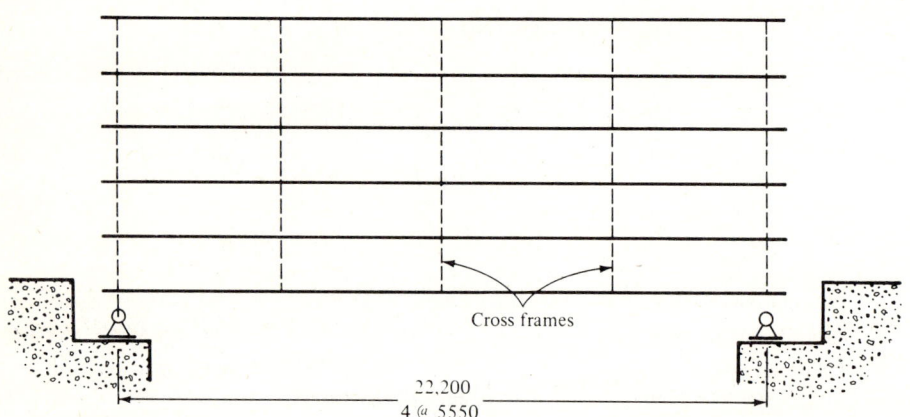

Figure E4-17a

of concrete, is

$$(0.230 + 0.040)(23.6 \text{ kN/m}^3) = 6.37 \text{ kPa}$$

Increase 10 percent for roadway debris, snow, etc.:

$$6.37 \times 1.10 = 7.0 \text{ kPa}$$

The dead-load moment is computed as

$$M_D = \frac{wL^2}{10} = \frac{7.0(1.823)^2}{10} = 2.33 \text{ kN} \cdot \text{m}$$

The impact factor for the slab is (using $L = s = 2.033$ m)

$$I = \frac{15}{L + 38} = 0.375 > 0.30 \qquad \text{use } I = 0.30$$

Since the $s/L = 2.033/22.4 = 0.091$ is so small, one-way principal slab reinforcement perpendicular to the traffic flow will be used in the slab design. For this case and five spans providing slab continuity across the roadway, the live-load moment is

$$M_L = 0.8\left(\frac{s + 0.61}{9.74}\right)(P)(1 + I)$$

$$= 0.8\left(\frac{1.823 + 0.61}{9.74}\right)(72)(1.30) = 18.7 \text{ kN} \cdot \text{m}$$

The design moment for the deck slab using strength design is

$$M_{\text{design}} = M_D + \frac{5M_L}{3} = 2.33 + \frac{5(18.7)}{3} = 33.5 \text{ kN} \cdot \text{m}$$

Take $d = D - 40$ (allowing 25 mm clear cover for bottom reinforcement in bridge decks AASHTO Art. 1.6.16):

$$d = 230 - 40 = 190 \text{ mm}$$

The workmanship factor $\phi = 0.90$.

$$M_u = \phi A_s f_y \left(d - \frac{a}{2}\right)$$

$$a = \frac{A_s f_y}{0.85 f'_c b} = \frac{415 A_s}{0.85(27.6)(1)} = 17.69 A_s$$

$$A_s \left(0.190 - \frac{17.69 A_s}{2}\right) = \frac{33.50}{0.9(415)(10^3)}$$

$$A_s^2 - 0.02148 A_s = -1.0140 \times 10^{-5}$$

Solve for A_s by completing the square:

$$A_s = \pm 0.0102576 + 0.0107405 = 0.0004829 \text{ m}^2/\text{m}$$

The maximum ratio of $A_s/A_c = 0.0214$, based on f'_c and f_y and taking 75 percent of the value to ensure initial steel yield. The actual steel percentage is

$$p = \frac{0.0004829}{0.19(1)} = 0.0025 \ll 0.0214 \quad \text{O.K.}$$

The minimum percentage is

$$p_{\min} = \frac{1.383}{f_y} = 0.0033 > 0.0025$$

Use $A_s = 0.0033(0.19 \times 1.0) = 0.000627 \text{ m}^2/\text{m}$.

Step 2. Design the steel stringers.

Note it was necessary to design the slab so that the dead load carried by the stringers could be computed. Assume unshored construction—the stringers must carry the dead weight of the deck slab until the concrete hardens.

Dead load of slab: (0.230 + 0.040)(23.6)(2.032)	= 13.00 kN/m
Haunch: 0.040(0.42)(23.6)	= 0.40 kN/m
Rolled beam assumed	= 3.60 kN/m
Miscellaneous, including debris, formwork, etc.	= 1.00 kN/m
Total	= 18.00 kN/m

The maximum dead-load moment at the center of span is

$$M_D = \frac{wL^2}{8} = \frac{18(22.4)^2}{8} = 1129 \text{ kN} \cdot \text{m}$$

$$V_D = \frac{wL}{2} = \frac{18(22.4)}{2} = 201.6 \text{ kN}$$

The maximum live-load moment, including impact, is based on the AASHTO truck and the stringer spacing.

$$I = \frac{15}{22.4 + 38} = 0.25$$

From Sec. 1-9, the live-load moment due to total truck load on span of this length is

$$M_L = \frac{W}{L}[(0.9L + 1.282)(0.5L + 0.71) - 3.41L](1 + I)$$

$$= \frac{180}{22.4}[(0.9(22.4) + 1.282)(0.5(22.4) + 0.71) - 3.41(22.4)](1.25)$$

$$= 1798 \text{ kN} \cdot \text{m}$$

The distribution factor for live load to a stringer of a bridge with two or more lanes is obtained from Table 4-1 as

$$\text{Factor} = \frac{S}{1.676} = \frac{2.033}{1.676} = 1.213 \text{ wheels}$$

Since 1.213 wheels = 1.213/2 = 0.6065 axle, the adjusted (for deck slab action) stringer bending moment is

$$M_{L(\text{design})} = 1798(0.6065) = 1090.5 \text{ kN} \cdot \text{m}$$

The maximum shear occurs when one of the rearmost truck wheel's [144-kN (32-kip) axle] load is at the beam end:

$$V = W\left(\frac{L - 5.1}{L} + 0.8\right) = 180\left(\frac{22.4 - 5.1}{22.4} + 0.8\right) = 283 \text{ kN}$$

This value must also be adjusted for plate action and impact:

$$V_{L(\text{design})} = 283.0(0.6065)(1.25) = 214.6 \text{ kN}$$

We note that M_D is at the center of the stringer span, whereas M_L is slightly off center. The error of summing $M_D + M_L$ for design is negligible and

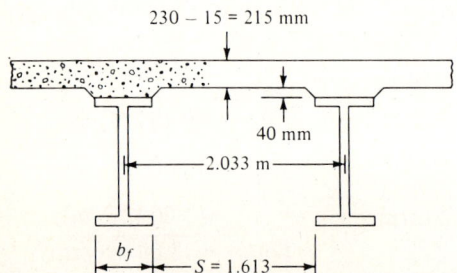

Figure E4-17b

DESIGN OF BEAMS FOR BENDING

conservative, so that we may check stresses at the center of span under M_D and add the composite effect of M_L (unshored construction sequence).

Step 3. Find the properties of the composite section (refer to Fig. E4-17b).

Assume that 15 mm of concrete is not usable in composite section because of wear and surface deterioration. Neglect the area of concrete in the haunch.

$$t' = 230 - 15 = 215 \text{ mm}$$
$$S = 2.033 - b_f = 2.033 - 0.42$$
$$= 1.613 \text{ m}$$
$$b = \frac{22.4}{4} = 5.6 \text{ m}$$
$$b = 0.42 + 1.613 = 2.033 \text{ m}$$
$$b = 0.42 + 16t = 3.86 \text{ m}$$

Use $b = 2.033$ m; for $f'_c = 27.6$ MPa, $n = 8$.

$$b' = \frac{2.033}{8} = 0.254 \text{ m}$$

It is usually sufficient to select a steel beam one or more sizes less than that required for noncomposite action. The section required for noncomposite action is approximately (and flange laterally supported)

$$S_x = \frac{M_T}{0.55 F_y} = \frac{1129 + 1090.5}{0.55(250)} = 16.14 \times 10^{-3} \text{ m}^3$$

Tentatively, try a W920 × 364.6/3.58:

$$I_x = 6701.3 \times 10^{-6} \text{ m}^4 \qquad S_x = 14.67 \times 10^{-3} \text{ m}^3$$
$$d = 916 \text{ mm} \qquad t_w = 20.3 \text{ mm}$$
$$t_f = 34.3 \text{ mm} \qquad b_f = 419 \text{ mm}$$
$$A = 46.52 \times 10^{-3} \text{ m}^2$$

Check the shear:

$$f_v = \frac{V_T}{dt_w} = \frac{201.6 + 283}{0.916(20.3)}$$
$$= 26.1 \text{ MPa} \ll \frac{F_y}{3} \qquad \text{O.K.}$$

Check the dead-load deflection:

$$\Delta = \frac{5wL^4}{384 EI} = \frac{5(0.018 \text{ MN/m})(22.4)^4}{384(200\ 000)(0.0067013)}$$
$$= 0.04402 \text{ m } (44.02 \text{ mm})$$

This much deflection in a 22.4-m span would tend to aggravate during placing the concrete to a level top surface (requires an additional 44 mm of concrete at the center of the span. This will require a temporary midspan shore; or, preferably, have the stringers cambered at the mill with a 44-mm camber (and being sure to place the cambered side up on the job or a real disaster would occur). The remainder of the beam properties (weight, b_f, etc.) are within the assumptions, so we can continue.

Compute I of the composite section (refer to Fig. E4-17c).

$$Ay = \Sigma M_{n.a.}$$
$$A = 46.52 + b'd'$$
$$= 46.52 + 0.254(215) = 101.13 \times 10^{-3} \text{ m}^2$$
$$101.13y = 54.61\left(\frac{916}{2} + 40 + \frac{215}{2}\right)$$
$$y = \frac{33\ 066}{101.13} = 327 \text{ mm}$$

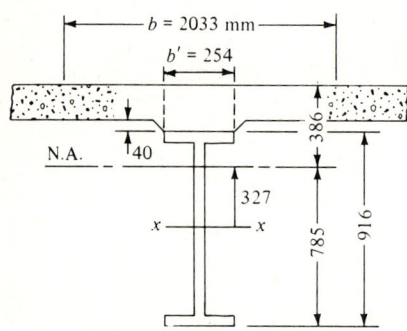

Figure E4-17c

which gives the dimensions shown in Fig. E4-17c.

$$I_{xx} = I_0 + \Sigma Ad^2$$
$$= 6701.3 + 46.52(0.327)^2 \times 10^3 + \frac{0.254(0.215)^3 \times 10^6}{12}$$
$$+ 54.61\left(0.131 + 0.040 + \frac{0.215}{2}\right)^2 \times 10^3$$
$$= 6701.3 + 4974.3 + 210.4 + 4235.7$$
$$= 16\ 121.7 \times 10^{-6} \text{ m}^4$$

The section modulus values are

$$S_{\text{top}} = \frac{16\ 121.7}{131 + 40 + 215} = 41.77 \times 10^{-3} \text{ m}^3$$
$$S_{\text{bot}} = \frac{16\ 121.7}{458 + 327} = 20.54 \times 10^{-3} \text{ m}^3$$

The stresses are:

Initial:
$$f_t = f_b = \frac{M_D}{S_x} = \frac{1129}{14.67} = 77.0 \text{ MPa}$$

DESIGN OF BEAMS FOR BENDING

Final:

$$f_{top} = \frac{M_L}{S_{top}} = \frac{1091}{41.77} = 26.1 \text{ MPa}$$

$$f_c = \frac{26.1}{8} = 3.26 \text{ MPa} \ll 0.4 f'_c \quad \text{O.K.}$$

$$f_{bot} = \frac{1091}{20.54} = 53.1 \text{ MPa}$$

The section stress profiles are shown in Fig. E4-17d. Check 100 percent overload stresses against 150 percent of allowable stresses:

$$f_c = \frac{2(1090.5)}{41.77} = 52.24 \text{ MPa}$$

$$f_{conc} = \frac{52.24}{8} = 6.5 \text{ MPa} < 1.5(0.4)f'_c \quad \text{O.K.}$$

$$f_t = \frac{2(1091)}{20.54} = 106.2 \text{ MPa}$$

$$f_{t(total)} = 77.0 + 106.2 = 183.2 \text{ MPa} < 1.5(0.55 F_y) \quad \text{also O.K.}$$

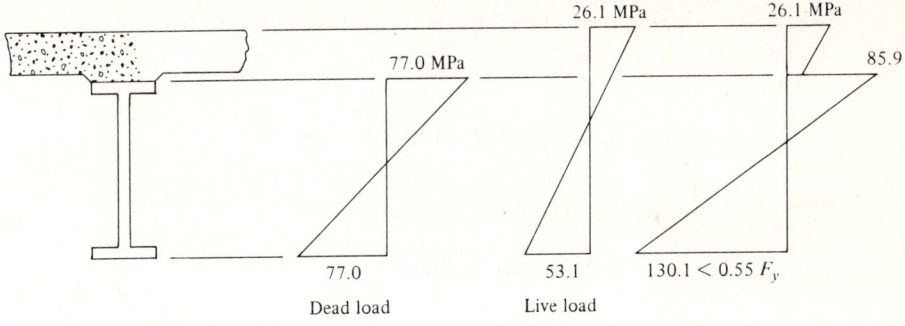

Figure E4-17d

Step 4. Design the shear connectors.
Use three 20-mm studs, as shown in Fig. E4-17e, with $L = 150$ mm. Thus the $L/d = 150/20 > 4$ is O.K. The studs must resist the smaller of

$$V_{hc} = \frac{0.85 f'_c b d'}{2} \quad \text{(neglect haunch)}$$

$$= \frac{0.85(27.6)(2.033)(0.215)}{2} = 5.13 \text{ MN}$$

$$= 5130 \text{ kN} \quad \text{controls (after computing } V_{hs}\text{)}$$

$$V_{hs} = \frac{A_s F_y}{2} = \frac{46.52(250)}{2} = 5815 \text{ kN}$$

206 STRUCTURAL STEEL DESIGN

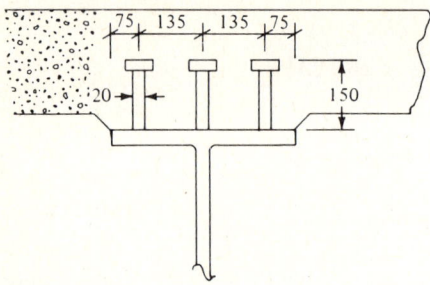

Figure E4-17e

The ultimate strength of the shear stud is

$$S_u = 0.4d^2\sqrt{f'_c E} \qquad E = 4740\sqrt{f'_c}$$
$$= 0.4(0.020)^2(27.6 \times 24\ 900)^{\frac{1}{2}} \text{ (w/SI adjustments)}$$
$$= 132.6 \text{ kN}$$

The minimum number of studs required from the end to the midspan (distance of $L = 22.4/2 = 11.2$ m) is

$$N_1 = \frac{5130}{132.6} = 38.7 \qquad \text{use 39 (multiples of 3)}$$

With studs in groups of three, there will be 13 groups. Now set the stud spacing based on fatigue.

The maximum range of live-load shear is 0 to 214.6 kN.

$$Z_r = Gd^2 = 0.0541(20)^2 = 21.6 \text{ kN}$$

$$S_r = \frac{V_r Q}{I}$$
$$= \frac{214.6(0.0541)(0.131 + 0.040 + 0.215/2)}{0.0161217}$$
$$= 200.6 \text{ kN/m}$$

$$\text{Pitch } p = \frac{nZ_r}{S_r} = \frac{3(21.6)}{200.6} = 0.323 \text{ m}$$

At $X = 3.5$ m from supports, the live-load shear plus impact is (refer to Fig. E4-17f)

$$22.4 R_a = 10.4(36) + 14.7(144) + 18.9(144) = 5212.8$$
$$R_a = \frac{5212.8}{22.4} = 232.7 \text{ kN}$$
$$V = 232.7(0.6065)(1.25) = 176.4 \text{ kN}$$

By proportion since $Q/I = $ constant,

$$p = 0.323\left(\frac{214.6}{176.4}\right) = 0.3929 \text{ m}$$

DESIGN OF BEAMS FOR BENDING **207**

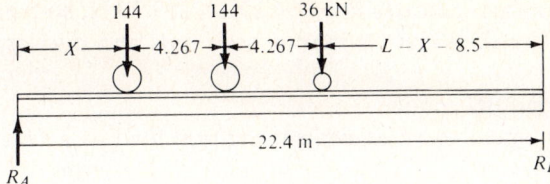

Figure E4-17f

At $X = 7$ m from supports, the live-load shear plus impact is

$$22.4 R_a = 4078.8$$
$$R_a = 182 \text{ kN}$$
$$V = 182(0.6065)(1.25) = 138 \text{ kN}$$
$$p = 0.323\left(\frac{214.6}{138}\right) = 0.5022 \text{ mm}$$
$$< 0.61 \text{ max. spacing allowed} \quad \text{O.K.}$$

Using the spacings above, the first 3.44 m requires 11 groups of studs, the next 3.43 m requires 9 groups, and the next 4.33 m requires 9 groups, for a total of 29 groups at 3 studs each, for a total of 87 studs vs. the 39 required for ultimate shear considerations.

Step 5. Design the diaphragms.

Lateral load:

Wind at 24 kPa × exposed surface area and, referring to Fig. E4-17a: $H = 460 + 225 + 270 + 916$ mm $= 1.871$ m

Curb load at 500 lb/ft for side opposite	$= 7.3$ kN/m
Wind $= 24(1.871)$	$=44.9$ kN/m
Total	$=52.2$ kN/m

Note: there is no wind load from the truck because the deck acts as a diaphragm.

With diaphragms spaced at 5.55 m,

$$P_{\text{diaph}} = 52.2(5.55) = 289.7 \text{ kN} \quad \text{(interior diaphragm)}$$
$$= \frac{289.7}{2} = 144.9 \text{ kN} \quad \text{(exterior diaphragms)}$$

Use all same-size diaphragms; also, the diaphragms must be at least one-third and preferably one-half of the girder depth. Arbitrarily select

$$W530 \times 65.5/0.64: \quad d = 525 \text{ mm} > \frac{916}{2} \quad \text{O.K.}$$
$$t_w = 8.9 \text{ mm} > 8.0 \text{ mm} \quad \text{O.K.}$$
$$A = 8.39 \times 10^{-3} \text{ m}^2$$
$$r_y = 32 \text{ mm} \quad L = 2.033 \text{ m}$$

A check (not shown) as a column indicates $P = 994$ kN $\gg 289.7$ kN, so the section is amply adequate. Use the same section for both ends and interior points.

Step 6. Check the live-load deflection.

Convert the live-load moment to an equivalent uniform load as a first approximation.

$$\frac{wL^2}{8} = 1090.5 \text{ kN} \cdot \text{m}$$

$$w' = \frac{1090.5(8)}{22.4^2} = 17.4 \text{ kN/m} = 0.0174 \text{ MN/m}$$

$$\Delta = \frac{5(0.0174)(22.4)^4}{384(200\ 000)(0.0161217)} = 0.01769 \text{ m}$$

Taking the maximum allowable deflection as

$$\frac{L}{1000} = \frac{22\ 400}{1000} = 22.4 \text{ mm} > 17.6 \text{ mm} \quad \text{O.K.}$$

If the designer assumes that the actual moment diagram resulting from three wheel loads is sufficiently close to a uniform load diagram, the deflection as computed is adequate; otherwise, a more exact analysis should be made.

///

4-14 BEAM DESIGN USING LOAD RESISTANCE FACTOR DESIGN (LRFD)

The design of a beam for bending using LRFD is relatively straightforward. It is necessary to have separate values for dead and live loads. The allowable bending stress is taken as

$$F_b = \phi F_y \qquad \phi = 0.86$$

$$M_u = \phi F_y Z$$

$$V_u = \phi \left(\frac{F_y}{\sqrt{3}} \right) dt_w$$

and M_u is computed using factored dead and live loads. This is illustrated in the following example.

Example 4-18 Given the floor system shown in Fig. E4-18 for an office building using simple framing. Take the loads as dead = 75 psf and live = 80 psf; use LRFD design and A-36 steel.

Make a preliminary design selection of a W shape with depth not a factor.

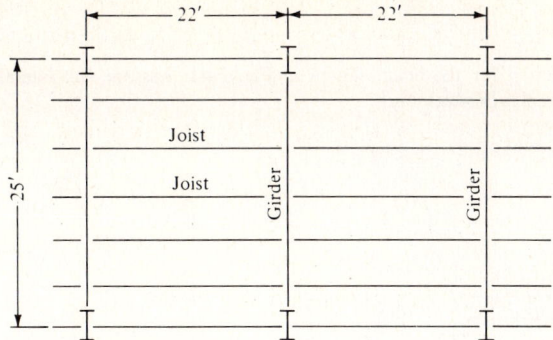

Figure E4-18

SOLUTION Since the girder is symmetrical on the floor plan and the joists are closely spaced, assume that the loads will give a uniform girder loading.

$$w_u = \psi(F_d D + F_L L)$$

Using load factors given in Sec. 3-7 (and Table 3-1) and noting that there is no live load reduction for floor area since the load factors are themselves statistical terms, we obtain

$$w_u = 1.1[1.1(22 \times 0.075) + 1.4(22 \times 0.080)]$$
$$= 1.1(1.815 + 2.464) = 4.71 \text{ kips/ft}$$

The design moment is

$$M_u = \frac{w_u L^2}{8} = \frac{4.71(25)^2}{8} = 367.97 \text{ ft} \cdot \text{kips}$$

$$M_u = \phi F_y Z \quad (\phi = 0.86 \text{ from Table 3-1})$$

$$Z = \frac{M_u}{\phi F_y} = \frac{367.97(12)}{0.86 \times 36} = 142.6 \text{ in}^3$$

From Table II-2 of SSDD, obtain a W21 × 62 with $Z = 144.4$ in^3, $d = 20.99$ in, and $t_w = 0.0400$ in.

Check the beam weight:

$$\Delta w_u = 1.1(1.1)(0.062) = 0.075 \text{ kip/ft}$$

By proportion,

$$\Delta Z = \frac{\Delta w_u Z}{w_u} = \frac{0.075(142.6)}{4.71} = 2.27 \text{ in}^3 \quad \text{O.K.}$$

Check the shear:

$$V_u = \frac{w_u L}{2} = \frac{4.71(25)}{2} = 58.9 \text{ kips}$$

$$= \phi \frac{F_y}{\sqrt{3}} dt_w = 0.86 \frac{36}{\sqrt{3}}(20.99)(0.40) = 180.3 \gg 58.9 \text{ kips} \quad \text{O.K.}$$

Use a W21 × 62 section. ///

PROBLEMS

4-1 Select the most economical W section for the beam shown in Fig. P4-1. Assume full lateral support. Use the AISC specifications and A-36 steel.
Answer: W24 × 76.

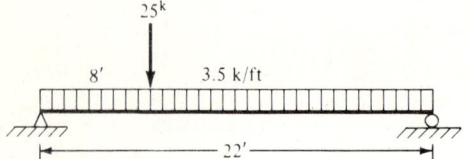

Figure P4-1

4-2 Select the most economical W section for the beam shown in Fig. P4-2. Assume full lateral support. Use the AISC specifications and A-36 steel.
Answer: W610 × 113.1.

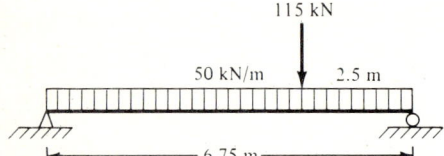

Figure P4-2

4-3 Select the most economical W section for the beam shown in Fig. P4-3. Use the AISC specifications and assume that the beam is adequately supported laterally between A and B but between B and C is supported only at the ends. Use $F_y = 50$ ksi.
Answer: W12 × 26.

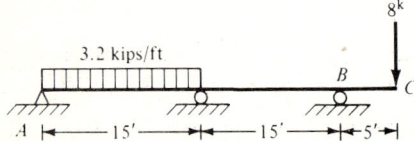

Figure P4-3

4-4 Select the most economical W section for the beam of Fig. P4-4. Use the AISC specifications and assume that the beam is adequately supported laterally between A and B, but B to C is supported only at the ends. Use $F_y = 345$ MPa.

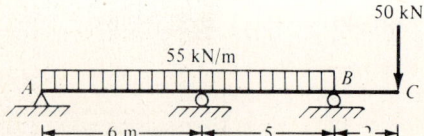

Figure P4-4

DESIGN OF BEAMS FOR BENDING **211**

4-5 A monorail hoist + trolley (Fig. P4-5) weighs 2050 lb with $S = 2.5$ ft and has a lifting capacity of $P = 12$ kips. It is designed to run along the bottom flange of a standard S shape. Design the section if the span is 20 ft, using the AISC specifications. Note that you should start with A-36 steel and use a greater F_y only if necessary.
   Answer: S18 × 54.7.

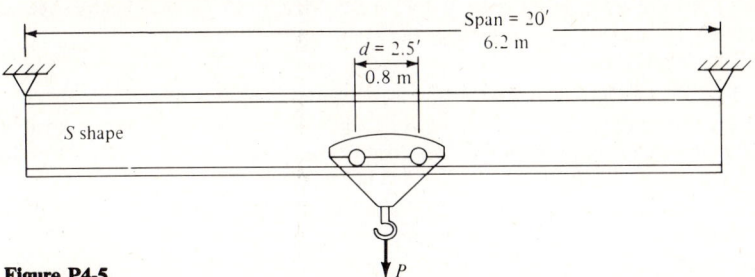

Figure P4-5

4-6 Referring to Fig. P4-5, make the design in SI units if $d = 0.8$ m, the trolley and hoist weigh 9.12 kN, the lift capacity $P = 50$ kN, and the span is 6.2 m. Use the AISC specifications and $F_y = 250$ MPa.

   For Probs. 4-7 to 4-18, refer to Fig. P4-7.

4-7 Make a tentative design for girders A1 and B1 using A-36 steel. Use fixed-end moments of $wL^2/12$ for the interior end and $wL^2/14$ for the exterior end. Take the wall as carried by A1 at 0.4 kip/ft (see Prob. 4-9).
   Answer: W18 × 65 for B1.

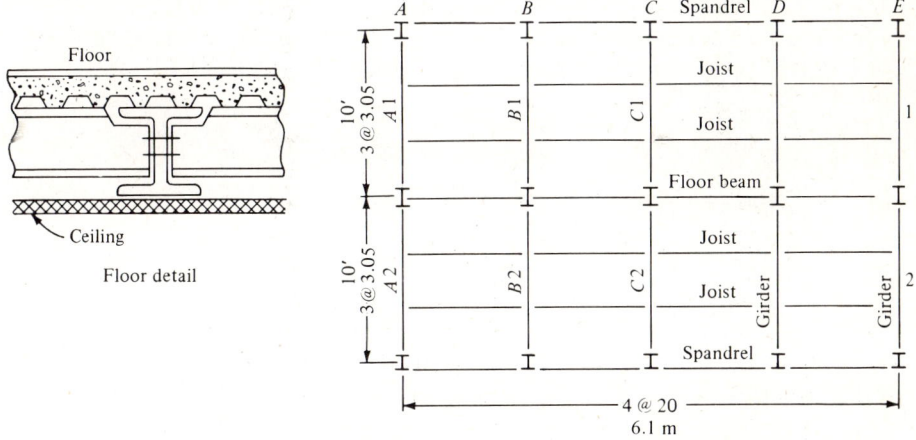

Other data: Floor + finish + metal deck = 80 psf   3.77 kPa
                        Ceiling = 10 psf   0.19 kPa
                        Joists (est.) = 3 psf   0.07 kPa
                        Live load = 80 psf   4.0 kPa

Live load reduction may be used as appropriate
Use Fixed-end-moments = $wL^2/10$ exterior; $wL^2/12$ interior
Limit approximate deflections to $L/360$
Limit d to 18 in or 460 mm (nominal depth)

Figure P4-7

4-8 Design the floor joists (beams) assuming simple supports and A-36 steel.

4-9 Design the spandrel beam AB assuming brick veneer (one brick thick) and a wall height to be carried by a spandrel of 8.5 ft. Assume that the metal deck is welded to the spandrel flange so that no twisting occurs due to the eccentricity of the brick. Use A-36 steel.

4-10 Design the floor beams spanning between columns assuming simple supports and A-36 steel.

4-11 Design girders A1 and B1 using SI units and $F_y = 345$ MPa. Take the wall load as carried by A1 at 5.90 kN/m. Use exterior end moments of $wL^2/14$ and $wL^2/12$ interior.
Answer: W410 × 59.8 for A1.

4-12 Design the floor joists (beams) assuming simple supports and $F_y = 345$ MPa.

4-13 Design spandrel AB assuming brick veneer (one brick thick) and a wall height to be carried by a spandrel of 2.60 m. Assume that the metal deck is welded to the spandrel flange so that no twisting occurs due to the eccentricity of the brick. Use $F_y = 345$ MPa.
Answer: W410 × 46.1.

4-14 Design the floor beams spanning between columns assuming supports and $F_y = 345$ MPa.
Answer: W410 × 46.1.

4-15 Do Prob. 4-8 using unshored composite construction. Use $f'_c = 3000$ psi.

4-16 Do Prob. 4-12 using composite unshored construction but use $F_y = 250$ MPa steel (instead of $F_y = 345$ MPa steel; why?). Use $f'_c = 21$ MPa.

4-17 Redo Example 4-16 using shored construction instead of the "unshored" construction of the example.

4-18 Redo Example 4-16 using a W18 × 46 and see if a workable solution can be obtained.

For Probs. 4-19 to 4-22, refer to the cross section shown in Fig. P4-19. The bridge span will be assigned by the instructor (36 to 46 ft or 11 to 14 m). If not assigned, use 40 ft or 12.5 m for the span. For exterior beams the maximum possible load due to truck is one-half of the truck load. The dead load is the cantilevered part plus the prorata part of the deck slab to adjacent interior stringer. Use an HS 20 truck and A-36 or $f_y = 250$ MPa steel.

4-19 Design the floor stringers noting full lateral support for the compression flanges and using A-36 steel. Use the AASHTO specifications.
Answer: W33 × 130.

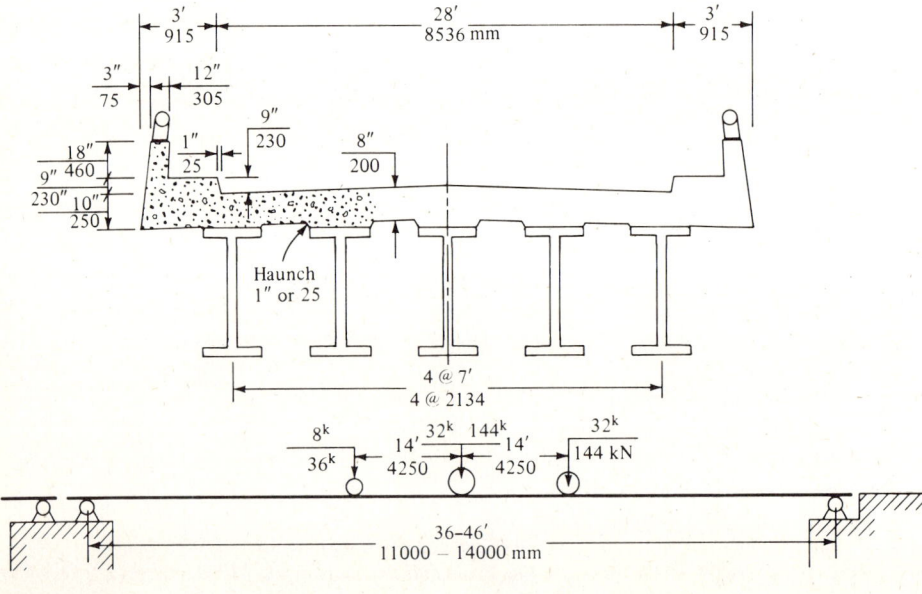

Figure P4-19

4-20 Design the floor stringers, noting full lateral support for the compression flange and using $F_y = 250$ MPa. Use the AASHTO specifications.

4-21 Design the interior floor stringers of Fig. P4-19 using composite design and unshored construction. Limit deflection under live load to $L/800$. Use $f'_c = 4$ ksi.

4-22 Design the interior floor stringers of Fig. P4-19 using composite design and unshored construction. Limit live-load deflections to $L/800$. Use $f'_c = 28$ MPa.

Following are miscellaneous beam problems for laterally unsupported spans and for shear considerations.

4-23 A column load of 160 kips is carried across an open work area as in Fig. P4-23. The laterally unsupported length is 39.5 ft. Select the lightest W shape with the deflection limited to $L/360$. Use any grade of steel if A-36 is not adequate. Note that you must assume your end conditions of fixity.
Answer: W36 × 300. (Simply supported.)

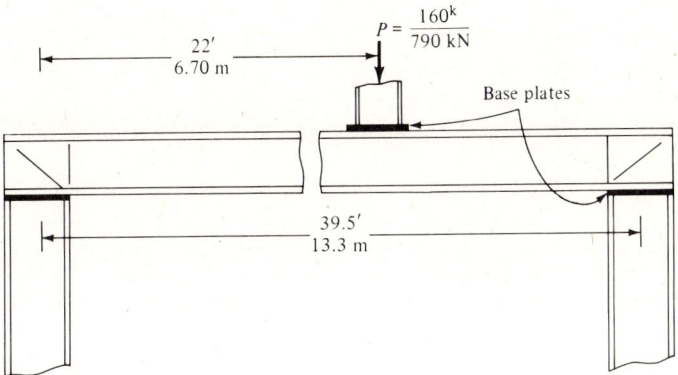

Figure P4-23

4-24 Redo Prob. 4-23 with $P = 790$ kN and the metric dimensions shown on Fig. P4-23.

4-25 If the column load in Prob. 4-23 is carried by a W8 × 48, what length of bearing distance N of Fig. 4-6 is required for the beam selected in Prob. 4-23?
Answer: $N = 8.50$ in.

4-26 If the beam selected in Prob. 4-24 supports the column load of 790 kN carried by a W200 × 86.3 column section, what length N of Fig. 4-6 is required?
Answer: $N = 222$ mm.

4-27 A W section is to be used in a span of 10 ft to carry a midspan concentrated load of 650 kips. Select the section, check for shear stresses, and compute the N values for both the reaction and the concentrated load, using A-36 steel and AISC specifications.

4-28 A W section is to be used in a span of 3.5 m to carry a midspan concentrated load of 2100 kN. Select the section, check for shear stresses, and compute the N values for both the reaction and the concentrated load using $F_y = 250$ MPa steel and the AISC specifications.
Answer: W840 × 299.1, $N = 260$ mm for reaction.

4-29 Redo Example 4-18 if $\phi = 0.90$ and $F_L = 1.5$.

4-30 Redo Example 4-18 if $F_y = 50$ ksi.
Answer: W21 × 50.

4-31 Redo Example 4-18 if $D = 3.75$ kPa, $L = 4.0$ kPa, $L = 7.75$ m (bay), and the joist length = 6.75 m. Use $F_y = 250$ MPa and LRFD with factors suggested in Sec. 3-7.
Answer: W530 × 92.3.

4-32 Redo Example 4-10 for the lightest W or M section.
Answer: W10 × 22 or M10 × 22.9.

4-33 Design the roof purlins for Prob. 5-20 (of the next chapter) using the lightest rolled W or C section and with a sag rod at midspan. Use the AISC specifications and A-36 steel.
Answer: W14 × 43.

4-34 Obtain the lightest W section for a beam span of 45 ft with two 20-kip concentrated loads at 15 ft from each end. Lateral support is available only at the ends and concentrated loads. Use A-36 steel and the AISC specifications.
Answer: W27 × 84.

4-35 Obtain the lightest W section for a 30-kips/ft uniform load on a 15-ft span. Also find the reaction distance N. Use A-36 steel and the AISC specifications.

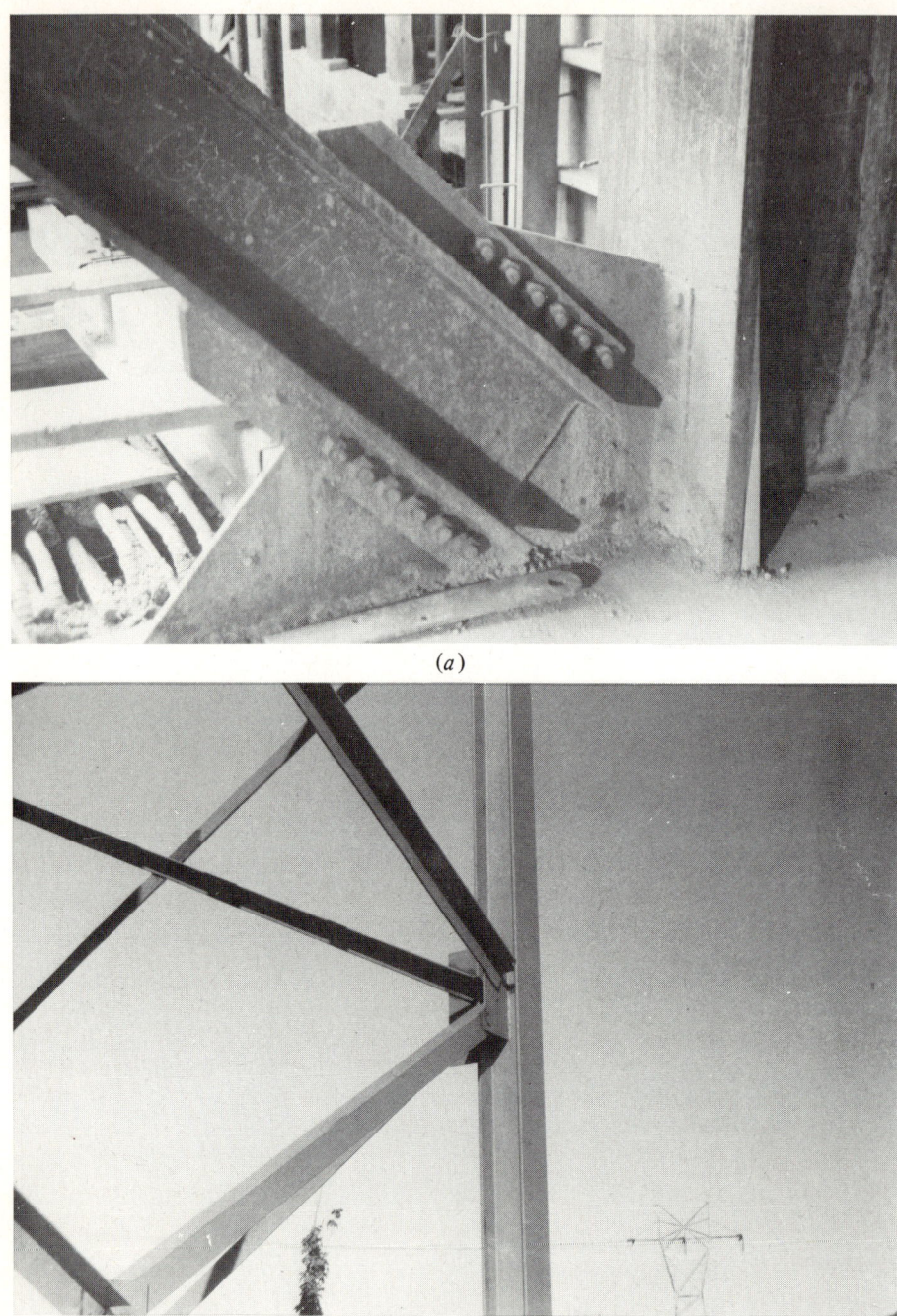

Figure V-1 Tension members. (*a*) Tension member in industrial building using high-strength bolts. (*b*) Tension member in small industrial building for wind. Erection bolt can be seen in lower member. Joint is welded.

CHAPTER
FIVE

DESIGN OF TENSION MEMBERS

5-1 TYPES OF TENSION MEMBERS

Those elements of a structure that carry tension loads are termed *tension members*. The bottom chords of roof and bridge trusses are classic examples of tension members. Certain of the truss web members may carry tension or may be a tension member for certain loading conditions and a compression member for other loading conditions.

Steel cables as used in suspension bridges and in cable-supported roofs are examples of steel tension members. Cables are also used to guy tall communication towers as well as power line poles where alignment changes occur.

Wind bracing in an X configuration is frequently used where the members are so flexible that "buckling" takes place under compression stresses developed by wind in one direction but functions as a tension member for the 180° wind. Other locations for tension members include selected web members of power transmission and communication towers, hangers for stair wells, elevator cables, hanger supports for curtain walls, and parts of hoisting equipment.

High-strength steel tie rods are used to strengthen existing structures by attaching to the bottom flange or chord of the structure and pretensioning to induce a compressive stress in the member which must then be overcome when a load is applied. Tie rods are sometimes used between the reactions of an arch or rigid frame to aid in resisting lateral base separation.

Considerable use is made of cables in cable-supported roof construction to produce an esthetically pleasing roof over a large, unobstructed floor area with a minimum amount of steel and other structural material required.

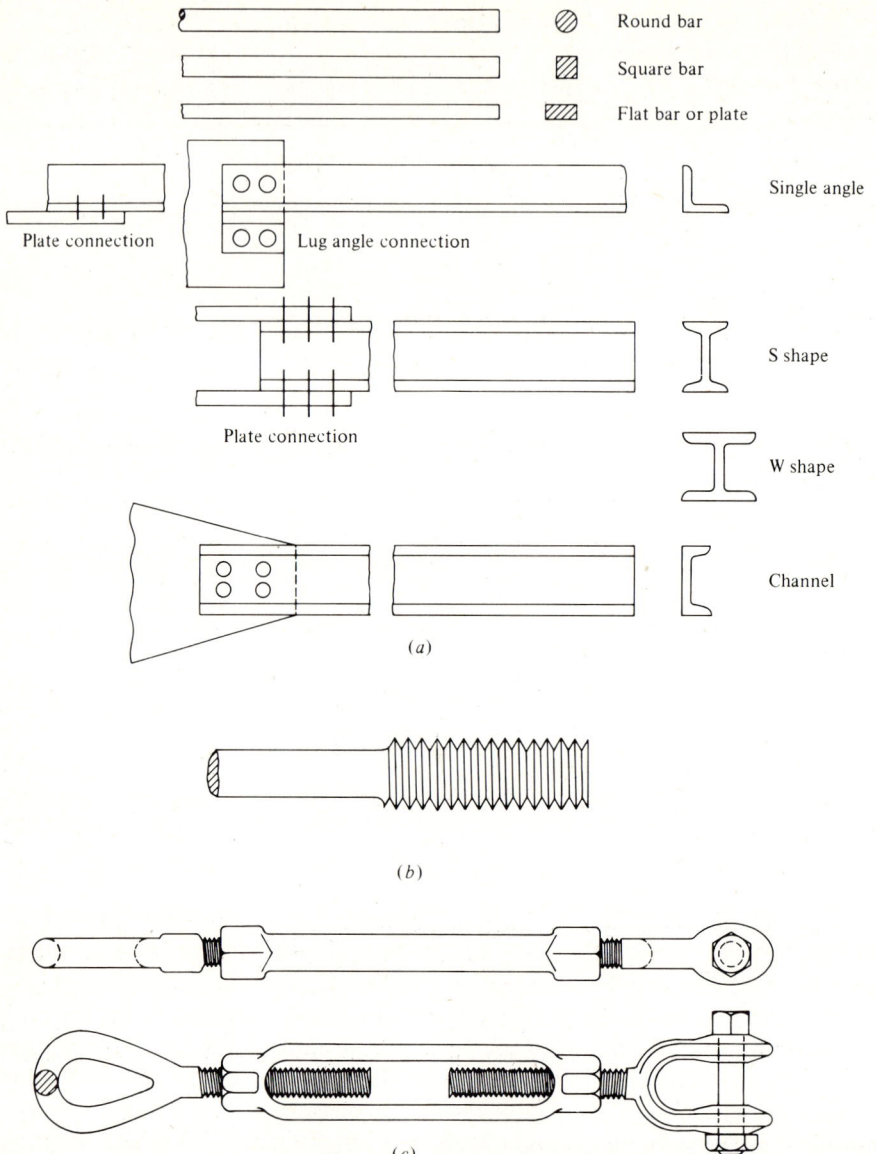

Figure 5-1 Tension members. See Fig. 5-2 for cables used as tension members. (*a*) Structural shapes used for tension members. (*b*) Upset bar. (*c*) Threaded bar and use of a turnbuckle to adjust bar length. Applicable for square and round bars.

In all these uses the tensile strength of the steel is used. In this stress configuration plate buckling or warping is not a consideration. In some instances, however, specifications will require a minimum amount of member stiffness for esthetic and safety reasons.

Generally, tension members may be categorized as rods and bars, rolled structural shapes, built-up members, and wires or cables. Several of these members are illustrated in Figs. 5-1 and following.

5-2 ALLOWABLE TENSION STRESSES

The AISC allowable tension stress of members, except eyebars, is limited to

$$F_t = 0.6 F_y \quad (\text{gross section area})$$
$$F_t = 0.5 F_u \quad (\text{net section area}) \tag{5-1}$$

The AASHTO and AREA allowable tension stress is somewhat more conservative, at

$$F_t = 0.55 F_y \tag{5-2}$$

The AASHTO specification further limits this basic stress to the lesser value of either Eq. (5-2) or

$$F_t = 0.46 F_u$$

but the *net* section is used for both these equations. For steel with F_y not over 80 ksi the basic tensile steel stress is governed by Eq. (5-2) for AASHTO design.

On the net section across the pin hole of an eyebar (see Fig. 5-3), the allowable AISC stress is

$$F_t = 0.45 F_y \tag{5-3}$$

Allowable tensile stresses for several steel grades are shown in Table 5-1, where the reader should note that established practice allows rounding of the values for A-36 steel to the values shown for both AISC and AASHTO/AREA.

In all cases, except eyebars, the tension stresses must be computed based on both the gross and net cross-sectional area when using AISC specifications. Only the net area is required for AASHTO and AREA specifications. The net area is the gross (total) area where welded connections are used. The net area is the least effective cross-sectional area for all other cases as where bolt or rivet holes are used for mechanical fasteners at the ends or where holes and/or area reductions occur along the member.

The effective net area at approximately the root of the thread of threaded tension members using the AISC specification is

$$A_e = 0.7854 \left(D - \frac{0.9743}{n} \right)^2 \tag{5-4}$$

where D = nominal outside diameter of threads
n = number of threads/in (or the SI equivalent of $n/25.4$)

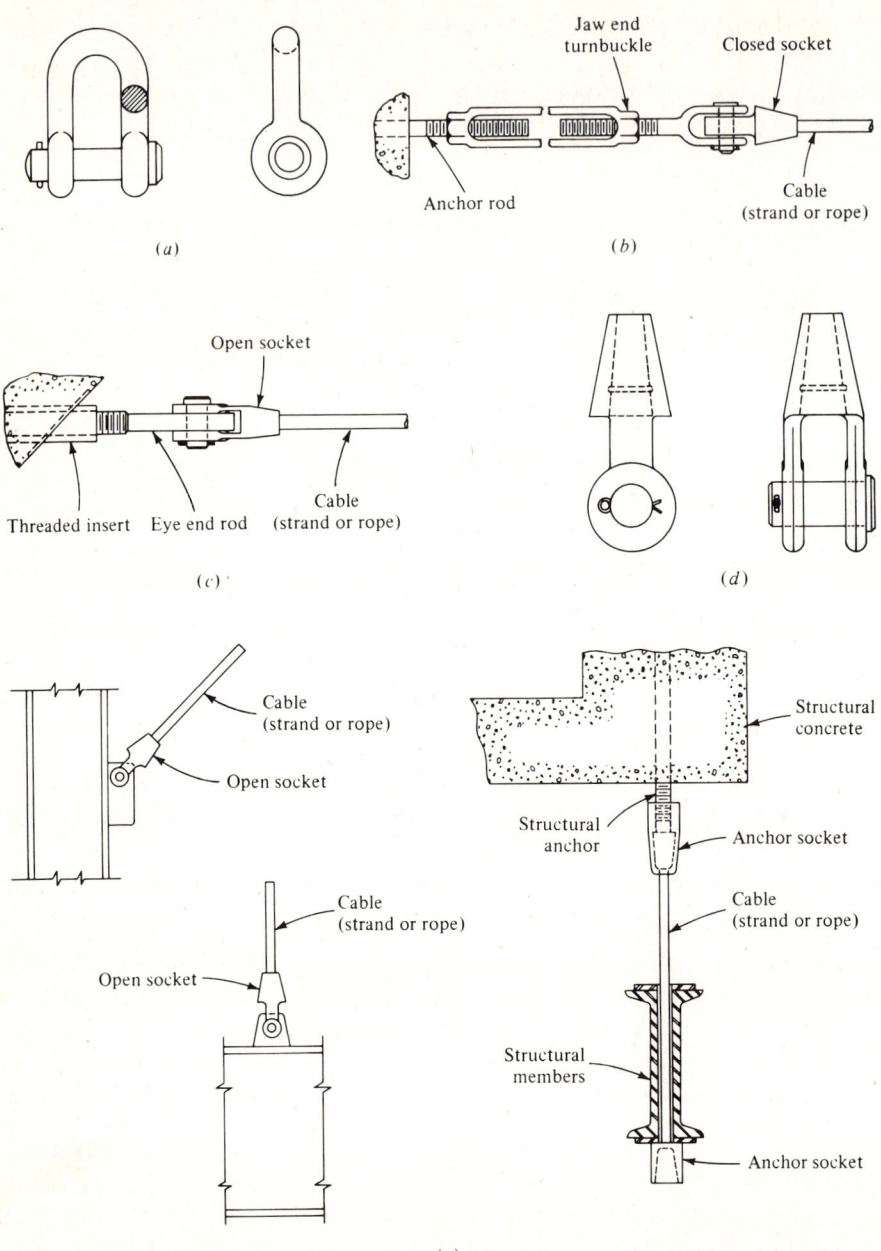

Figure 5-2 Cable tension members and attachment devices. Note that in many of these situations one may use a round or square bar. (*a*) Clevis. (*b*) Cable socket and turnbuckle. (*c*) Method of using cable socket with structural anchorage. (*d*) Socket details. (*e*) Three situations using cables for the tension member.

Table 5-1 Allowable tensile stresses for specifications and equations shown

		AISC					AASHTO and AREA			
		Not at pinholes[a]		At pinholes[b]		Net section		Net section		
F_y		F_u		$0.6F_y$		$0.45F_y$		$0.5F_u$	$0.55F_y$	$0.46F_u$
ksi	MPa	ksi	MPa	ksi	MPa	ksi	MPa	ksi	MPa	ksi	MPa
36	250	58	400	22.0	150	16.2	112	29	200	20	138	26.7	184
42	290	63[c]	435	25.2	174	18.9	130	31.5	218	23.1	160	29.0	200
46	315	67[c]	460	27.6	189	20.7	142	33.5	230	25.3	173	30.8	212
50	345	70[c]	485	30.0	207	22.5	155	35.0	242	27.5	190	32.2	223
60	415	75[d]	520	36.0	216	27.0	187	37.5	260	33.0	198	34.5	239
65	450	80[d]	550	39.0	270	29.3	202	40.0	275	35.8	248	36.8	253

[a] On gross section not at pinholes.
[b] On gross section at pinholes.
[c] F_u varies with ASTM designation; values shown are for A-588 steel.
[d] For A-572 steel.

The thread root location will be the critical area for standard bars that are threaded and effectively produces considerable wasted material. This can be avoided by the use of upset-ended bars (Fig. 5-1b), where the threaded ends are forged to a larger diameter prior to threading so that application of Eq. (5-4) produces $A_e \geq A_{bar}$. There is not a total savings, however, since there is an extra fabrication charge for producing an upset end.

5-3 GENERAL DESIGN CONSIDERATIONS

The design of tension members requires consideration of several factors, including connections, types of shapes available or required, and shear flow. These will be considered in the following sections.

5-3.1 Connections

A principal factor is how to affect a connection of the tension member to the remainder of the structural system (i.e., other members or to the foundation). Figures 5-1, 5-2, and 5-4 illustrate several tension member connections. The simplest connection is the pin-connected eyebar, but this type of member has had a poor service history and requires special dimensioning, as shown in Fig. 5-3. Next in simplicity would be some kind of threaded bar or cable. Here

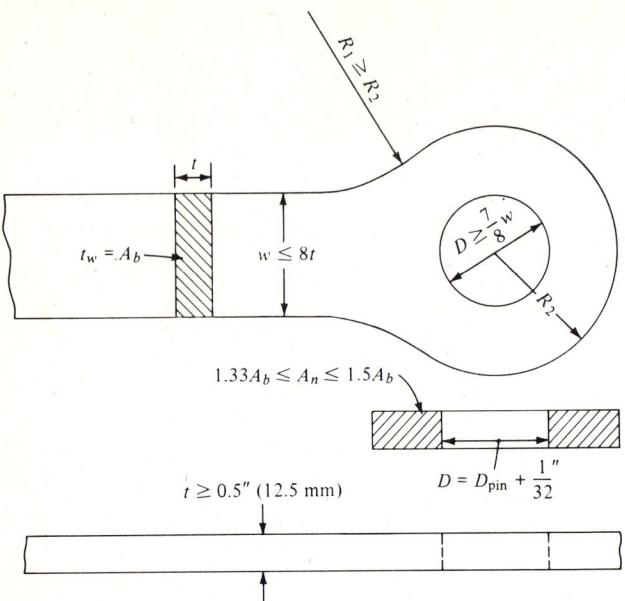

Figure 5-3 AISC eyebar dimensions.

several problems develop, including the area in tension for the bar [Eq. (5-4)] and fitting the member into the structure. This problem is usually solved by use of a turnbuckle or by having an extra threaded distance on one end to take up the slack. Bars and cables used as tension members are generally given a small initial tension when installed to eliminate any tendency to sag and to avoid rattling when the structure vibrates under service loads. The initial tension is sometimes useful in "tightening" up the remainder of the structure.

Cables may be strands of wire rope, with the terms "bridge strand" or "bridge rope" being used to specify the structural quality of the cables. Single wires are not used in structural applications; rather, strands that are arrangements of 7 to 61 or more single wires are wrapped around a central core wire to produce a symmetrical section. Wire rope is produced by laying several strands helically around a wire core. Commonly wire rope consists of 6 to 37 strands. Table 5-2 gives design data for selected sizes of both bridge strand and wire rope.

Cable connections commonly take the form of one of the configurations illustrated in Fig. 5-2. In the connections shown, the end of the cable is carefully cleaned and then fed through the opening in the connector. The cable end is then broomed (strands separated somewhat), carefully cleaned, and molten zinc at about 850°F is poured into the wire matrix. After the zinc cools, and the connection is again cleaned and assembled, the member is ready to be installed. This method of attaching the end connector produces a joint at least as strong as the cable. Occasionally (but not shown), the cable can be inserted into a longer connector which is squeezed (termed *swaging*) to produce a friction connection.

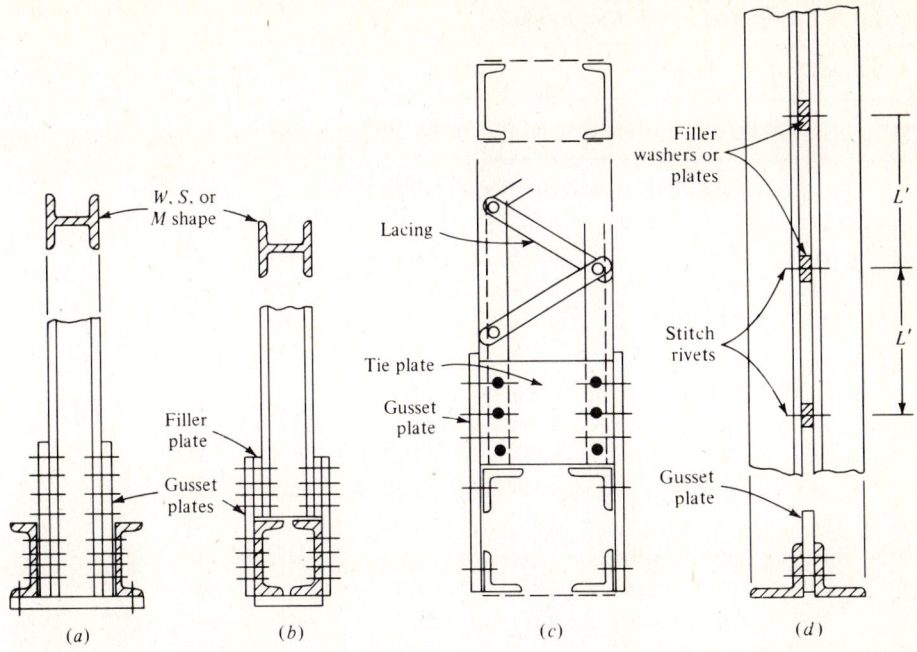

Figure 5-4 Tension member connections. (a) Web member dimension controls spacing of channels for chord. (b) Use of filler plates to make connection. (c) Framing of a built-up member to a built-up member using gusset plates. (d) Pair of angles with gusset plate. Generally requires stitch rivets (or bolts) to maintain L'/r_e between stitch bolts $\leq L/r_m$ of entire member. Note necessity of using r_z of element to compute L'/r between stitch bolts.

The modulus of elasticity of strands and ropes is needed for computation of elongation under stress. The modulus of elasticity of wire rope is not the same as for the basic high-strength steel used in the manufacture of the individual strands. The modulus of elasticity is determined after forming the strand or rope and stretching it so that the component parts are fitted together.

The modulus of elasticity of bridge strand and rope may be taken as follows:

Diameter or class		E	
Bridge strand (diameter)			
fps, in	SI, mm	fps, ksi	SI, MPa
$\frac{1}{2}$ to $2\frac{9}{16}$	12.5 to 65	24 000	165 000
$2\frac{5}{8}$	67 (and larger)	23 000	158 000
Bridge rope (class)			
6 × 7 (6 strands with 7 wires/strand)		20 000	138 000
6 × 19 (6 strands with 19 wires/strand)		20 000	138 000
6 × 37 (6 strands with 37 wires/strand)		20 000	138 000

Table 5-2 Selected cable design data

Diameter		Weight		Area		P_u	
in	mm	lb/ft	kN/m	in²	m² (× 10⁻³)	kips	kN
Bridge Strand (single strand with multiple wires)							
$\frac{1}{2}$	13	0.52	0.008	0.15	0.0968	29.0	129
$\frac{5}{8}$	16	0.82	0.012	0.234	0.1510	46.6	207
$\frac{11}{16}$	18	0.99	0.014	0.284	0.1832	56.2	250
$\frac{3}{4}$	19	1.18	0.017	0.338	0.2181	66.0	294
$\frac{7}{8}$	22	1.61	0.023	0.459	0.2961	89.2	397
1	25	2.10	0.031	0.600	0.3871	118.4	527
$1\frac{1}{8}$	28	2.66	0.039	0.759	0.4897	151.4	673
$1\frac{1}{4}$	32	3.28	0.048	0.938	0.6052	188.2	837
$1\frac{3}{8}$	35	3.97	0.058	1.130	0.7290	228.0	1014
$1\frac{1}{2}$	38	4.73	0.069	1.350	0.8710	270.0	1201
$1\frac{5}{8}$	41	5.55	0.081	1.59	1.0258	318.0	1414
$1\frac{3}{4}$	44	6.43	0.094	1.84	1.1871	368.0	1637
$1\frac{7}{8}$	48	7.39	0.108	2.11	1.3613	424.0	1886
2	50	8.40	0.123	2.40	1.5484	482.0	2144
$2\frac{1}{2}$	64	12.8	0.187	3.75	2.4194	740.0	3292
4	100	33.6	0.490	9.60	6.1935	1822.0	8105
Bridge Rope [6 × 7 (6 strands of 7 wires/strand)]							
$\frac{3}{8}$	10	0.24	0.004	0.065	0.0419	13.0	58
$\frac{1}{2}$	13	0.42	0.006	0.119	0.0768	23.0	102
$\frac{5}{8}$	16	0.65	0.009	0.182	0.1174	36.0	160
$\frac{3}{4}$	19	0.95	0.014	0.268	0.1729	52.0	231
$\frac{7}{8}$	22	1.28	0.019	0.361	0.2329	70.0	311
1	25	1.67	0.024	0.471	0.3039	91.4	407
Bridge Rope [6 × 19 (6 strands of 19 wires/strand)]							
$1\frac{3}{4}$	44	5.24	0.076	1.47	0.9484	286.0	1272
2	50	6.85	0.100	1.92	1.2387	372.0	1655
$2\frac{1}{2}$	64	10.60	0.155	2.97	1.9161	576.0	2562
Bridge rope [6 × 37 (6 strands of 37 wires/strand)]							
3	75	15.1	0.220	4.25	2.7419	824.0	3665
$3\frac{1}{2}$	90	21.0	0.306	5.83	3.7613	1110.0	4938
4	100	27.0	0.394	7.56	4.8774	1460.0	6494

High-strength steel (F_y on the order of 160 ksi) is used in the manufacture of bridge strand and rope. The manufacturer guarantees the breaking strength (P_u given in Table 5-2), but it is up to the designer to specify the safety factor. The SF should not be less than 1.7 and may be as high as 3.0, depending on the project and design uncertainties.

5-3.2 Shear Flow and Maximum Effective Cross-sectional Areas

Figure 5-5 qualitatively illustrates the problem of shear flow or shear lag as load is transferred to gusset plates from the main member. The problem is even more critical when single angles are used and only one leg is connected to transfer the load. In this case lug angles are sometimes used (Fig. 5-1a). The AISC specifications (Sec. 1-14.2) consider this problem as follows.

Any case where tension members have a profile with not all the segments in a common plane (as the outstanding leg of an angle, webs of beams with flange connections, flanges of channels when the web is connected, and similarly for built-up members) will have an effective area A_e computed as follows:

1. Equal or unequal leg angles connected by one leg to a plate (singly or in pairs), $A_e = 0.90 A_n$.
2. For W, M, or S shapes with $b_f \geq 0.67d$ or structural tees cut from these shapes and fastened by the flange only, $A_e = 0.90 A_n$.
3. For all other shapes, including built-up shapes, with at least three fasteners in a line (Fig. 5-5 has five fasteners in line), $A_e = 0.85 A_n$.
4. Any tension members with only two fasteners in a line, $A_e = 0.75 A_n$.

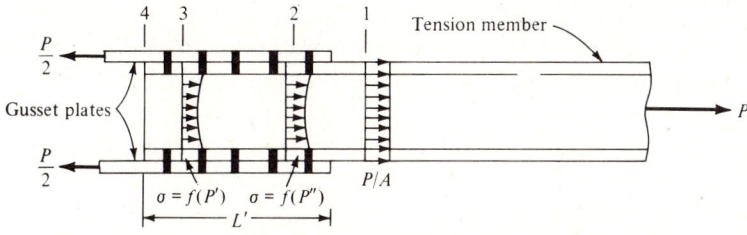

At point 1: Uniform stress distribution assumed in analysis.
        2: Some load transferred to gusset plates leaving P'. Load transfer is at flanges which results in qualitative stresses across section shown. Interior web stresses will be lower than P/A, due to shear lag. Thus, in long joints the web may tear due to large differential strains resulting in a progressive tension failure at the forward end of joint.
        3: Additional load transferred to gusset plates leaving P''.
        4: All load transferred to gusset plates and tension member stresses are zero.

Figure 5-5 Shear flow at the joint of a tension member during load transfer to gusset plates.

We note that the effective area A_e is always less than the net area when shear lag is to be considered.

Tests on large numbers of fabricated tension connections have shown that welded connections may be close to 100 percent efficient ($A_e \to A_n$). No connections using mechanical fasteners were found to be much over 90 percent efficient, so to provide an adequate statistical coverage of the tests, the AISC specifications limit the efficiency of any connection using mechanical fasteners to a maximum net cross-sectional area of

$$A_n \leq 0.85 A_g \tag{5-5}$$

With welded connections without holes, the net area A_n is equal to the total or gross section area A_g. The author suggests that

$$A_e = 0.90 A_n$$

be used for tension members that are continuous through the joint, to compute the effective tension area for double angles, channels, and structural tees used as bottom chords of trusses and continuous across several panels.

5-3.3 Use of Structural Shapes as Tension Members

The use of a particular rolled or built-up shape will be dictated by the remainder of the structure and, of course, the axial load to be resisted. When the member frames into a joint that has sufficient width, the use of W, M, or C shapes may be desirable, as illustrated in Fig. 5-4. The joint length L' will have to be sufficient that the tensile stress in the full cross section can be transferred into the joint via "shear flow," as qualitatively illustrated in Fig. 5-5.

More commonly, tension members are either single or, preferably, double angles, as shown in Fig. 5-6. The use of double angles is generally preferred because the joint will be more symmetrical both in and out of plane, as opposed to using a single angle, which will always have an out-of-plane eccentricity. The reader should note the relative ease of joint fabrication using angles via use of the stem of the WT in Fig. 5-6a and the use of a single gusset plate in Fig. 5-6b as opposed to the connections of Fig. 5-4. Somewhat offsetting this is the recommended use of spacer plates with bolts or welds at 1.5- to 2-m intervals along the pair of angles, to reduce vibration noise.

A decision to use an S versus the W or M shape requires taking into account the fact that W shapes are better balanced with respect to rigidity (r_x and r_y) about both axes than are the S shapes. The actual and nominal depths of S shapes are the same, whereas for the W and M shapes, the nominal and actual depths differ so that the connections shown in Fig. 5-4 will generally require spacer plates to build the width so that the connection can be fabricated.

5-3.4 Use of Built-up Tension Members

Built-up members are required where use of a single or a pair of angles, or any one of the standard rolled shapes, either does not have a sufficient cross section

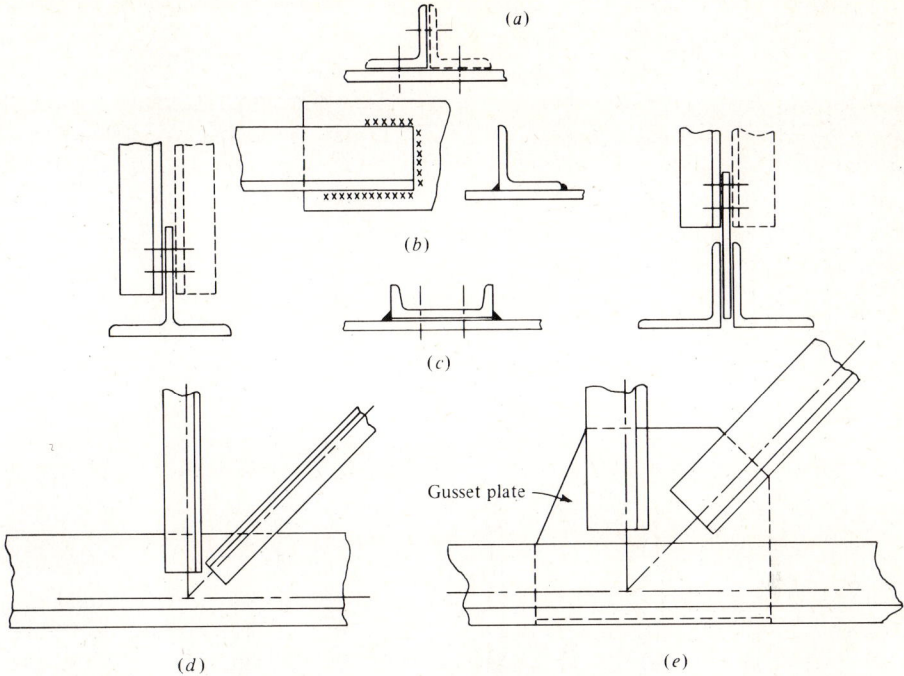

Figure 5-6 Joints for tension (and compression) members where the axial loads are small, as in industrial building roof trusses or transmission towers. (*a*) Single or double angles with bolts or rivets. (*b*) Welded single angle. (*c*) Welded or bolted channel. (*d*) T section with either single or double angles. (*e*) Double-angle chord with gusset plate and single- or double-angle vertical and diagonal members.

or rigidity (measured by L/r), or the joint will be impractical to fabricate. The use of a pair of angles back to back is not considered to be a built-up member.

Several possible built-up sections are shown in Fig. 5-7. These members are made up from two or more rolled shapes sufficiently connected that the resulting section acts as a single member. The stiffness of a built-up member, as measured by the radius of gyration r, can be controlled by the designer so that the use of these sections in locations where stress reversals occur (member carries both tension and compressive loads) is one of the principal advantages, which offsets somewhat the considerable fabrication costs, unless a number of members with the same section is used (e.g., in very long span truss bridges).

Besides load, rigidity, and connection requirements, other considerations in the use of built-up members are minimum weight, ease of fabrication, and where the use is in a hostile environment, such as in bridges, corrosion protection and maintenance/repairs. Bridge and other members exposed to harsh environmental conditions, including both weather effects and birds, must be constructed so that accumulations of bird droppings and water (particularly if contaminated by carbon and sulfur oxides, which create a weak acid and rapid corrosion) does

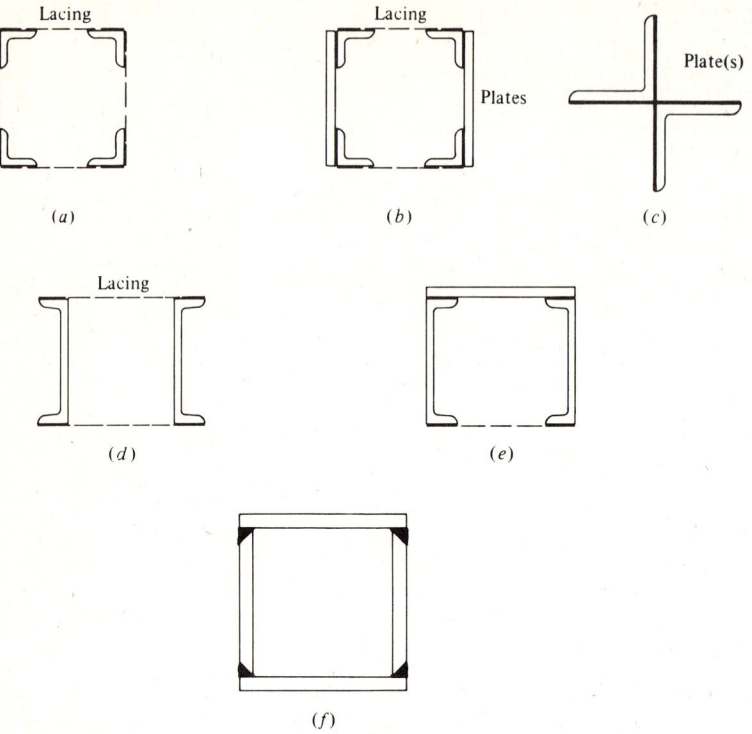

Figure 5-7 Cross sections of several built-up sections. General cross-section configuration is limited only by designer's need and ingenuity and may include W and S shapes with lacing and/or plates. Where additional area is needed, more plates may be added to any of the above sections. (*a*) Four angles with lacing. (*b*) Four angles with both plates and lacing. (*c*) Two angles and one or two plates. (*d*) Two channels with lacing. (*e*) Two channels with flanges reversed from (*d*) and both plate and lacing. (*f*) Four plates welded to form box section.

not occur. It is also necessary that the member(s) be constructed so that painting of the complete member can be affected. This requirement generally precludes use of fully enclosed box or other built-up sections with enclosed cavities. Instead, the built-up sections are open on one or more sides, with continuity being obtained on those sides by use of lacing bars or by use of perforated cover plates. Either of these configurations allows maintenance of the interior as well as the exterior of the member, as shown in Fig. 5.8.

5-4 STRESSES DUE TO AXIAL LOAD ON THE NET SECTION

The presence of a hole at a cross section obviously creates a "notch" effect, as was discussed briefly in Sec. 1-6. This is true even if the hole is occupied by a rivet or bolt. An exception may occur if the fastener clamps the two (or more)

DESIGN OF TENSION MEMBERS 229

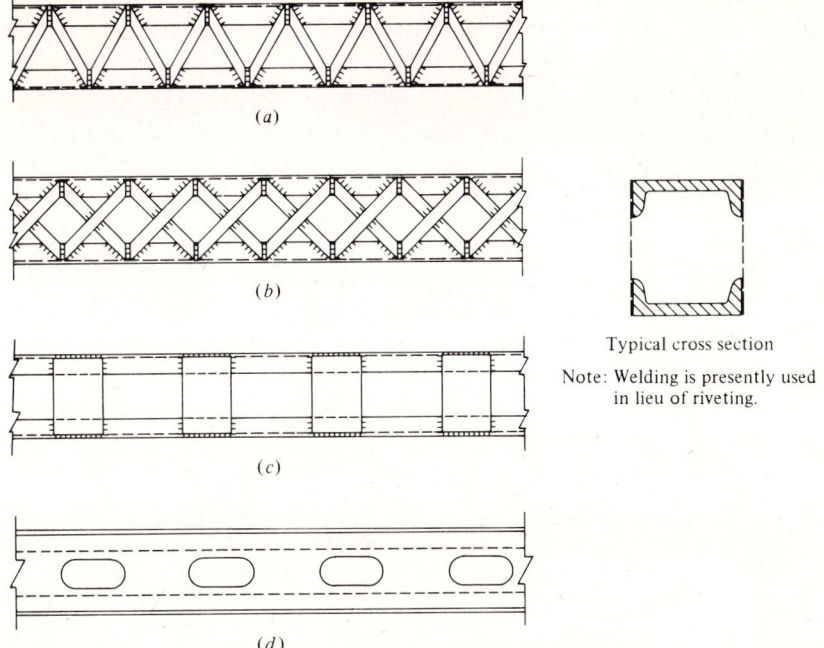

Typical cross section
Note: Welding is presently used in lieu of riveting.

Figure 5-8 Lacing and other means of producing a built-up member with access to interior. Present fabrication shop practice gives preference to use of batten or perforated cover plates which are welded. (*a*) Single lacing. (*b*) Double lacing. (*c*) Batten plates. (*d*) Perforated cover plates.

parts, making the connection sufficiently strong that no relative movement can take place across the hole. Tension tests using plastic and photoelastic techniques indicate stress concentrations at the edges of the bolt hole which may have a maximum value on the order of two to three times the average stress across the section. Since steel stresses are redistributed across the section when $f_t = F_y$, it is reasonable to assume a uniform stress distribution across any cross section, at least for static stress conditions. This means that at working stresses the actual stress distribution in the vicinity of the connection with holes is indeterminate, but should the working load be inaccurately estimated so that yield stresses are produced, the stresses are redistributed until the load is sufficient to produce $f_t = F_y$ across the net section.

When a member is subject to either a large number of stress reversals or to cyclic stresses, the result is application of a stress range to the member. The allowable stresses will need to be reduced to avoid a fatigue failure if the number of stress cycles and/or the stress range is large.

It is usual to assume that fatigue can be neglected in normal building construction. We should note, however, that with 20 load cycles/day, the total cycles over a period of 20 years is only

$$N_{\text{cycles}} = 20(365)(20) = 146\ 000$$

Special situations exist, of course, where fatigue stresses will have to be considered (e.g., crane runway girders and their connections), but the designer should be alert to this in the course of the design. The AISC specifications consider the fatigue problem in Appendix B of the steel manual.

5-5 DESIGN OF TENSION RODS

The design of a tension rod proceeds directly from the determination of the rod load-to-area selection based on the allowable tension stress of $0.6F_y$. Threaded rods are commonly used as sag rods to support purlins in industrial buildings with sloping roofs as well as to support girts in wall construction, and to support balconies and stair landings. The bars will have standard threading, as shown for selected bar diameters in Table 5-3. The design of a sag rod will be illustrated in the following example.

Table 5-3 Design data for threading of standard bar diameters shown

Bar diameter		Root diameter		Number of threads
in	mm	in	mm	per inch or per 25 mm
$\frac{1}{2}$	12.7	0.400	10.2	13
$\frac{5}{8}$	15.9	0.507	12.9	11
$\frac{3}{4}$	19.0	0.620	15.7	10
$\frac{7}{8}$	22.2	0.731	18.6	9
1	25.4	0.838	21.3	8
$1\frac{1}{4}$	31.8	1.064	27.0	7
$1\frac{1}{2}$	38.1	1.283	32.6	6
$1\frac{3}{4}$	44.4	1.490	37.8	5
2	50.8	1.711	43.5	$4\frac{1}{2}$
$2\frac{1}{4}$	57.2	1.961	49.8	4
$2\frac{1}{2}$	63.5	2.175	55.2	4
$2\frac{3}{4}$	69.9	2.425	61.6	4
3	76.2	2.675	67.9	4

Example 5-1 Design the sag rods for the purlins of Example 4-10 (the roof of the side sheds of the industrial building of Example 2-5) (see Fig. E5-1a). Use A-36 steel.

SOLUTION From Example 4-10 the following purlin loading was developed, with the load condition of $D + S + W$ producing the most critical loading. Note, however, that the 0.066-kip/ft loading is due only to $D + S$. Taking this into account, we will use $F_t = 0.6F_y$ and not $F_t = 0.6F_y(1.33)$. From Table IV-5, obtain the sag rod force as the center reaction of a two-span

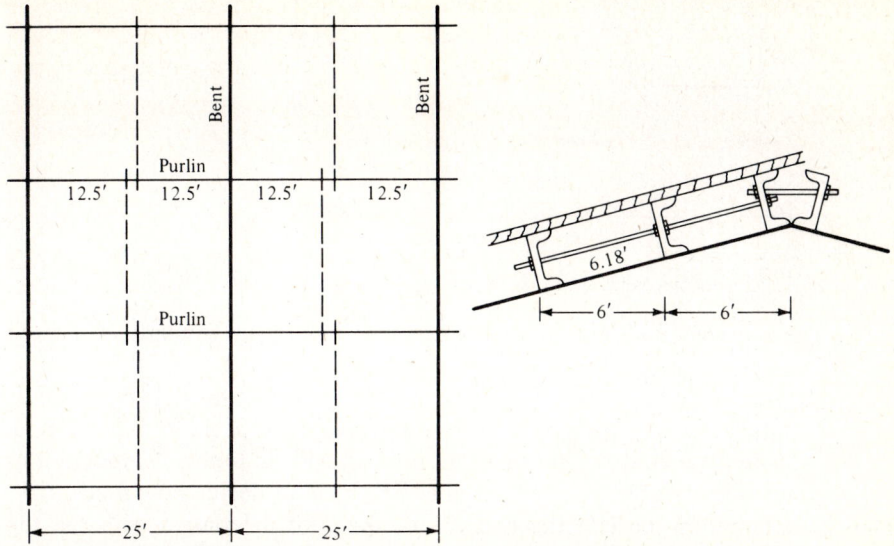

Figure E5-1a

symmetrical continuous beam (Fig. E-51b):

$$P = 2\frac{5wL}{8} = 1.25(0.066)\frac{25}{2} = 1.03 \text{ kips}$$

For 12 purlins (Fig. E2-5a and Example 4-10) the top value = 12 × 1.03 = 12.36 kips.

$$A_{\text{reqd}} = \frac{12.36}{22} = 0.562 \text{ in}^2$$

The AISC specifications require a maximum L/r for bracing and secondary tension members (Sec. 1-8.4) of 300 but not for bars and rods. We will use this as a guide, however, so that the member is not overly small. The minimum force of 6 kips in AISC Sec. 1-15.1 does not apply to sag rods.

$$r \text{ for round bar} = \sqrt{I/A} = \sqrt{\frac{\pi d^4/64}{\pi d^2/4}} = \frac{d}{4}$$

$$\frac{L}{r} = 300 \qquad d = \frac{4(6.18 \times 12)}{300} = 0.988 \text{ in}$$

Check the 1-in-diameter rod (see Table 5-3) using $n = 8$ threads/in:

$$A_e = 0.7854\left(1.00 - \frac{0.9743}{8}\right)^2 = 0.605 > 0.562 \text{ in} \qquad \text{O.K.}$$

232 STRUCTURAL STEEL DESIGN

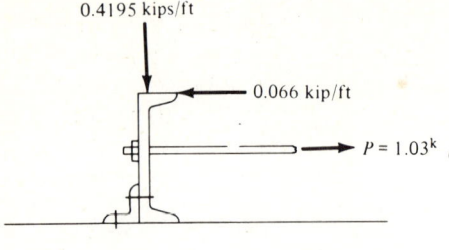

Figure E5-1b

Notes:

1. Tie rod size could be reduced in the lower-half group of purlins for a rod force of $6 \times 1.03 = 6.18$ kips.
2. A rafter, substantial girt, or other member spanning the 25-ft bays will be required to carry the concentrated midspan load of 12.18 kips without excessive outward deflection. The roofing will undoubtedly reduce this effect considerably, but the designer will have to decide how much.
3. Computations neglect the benefits of the roof in contact with the top flange of the purlins, and may reduce the sag rod force 50 percent or more. ///

5-6 NET SECTIONS

When the cross section contains a row of holes, the critical section will occur through one of the holes, as shown in Fig. 5-9. When there is more than one row of holes, the designer must determine a failure section that will yield the minimum area. The path across the section producing the minimum area is the critical net section.

In Fig. 5-9b, which is a portion of a repeating bolt or rivet pattern as might be obtained in a large plate assembly (water tank, steam boiler, or other similar structure), the critical section is not directly obvious, as the plate may tear along

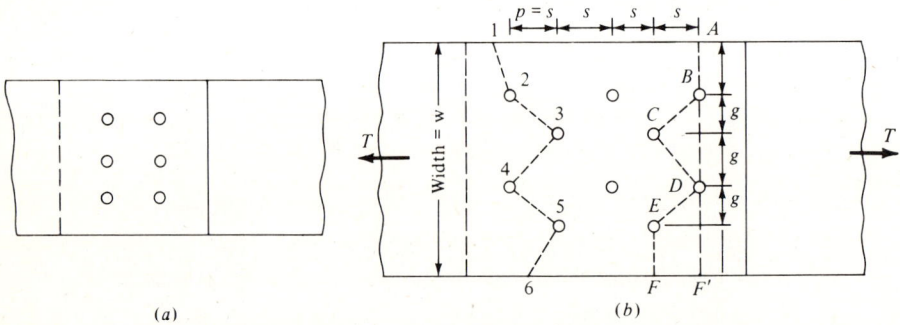

Figure 5-9 Critical net section for tension members with holes. (*a*) One or more rows of holes where the net section is obvious. (*b*) Hole pattern where net section must be found by trial.

diagonal lines such as dashed line 1-2-3-4-5-6. Theoretical solutions have been developed for failure along some possible path as just cited, which allows for diagonal tension material behavior along irregular failure lines. The theoretical solutions are too complicated for design and an approximation is allowed. The approximation is based on a procedure by Cochrane (*Engineering News-Record*, Vol. 89, 1922, pp. 847–848) and is almost universally used. The method is as follows:

1. Take any reasonable (and possible) path across a chain of holes and deduct one hole width for each hole encountered.
2. For each change in direction from one hole to the next hole, add back the quantity

$$\frac{s^2}{4g}$$

where $s=$ pitch, or longitudinal distance between adjacent holes
$g=$ gage distance between adjacent holes across the width

This method of predicting the capacity of a connection (based on net area) in tension is in reasonable agreement with tests, so there is no reason not to continue using it since it is so simple and direct to use.

It is evident that if s or g is of the proper value, we may effectively add back a "hole" on the diagonal, thus canceling that hole reduction on the net width.

The effective hole diameter (except for slotted holes) is taken as

$D + \frac{1}{8}$ in (or 3 mm)   AISC specifications

$D + \frac{1}{8}$ in (or 3 mm)   AASHTO and AREA specifications

The hole is always larger than the fastener by at least $\frac{1}{16}$ in (or 1.5 mm) to allow a slight misfit (including burrs on either the hole or fastener and/or large changes in temperature between punching and erection) so that erection can proceed rapidly. When the hole is punched, an additional $\frac{1}{16}$ in (or 1.5 mm) should be deducted to allow for possible metal damage around the hole, thus producing the effective hole diameter as $D + \frac{1}{8}$ in or $D + 3$ mm.

Example 5-2 What is the critical net section of the hole pattern shown in Fig. E5-2? Take holes for $\frac{7}{8}$-in-diameter bolts. Note that this pattern is more academic than practical, but is used to illustrate the method of using $s^2/4g$. Use the AISC specifications for the hole diameter.

SOLUTION

$$\text{Total width } w = 2(1.5) + 3 + 2(2)$$
$$= 10 \text{ in}$$
$$D = \tfrac{7}{8} + \tfrac{1}{8} = 1.00 \text{ in}$$

Net width along path $ABDF'$:

$$w = 10 - 2(1.00) = 8.00 \text{ in}$$

234 STRUCTURAL STEEL DESIGN

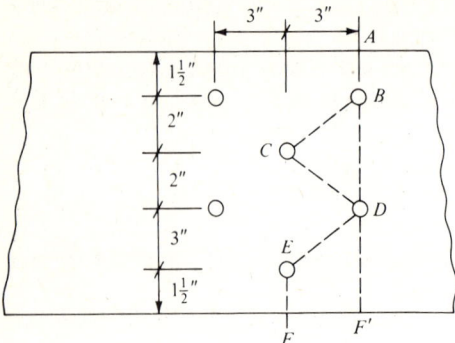

Figure E5-2

Net width along path $ABCDEF$: $w_{net} = w - 4$ holes $+ s^2/4g$ terms for B to C, C to D, and D to E.

$$w_{net} = 10 - 4(1.0) + \frac{3^2}{4(2)} + \frac{3^2}{4(2)} + \frac{3^2}{4(3)} = 9.00 \text{ in}$$

Path $ABDEF$:

$$w_{net} = 10 - 3(1.0) + \frac{3^2}{4(3)} = 7.75 \text{ in} \qquad \text{controls}$$

$$w_{0.85} = 0.85(10) = 8.50 \text{ in}$$

The critical net area $= 7.75(t)$ in^2.

Question: What value of pitch s will produce a net section equivalent to $ABDF'$?

Solution: The critical path is along $ABDEF$, and we have

$$w_{net} = 10 - 3(1.0) + \frac{s^2}{(4)(3)} = 10 - 2.00 = 8.00 \text{ in}$$

$$\frac{s^2}{12} = 8.000 - 7.000$$

$$s^2 = \sqrt{(12)(1.0)}$$

$s = 3.46$ in   (say 3.5 in, since anything over 3.46 is O.K.)

Back substitution of this value of s readily verifies that the new critical section will be 8.00 (within roundoff tolerances).   ///

5-6.1 Net Section of Rolled Sections

The method of producing the critical net section of rolled sections is slightly different from the plate solution just presented, in that we must directly consider the area ($w \times t$). This is done by deducting the area across each hole ($D \times t$) and adding back the partial hole area ($s^2/4g \times t$) for each change in direction of the path. This will be illustrated for a channel section being used as a tension member. Note in this example that the holes are located using the gage distances

DESIGN OF TENSION MEMBERS

g_1 shown in Tables I-3 and V-3. These distances are essentially standard for ease in maintaining templates in the fabrication shop to produce holes located such that minimum interference with the flanges is produced when installing either bolts or rivets. Not all sizes of bolts and/or rivets can be easily installed using these gage distances, but the spacing is such that most sizes can be accommodated.

Example 5-3 What is the critical tension area for the C250 × 37.2 channel shown in Fig. E5-3? The bolts are 22 mm (Table VI-7). Note the g_1 distance used to locate holes B and D with respect to the flanges and given in Table V-6. Use the AISC specifications.

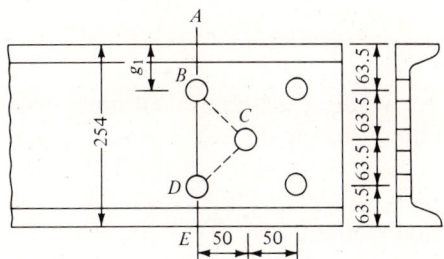

Figure E5-3

SOLUTION From Table V-6, obtain

$$A_t = 4.74 \times 10^{-3} \text{ m}^2$$

$$t_w = 13.4 \text{ mm}$$

$$D = 22 + 3.0 = 25.0 \text{ mm}$$

Net area along path $ABDE$:

$$A_n = 4.74 - 2(0.025)(13.4) = 4.07 \times 10^{-3} \text{ m}^2$$

Net area along path $ABCDE$ [area − 3 holes + $2(s^2/4g)$]:

$$A_n = 4.74 - \left[3(0.025) - 2\left(\frac{0.050^2}{4(0.0635)} \right) \right] \times 0.0134 \times 10^3$$

$$= 4.74 - 0.740 = 4.000 \times 10^{-3} \text{ m}^2 \quad \text{found to be critical}$$

$$A_{0.85} = (4.74)(0.85) = 4.029 \times 10^{-3} \text{ m}^2$$

For a channel shape with flanges out-of-plane and not connected, and referring to Sec. 5-3.2, the effective area A_e (w/3 lines of fasteners) is

$$A_e = 0.85 A_n = 0.85(4.00) = 3.40 \times 10^{-3} \text{ m}^2 \qquad ///$$

5-6.2 Net Sections of Angles with Staggered Holes

The critical section of angles with staggered holes is similar to that of plates, as illustrated in Fig. 5-10. The locations of holes in angles are standardized as in Table I-13 or V-13. The critical section is based on the net width less the

236 STRUCTURAL STEEL DESIGN

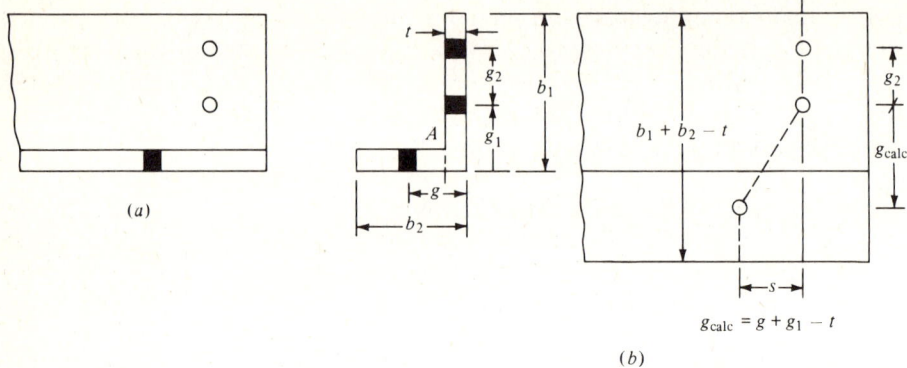

(b)

Figure 5-10 Net section of angle used as tension member with staggered holes in both legs.

holes $+ s^2/4g$, as before. Note that the net width of the equivalent unpunched plate $= b_1 + b_2 - t$, as obtained by cutting the angle along line AB and repositioning it as in Fig. 5-10b.

Example 5-4 What is the critical section of the $8 \times 6 \times \frac{3}{4}$ angle with three $\frac{3}{4}$-bolt holes as shown in Fig. E5-4? Use the AISC specifications.

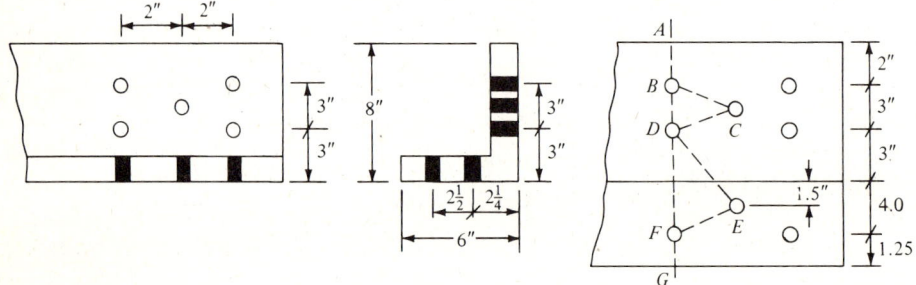

Figure E5-4

SOLUTION
$$D = 0.75 + 0.125 = 0.875 \text{ in}$$
Net width along path $ABDFG$:
$$w = 8 + 6 - 0.75 - 3(0.875) = 10.625 \text{ in}$$
Net width along path $ABCDEFG$:
$$w_n = w - 5 \text{ holes} + s^2/4g \text{ for } B\text{-}C, C\text{-}D, D\text{-}E, \text{ and } E\text{-}F$$
$$w_n = 13.25 - 4.375 + \frac{2^2}{4(1.5)} + \frac{2^2}{4(1.5)} + \frac{2^2}{4(3 + 1.5)} + \frac{2^2}{4(2.5)}$$
$$= 13.25 - 4.375 + 0.667 + 0.667 + 0.222 + 0.400$$
$$= 10.83 \text{ in}$$

Net width along path *ABCDFG*:
$$w_n = 13.25 - 4(0.875) + 0.333 + 0.333 = 10.42 \text{ in}$$
Net width along path *ABDEFG*:
$$w_n = 13.25 - 4(0.875) + 0.222 + 0.400 = 10.37 \text{ in}$$
Check the 85 percent requirement:
$$w_n = 13.25(0.85) = 11.26 \text{ in}$$

Based on the critical net width of 10.37 in (note that both legs are connected, so a shear lag reduction is not required) and $t = 0.75$ in:

$$A_e = 10.37(0.75) = 7.78 \text{ in}^2$$
$$P_{\text{allow}} = A_g F_t = 13.25(0.75)(0.6 F_y)$$
or
$$P_{\text{allow}} = A_e F_t = 7.78(0.5 F_u)$$

whichever is smaller. ///

5-7 DESIGN OF AISC TENSION MEMBERS

The design of tension members requires computations to determine the load and additional computations to determine the critical net section where holes are used for mechanical fasteners. Member selection proceeds based on satisfying the gross and net area requirements, including the 85 percent maximum when holes are in the member, and any reduction in net area from shear flow considerations.

Minimum L/r requirements must be met in all the specifications, and all three specifications (AISC, AASHTO, and AREA) have minimum connection capacity requirements. The AISC minimum connection requirement is found in Sec. 1-15.1 and requires at least 6 kips.

Example 5-5 Select the lightest single angle section for the vertical members of the side shed truss of the industrial building of Example 2-6. It is desirable to use the same size angle section for all the verticals, a single size section for all the diagonals, and similarly, top and bottom chords designed using constant section sizes. Only the vertical web members will be designed in this example. Use $F_y = 250$ MPa and the AISC specifications.

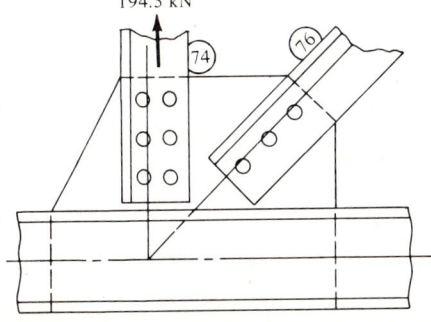

Figure E5-5

SOLUTION From inspection of the computer output (part of which is not shown in Example 2-6), the following values are obtained from the two load conditions used:

	Right side			Left side	
Member	LC-1 (kN)	LC-2 (kN)	Member	LC-1 (kN)	LC-2 (kN)
4	0.0	0.0	74	124.73	157.57
8	31.18	48.62	78	93.56	118.18
12	62.36	97.26	82	62.37	78.80
16	93.54	145.88	86	31.18	39.40
20	124.72	194.50	90	0.0	0.0

Since the signs are (+), all the vertical members have only tension forces for either load condition. Note that the computer program automatically adjusts output with wind by the factor 0.75 so that all loads are on the same design basis. Member 20 has the largest axial tension force with wind from the left; wind from the right would produce this design value in member 74. Member 20 is the longest vertical of the side shed members and has $L = 4.5$ m. If one of the other members had been longer, that member length would control the L/r—in this case both the controlling length and the maximum tension force are in the same member. Design $P = 194.5$ kN. The maximum $L/r = 240$ (per AISC, Sec. 1-8.4, and assuming that this is a main member with such a large axial force). The minimum radius of gyration is

$$r_{min} = \frac{4.5(1000)}{240} = 18.75 \text{ mm}$$

A preliminary side computation indicates that the bolt pattern shown in Fig. E5-5 can probably be used, as L/r rather than stress is likely to control. This pattern may be able to use standard gage distances and be most economical. For two bolts at the section and using 22-mm A-325 bolts,

$$D = 22 + 3.0 = 25.0 \text{ mm}$$

The effective angle area using AISC criteria for shear lag (see Sec. 5-3.2) is

$$A_e = 0.90 A_n$$

Based on $F_t = 0.6 F_y$, the gross angle section must be at least

$$A_g = \frac{194.5}{0.6(250)} = 1.2967 \times 10^{-3} \text{ m}^2$$

Based on $F_t = 0.5 F_u$ (use Table 5-1 for F_u), the effective net area A_e must be

$$A_e = \frac{194.5}{200} = 0.9725 \times 10^{-3} \text{ m}^2$$

This effective net area A_e must be obtained from a gross section of at least

$$A_g = \frac{0.9725}{0.85(0.90)} = 1.2712 \times 10^{-3} \text{ m}^2 < 1.2967$$

DESIGN OF TENSION MEMBERS 239

We cannot generally tell which angle section will be best until we find one or more that are adequate and then select the lightest. This is best done by using a table as follows (also note that $r_{min} = r_z$ for single angles):

t, mm	A_{holes}, m^2	A_{net}, m^2	Section	r_z, mm	A_{furn} ($\times 10^{-3}$), m^2
6.3	0.315	1.175 > 1.081	L152 × 89 × 6.3	19.7	1.490 > 1.2967
7.9	0.395	1.153 > 1.081	L102 × 102 × 7.9	20.1	1.548

In the table above,

$$A_{holes} = 2Dt = 2(25.0)(0.0063) = 0.315 \times 10^{-3} \text{ m}^2$$

$$A_{net} = A_g - A_{holes} \geq \frac{0.9725}{0.90} = 1.081 \times 10^{-3} \text{ m}^2$$

These are the only two angles in Tables V-8 and V-9 that need to be considered. Here the radius of gyration controls the selection of the section more than the area requirements. Tentatively use: L152 × 89 × 6.3 (as the lightest). We must select a section that can be used in the joint as shown. Check if it is possible to get two 22-mm bolts side by side in the long leg of the angle. Use Table V-13 and obtain:

    Standard gage distances: $g_1 = 57.2$ mm $g_2 = 63.5$ mm

    AISC specifications: center to center of hole = $2.67D$ ($3D$ preferred)

    AISC specifications: center of hole to edge of angle leg (see Table
        1.16.4 SSDD) = 28 mm $\left(\text{or } 1\frac{1}{8} \text{ in}\right)$

$$g_2 = 63.5 > 2.67D \quad \text{O.K.}$$

    Actual hole diameter = 22 + 1.5 = 23.5 mm
    Edge distance = 152 − 57.2 − 63.5 = 31.3 > 28 mm   O.K.

Use an L152 × 89 × 6.3 section.   ///

Example 5-6 Select the lightest pair of angles for the vertical members of the main roof truss of Example 2-6. Use the AISC specifications and $F_y = 250$ MPa. The reader should note that the members in this truss will have to be designed for both tension and compression (and possibly for fatigue). From the computer output, members 39 and 47 will be critical for axial forces.

Member	LC-1 (kN)	LC-2 (kN)	L, m
39	−63.65	+20.39	6.76
47	+70.18	−37.60	8.2 (longest of verticals)

$$P_{sr} = 70.18 - (-37.60) = 107.78 \text{ kN}$$

Member 47 is critical for tension design.

$$\text{Minimum } r = \frac{8.2(1000)}{240} = 34.17 \text{ mm}$$

Use 25-mm bolts: $D = 25 + 3.0 = 28.0$ mm. Use two L sections with a 12-mm gusset plate as in Fig. E5-6. Assume two lines of holes, since P is only 70.18 kN, which gives $A_e = 0.90A_n$.

$$A_g = \frac{70.18}{150} = 0.4679 \times 10^{-3} \text{ m}^2$$

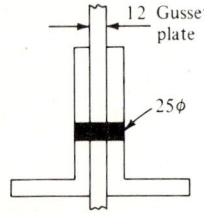

Figure E5-6

For $A_e = 0.75A_n$, the gross angle area is at least

$$A_g = \frac{70.18}{(200)(0.85)(0.75)} = 0.5504 \times 10^{-3} \text{ m}^2$$

Use long legs of unequal leg angles back to back. (Why?). Again set up a table of double angles using Tables V-10 and V-11 of SSDD [A_e = smaller of (A_n or $0.85A_g$) × 0.90]:

t, mm	$A_{2\,holes}$, m²	A_{net}, m²	A_e, m²	Section	r, mm	A_{furn} (×10⁻³), m²
6.3	0.353	2.307	2.03	2L127×89×6.3	37.5	2.660
7.9	0.442	3.468	2.991	2L127×127×7.9	39.9	3.910

There is nothing lighter than these two double angles. Note that minimum L/r controls the design. Now checking stress range and using A_e, we have

$$f_{sr} = \frac{P_{sr}}{A_e} = \frac{107.78}{2.03} = 54.0 \text{ MPa}$$

This value of 63.6 is much smaller than any value of F_{sr} in Table 1-4 up to 2000×10^3 stress cycles for "base metal" at mechanically fastened joints. Use two L127 × 89 × 6.3 sections. ///

5-8 DESIGN OF BRIDGE TENSION MEMBERS

The design of bridge tension members is similar to that using the AISC specifications except that fatigue will have to be considered, as outlined in Secs. 5-4 and 1-9. This will be illustrated by the following example.

Example 5-7 A portion of a highway bridge truss is shown in Fig. E5-7. It is required to select the lightest W12 section for member 9 using A-36 steel and the AASHTO specifications. Assume 2×10^6 load cycles for the service life of the structure.

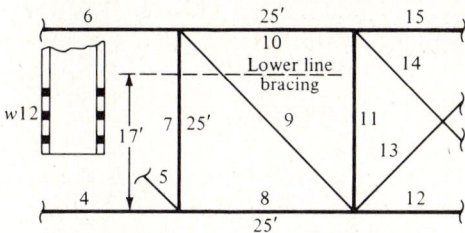

Figure E5-7

SOLUTION From the computer output, we obtain the following (includes impact in live loads):

$$\text{Dead load} = +80.3 \text{ kips}$$
$$\text{Maximum live load} = +60.2 \text{ kips}$$
$$\text{Minimum live load} = -22.8 \text{ kips}$$

From which

$$P_{\max} = 80.3 + 60.2 = 140.5 \text{ kips}$$
$$P_{\min} = 80.3 - 22.8 = 57.5 \text{ kips}$$

The force range (analogous to stress) is

$$P_{sr} = 140.5 - 57.5 = 83 \text{ kips}$$

We will assume four holes in the flange at any net section, as shown in the insert of Fig. E5-7. Use $\frac{7}{8}$-in-diameter bolts, so that the effective hole diameter $D = \frac{7}{8} + \frac{1}{8} = 1.0$ in.

$$A_{\text{holes}} = 4(1)t_f$$

High-strength bolts will be used to connect the member to the joint, producing stress range conditions for "base metal at friction fasteners." The data given in Sec. 1-12 are based on an allowable stress range when a catastrophic failure does not immediately occur when the member fails. In this case it is not likely the truss would collapse if member 9 fails. With this consideration we obtain the allowable stress range:

$$F_{sr} = 16 \text{ ksi} \quad \text{(first column and checking footnote)}$$

We will use lateral and sway bracing across the top and spanning between the two trusses to satisfy stability, but this will not reduce the unbraced length of the several web members with respect to the Y axis. (AASHTO requires the effective depth of this bracing to be at least 5 ft or 1.8 m.) Joint fabrication requires orienting the X axis in the plane of the truss for rolled sections. Note that commonly the transverse floor beams are somewhat below the point of intersection of the truss web members and the bottom chord, but this should not affect the unbraced length L significantly.

The AASHTO limitation for $L/r = 200$ for main tension members and no stress reversals (Sec. 1-7.5), and for a $P = 140.5$ kips, this would seem to qualify. The net area, assuming that $F_a = 0.55 F_y$, is

$$A_{net} = \frac{P}{F_a} = \frac{140.5}{20} = 7.025 \text{ in}^2$$

$$r_{min} = \frac{25(1.414)(12)}{200} = 2.12 \text{ in (for } r_y)$$

We will arbitrarily use the AISC 85 percent requirement, so A_g is at least

$$A_g \geq \frac{7.025}{0.85} = 8.26 \text{ in}^2$$

By trial set up the following table, where $A_{holes} = 4(1.0)t_f$; $A_{reqd} = A_g - A_{holes}$.

Section	t_f, in	A_{holes}, in^2	A_{reqd}, in^2	A_{furn}, in^2	r_y, in	
W12×53	0.575	2.30	13.30	15.60 > 8.26	2.48	O.K.
W12×50	0.640	2.56	12.14	14.70	1.96	N.G.

The W12 × 53 section is the lightest W12 that is satisfactory for both area and L/r requirements. This section is also selected so that the connection can be more easily fabricated when the vertical member (No. 7) is designed in Chap. 6.

Check the stress range for the W12 × 53 section:

$$f_{sr} = \frac{P_{sr}}{A_n} = \frac{83.0}{13.30} = 6.24 \text{ ksi} \ll 16 \text{ ksi} \quad \text{O.K.}$$

Although fatigue is not a controlling parameter for this member, all the truss members should be checked similarly. Use a W12 × 53 section. ///

5-9 CABLE DESIGN

A moment summation about a convenient location with respect to the parabolic cable geometry as shown in Fig. 5-11 gives

$$H = \frac{wL^2}{8h} \tag{5-6}$$

where h = midspan sag, as shown in Fig. 5-11
 w = uniform cable loading/unit of length; there will always be a uniform load caused by the cable weight
 L = span length (the cable length is always somewhat longer)

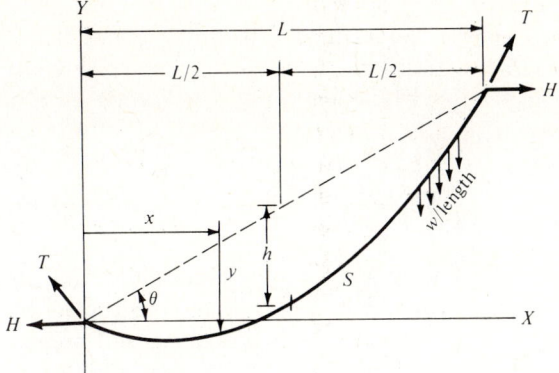

Figure 5-11 Cable geometry for developing design equations. Note that in this configuration the T at the origin is larger than the T at the other end of the cable.

In general, the cable sag y at any point is

$$y = \frac{4hx}{L^2}(x - L) + x \cdot \tan\theta \qquad (a)$$

Differentiating, we obtain

$$\frac{dy}{dx} = \frac{8hx}{L^2} - \frac{4h}{L} + \tan\theta \qquad (b)$$

Since

$$ds = \left[1 + \left(\frac{dy}{dx}\right)^2\right]^{1/2} \quad \text{and} \quad T_x = \frac{H}{\cos\theta} = H\left(\frac{ds}{dx}\right)$$

we obtain a general equation for the tension force in the cable at any point as

$$T_x = H\left[1.0 + \frac{64h^2x^2}{L^4} + 16\left(\frac{h}{L}\right)^2 + \tan^2\theta \right.$$
$$\left. - \frac{64h^2x}{L^3} + \frac{16hx}{L}\tan\theta - 8\frac{h}{L}\tan\theta\right]^{1/2} \qquad (5\text{-}7)$$

Noting that all the terms under the square root are relatively insignificant except the first, third, and fourth, we may simplify to obtain

$$T_x = H\left[1 + 16\left(\frac{h}{L}\right)^2 + \tan^2\theta\right]^{1/2} \qquad (5\text{-}8)$$

When the ends of the cable are at greatly differing elevations, Eq. (5-7) rather than Eq. (5-8) should be used, because the tension at the upper end of the cable will be considerably different (it is "carrying" the weight of the cable + any additional cable loading). For horizontal cables T has the same value at both ends and is directly computed as

$$T = H\left[1 + 16\left(\frac{h}{L}\right)\right]^{1/2} \tag{5-9}$$

The cable length is approximately given by

$$S = L\left[1 + \frac{8}{3}\left(\frac{h}{L}\right)^2 - \frac{32}{5}\left(\frac{h}{L}\right)^5\right] \tag{5-10}$$

Cables have been used to support large-span roof structures as well as bridges and guys for towers. In buildings the cable roof is constructed by stringing cables across the open space at sufficiently close spacing and pretension force T to produce the desired sag based on Eq. (5-9). The sag in buildings is on the order of $\frac{1}{12}$ to $\frac{1}{16}$ (bridges may be as large as $\frac{1}{6}$ to $\frac{1}{8}$). In round structures the cables may be attached to a large compression ring at the building perimeter and terminate in the center in a tension ring. This configuration is the most desirable, since the cable tension must be carried by some kind of anchorage. The compression ring is the most desirable, since large compressive stresses can be used if the ring is made of steel. The use of prestressed concrete roofing "plank" directly on the cables produces the necessary roof and at the same time tends to reduce the vibrations, since the concrete planking is rather heavy and develops a large system mass.

Where the use of concrete planking is not sufficient to restrict damping, other means must be employed. Damping tends to modify the natural amplitude of a vibrating system and if reliably introduced, eliminates resonance amplitudes of vibration. The natural frequency of a cable system (the same as any other type of natural frequency computation) is

$$f_n = n\frac{\pi}{L}\left(\frac{T}{m}\right)^{1/2} \tag{5-11}$$

where $m = w/g$ = mass of the entire cable (including any attachments, such as a roof)
n = any integer, such as 1, 2, or 3, used to obtain the fundamental modes; the value at $n = 1$ is of primary interest, but we may need values for $n = 2, 3,$ and perhaps 4
T = cable tension; can be written as $T + \Delta T$, so that it is evident that a change in T produces a new f_n

A convenient way to dampen a cable to control vibration is to attach it to a second cable, as illustrated in Fig. 5-12, which has a different natural frequency. This can be accomplished by use of Eq. (5-11), which indicates that changes in T produce different f_n values. It will then be necessary to install the cables using vertical struts to a tension sufficient to satisfy design forces and at T values such that the f_n values at any of the fundamental modes does not match, which would produce resonance (very large vibration amplitudes that would likely cause a structural collapse).

DESIGN OF TENSION MEMBERS 245

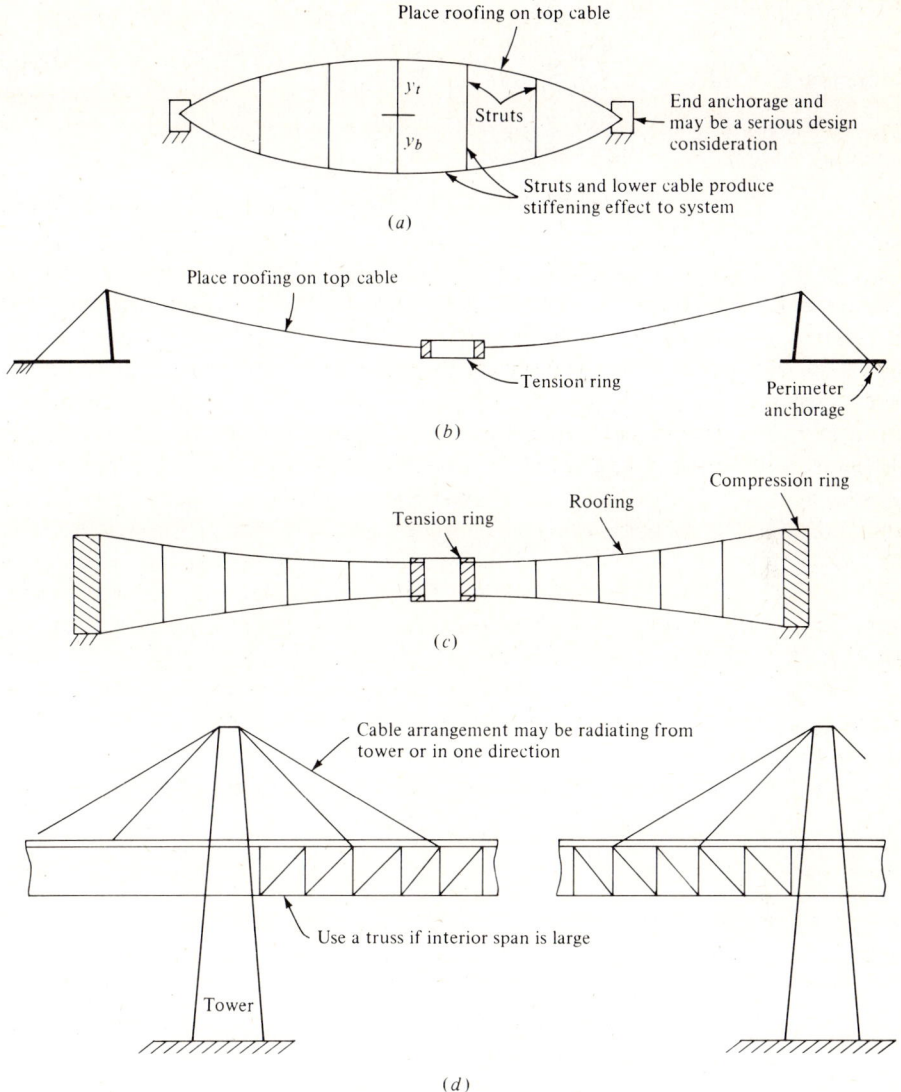

Figure 5-12 Several configurations of cables in building construction. (*a*) Cables spaced across span for a rectangular building plan. (*b*) Single cable using a central tension ring and for circular building. (*c*) Double-cable system for increased stiffness and vibration control. For round building plans. (*d*) Cables used in roof support system. If cables radiate, the towers may be essentially self-supporting.

The safety factors when using cables have been suggested by the AISI as being from 2.5 to 3.0; compare the factored ultimate tension to P_u as given in tables such as Table 5-2. The factored load may be

$$U = 1.5 \text{ dead} + 3 \text{ live}$$
$$U = 2.5(D + L)$$
$$U = 2(D + L + W \text{ or } E)$$
or
$$U = 2.0 \times \text{erection loads}$$

Example 5-8 Design the guy cables at one level for a 250-m TV antenna. The horizontal forces for design are shown in Fig. E5-8. Four cables will be used, but each is to be designed for the force shown.

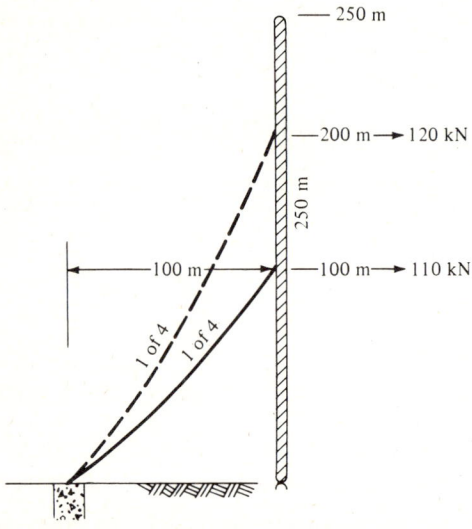

Figure E5-8

SOLUTION For the 100-m level, using Eq. (5-6), we obtain

$$\frac{wL^2}{8h} = 110 \text{ kN} \qquad \frac{w}{h} = \frac{8(110)}{100^2} = 0.088$$

Also, the cable tension is (Eq. (5-7)):

$$T = H\left[1 + 16\left(\frac{h}{L}\right)^2 + \tan^2\theta - 8\left(\frac{h}{L}\right)\tan\theta\right]^{1/2}$$

obtained with slight simplification of terms. By trial and using SF = 2.5, $\theta = 45° \rightarrow \tan\theta = 1.0$:

Trial	$h = w/0.088$	T	T_u	Cable w, kN/m	Diameter, mm
1	0 (assumed)	155.6	389	0.024	25
2	0.272 m	154.7	386		Close enough to initial

Use 25-mm bridge rope 6 × 7 at $P_u = 407 > 386$ kN. The anchorage would have to be designed for a pullout force of T_{min} computed as

$$T_{min} = H\left[1 + 16\left(\frac{h}{L}\right)^2 + \tan^2\theta - 8\frac{h}{L}\tan\theta\right]^{1/2} = 110(1.4065) = 154.7 \text{ kN}$$

With SF = 3, the anchorage resistance would be $3 \times 154.7 = 464$ kN.   ///

Example 5-9 It is desired to produce a domed roof over a building used as a sports arena (Fig. E5-9). The building is roughly rectangular at 350 × 700 ft in plan. Cables will run across the narrow dimension (span 350 ft) in the profile shown, using two cables with struts to control damping. The roofing will be placed directly on the top cable. The strut rods will be located on 20-ft centers and will be in compression. The design of the struts will be considered in Chap. 6.

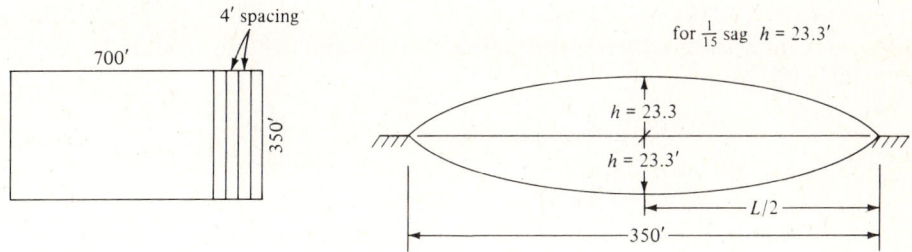

Figure E5-9

SOLUTION

$$\begin{aligned}
&\text{Roofing dead load, including weight of cables} \\
&\text{(horizontal projection)} && = 60 \text{ lb/ft} \\
&\text{Live load at 40 psf:} \quad 40 \times 4 && = 160 \text{ lb/ft} \\
&\text{Total load} && = 220 \text{ lb/ft}
\end{aligned}$$

Since the weight of the top cable load tends to depress the cable, we will put an arbitrary tension in the top to produce an equivalent sag of 23.3 ft, as shown, and this tension will be [combine Eqs. (5-6) and (5-8) to obtain that shown]

$$T = \frac{wL^2}{8h}\left[1 + 16\left(\frac{h}{L}\right)^2\right]^{1/2}$$

Taking 1/2 the roof load as w,

$$T = \frac{0.220(350)^2}{2(8)(23.3)}\left[1 + 16\left(\frac{23.3}{350}\right)^2\right]^{1/2} = 74.8 \text{ kips}$$

The bottom cable will carry the uniform load, producing the 23.3-ft sag of the upper cable plus the total roof load:

$$T = \frac{(0.220/2 + 0.22)(350)^2}{8(23.3)}(1.035) = 224.4 \text{ kips}$$

The required cable areas using $F = 3.0$:

Top cable: $P_u = 3(74.8) = 224.4$ kips → use $1\frac{1}{2}$-in-diameter bridge strand

From Table 5-2, $w = 4.73$ lb/ft.

Bottom cable: $P_u = 3(224.4) = 673.2$ → use $2\frac{1}{2}$-in-diameter bridge strand

From Table 5-2, $w = 12.8$ lb/ft.

The struts will carry a compression load based on q producing an h value in the upper cable of 23.3 ft. This gives an equivalent uniform strut "or diaphragm" load of $0.220/2 = 0.110$ kip/ft. At a 20-ft spacing, the strut load is $20(0.110) = 2.2$ kips.

Check the natural frequency of the top and bottom cables: First find the cable length:

$$L = 350\left[1 + \frac{8}{3}\left(\frac{23.3}{350}\right)^2 - \frac{32}{5}\left(\frac{23.3}{350}\right)^5\right]$$

$$= 354.13 \text{ ft}$$

With dead and live loads in contact with the top cable and using 350 ft as the contribution span for these loads, the mass is

$$m = \frac{W}{g} = \frac{0.220}{32.2}\frac{350}{354} = 0.00676 \text{ lb} \cdot \text{s}^2/\text{ft}$$

$$f_n = n\frac{\pi}{L}\left(\frac{T}{m}\right)^{1/2} = n\frac{\pi}{350}\left(\frac{77.4}{0.00676}\right)^{1/2} = 0.96n$$

For $n = 1, 2$, and 3, we obtain $f_n = 0.96, 1.92$, and 2.88 Hz.

For the bottom cable, only the bottom cable weight will be used, since only the struts make continuity with the top cable:

$$f_n = n\frac{\pi}{354}\left[\frac{232.1(32.2)}{0.0128}\right]^{1/2} = 6.78n$$

For $n = 1, 2$, and 3, we obtain $f_n = 6.78, 13.6$, and 20.3 Hz.

Since the natural frequencies f_n of the two cables are considerably different, no resonance is likely to occur. When one cable is at resonance, the other is at a different frequency, which acts to dampen the resonance vibrations so that the total vibration amplitude is kept small. ///

5-10 DESIGN OF TENSION MEMBERS USING LRFD

The design of tension members using LRFD is relatively straightforward. Again the dead and live loads must be identified so that the appropriate load factor can be used. Once this is done and the ultimate axial tension load P_u is obtained, we have

$$P_u = A_n \phi F_y \quad (\phi = 0.88, \text{ Table 3-1})$$

$$P_u = A_n \phi F_u \quad (\phi = 0.74)$$

Example 5-10 Given a roof truss member with a length of 11.5 ft in a building frame with a dead load of 18.52 kips and a live load of 22.54 kips (snow load). Use the LRFD method, A-36 steel, and $\frac{7}{8}$-in-diameter A-325 high-strength bolts. Design the member using the lightest C shape possible.

SOLUTION We will use a bolt pattern as shown in Fig. E5-10 so that the AISC shear lag reduction factor will be 0.85 (at least three fasteners in the line of stress).

$$\frac{L}{r} = 240 \quad \text{(AISC maximum for a main member)}$$

$$r_{min} = \frac{11.5(12)}{240} = 0.575 \text{ in}$$

Figure E5-10

With two holes out at the critical section, the net area is

$$A_{net} = A_g - 2(\tfrac{7}{8} + \tfrac{1}{8})t_w$$

From Sec. 3-7, obtain (ϕ factors from Table 3-1)

$$P_u = 1.1(1.1D + 1.5S)$$
$$= 1.1[1.1(18.52) + 1.5(22.54)] = 59.6 \text{ kips}$$

$$A_n \geq \frac{P_u}{\phi F_y} = \frac{59.6}{0.88(36)} = 1.88 \text{ in}^2$$

$$A_n \geq \frac{P_u}{\phi F_u} = \frac{59.6}{0.74(58)} = 1.39 \text{ in}^2$$

Using the largest A_n, the gross section using the AISC efficiency factor and the shear lag factor gives the gross area as at least

$$A_g = \frac{1.88}{0.85(0.85)} = 2.602 \text{ in}^2$$

Try a C7 × 9.80:

$$A_g = 2.87 \text{ in}^2 \quad r_y = 0.625 \text{ in} > 0.575 \quad \text{O.K.}$$

Now check if we can get two bolts in the web, as illustrated in Fig. E5-10.

$$g_1 = 2.50 \text{ in} \quad \text{and} \quad 2g_1 = 2(2.50) = 5.00 \text{ in}$$

This leaves a bolt hole spacing of

$$7.00 - 5.00 = 2.00 < 2.67(7/8) \quad \text{N.G.}$$

Try a C8 × 11.50:

$$A = 3.38 \text{ in}^2 \quad r_y = 0.625 \text{ in} > 0.575 \quad \text{O.K.}$$
$$g_1 = 2.50 \text{ in} \quad t_w = 0.220 \text{ in}$$

This leaves a center-to-center bolt spacing of

$$8 - 5 = 3 \text{ in} > 3D \quad \text{O.K.}$$

Check the net area: with two holes out:

$$A_n = 3.38 - 2\left(\frac{7}{8} + \frac{1}{8}\right)0.22 = 2.940 \text{ in}^2$$

Check the shear lag:

$$A_e = 2.940(0.85) = 2.499 > 1.88 \text{ in}^2 \quad \text{O.K.}$$

Joint efficiency and shear lag:

$$A_e = 3.38(0.85)(0.85) = 2.44 > 1.88 \quad \text{O.K.}$$

Use a C8 × 11.50 member. ///

PROBLEMS

For all problems, assume adequate fastener strength so that only the net/gross section requirements control.

5-1 Design an eyebar to carry a tension load of 40 kips using a 1-in-diameter pin. Use A-36 steel and the AISC specifications. Use $t \geq \frac{3}{4}$ in and w in multiples of $\frac{1}{8}$ in.

5-2 Design an eyebar to carry a tension load of 200 kN using a 25-mm-diameter pin. Use $F_y = 250$ MPa and the AISC specifications. Use $t \geq 15$ mm and w in multiples of 3 mm.

5-3 What is the net width of the plate shown in Fig. P5-3 using the given fps units?
Answer: 6.75 in.

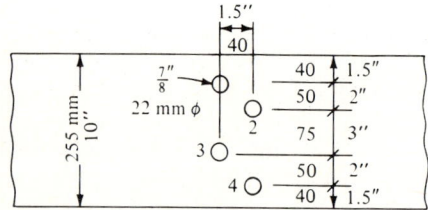

Figure P5-3

5-4 What is the net width of the plate shown in Fig. P5-3 using the given SI units?
Answer: 176.3 mm.

5-5 What is the bolt pitch in Fig. P5-3 so that the critical net section is at least 8 in?

5-6 What is the bolt pitch in Fig. P5-3 so that the critical net section is at least 205 mm?
Answer: 62 mm.

5-7 What is the allowable plate capacity of Fig. P5-3 if A-36 steel is used and the plate thickness is $\frac{1}{2}$ in?
Answer: 97.9 kips.

5-8 What is the allowable plate capacity of Fig. P5-3 if $F_y = 250$ MPa steel is used and the plate thickness is 12 mm?
Answer: 423 kN.

5-9 What is the allowable tensile load for the plate shown in Fig. P5-9 using $F_y = 345$ MPa steel, 20-mm-diameter bolts, and the AISC specifications?

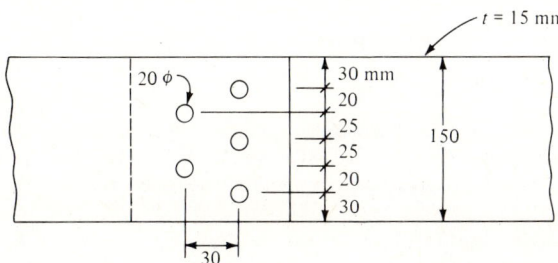

Figure P5-9

5-10 What pitch is necessary in Prob. 5-9 so that only three bolt holes are deducted from the width to produce the net section?
Answer: 32 mm.

5-11 Select the lightest single angle for a tension load of 50 kips. The length is 6 ft and $\frac{7}{8}$-in-diameter bolts will be used as shown in the pattern on Fig. P5-11. Use A-36 steel and the AISC specifications.
Answer: L6 × 4 × $\frac{1}{4}$.

Figure P5-11

5-12 Select the lightest single angle for a tension load of 210 kN. The length is 1.9 m and 22-mm-diameter bolts will be used in a pattern similar to that shown in Fig. P5-11. Use $F_y = 250$ MPa steel and the AISC specifications.

5-13 Select the lightest pair of angles back to back to carry a tensile load of 400 kN. Use 22-mm-diameter bolts, a 12-mm gusset plate, and $F_y = 345$ MPa steel. The member length is 3.2 m. Use the bolt pattern of Fig. P5-13 and the AISC specifications.
Answer: L102 × 76 × 6.3.

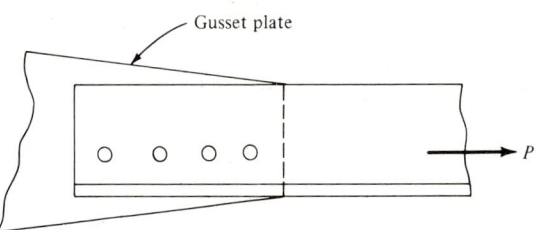

Figure P5-13

5-14 Select the lightest pair of angles back to back to carry a tensile load of 92.5 kips. Use $\frac{7}{8}$-in-diameter bolts, a $\frac{1}{2}$-in gusset plate, and $F_y = 50$ ksi steel. The member is 8.375 ft long. Use the bolt pattern of Fig. P5-13 and the AISC specifications.

5-15 Select the lightest single angle for a tension load of 68 kips, assuming one $\frac{3}{4}$-in-diameter bolt at the critical section. The member is 7.5 ft long. Use A-36 steel, the bolt pattern shown in Fig. P5-15, and both the AISC and AASHTO specifications. Assume no stress reversals for AASHTO.
    Answer: By AASHTO, L7 × 4 × $\frac{7}{16}$.

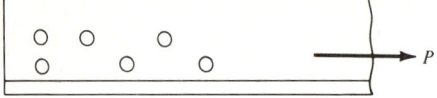

Figure P5-15

5-16 Select the lightest single angle for a tension load of 220 kN assuming one 20-mm-diameter bolt at the critical section. The member is 4.3 m long. Use $F_y = 345$ MPa steel, the bolt pattern shown in Fig. P5-15, and both the AISC and AASHTO specifications. Assume no stress reversals for AASHTO.

5-17 Design the bottom chord members to satisfy tension for the side shed truss of Example 2-5 using WT (structural tee), $\frac{7}{8}$-in-diameter bolts, A-36 steel, and the AISC specifications. Assume two bolts at the critical section is the web of the tee. The tee is continuous across the critical joint.
    Answer: WT9 × 27.5.

5-18 Design member 5 of Fig. E5-7 using a W12 section if the computer output (including impact) is

$$\text{Dead load} = 160.51 \text{ kips}$$
$$\text{Live load} = 77.17 \text{ kips (maximum)}$$
$$\text{Live load} = -5.81 \text{ kips (minimum)}$$

Use the AASHTO specifications and A-36 steel.
    Answer: W12 × 53.

5-19 Design the guy cable for the 200-m level of the TV antenna of Example 5-8.

5-20 Design sag rods for the purlins shown in Fig. P5-20. The purlin span is 28 ft and spaced on 8-ft centers. The sag rod is at the midspan of the purlin. The roof slope is as shown.

$$\text{Dead load} = 25 \text{ psf of roof surface}$$
$$\text{Live load} = 45 \text{ psf (horizontal project)}$$

Use A-36 steel and the AISC specifications.
    Answer: Diameter = $1\frac{1}{8}$ in.

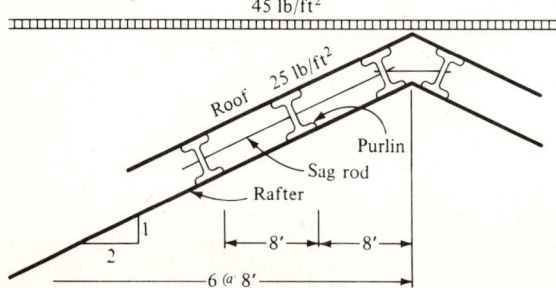

Figure P5-20

5-21 What is the maximum allowable tensile load for a C12 × 25 connected to a $\frac{1}{2}$-in gusset plate as shown in Fig. P5-21. Use A-36 steel and the AISC specifications. Also find the minimum width of the gusset plate at the last bolt row in the channel.
Answer: $P = 154$ kips.

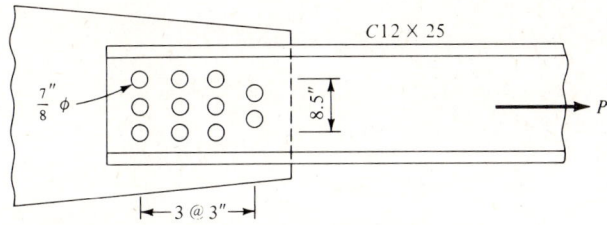

Figure P5-21

5-22 What is the maximum allowable tensile load for a W310 × 52.1 connected to a pair of 12-mm gusset plates as shown in Fig. P5-22 with four holes at each critical section. Use $F_y = 250$ MPa and the AISC specifications.
Answer: 978.5 kN.

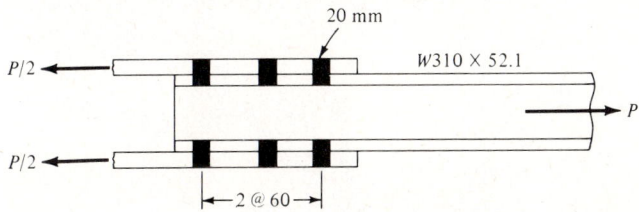

Figure P5-22

5-23 Do Prob. 5-21 using the AASHTO specifications. Do not consider fatigue.

5-24 Do Prob. 5-22 using the AREA specifications. Do not consider fatigue.
Answer: 747.4 kN.

5-25 Redo Example 5-10 using a single angle.
Answer: L$3\frac{1}{2}$ × 3 × 7/16.

5-26 Redo Example 5-10 if the truss member length is 14.5 ft instead of 11.5 ft.

5-27 Given the bottom chord of a truss using a pair of C200 × 17.11 back to back with a 15-mm gusset plate between them. Using two 25-mm-diameter A-325 bolts at the critical section and for a dead-load bar force of 120 kN, what is the maximum live-load bar force that is allowed using $F_L = 1.67$? Use $F_y = 250$ MPa and a panel length of 5.1 m. Assume at least three fasteners in the line of stress and the AISC specifications as applicable.
Answer: LL = 365 kN.

5-28 Design the bottom chord member (No. 12) of the highway bridge truss of Example 5-7 (refer to Fig. E5-7) using a built-up section. Use Examples 6-7 and 8-3 as a guide in selecting the rolled sections to make up the cross section. Loads: dead = 336.9 kips (tension); live maximum = 123.8 kips (tension); live minimum = 0.0 kip. Use the AASHTO specifications, $\frac{7}{8}$-in-diameter high-strength bolts, and A-36 steel. Panel length = 25 ft, as shown in Fig. E5-7.

Figure VI-1 Bridge end-post compression member. End post is built up and uses all-welded construction. Joints are field-fabricated using high-strength bolts.

CHAPTER SIX

AXIALLY LOADED COLUMNS AND STRUTS

6-1 INTRODUCTION

The vertical compression members in a structure are commonly identified as columns (sometimes stanchions in foreign literature). Sometimes vertical compression members are called *posts*. The diagonal compression members comprising the top chord of bridge approaches are *end posts*. The diagonals of a truss or members used in wind bracing may be called *struts*. Short compression members at the junction of columns and roof trusses or beams may be called *knee braces*. In all cases, however, the member under consideration is carrying a compressive load.

A structural member carrying a compression load is termed a *column* if the length is sufficiently great. For lesser lengths the member may be called a *compression block*. The length which divides these two classifications is such that it affects the maximum compressive stress which can be developed under the load. The length is seldom used alone in describing column behavior. Rather, as an offshoot of the Euler column formula developed in the next section, the ratio of column length to radius of gyration (L/r) is used.

Material	Limiting L/r (approximately)
Steel	60
Aluminum	30
Wood	10

256 STRUCTURAL STEEL DESIGN

Tests indicate that when the L/r ratio is at or above the limiting values given, the member exhibits column behavior; below this L/r, the member is a compression block. A member with an L/r ratio of 60 has substantial length, so the term "compression block" is somewhat of a misnomer.

Column design is considerably less exact than beam design for several reasons, which include the following:

1. The difficulty of determining the exact point of demarcation between compression blocks and columns.
2. Columns, although appearing straight and homogeneous, may have small imperfections and always have residual stresses from mill operations, such as rolling, cooling, and so on. Any small imperfections will result in a net eccentricity about one (or both) of the axes and produce lateral deflections (buckling) due to the bending moment that is produced as the product of load x eccentricity.
3. It is often difficult to apply a load through the center of area (i.e., apply a truly axial load).
4. The character of the end restraints markedly affects column behavior. For example, the top chords of trusses are usually compression members. Trusses are usually analyzed as having pinned joints. The reader should ask himself if he has ever seen a pinned truss joint. He no doubt has seen many joints consisting of large numbers of bolts or rivets through gusset plates and into the member to produce a joint. Further, one must ask what effect is produced on the compression member if all the force vectors at a joint are not coincident at the intended intersection location.

From these several considerations it is evident that if an ideal, isotropic, axially loaded column is produced, it is accidental. Nevertheless, it is convenient to analyze a large class of structural members as "axially loaded" columns, struts, posts, or whatever the local terminology used to identify the compression member.

6-2 THE EULER COLUMN FORMULA

One of the most popular column formulas ever proposed (and there are a very large number) was derived in about 1759 by Leonhard Euler, a Swiss mathematician. The formula is readily derived (refer to Fig. 6-1) as follows. Writing the fundamental bending equation for moment,

$$EI\frac{d^2x}{dy^2} = -M$$

we have for the Euler case that the moment is Px, so

$$EI\frac{d^2x}{dy^2} = -Px \qquad (a)$$

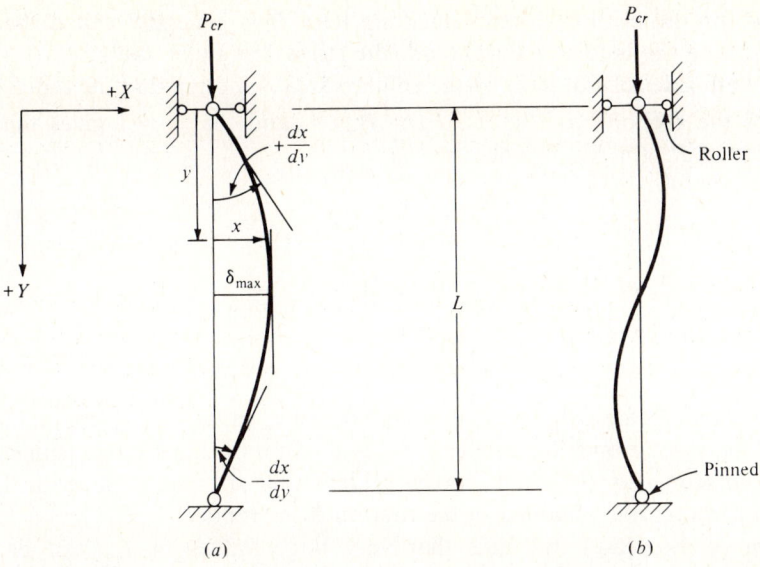

Figure 6-1 Ideal column under Euler column buckling conditions. (*a*) General case with $n = 1$. (*b*) Second buckling ($n = 2$) mode.

where the negative sign for M results from going from $+dy/dx$ at the origin to $-dy/dx$ at $y = L$, so that the change of slope must have a negative value. The classical general solution for the differential equation in the form of Eq. (*a*) is

$$x = A \sin ky + B \cos ky \qquad (b)$$

where

$$k = \left(\frac{P}{EI}\right)^{1/2} \qquad (c)$$

With boundary conditions of $x = 0$ at $y = 0$, we obtain $B = 0$ and Eq. (*b*) becomes

$$x = A \sin ky \qquad (6\text{-}1)$$

Since $x = 0$ at $y = L$, either $A = 0$ or $\sin ky = 0$. With $A = 0$, there is no deflected shape by inspection of the equation, so the solution must be in $\sin ky = 0$. This is only possible for values of kL as follows:

$$kL = \pi, 2\pi, 3\pi, \ldots, n\pi \qquad (d)$$

and in general

$$k = \frac{n\pi}{L} \qquad (e)$$

Using Eqs. (*e*) and (*c*) to obtain P, we obtain

$$P = \frac{n^2 \pi^2 EI}{L} \qquad (6\text{-}2)$$

This equation for the critical column buckling load P is generally called the *Euler equation* (and the load, the Euler load; the stress, the Euler stress).

Dividing both sides of Eq. (6-2) by the column area A, noting that the radius of gyration of the section $r = \sqrt{I/A}$, $F_{cr} = P_{cr}/A$, and that $n = 1$ gives the minimum value of P_{cr} (or F_{cr}), we obtain

$$F_{cr} = \frac{\pi^2 E}{(L/r)^2} \tag{6-3}$$

We note that if $n = 2$, we obtain from Eq. (6-2),

$$F_{cr} = \frac{4\pi^2 E}{(L/r)^2}$$

which is equivalent to the column containing two sine waves in the length L. This is called the second buckling mode, $n = 1$ is the first buckling mode (single sine wave), and from Eq. (6-2) it becomes evident that the minimum critical buckling load (or stress) is obtained in the first buckling mode.

Inspection of Eq. (6-3) indicates that very large values of F_{cr} can be obtained by using $L/r \to 0$. Implicit in writing the differential equation of bending [Eq. (a)], however, is stress being proportional to strain. Thus the upper limit of validity is the proportional limit, which is often taken as $F_{cr} \to F_y$.

6-3 COLUMNS WITH END CONDITIONS

Rotation of the ends of columns in buildings is restrained by the beams that frame into them. The ends of truss members are similarly restrained. In both cases the design *may* be made on the basis of pinned ends. Note that the Euler equation derivation was for a perfectly straight, pin-ended column. The derivation for the critical buckling load for columns with various end restraints can be done in a manner similar to that for the Euler case. This will be illustrated for the fixed-end column shown in Fig. 6-2a. The differential equation for bending now becomes

$$EI\frac{d^2x}{dy^2} = -Px + M_0 \tag{f}$$

Using the standard method of solution as before and noting that we have simply added a constant, we obtain

$$x = A \sin ky + B \cos ky - \frac{M_0}{P} \tag{g}$$

where $k = (P/EI)^{1/2}$, as before. With boundary conditions of $x = 0$ at both $y = 0$ and L, we obtain the constants A and B:

$$x = \frac{M_0}{P}\left(\frac{1 - \cos kL}{\sin kL} \sin ky + \cos ky - 1\right) \tag{h}$$

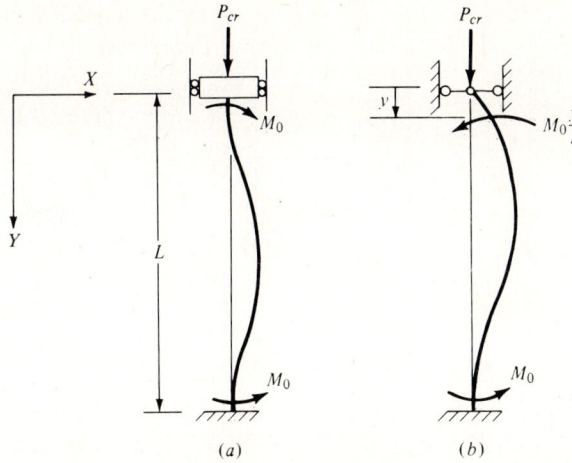

Figure 6-2 Column with end moments (*a*) on both ends; (*b*) pinned one end and fixed on base end.

Taking the derivative dx/dy for the slope, we obtain

$$\frac{dx}{dy} = \frac{kM_0}{P}\left(\frac{1-\cos kL}{\sin kL}\cos ky - k\sin ky\right) \quad (i)$$

Since $dx/dy = 0$ at $y = 0$, we have

$$\frac{kM_0}{P}\frac{1-\cos kL}{\sin kL} = 0$$

The smallest root of this equation gives $kL = 2\pi$ and with $k = (P/EI)^{1/2}$, we have

$$P = \frac{4\pi^2 EI}{L^2} = \frac{\pi^2 EI}{(0.5L)^2} = \frac{\pi^2 EI}{(KL)^2}$$

or the effective column length KL is $L/2 = 0.5L$ and $K = 0.5$. We have now introduced one of the first attempts to adjust the column length for end conditions. The effective length KL or some equivalent is used in nearly all design formulas in all building specifications.

When the top of the column is fixed against translation and the base fixed against both translation and rotation as in Fig. 6-2*b*, we can rewrite Eq. (*f*) as

$$EI\frac{d^2x}{dy^2} = -Px + M_0\frac{y}{L}$$

and in a similar manner obtain

$$P_{cr} = \frac{\pi^2 EI}{(0.7L)^2}$$

or $K = 0.7$. Several of the more common cases of column end fixity are shown in Fig. 6-3.

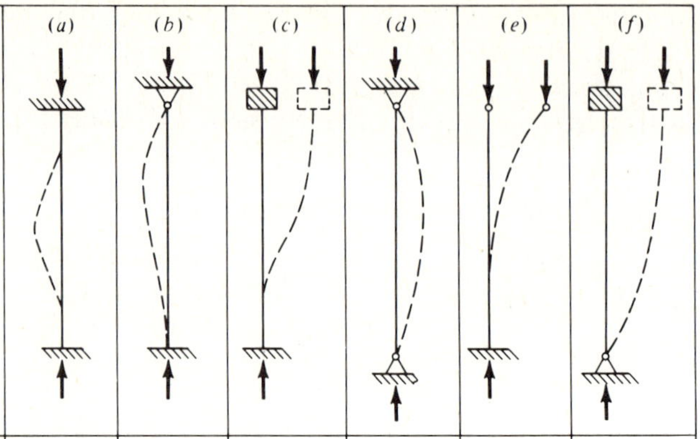

260 STRUCTURAL STEEL DESIGN

	(a)	(b)	(c)	(d)	(e)	(f)
Theoretical K	0.5	0.7	1.0	1.0	2.0	2.0
Recommended design K	0.65	0.80	1.2	1.0	2.1	2.0

End conditions:
- Rotation fixed, Translation fixed
- Rotation free, Translation fixed
- Rotation fixed, Translation free
- Rotation free, Translation free

Figure 6-3 Theoretical and design values of K for columns with end conditions shown.

Strictly, according to the derivation of these equations for critical buckling load, the tangent modulus of elasticity E_t should replace the elastic modulus in Eq. (6-3) when the column stress is greater than the proportional limit. The tangent modulus concept was introduced in Sec. 3-7.

6-4 ALLOWABLE STRESSES IN STEEL COLUMNS

The column stress obtained with the Euler equation of

$$F_{cr} = \frac{\pi^2 E}{(KL/r)^2}$$

is the equation of a parabola, and thus it would seem appropriate to use a parabolic equation form to develop the allowable column stress. It would similarly seem appropriate that KL/r would be one of the significant parameters. Further, since the Euler stress is a buckling stress, it would be necessary to apply a safety factor which could account for this stress level and additionally account for eccentricity, residual stresses, and the several other factors that complicate the theory. It would also be appropriate to use a variable safety factor that accounts for a lessening effect of some of these variables as the length (or KL/r) decreases. For example, a small eccentricity is not as critical for a short column as for a long slender one, and so on. With these considerations, let us investigate the AISC column equation for axially loaded columns.

6-4.1 AISC Axially Loaded Column Design Stresses

We develop the AISC design stresses by taking the Euler equation for critical stress, and with $n = 1$ we have

$$F_{cr} = \frac{\pi^2 E}{(KL/r)^2}$$

and differentiating obtain

$$\frac{d(F_{cr})}{d(KL/r)} = \frac{-2\pi^2 E}{(KL/r)^3} \qquad (a)$$

Since the maximum value of the Euler stress or any critical buckling stress is limited to F_y, let us also define the critical stress in any region where the Euler stress is not valid (such as small KL/r values):

$$F_{cr} = F_y - m\left(\frac{KL}{r}\right)^p \qquad (b)$$

This equation can be differentiated to obtain

$$\frac{d(F_{cr})}{d(KL/r)} = -mp\left(\frac{KL}{r}\right)^{p-1} \qquad (c)$$

We will take the slope $d(F_{cr})/d(KL/r) = 0$ at $KL/r = 0$. Also, at some point where we will arbitrarily define a parameter

$$\frac{KL}{r} = C_c$$

the slopes of Eqs. (a) and (c) will be equal (i.e., the two curves defined by these separate equations will have a common tangent). Also, experimental column test data indicate that taking $p = 2$ is adequate. Now equating slopes at $KL/r = C_c$ and for $p = 2$, we obtain

$$-2m(C_c)^2 = \frac{-2\pi^2 E}{(C_c)^3}$$

from which obtain m as

$$m = \frac{\pi^2 E}{C_c^4} \qquad (d)$$

Now from rearranging Eq. (b), inserting $KL/r = C_c$, $p = 2$, and using the Euler value for F_{cr}, we obtain

$$F_y = \frac{\pi^2 E}{C_2^2} + \frac{\pi^2 E C_c^2}{C_c^4} \quad (e)$$

and solving, we obtain for C_c,

$$C_c = \left(\frac{2\pi^2 E}{F_y}\right)^{1/2} \quad (6\text{-}4)$$

If the value of m, p, and C_c are placed in Eq. (b), we obtain

$$F_{cr} = F_y - \frac{\pi^2 E C_c^2}{C_c^4} = F_y - 0.5 F_y = 0.5 F_y$$

Thus the critical buckling stress becomes $0.5 F_y$ at $KL/r = C_c$ under our assumptions. In general, we have the buckling stress from Eq. (b):

$$F_{cr} = F_y \left[1.0 - \frac{0.5(KL/r)^2}{C_c^2}\right] \quad (f)$$

and the allowable stress is obtained for $KL/r \leq C_c$ from Eq. (f):

$$F_a = \frac{F_{cr}}{SF} = \frac{F_y}{SF}\left[1.0 - \frac{0.5(KL/r)^2}{C_c^2}\right] \quad \left(\frac{KL}{r} \leq C_c\right) \quad (6\text{-}5)$$

The AISC has used the following variable safety factor since 1963:

$$SF = \frac{5}{3} + \frac{3}{8}\frac{KL/r}{C_c} - \frac{1}{8}\frac{(KL/r)^3}{C_c^3} \quad (6\text{-}6)$$

for all values of $KL/r \leq C_c$. For $KL/r > C_c$, use a constant value of safety factor based on using $KL/r = C_c$ in the equation above to give

$$SF = \frac{23}{12} = 1.92$$

When $KL/r > C_c$, the Euler equation with $SF = 23/12$ is used to obtain the allowable column stress as

$$F_a = \frac{\pi^2 E}{SF(KL/r)^2} \quad \left(\frac{KL}{r} > C_c\right) \quad (6\text{-}7)$$

With standard values of $\pi^2 E$, we obtain:

$$\text{fps:} \quad F_a = \frac{149\,000}{(KL/r)^2} \quad \text{ksi}$$

$$\text{SI:} \quad F_a = \frac{1.03 \times 10^6}{(KL/r)^2} \quad \text{MPa}$$

Equations (6-5) and (6-7) are used for main compression members. Secondary compression members may be designed for an allowable stress based on the following amplification factor using L/r ($K = 1$) *when L/r exceeds 120*:

$$\Psi = \frac{1}{1.6 - L(200r)} \tag{6-8}$$

We obtain the allowable design stress using either Eq. (6-5) or (6-7):

$$F'_a = F_a \times \Psi$$

Equations (6-5) to (6-7) are somewhat awkward for routine computations (even with the programmable desktop calculators), and it is convenient to write a computer program to produce a table of F_a vs. KL/r for various values of F_y shown in Tables II-5, II-6, and VI-5, VI-6. The AISC manual has more complete tables using the several commonly used grades of F_y, including the amplification factor in Eqs. (6-5) and (6-7) for secondary members when $L/r > 120$.

6-4.2 AASHTO Axially Loaded Column Design Stresses

The AASHTO formulas for axially loaded columns are derived similarly to the AISC values, but the SF tends to be somewhat more conservative, since the members of the bridge structure are, in general, in a more hostile environment than building members. The AASHTO formulas are as follows:

For $KL/r \leq C_c$:

$$F_a = \frac{F_y}{SF}\left[1 - 0.5\frac{(KL/r)^2}{C_c^2}\right] \tag{6-9}$$

For $KL/r > C_c$:

$$F_a = \frac{\pi^2 E}{SF(KL/r)^2} \tag{6-10}$$

The SF = 2.12 for the AASHTO specifications and the values for the length factor K and C_c are computed the same as for the AISC specifications. Table 6-1 gives C_c for the several values of F_y commonly used for columns.

Table 6-1 Values of C_c according to the AISC and AASHTO specifications for several values of F_y; $C_c = (2\pi^2 E/F_y)^{1/2}$

F_y		
fps, ksi	SI, MPa	C_c
36	250	126.1
42	290	116.7
46	315	111.6
50	345	107.0
60	415	97.7
65	450	93.8

6-4.3 AREA Axially Loaded Column Design Stress

The AREA allowable column design stress formulas are somewhat similar to the AASHTO and AISC formulas and are given as follows:

Limitation	Allowable stress	
$\dfrac{kL}{r} \leq \dfrac{107}{\sqrt{F_y}}$	$F_a = 0.55 F_y$	(6-11)
$\dfrac{107}{\sqrt{F_y}} < \dfrac{kL}{r} \leq \dfrac{857}{\sqrt{F_y}}$	$F_a = 0.6 F_y - \left(\dfrac{F_y}{166.2}\right)^{1.5} \cdot \dfrac{kL}{r}$	(6-12)
$\dfrac{kL}{r} > \dfrac{857}{\sqrt{F_y}}$	$F_a = \dfrac{147\,000}{(kL/r)^2}$	(6-13)

In the AREA specifications,

F_y = ksi

k = 0.75 for riveted, bolted, or welded compression member end connections (and k, not K)

  = 0.875 for pinned-end members

6-4.4 Net versus Gross Column Cross-sectional Area

The net area (= gross area − loss for holes) is used in tension member design. In any connection design using mechanical fasteners (rivets or bolts), it is assumed that the fastener completely fills the hole. This assumption is very nearly met in riveted work, where the head fabrication enlarges the rivet shank, and very nearly occurs in bolted work since the hole is only about 1.5 mm larger than the nominal bolt diameter. Under axial compression, although there are stress concentrations at the hole, it can be safely assumed that no loss of net area occurs when a mechanical fastener fills the hole. When an intermediate open hole is in a compression member (as for utilities, erection, etc.), the designer must exercise judgment as to whether to use the net or the gross area. The AISC specifications give the allowable stresses as "On the gross section of axially loaded compression members . . . ." Undoubtedly, there will be adequate arching to transmit the load around the hole if there is only a small amount of area lost at any section due to the holes. The most conservative procedure would be to use the gross section when the hole has a mechanical fastener that fills, or nearly fills, it and the net section for all other cases.

6-4.5 Column Design

It is necessary to use an iterative process in the design of compression members using any of the AISC, AASHTO, or AREA allowable stress equations. The usual design problem involves the following steps:

1. Determine column loads (unless the problem only involves checking a section for adequacy).
2. Determine the effective column length KL (or k for AREA).
3. Make a tentative section selection (sometimes use can be made of tables such as Table II-4 or Table VI-4 of SSDD).
4. Compute KL/r for the section selected (now that r is known) and use the appropriate stress equation to compute F_a (or use tables such as Table II-5 of SSDD, which gives F_a for several KL/r ratios).
5. Compute the allowable column load $P_a = A \times F_a \geq P_{\text{design}}$.
6. Revise the section until P_a is reasonably close (and slightly larger) than P_{design}. The lightest section is obtained (not necesarily the most economical) when $P_a = P_{\text{design}}$.

Two additional factors should be given consideration:

1. Commonly, only the W8, W10, W12, and W14 sections and rectangular tube and round pipe sections are used for columns, since the critical radius of gyration is with respect to the Y axis. These sections have the best r_y values (and corresponding r_x/r_y ratios). In building design, practical considerations often necessitate use of a given nominal column size throughout the building. It is usual in building construction to run a single column through at least two, and often three or more floors to avoid field column splices. The labor savings more than offset the increased weight of steel.
2. When $KL/r > C_c$, the AISC specification requires use of Eq. (6-7), the AASHTO specification requires Eq. (6-10), and AREA has a somewhat similar requirement for Eq. (6-13). In all these equations F_a is independent of F_y. Therefore, in column design one should use A-36 steel for all cases where KL/r exceeds C_c or the AREA limitation, and even if $F_y > 36$ ksi is being used for some of the other members. For example, if we are using $F_y = 60$ ksi and a section is found where $KL/r > 97.7$ (refer to Table 6-1), we should try to specify A-36 steel instead of the more expensive 60-ksi steel. The slightly larger safety factor in Eq. (6-7) allows for the transition between $KL/r = C_c = 126.1$ (for A-36 steel) and the lower values of C_c for the higher-strength steels.

Advantage should be taken of any available design aids in making a preliminary column selection. Tables such as Table II-4 or VI-4 of SSDD can often be used to make the final design. Tables II-5 and II-6 and the corresponding SI tables (Tables VI-5 and VI-6) can be used to advantage for any other

266 STRUCTURAL STEEL DESIGN

(and including W) shapes to quickly obtain F_a when KL/r is computed—particularly when $KL/r < C_c$, since both stress and safety factor now depend on KL/r.

6-4.6 Design Examples

The design of simple axially loaded columns and struts will be illustrated in the following examples.

Example 6-1 Design a column to be used in a one-story discount department store building. Columns are spaced 20 ft on center both ways. The roof load is taken as 30 psf dead load and 60 psf snow. This gives a column load of $20(20)(0.090) = 36$ kips. Use A-36 steel ($C_c = 126.1$).

SOLUTION From inspection of Fig. E6-1, take $K = 1$. By using Table II-4 as a guide, the lightest W section (W8 × 18) for $KL = 14$ ft can carry 41 kips. This could be a solution but is not very economical. We will first check this W8 section and then compare to using a tube column.

Check the a W8 × 18 section:

$$r_y = 1.23 \qquad A = 5.26 \text{ in}^2$$

$$\frac{KL}{r_y} = \frac{14(12)}{1.23} = 136.6 > C_c \qquad [\text{use Eq. (6-7)}]$$

$$F_a = \frac{149\,000}{(136.6)^2} = 7.99 \text{ ksi}$$

$$P_a = 5.26(7.99) = 42.0 > 36 \text{ kips} \qquad \text{O.K.}$$

Use a W8 × 18 section (tentatively); the Table II-4 value of 41 kips is due to computer roundoff.

Let us also investigate a pipe section (see Table I-14). Now that we have some "experience" from just checking a W shape, let us investigate a 4-in-diameter pipe:

$$r_x = r_y = 1.51 \text{ in} \qquad A = 3.17 \text{ in}^2 \qquad wt = 10.79 \text{ lb/ft}$$

$$\frac{KL}{r} = \frac{14(12)}{1.51} = 111.25 < C_c$$

From Eq. (6-5),

$$F_a = \frac{36}{SF}\left[1.0 - 0.5\left(\frac{111.25}{126.1}\right)^2\right] = \frac{21.98}{SF}$$

AXIALLY LOADED COLUMNS AND STRUTS **267**

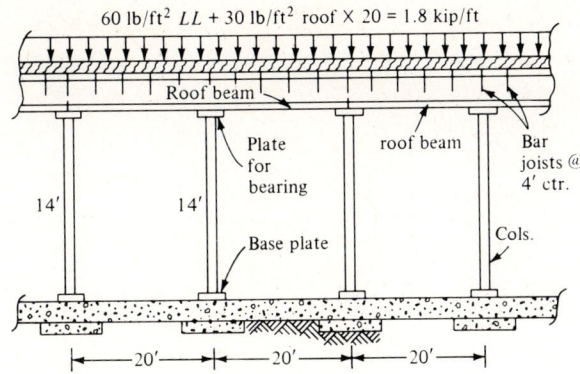

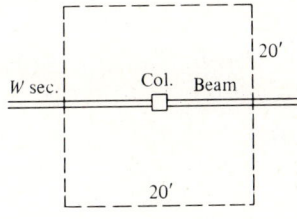

Figure E6-1

From Eq. (6-6), the safety factor is computed as

$$SF = 1.67 + 0.375\left(\frac{111.25}{126.1}\right) - 0.125\left(\frac{111.25}{126.1}\right)^3$$

$$= 1.915$$

$$F_a = \frac{21.98}{1.915} = 11.5 \text{ ksi (vs. 11.54 in Table II-5)}$$

$$P_a = 11.5(3.17) = 36.45 > 36 \text{ kips} \qquad \text{O.K.}$$

Use steel pipe 4 in × 10.79 lb/ft.

Note that pipe and any other sections where $r_x/r_y \to 1$ are generally the most economical shapes to use when $KL_x = KL_y$. ///

Example 6-2 Design a W section for the conditions shown in Fig. E6-2. Use $F_y = 50$ ksi steel and the AISC specifications.

SOLUTION

$$KL_x = 16 \text{ ft}$$
$$KL_y = 8 \text{ ft}$$

We may use Table II-4 to obtain an initial estimate of column size. Since Table II-4 is based on $F_y = 36$ ksi, use a ratio as

$$P_{equiv} = (750)\frac{36}{50} = 540 \text{ kips}$$

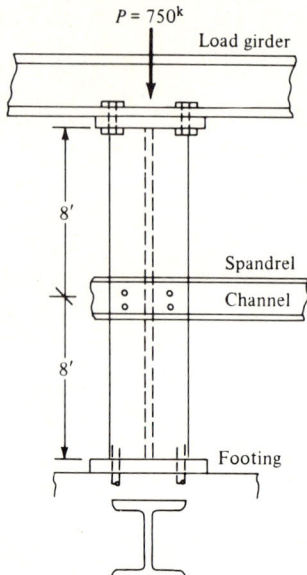

Figure E6-2

Since the table is based on KL/r_y, we note that $r_x/r_y \simeq 1.6$ in the column sizes approaching $P = 540$ kips; therefore, look for $L = 1.6 \times 8 = 13$ ft column length and 540 kips.

Try W14 × 99:

$$A = 29.10 \text{ in}^2 \qquad r_y = 3.71 \text{ in} \qquad \frac{r_x}{r_y} = 1.66$$

Compute

$$r_x = 1.66 r_y = 1.66(3.71) = 6.16 \text{ in}$$

$$\frac{KL}{r_x} = \frac{16(12)}{6.16} = 31.2 \qquad \text{controls since it is largest}$$

$$\frac{KL}{r_y} = \frac{8(12)}{3.71} = 25.9$$

$$C_c = 107 \qquad \text{(Table 6-1)}$$

$$F_a = \frac{50}{SF}\left[1.0 - 0.5\left(\frac{31.2}{107}\right)^2\right] = \frac{47.9}{SF}$$

$$SF = 1.67 + 0.375\left(\frac{31.2}{107}\right) - 0.125\left(\frac{31.2}{107}\right)^3 = 1.78$$

$$F_a = \frac{47.9}{1.78} = 26.9 \text{ ksi} \qquad (\simeq 27.0 \text{ in Table II-6})$$

$$P_a = 26.9(29.10) = 782.8 > 750 \text{ kips} \qquad \text{O.K.}$$

This appears to be the lightest rolled section possible for this loading situation. ///

Example 6-3 Check the pair of angles selected in Example 5-6 to be used as vertical members in the main roof truss of Example 2-6. Use A-36 steel and the AISC specifications.

Member	LC-1 (kN)	LC-2 (kN)	L, m
39	−63.65	+20.39	6.76
47	+70.18	−37.60	8.2

The section selected for tension considerations was

$$2L127 \times 89 \times 6.3: \quad A = 2.66 \times 10^{-3} \text{ m}^2 \quad r_{min} = 37.5 \text{ mm}$$

SOLUTION Take $K = 1$. For member 39:

$$\frac{KL}{r} = \frac{6.76(1000)}{37.5} = 180 < 200 \quad \text{O.K.}$$

For member 47:

$$\frac{KL}{r} = \frac{8.2(1000)}{37.5} = 218.7 > 200 \text{ max. (AISC Sec. 1-8.4)} \quad \text{N.G.}$$

Revise the section for minimum r:

$$r_{min} = \frac{8.2(1000)}{200} = 41 \text{ mm}$$

Note that a 12-mm gusset plate is used between angles at the connection. Possible angles are:

$2L152 \times 152 \times 7.9:$ $r_{min} = 48.0$
$A = 4.71 > 0.5504$ (required for tension)

$2 L152 \times 102 \times 6.3:$ $r_{min} = 41.5$ mm
$A = 3.15 > 0.5504$

We will check the pair of angles $152 \times 102 \times 6.3$, since they are the lightest.
Member 47:

$$\frac{KL}{r} = \frac{8.2(1000)}{41.5} = 197.6 \rightarrow F_a = 26.3 \text{ MPa} \quad \text{(Table VI-5)}$$

Member 39:

$$\frac{KL}{r} = \frac{6.76(1000)}{41.5} = 162.9 \rightarrow F_a = 38.8 \text{ MPa}$$

$$P_{39} = 38.8(3.15) = 122.2 \text{ kN} > 63.65 \quad \text{O.K.}$$

$$P_{47} = 26.3(3.15) = 82.8 \text{ kN} > 37.6 \quad \text{O.K.}$$

Use two L152 × 102 × 6.3. We note that L/r controlled the design in both tension and compression. ///

Example 6-4 Design member 7 of the highway bridge truss of Fig. E6-4 (same truss as Example 5-7).

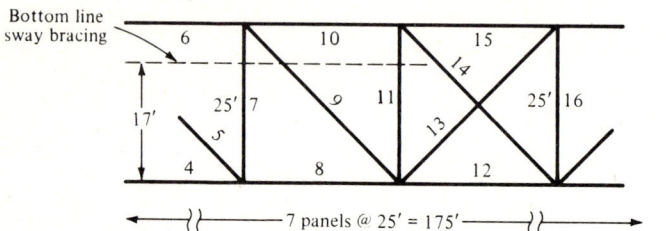

Figure E6-4

SOLUTION Computer output:

$$\text{Dead load} = -56.75 \text{ kips}$$
$$\text{Live load} = -42.56 \text{ kips}$$
$$\text{Live load} = +16.11 \text{ kips}$$

Note that this member has an effective out-of-plane length of 17 ft (as in Example 5-7) due to the framing of lateral and sway bracing across the top of the two trusses making up the bridge. The impact factor has been included in the live load but is computed as follows:

$$I = \frac{50}{L + 125} = \frac{50}{300} = 0.17$$

Maximum column force = 56.75 + 42.56 = 99.3 kips (compression)
Minimum column force = 56.75 − 16.11 = 40.6 kips (compression)
$P_{sr} = P_{max} - P_{min} = 99.3 - (+40.6) = 58.7$ kips (stress range)

The AASHTO specifications (Sec. 1-7.5) limits KL/r for compression members to $KL/r = 120$ for main members or members with both dead- and live-load stresses. The minimum r is

$$r_{min} = \frac{25(12)}{120} = 2.50 \text{ in}$$

Try a W12 section, since we have used a W12 × 53 in Example 5-7 for member 9, which frames into the joint on one end of this member. A gusset plate can cover both members with a minimum of filler material if all the web members have the same nominal depth.

We further note that Eq. (6-9) always determines the column stress for members using A-36 steel since $KL/r \le 120$ and $C_c = 126.1 > 120$.

AXIALLY LOADED COLUMNS AND STRUTS **271**

By trial [and programming Eq. (6-9) on a pocket programmable calculator]:

Section	A_{furn}, in^2	r_y, in	F_a, ksi	$P_{\text{allow}} = AF_a$, kips
W12×53	15.60	2.48		N.G., since r_y too small
W12×58	17.00	2.51	9.35	159.0 > 99.3 O.K.

The r_x value for the W12 × 58 is 5.28 > 2.5, so the section is satisfactory. Also, this section is the same size as member 9, so the joint will be easy to fabricate. Use a W12 × 58 section. ///

Example 6-5 Design the struts for the cable-supported roof of Example 5-9. The strut load is 2.2 kips based on cables at 4 ft on centers and struts as shown in Fig. E6-5. The maximum strut length is 2(23.3) = 46.6 ft.

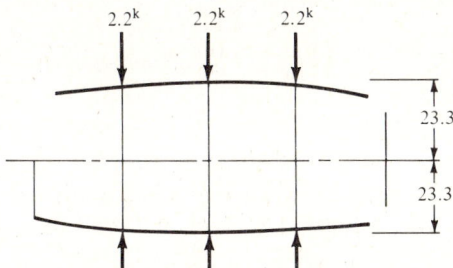

Figure E6-5

SOLUTION The maximum KL/r for compression members is limited to 200. According to the AISC specifications, this would require a radius of gyration r of at least

$$r = \frac{1(46.6)(12)}{200} = 2.80 \text{ in}$$

Commonly, pipe struts are used with diameters ranging from 4 to 6 in. Here we have a rather large unbraced length, so examination of Table 1-14 (SSDD) indicates that we can use:

Extra strong pipe: 8-in diameter: $r = 2.88 > 2.80$ in

$A = 12.8$ in^2

For $KL/r = 46.6(12)/2.88 = 194.2$,

$F_a = 3.95$ ksi   (Table II-5)

$P_{\text{allow}} = AF_a = 12.8(3.95) = 50.6 \ggg 2.2$ kips

Use 8-in-diameter extra-strong pipe. It may be advantageous to use a smaller section in the outer one-fourth of the span, where the value of $L = 2h$ is less than 46.6 ft.   ///

6-5 DESIGN OF BUILT-UP COMPRESSION MEMBERS

A built-up section is a more practical design than using a rolled shape in many situations. This is particularly true when there is a very long unsupported column length involved such that to meet the L/r requirements would require one of the heavier rolled shapes. Another factor of primary importance is that the radius of gyration of built-up members can be controlled (see Table 6-2 for selected examples so that the value of r_x can be made more nearly equal to r_y, to produce maximum section efficiency. This efficiency cannot be obtained using the standard rolled W shapes, where the ratio of r_x/r_y is often 1.5 to 5 or more, unless bracing is provided with respect to the Y axis.

Table 6-2 Approximate radius of gyration for several built-up shapes

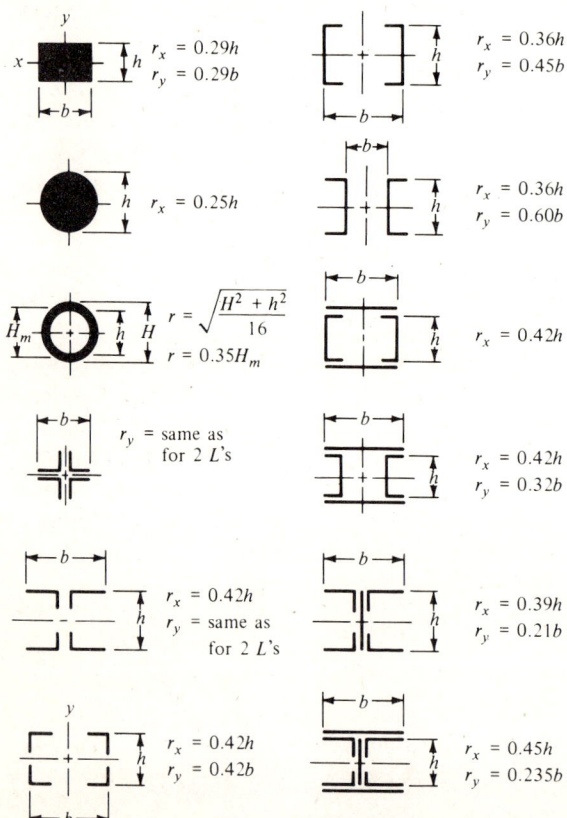

Built-up sections are very commonly used for bridge trusses and in columns for water towers. Antennas are essentially built-up columns, although not directly considered as such. In any case, where compression (and tension) members are used in large unsupported spans, a built-up member may need to be considered. Any of the sections considered in Chap. 5 may be used (refer to Figs. 5-4, 5-7, and 5-8) in addition to any other section configuration, which may, or could, be made appropriate for the design problem.

It is somewhat more difficult to produce an optimum, or least-weight, built-up section since there are several design parameters to satisfy, including:

1. Types of members to use, including rolled angles; channels; W, S, or M shapes, as well as plate segments.
2. Arrangement of the basic members, including any size limitations for overall section dimensions.
3. The resulting computed values of I_x, I_y, r_x, and r_y and KL/r_{max}, which produce the allowable compression stress.
4. Producing an acceptable section area based on the allowable stress from step 3, which is not known until the area has already been established.

One or more iterations are usually required to finally develop a satisfactory section. The number of iterations obviously will depend on

1. Engineering design versus material cost.
2. Number of sections to be fabricated; if 100 sections are to be fabricated, the material savings can be considerable, whereas in the fabrication of only four or five sections, the cost of producing a refined design might cost more than the material saved.
3. Reliability of load data and intended use of the built-up members.

In general, a 5 to 15 percent "overdesign" of a built-up member will be satisfactory.

Built-up sections may be built using rolled shapes as in Fig. 6-4, but more commonly are constructed using lacing, perforated cover plates, or batten plates, as shown in Fig. 5-7. A larger radius of gyration and more efficient use of steel is obtained by separating the load-carrying parts as much as practical (and as done with the flanges of W shapes). Where this is done, it is necessary to interconnect the parts so that these several parts act as a load-carrying unit. Several methods

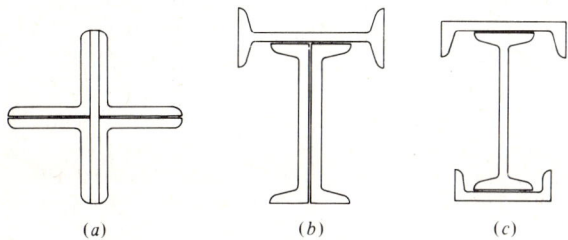

Figure 6-4 Built-up shapes using combinations of rolled shapes. Section geometry limited only by imagination, load and clearances. (*a*) Four angles and plate. (*b*) Two channels and S shape. (*c*) Two channels and S shape.

of using lacing, single and double batten, and perforated plates (called, collectively, cover plates) are usually used. Where the steel is located inside a building, the cover plates may be solid and their use could reduce fabrication costs. In exterior environments, where corrosion is always a problem, it is necessary to have access to the interior of the section for maintenance and inspection; otherwise, the interior must be completely sealed. The "open" cover plates and lacing allow access to the interior of the section for cleaning and painting without the careful fabrication required to completely seal the interior. Presently, it appears that the economics of fabrication favors perforated cover plates to lacing, since automatic gas-cutting methods allow rapid cutting of the plate openings in a length of cover plate.

The design of lacing and batten plates, in particular, requires attention to several details. Lack of proper attention to lacing design was believed to have caused the first Quebec Bridge in Canada to fail in 1907. It is standard practice to allocate a portion of the axial load as the shear developed in the lacing or batten plate when the compression member "buckles," as shown in Fig. 6-5.

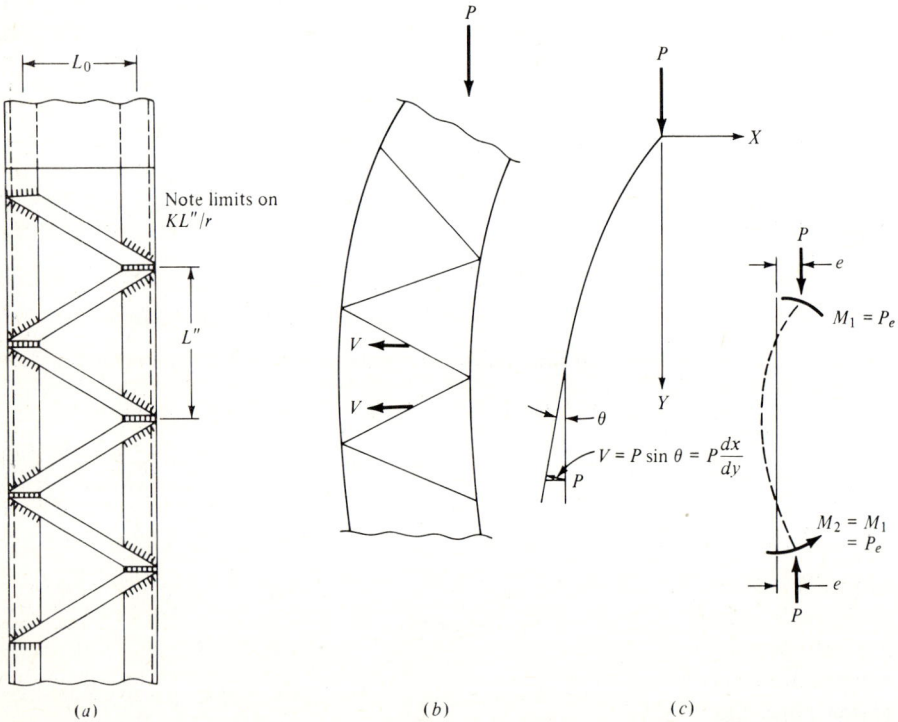

Figure 6-5 Shear development for a laced (or battened) compression member. (*a*) Laced column. (*b*) Identification of laced column shear V. For lacing on both through faces, divide V equally on both lacing bars; for four-side lacing obtain 90° as above. (*c*) Basic concept of shear in lacing of built-up section.

If we assume equal end moments, as shown in Fig. 6-5c, and use the differential equation for bending as used to develop the Euler column equation, and allow for boundary conditions, we obtain

$$x = e\left(\tan\frac{kL}{2}\sin ky + \cos ky - 1\right)$$

The derivative at $y = 0$ is

$$\frac{dx}{dy} = ke\tan\frac{kL}{2}$$

Now referring to Fig. 6-5b, we obtain for the shear in the lacing,

$$V = P\left(ke\tan\frac{kL}{2}\right)$$

where $k = P/EI$ as in the Euler equation. AISC simply takes

$$ke\tan\frac{kL}{2} = 0.02$$

The AASHTO and AREA specifications make the assumption that the end eccentricity e shown in Fig. 6-5c is equal and opposite on the two ends of the column (shown equal and with the same sign in Fig. 6-5c). With some additional simplification of the preceding equation for V, we obtain

$$V = \frac{P}{100}\left(\frac{100}{L/r + 10} + \frac{L/r}{C/F_y}\right) \qquad (6\text{-}14)$$

where F_y = steel yield stress, ksi or MPa
L/r = value for entire member with respect to an axis perpendicular to the plane of lacing or perforated cover plate as follows:

AASHTO		AREA	
fps	SI	fps	SI
3300	22 920	3600	25 000

AASHTO and AREA specifications also require that V be increased for any additional shear on the member, such as section weight or other transverse loadings. Wind on bridge trusses would also contribute an increase in V under this interpretation. The value of V obtained for either the AASHTO or AISC computation may be either a tension or a compression value, and the lacing or batten plate should be so designed.

The spacing of lacing and batten plates must be such that the L/r of the main elements between fasteners is not greater than KL/r of the entire member; otherwise, local buckling might develop, particularly where angles are used, and L/r_z between fastener points may be critical. The AASHTO and AREA specifications actually limit the fastener distance to

$$\frac{L}{r} \leq \frac{2}{3} \frac{KL}{r}_{\text{entire member}}$$

or

$$\frac{L}{r} \leq 40$$

whichever is less. It may be necessary to terminate lacing at interior points where gusset plates are used to connect transverse members. At the ends, tie plates (called stay plates by AASHTO and AREA) are usually used because of the gusset plates used to connect the member to the rest of the structure. Several of the lacing and tie or stay plate requirements for the three specifications are as follows:

	AISC	AASHTO	AREA
K for: single lacing	1	1	1
double lacing	0.7	0.7	0.7
Single lacing $L' \leq$ (Fig. 6-5a) Double lacing $L' >$		15 in or 380 mm 15 in or 380 mm	
Single lacing angle α Double lacing angle α		$\simeq 60°$ $\simeq 45°$	
t of lacing bar	As required for L/r	Single $\leq L/40$ Double $\leq L/60$ Decrease 25 percent for secondary members	
w of lacing bar	As required	$3 \times$ diameter of fastener	

The general provisions for the design of perforated cover plates by the various design specifications are summarized in Fig. 6-6.

The design of a built-up member requires at least the following steps:

1. General outside dimensions and load to be carried.
2. Estimate the tentative compression area based on assuming an F_a between 15 and 20 ksi or 100 to 140 MPa (based on $F_y = 36$ ksi). This allows a modest reduction in the allowable stress from $0.6F_y$ due to the KL/r of the built-up shape.
3. Decide on lacing, batten, or perforated cover plates (Fig. 5-7) or if section is to be something like that of Fig. 6-4.
4. Compute area, I_x, I_y, r_x, and r_y. The moment of inertia of the built-up section is

$$I_i = \Sigma I_{0(j)} + \Sigma A_j d_{j(i)}^2$$

AXIALLY LOADED COLUMNS AND STRUTS **277**

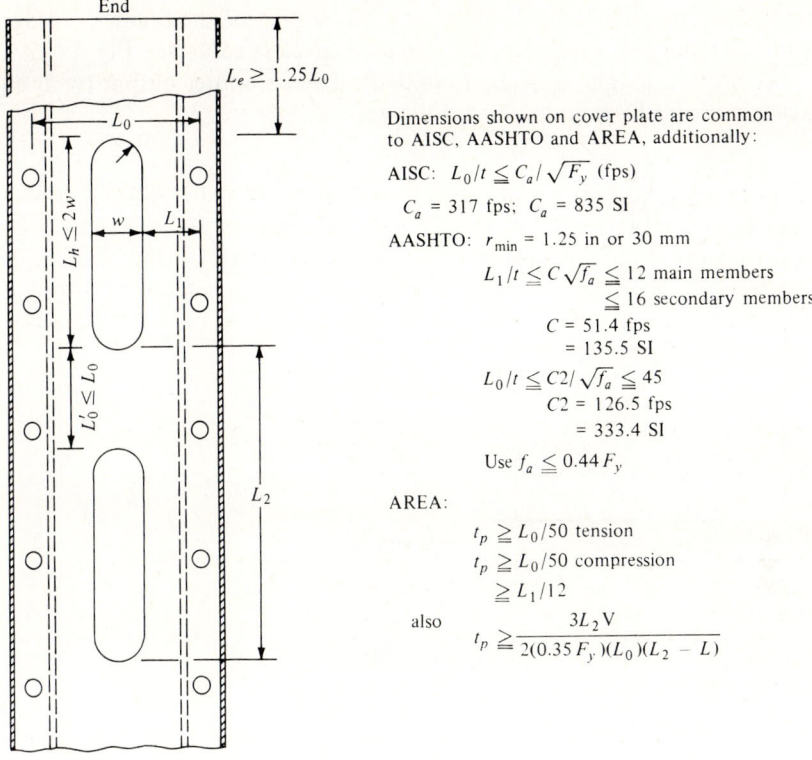

Dimensions shown on cover plate are common to AISC, AASHTO and AREA, additionally:

AISC: $L_0/t \leq C_a/\sqrt{F_y}$ (fps)

C_a = 317 fps; C_a = 835 SI

AASHTO: r_{min} = 1.25 in or 30 mm

$L_1/t \leq C\sqrt{f_a} \leq$ 12 main members
$\leq$ 16 secondary members
C = 51.4 fps
= 135.5 SI

$L_0/t \leq C2/\sqrt{f_a} \leq$ 45
$C2$ = 126.5 fps
= 333.4 SI

Use $f_a \leq 0.44 F_y$

AREA:

$t_p \geq L_0/50$ tension
$t_p \geq L_0/50$ compression
$\geq L_1/12$

also $t_p \geq \dfrac{3L_2 V}{2(0.35 F_y)(L_0)(L_2 - L)}$

Figure 6-6 Design of perforated cover plates.

where $I_{0(j)}$ = moment of inertia of the jth part with respect to the parallel axis i and through the centroid of the jth part
A_j = cross-sectional area of the jth part
$d_{j(i)}$ = perpendicular distance from the centroid of the jth area to the i axis

5. Compute the radius of gyration with respect to both axes.

$$r_i = \sqrt{\dfrac{I_i}{A}}$$

6. Compute KL/r_x and KL/r_y and obtain the allowable compressive stress based on the largest KL/r.
7. Check $P_{allow} = AF_a > P_{design}$ and iterate as necessary.
8. Design lacing, perforated cover plates, batten plates, and/or tie (or stay) plates.

This procedure will be illustrated by the following examples.

Example 6-6 Design a laced section for the end post of the highway bridge truss of Example 6-4, which has seven panels at 25 ft each (see Fig. E6-6a). The unsupported length $L_y = L_x$ is 24.04 ft. The computer output (with an impact factor of 0.17 included) is as follows:

Member	LC-11, kN	Dead load, kips
2	−94.13	−240.77

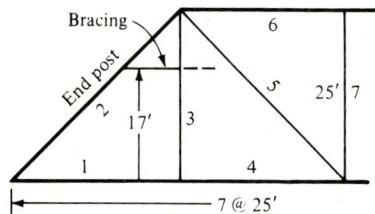

Figure E6-6a

SOLUTION

$$\text{Total design load} = 240.77 + 94.13$$
$$= 334.9 \text{ kips}$$

Assuming that $F_a \simeq 14$ ksi the area required in the section will be approximately

$$A \simeq \frac{334.9}{14} = 23.92 \text{ in}^2$$

Let us try two channels with a solid cover plate and lacing as shown in Fig. E6-6b. This configuration, with solid cover plate up, will provide some protection to the interior of the built-up section, and lacing will allow access for painting and cleaning. The spacing and configuration will be such that a reasonably easy framing of the W12 web sections can be made, as shown, using a pair of gusset plates. We note that filler plates will be required, since the W12's have depths greater than 12 in.

C12 × 30 data:

$$A = 8.82 \text{ in}^2 \qquad b_f = 3.17 \text{ in}$$
$$t_f = 0.501 \text{ in} \qquad t_w = 0.51 \text{ in}$$

Both t_f and $t_w > 0.23$ in (Sec. 1-7.7)

$$I_x = 162 \text{ in}^4 \qquad r_x = 4.29 \text{ in}$$
$$I_y = 5.14 \text{ in}^4 \qquad r_y = 0.763 \text{ in}$$
$$\bar{x} = 0.674 \text{ in}$$

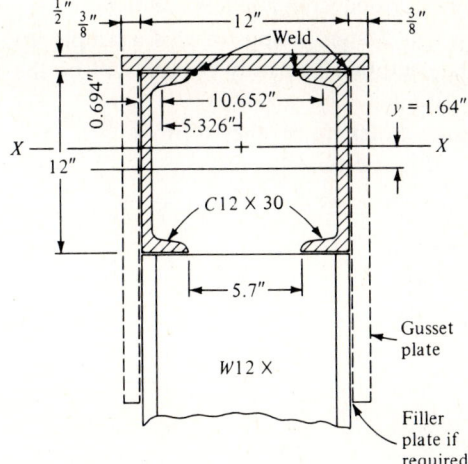

Figure E6-6b

Compute the radius of gyration about the X and Y axes:
About the X axis: locate a new X axis:

$$(2 \times 8.82 + 0.5 \times 12.75)y = 0.5(12.75)\left(\frac{12}{2} + 0.25\right)$$

$$y = \frac{39.84}{24.02} = 1.66 \text{ in}$$

$$I_{xx} = 2I_{x0} + 2A_c d^2 + A_p d^2$$

$$= 2(162) + 2(8.82)(1.66)^2 + 6.375(4.59)^2$$
$$= 324 + 48.6 + 134.4$$
$$= 507.0 \text{ in}^4$$

$$r_x = \left(\frac{507.0}{24.02}\right)^{1/2} = 4.59 \text{ in} \qquad \text{controls (after computing } r_y \text{ below)}$$

About the Y axis: $I_{yy} = 2I_y + 2A_c d^2 + I_p$

$$= 2(5.14) + 2(8.82)(5.326)^2 + \frac{0.5(12.75)^3}{12}$$
$$= 10.3 + 500.4 + 86.4 = 597.1 \text{ in}^4$$

$$r_y = \left(\frac{597.1}{24.02}\right)^{1/2} = 4.99 \text{ in} \qquad \frac{KL}{r_x} = \frac{24.04(12)}{4.59} = 62.8$$

$$F_a = \frac{36\left[1 - 0.5\left(\frac{62.8}{126.1}\right)^2\right]}{2.12} = 14.9 \text{ ksi} \quad [\text{using Eq. (6-9)}]$$

$$A_{\text{reqd}} = \frac{334.9}{14.9} = 22.5 < 24.02 \text{ in}^2 \text{ furnished} \qquad \text{O.K.}$$

280 STRUCTURAL STEEL DESIGN

Current practice is to use perforated cover plates welded to the rolled sections rather than lacing. We will design lacing for this example and use welding to attach it to the rolled sections. The only bolting will be the field connections of the joints.

The distance between flange holes (and the approximate center-to-center weld distance) is shown in Fig. E6-6c.

$$L' = 12 - 2(1.75) = 8.5 < 15 \text{ in} \qquad \text{(also AISC)}$$

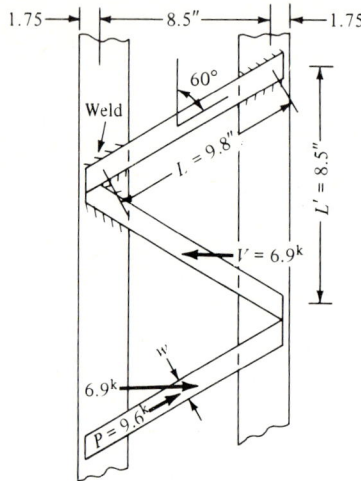

Figure E6-6c

Use single lacing bars at an angle of 60° to the member axis, as shown in the figure (refer also to Fig. 6-5a):

$$L = \frac{8.5}{\cos 30°} = 9.81 \text{ in}$$

$$L' = 2(8.5)\sin 30° = 8.5 \text{ in}$$

$$\frac{L'}{r_{y(\text{of channel})}} = \frac{8.5}{0.763} = 11.1 \ll 40$$

Also, 11.1 is less than $0.67 \times L/r$ (= 41.9). Limit the L/r of the lacing to 130. The radius of gyration of a flat bar is

$$r = \left[\frac{bt^3}{12(bt)}\right]^{1/2} = 0.288t$$

$$t = \frac{9.81}{130(0.288)} = 0.262 \text{ in}$$

$$t \geq \frac{L}{40} = \frac{9.81}{40} = 0.245 \text{ in}$$

Try $t = \frac{5}{16}$ in (0.313 in also minimum AASHTO):

$$\frac{L}{r} = \frac{9.81}{0.288(0.313)} = 109 \quad \text{O.K.}$$

Compute the lacing bar force. The bar force component perpendicular to the member axis is computed using Eq. (6-14):

$$V = \frac{P}{100}\left(\frac{100}{L/r + 10} + \frac{L/r}{3300/F_y}\right)$$

$$= \frac{334.9}{100}\left[\frac{100}{62.8 + 10} + \frac{62.8(36)}{3300}\right] = 3.349(2.06) = 6.9 \text{ kips}$$

We will increase this value 20 percent (author's decision) to allow for member weight, wind, and any other factors; thus V_{design} is

$$V_d = 1.2(6.9) = 8.3 \text{ kips}$$

The axial force is

$$P_d = \frac{8.3}{\cos 30°} = 9.6 \text{ kips}$$

$$F_a = \frac{36[1 - 0.5(109/126.1)^2]}{2.12} = 10.64 \text{ ksi}$$

$$bt F_a = 9.6 \text{ kips}$$

$$b = \frac{9.6}{0.313(10.75)} = 2.88 \text{ in}$$

Let us revise t to 7/16 in; $F_a = 13.78$ ksi:

$$b = \frac{9.6}{0.44(13.78)} = 1.58 \text{ in} \quad \text{use } 1\frac{3}{4}\text{-in plate}$$

The final lacing bar dimensions will be taken as ($L = 9.81 + 1.69$ in)

$$1\tfrac{3}{4}w \times \tfrac{7}{16}t \times 11.5L$$

Design the end tie plates (AASHTO calls these stay plates). The AASHTO requirements are:

$$L \geq 1.25 L_0 \quad t \geq \frac{L_0}{50}$$

Use a minimum of three fasteners (or equivalent weld) each side. Translated into design (Fig. E6-6d):

$$L = 1.25(12) = 15 \text{ in}$$

$$t = \frac{12}{50} = 0.24 \quad \text{use } \tfrac{7}{16}\text{-in, to match lacing bars}$$

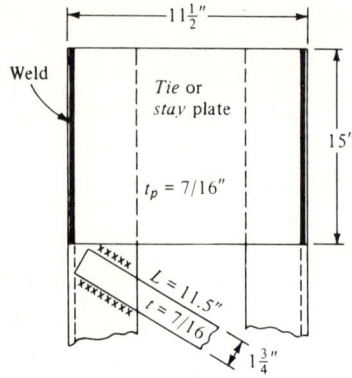

Figure E6-6d

///

Example 6-7 Design a built-up section using perforated cover plates for use as a column in a water tower (see Fig. E6-7a). The unsupported column length is 24.7 m and the axial load for design is 1350 kN. Use $F_y = 250$ MPa and the AISC specifications (noting that a water tower is not a "building" and may be located where a collapse is more of an expensive nuisance than a hazard to people, so it may not be necessary to use any specifications), since adherence to these specifications, although not necessary, will ensure a safe design. There is usually some wind bracing in water towers, but we will assume that the bracing is not sufficient to develop restraint against column buckling.

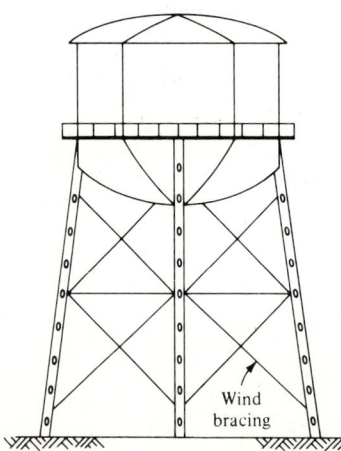

Figure E6-7a

SOLUTION The minimum KL/r will be taken as 200 for main members. Therefore,

$$r_{min} = \frac{24.7(1000)}{200} = 123.5 \text{ mm}$$

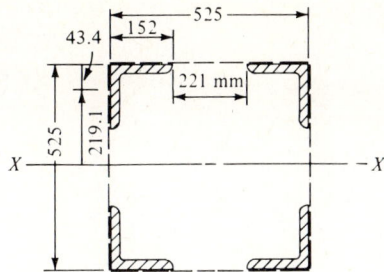

Figure E6-7b

For this L/r value of 200, the allowable stress $F_a = 25.7$ MPa. Therefore, we must use an r greater than this or the steel area will be excessive. For an $L/r = 100$, the value of $F_a = 90$ MPa and the area of steel will be approximately $1350/100 = 13.5 \times 10^{-3}$ m^2.

Try 4 L152 × 152 × 14.3 (somewhat arbitrary choice). The data are:

$$A = 4.184 \times 10^{-3} \text{ m}^2/\text{angle}$$

$$I_y = I_x = 9.199 \times 10^{-6} \text{ m}^4$$

$$\bar{x} = \bar{y} = 43.4 \text{ mm}$$

Place the angles in a symmetrical section with spacing as shown in Fig. E6-7b. Compute $I_x = I_y$ and $r_x = r_y$ using section data:

$$I_x = 4I_0 + 4Ad^2$$

$$I_x = 4(9.199) + 4(4.184 \times 10^{-3})(0.2191)^2 = 36.796 + 803.4$$

$$= 840.2 \times 10^{-6} \text{ m}^2$$

$$r_x = r_y = \sqrt{\frac{840.2 \times 10^{-6}}{4 \times 4.184 \times 10^{-3}}} = \sqrt{50.2 \times 10^{-3}} = 224 \text{ mm}$$

$$\frac{L}{r} = \frac{24.7(1000)}{224} = 110.26 \quad \text{and from Table VI-5 obtain } F_a = 81 \text{ MPa}$$

$$P_{\text{allow}} = 81(4.184 \times 4) = 1356 > 1350 \text{ kN required} \quad \text{O.K.}$$

This cross section will be considered adequate and we will proceed to design the perforated cover plates.

The cover plate design is not so much "design" as satisfying selected criteria and producing a hole spacing that will fit the column length. Referring to Figs. E6-7c and 6-6, we obtain:

$$L_0 = 221 \text{ mm}$$

$$r = 75 \text{ mm (arbitrary selection of hole radius)}$$

$$w = 2r = 150 \text{ mm}$$

Use $L_1 \geq 1.25 L_0 \to 0.45$ m. Length of cover plate = $24.7 - 2(0.45) = 23.8$ m. This choice is made so that with a hole length of $L_h = 300$ mm and

284 STRUCTURAL STEEL DESIGN

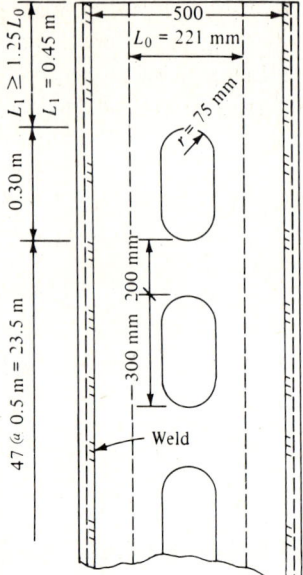

Figure E6-7c

$L'_0 = 200$ mm the length of 23.8 m gives

$$48 \text{ holes} = 48(0.30) = 14.4 \text{ m}$$
$$47 \text{ spaces} = 47(0.20) = 9.4 \text{ m}$$
$$\text{Total} \qquad\qquad = \overline{23.8 \text{ m}}$$

Find the cover plate thickness (refer to Fig. 6-6):

$$\frac{L_0}{t_p} \le \frac{835}{\sqrt{250}} = 52.8$$

$$t_p = \frac{221}{52.8} = 4.19 \text{ mm}$$

Use 6 mm (1/4 in). Use welding and locate the plate as shown. Welding may be intermittent. If the welding is adequate to allow the cover plate and angles to act as a unit, AISC allows a contribution of the cover plate based on

$$(L - 2r)t_p = [500 - 2(75)]0.006 = 2.1 \times 10^{-3} \text{ m}^2$$

of area to be used to increase the axial capacity of the section.   ///

6-6 COLUMN BASE PLATES

Steel columns are placed on some type of supporting member to interface the column and support. The supporting member may be a concrete column in composite building construction, but more commonly the column is terminated on a footing, pedestal, or pilaster. A *pedestal* is used to support the metal column above the ground to prevent corrosion when the footing is below ground level. A *pilaster* is an enlarged section of the basement wall used to transmit the column load through the wall zone to the footing. Sometimes, but not commonly, the column terminates directly on the footing.

A base plate is necessary when a steel column terminates on any type of masonry to spread the high intensity of stress in the steel to a value that can be safely carried by the masonry material. Masonry is here defined as concrete, concrete block, brick, and tile; concrete is most often used and will be the only material considered here.

The base plate and the mating end of the column may be planed to affect load transfer by direct bearing. The base plate is seated on the foundation using a cement grout (thick sand–cement mixture, often with an expansive agent to produce an intimate contact, since cement paste tends to shrink on drying). Grout can take out up to about 1 in (25 mm) of footing-to-column mismatch as long as it is a "fill-in" discrepancy.

Angles may be used to bolt or weld the base plate to the column. However, present practice tends to the use of a base plate that is shop-welded to the column. Slightly oversize anchor bolt holes are drilled in the base plate to fit over anchor bolts placed in the foundation/footing element during field construction. The oversized holes allow some misalignment of the anchor bolts without redrilling the base plate holes or taking out and resetting the anchors. The general method of fastening the column to the foundation is illustrated in Fig. 6-7.

The required base plate dimensions are based on the allowable unit contact pressure of the footing. The base plate thickness is based on the base plate contact pressure, producing bending on the critical section with dimensions as shown in Fig. 6-8c. When the column base resists a moment, the plate dimensions must be adjusted so that

$$F_p \geq f_p = \frac{P}{A} \pm \frac{M}{S}$$

where S is the section modulus of the rectangular base plate with respect to the moment axis. The plate thickness for this case is also shown in Fig. 6-8c.

From Fig. 6-8a the area of the column base plate is

$$B \times C = \frac{P}{F_p}$$

A number of combinations of side dimensions B and C can be obtained, but that combination should be used which produces $m \simeq n$. Fabrication practice favors B and C in integers.

The thickness of the base plate is obtained by considering bending on a critical section the distance m or n from the corresponding free edge (Fig. 6-8c).

286 STRUCTURAL STEEL DESIGN

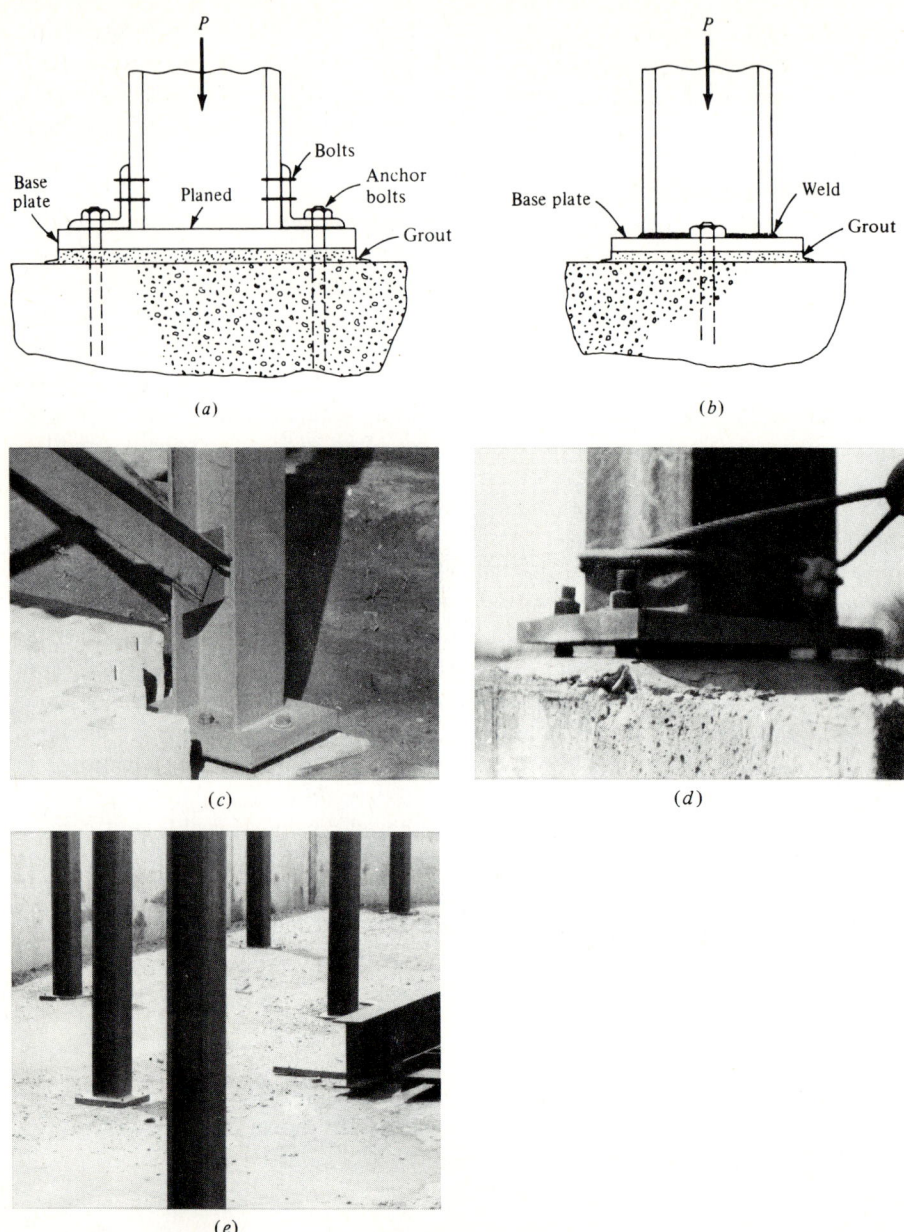

Figure 6-7 Column to foundation interfacing using base plates. (*a*) Use of angles to attach column to foundation. Method not widely used at present due to extra fabrication (cutting two angles and carefully planing column end and base plate). (*b*) Widely used shop-welded column-to-base plate method for attaching columns to foundations [see also field photographs in (*c*), (*d*), and (*e*)]. (*c*) Corner column using shop-welded base plate. Gap for grouting to final grade can be easily seen. There are four anchor bolts being used to attach column to footing. Diagonal member is a field-welded wind bracing element. (*d*) Column also using shop-welded base plate. Grouting space can be easily seen. The cable is being used to align frame. (*e*) Series of interior columns fastened to footings directly. Note again the use of shop-welded base plates.

AXIALLY LOADED COLUMNS AND STRUTS **287**

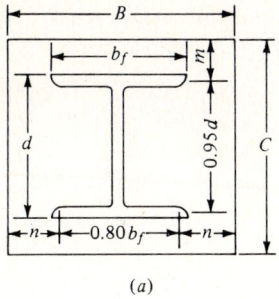

(a)

Allowable AISC stresses Contact material	F_p	
	fps: kips/in^2	SI, MPa
Sandstone & limestone	0.4	2.8
Brick in cement matrix	0.25	1.75
Full area of concrete foundation	$0.35 f'_c$	
Less than full area on foundation	$0.35 f'_c \sqrt{A_2/A_1} \leq 0.7 f'_c$	

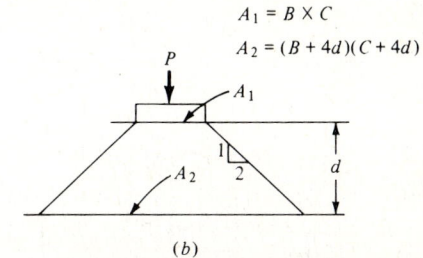

$A_1 = B \times C$
$A_2 = (B + 4d)(C + 4d)$

(b)

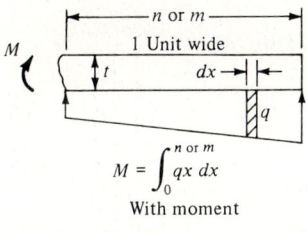 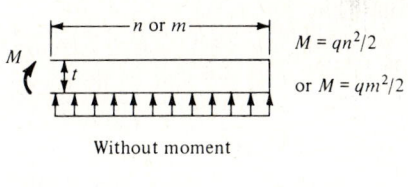

(c)

Figure 6-8 General base plate dimensions and other design criteria. (a) Base plate dimensions. (b) Allowable stresses F_p. (c) Base plate moment to compute base plate thickness.

For a uniform pressure and a strip m or $n \times 1$ unit wide $\times t$ thick, we have

$$M = q(m)\frac{m}{2} \quad \text{or} \quad M = q(n)\frac{n}{2}$$

Using the largest value of M (and noting that if $m = n$ they are equal), we have

$$f_b = F_b = \frac{M}{S} = \frac{6M}{t^2} \quad \text{(for a strip one unit wide)}$$

Solving for t, we obtain

$$t = \left(\frac{6M}{F_b}\right)^{1/2} = \left[\frac{3 \times q \times (m^2 \text{ or } n^2)}{F_b}\right]^{1/2}$$

where $q =$ actual contact pressure
$F_b = 0.75 F_y$ (AISC Sec. 1-5.1.4.3, based on bending on rectangular section)

Example 6-8 Design the base plate for a column as shown in Fig. E6-8. Use $F_y = 250$ MPa, $f'_c = 20.7$ MPa, and the AISC specifications.

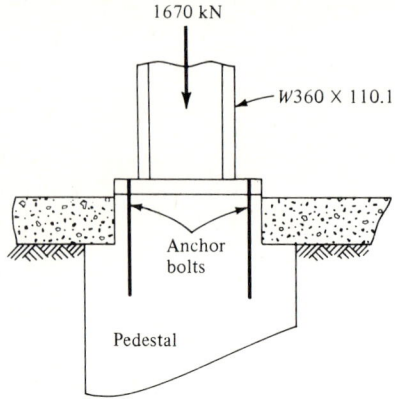

Figure E6-8

SOLUTION The pedestal floor line dimensions will be the same as the base plate. Thus

$$F_p = 0.35 f'_c = 7.245 \text{ MPa} \quad (\text{Fig. 6-8}b)$$

$$B \times C = \frac{1670}{7.245} = 230.5 \times 10^{-3} \text{ m}^2$$

Let us make $m \cong n$:
From Table V-3 obtain $d = 360$ mm; $b_f = 256$ mm

$$B = 0.80(256) + 2n = 205 + 2n = 205 + 2m$$
$$C = 0.95(360) + 2m = 342 + 2m$$
$$(0.205 + 2m)(0.342 + 2m) = 0.2305 \text{ shifting decimal}$$
$$m^2 + 0.2735m = 0.04009$$
$$m = 0.106 \text{ m}$$

Check:
$$B = 205 + 2(106) = 417 \text{ mm}$$
$$C = 342 + 2(106) = 554 \text{ mm}$$
$$B \times C = 0.417 \times 0.554 = 0.2310 \text{ m}^2 > 0.2305 \quad \text{O.K.}$$

The actual contact pressure q is

$$q = \frac{1670}{231.0} = 7.229 \text{ MPa}$$

$n = m = 0.106 \text{ m}$

$$M = \frac{7.229(0.106)^2}{2} = 0.08123 \text{ MN} \cdot \text{m} \qquad t = \left[\frac{6(0.08123)}{0.75(250)} \right]^{1/2} = 0.0509 \text{ m}$$

$$= 50.9 \text{ mm} \qquad \text{say } 52 \text{ mm}$$

Use a column base plate $417 \times 554 \times 52$ mm thick. ///

Example 6-9 Redesign the column base plate of Example 6-8 to resist a bending moment of 265 kN · m in addition to the axial load (Fig. E6-9a).

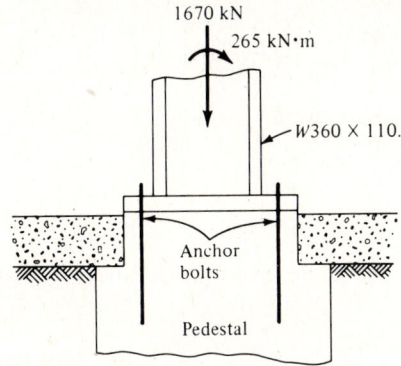

Figure E6-9a

SOLUTION We will design the column-to-base plate weld in Example 9-7.

$$F_p = 7.245 \text{ MPa} \leq q$$

$$q = \frac{1670}{BC} \pm \frac{6(265)}{BC^2}$$

After several side computations, let us assume that $C = 780$ mm:

$$B = \frac{1670}{7245(0.78)} + \frac{6(265)}{7245(0.78^2)}$$

$$= 656 \text{ mm}$$

Figure E6-9b illustrates data so far including $q = 7.248$ and -0.724 obtained from using B and C in the preceding equation for q.

Along line x–x:

$$q = 7.248 - 10.22x$$

$$V = 7.248x - \frac{10.22x^2}{2} \quad \text{(integrating)}$$

$$M = \frac{7.248x^2}{2} - \frac{10.22x^3}{6}$$

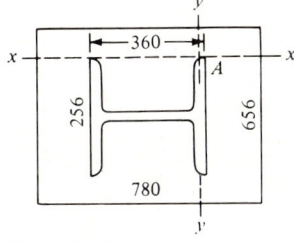
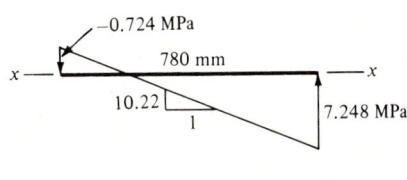

Figure E6-9b

For $x = 0.219$ m:

$$M = \frac{7.248(0.219)^2}{2} - \frac{10.22(0.219)^3}{6} = 0.1559 \text{ MN} \cdot \text{m}$$

The corresponding thickness is

$$t = \left(\frac{6 \times 0.1559}{187.5}\right)^{1/2} = 0.0706 \text{ m}$$

$$= 70.6 \text{ mm} \quad \text{say 75 mm}$$

From the other direction (line $y-y$) at point A:

$$n = \frac{656 - 0.80(256)}{2} = 226 \text{ mm}$$

$$q = 7.248 - 10.22(0.219) = 5.01 \text{ MPa} \quad \text{(average for a unit width)}$$

$$M = \frac{5.10(0.226)^2}{2} = 0.1302 \text{ MN} \cdot \text{m}$$

$$t = \left(\frac{6 \times 0.1302}{187.5}\right)^{1/2} = 0.0645 \text{ m}$$

$$= 65 \text{ mm}$$

Use the largest thickness of 75 mm. The final column base plate dimensions are 780 × 656 × 75 mm thick.

For anchor bolts as shown in Fig. E6-9c, assume that the bolts will carry the full moment even though the axial force will reduce the effect of the moment considerably. This assumption provides some reserve capacity of the anchoring system to resist a considerable lateral force (column shifting laterally).

$$T = \frac{265}{0.570} = 465 \text{ kN}$$

Use $F_y = 345$ for anchor bolts.

$$F_t = 0.6 F_y = 207 \text{ kN}$$

$$A_{\text{reqd}} = \frac{465}{207} = 2.246 \times 10^{-3} \text{ m}^2$$

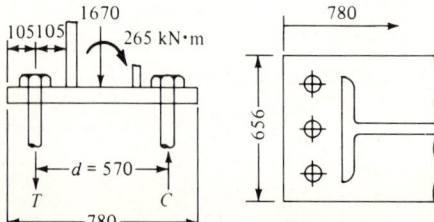

Figure E6-9c

Using three anchor rods for each side, the diameter is

$$D = \left[\frac{0.002246}{3(0.7854)} \right]^{1/2} = 0.03087 \text{ m}$$

Use three 32-mm-diameter anchor rods. ///

6-7 LATERAL BRACING OF COLUMNS

A common and widely used empirical rule for lateral bracing for both the compression flanges of beams and columns is to provide a bracing (also a compression member) element to carry a lateral brace force of

$$P_b = 0.02P$$

where P is the axial force in the compression member being braced; that is, $P = A_f f_b$ for bending members, where A_f = area of compression flange and f_b = average (or maximum) bending compressive stress. For columns, use P = average axial force in column. This recommendation is given by the Structural Stability Research Council in *Guide to Stability Design Criteria for Metal Structures*, 3rd ed., edited by Johnston.

A series of tests at Cornell University by Winter (see "Lateral Bracing of Columns and Beams," *Transactions, ASCE*, Vol. 125, 1960) indicates that very little lateral bracing is required to allow the compression element to develop the allowable design stress. This restraint could generally be developed by the weight of the floor system onto beams when full-length contact with the compression flange is made. Because of the variable nature of flooring (metal decking, concrete-to-beam, wood-to-beam, and so on), it is suggested that the 2 percent criterion be used. Winter also derived an analytical expression for the bracing requirement based on both restraint and the relative stiffness of the column and brace. If a SF of approximately 2.5 is used with this derived expression, the empirical rule of 2 percent can be obtained.

Example 6-10 Determine the minimum-size spandrel (or girt) to brace the W section with respect to the Y axis for the 750-kip axial load of Example 6-2. The distance between columns may be taken as 18 ft. F_y for column = 50 ksi.

SOLUTION The axial force in the channel section used for the brace is

$$P_b = 0.02P = 0.02(750) = 15 \text{ kips}$$

The maximum L/r (AISC) = 200 (compression member). The minimum radius of gyration, r_y = 18(12)/200 = 1.08 in. For an L/r ratio = 200, the allowable axial stress F_a = 3.73 ksi (Table II-5, SSDD), applicable for all F_y.

$$A_{\text{reqd}} = \frac{P}{F_a} = \frac{15}{3.73} = 4.02 \text{ in}^2$$

Simply search Tables I-6 and I-7 for this combination of A and r_y and find

$$\text{MC10} \times 28.5: \quad A = 8.37 \text{ in}^2$$
$$r_y = 1.17 \text{ in}$$

If bending or other requirements are also satisfied, this section can be used for the girt (or spandrel). ///

6-8 COLUMN AND STRUT DESIGN USING LRFD

The use of LRFD requires the separation of dead and live loads on the column. Once this is done, obtain the ultimate column load in the following form:

$$P_u = \psi(F_d D + F_L L \cdots)$$

Also,

$$P_u = A_g \phi F_{cr}$$

Here we need Table 3-1, since the value of ϕ ranges from 0.86 to 0.65 depending on η, which in turn depends on KL/r of the column as well as F_y. We may also note that the value of F_{cr} depends on the value of η as follows:

$$F_{cr} = F_y(1 - 0.25\eta^2) \qquad \eta \leq \sqrt{2}$$

$$F_{cr} = \frac{F_y}{\eta^2} \qquad \eta > \sqrt{2}$$

where η = value given in Table 3-1 and is repeated here:

$$\eta = \frac{KL}{\pi r}\sqrt{\frac{F_y}{E}}$$

Example 6-11 Given the columns of Example 6-1 spaced on 20 × 20 centers, dead load = 30 psf, live load = 60 psf (snow), column length = 14 ft, and $K = 1$ (as in Example 6-1 and Fig. 6-3). Use A-36 steel and the LRFD method. (A 4 in × 10.79 lb/ft pipe column was selected in Example 6-1.) Redesign the column using LRFD as given in Sec. 3-7.

SOLUTION

$$P_u = 1.1(1.1D + 1.5S)A$$
$$= 1.1[1.1(0.030) + 1.5(0.060)](20 \times 20) = 54.12 \text{ kips (vs. 36 kips } P_w\text{)}$$

Also,

$$\eta = \frac{1(14 \times 12)}{\pi r}\left(\frac{36}{29\,000}\right)^{1/2} = \frac{1.884}{r}$$

Using Example 6-1 as a guide, try a 4 in × 10.79 lb/ft pipe column:

$$A = 3.17 \text{ in}^2$$
$$r = 1.51 \text{ in}$$
$$\eta = \frac{1.884}{1.51} = 1.25 < \sqrt{2}$$
$$F_{cr} = F_y(1 - 0.25\eta^2) = 36[1 - 0.25(1.25)^2] = 21.94 \text{ ksi}$$
$$\phi = 0.65 \quad (\eta > 1.0)$$
$$A_{reqd} = \frac{P_u}{\phi F_{cr}} = \frac{54.12}{0.65(21.94)} = 3.79 \text{ in}^2 > 3.17 \quad \text{N.G.}$$

Try a 5 in × 14.62 lb/ft pipe:

$$A = 4.30 \text{ in}^2$$
$$r = 1.88 \text{ in}$$
$$\eta = \frac{1.884}{1.88} = 1.002$$
$$F_{cr} = 36[1 - 0.25(1.002)^2] = 26.96 \text{ ksi}$$
$$\phi = 0.65$$
$$A_{reqd} = \frac{54.12}{0.65(26.96)} = 3.09 \text{ in}^2 < 4.30 \quad \text{O.K.}$$

Use a pipe column 5 in × 14.62 lb/ft. ///

PROBLEMS

6-1 Determine the allowable load that can be carried by a W14 × 211 column using $F_y = 50$ ksi steel and the AISC specifications if:
 (a) $KL = 16$ ft.
 (b) $KL = 42$ ft.
 (c) $KL_x = 68$ ft and $KL_y = 44$ ft.
Make appropriate comments.
Answer: (b) 602.7 kips. (c) 549.1 kips.

6-2 Determine the allowable load for a W360 × 346.7 rolled section using $F_y = 415$ MPa and the AISC specifications if:
 (a) $KL = 5.1$ m.
 (b) $KL = 13.2$ m.
 (c) $KL_x = 20.5$ m and $KL_y = 13.6$ m.
Make appropriate comments.
Answer: (c) 2669 kN.

6-3 What is the lightest square tube section (see Table I-15, SSDD) for a column loading of 134 kips and an unsupported length of 12.4 ft? Use the AISC specifications and A-36 steel.
Answer: 6 × 6 × 0.375.

6-4 What is the allowable column load using the AISC specifications and $F_y = 345$ MPa for a rectangular tube section 300 × 200 × 9.52 mm wall (Table V-16, SSDD) for an unsupported length of 4.8 m?

6-5 What is the allowable load using the AREA specifications for a W14 × 145 used as a compression member in a bolted end connection for a bridge truss? The member is 15.5 ft long and uses A-36 steel.

Answer: 771.6 kips.

6-6 What is the allowable load using the AREA specifications for a W310 × 178.6 rolled section used as a compression member with a bolted end connection for a bridge truss? The member is 4.75 m long and uses $F_y = 345$ MPa steel.

Answer: 3163 kN.

6-7 What is the allowable load for the built-up section shown in Fig. P6-7 using the AREA specifications and A-36 steel? Assume a bolted end connection and a length of 5.25 m.

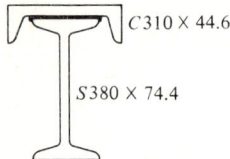

Figure P6-7

6-8 What is the allowable load for the built-up member shown in Fig. P6-8, using the AASHTO specifications, $F_y = 50$ ksi, and an unsupported length of 18.7 ft?

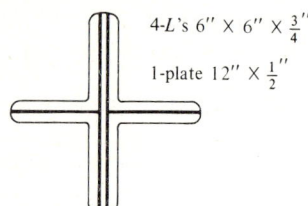

Figure P6-8

6-9 Referring to Fig. P6-7, place a second channel C310 × 44.64 on the bottom of the S shape to make it symmetrical. What is the allowable column load using the AASHTO specifications if $L = 6.6$ m and using A-36 steel?

Answer: 1962 kN.

6-10 What is the allowable column load for the built-up section shown in Fig. P6-10 if the member is of A-36 steel and the length is 14.5 ft? Use the AASHTO specifications.

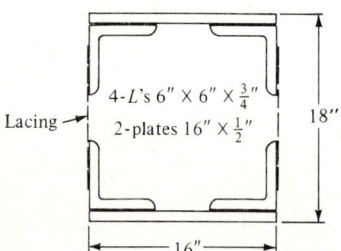

Figure P6-10

6-11 What is the allowable column load for the built-up section shown in Fig. P6-11? The length is 5.3 m. Use the AISC specifications and A-36 steel. Neglect the contribution of the perforated cover plate.

Answer: 3090.3 kN.

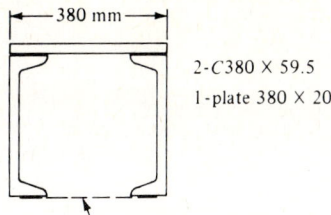

2-C380 × 59.5
1-plate 380 × 20

Perforated C.P.

Figure P6-11

6-12 Design the lacing for the allowable load found in Prob. 6-10, allowing a 20 percent increase in V for wind, member weight, etc.

6-13 Design the perforated cover plates for the allowable load and section used in Prob. 6-11.

6-14 Redo Example 6-1 if the contributory column area is 30×20 instead of 20×20 but all the other data are the same.
Answer: 5 in at 14.6 lb/ft.

6-15 Design a laced column section (refer to Example 6-5) for a highway bridge truss end post. The length is $7.2\sqrt{2} = 10.18$ m. The truss span is 50.4 m and the loads are dead load $= -1120$ kN and live load $= -422$ kN (without impact). Use $F_y = 250$ MPa and the AASHTO specifications.
Answer: Try two C380 × 50.4, 300 × 15 mm cover plate.

6-16 Redesign the truss end post of Example 6-6 using a perforated cover plate for both sides of the channels. Note that AASHTO allows use of the net area of the perforated cover plate in computing the total section area and column capacity.
Answer: 2C12 × 25, $A_{furn} = 23.7$, $r_{min} = 4.58$ in, includes two 12 in × $\frac{1}{2}$ plates with 3-in holes.

6-17 Design a column base plate for the maximum capacity of a W12 × 170 column with an unbraced length of 12.0 ft. Assume that $K_x = K_y = 1.0$. Use $F_y = 50$ ksi, $f'_c = 4$ ksi, and the AISC specifications. The column is interfaced to a concrete pedestal.

6-18 Design the column base plate for a W14 × 120 section that carries an axial load of 500 kips and a base moment of 200 ft · kips. Use A-36 steel, $f'_c = 3$ ksi, and the base plate interfaces the column directly to the footing, which has a total depth of 21 in.
Answer: $24\frac{3}{4} \times 22\frac{3}{4} \times 2\frac{1}{2}$.

6-19 Design a column base plate for the maximum capacity of a W310 × 117.6 rolled section with an unbraced length of 4.1 m. Assume that $K_x = K_y = 1.0$. Use $F_y = 345$ MPa, $f'_c = 28$ MPa, and the AISC specifications. The column is interfaced to a concrete pedestal.

6-20 Redo Example 6-11 for the lightest available W8 section.

6-21 Redo Example 6-11 if the loads are as follows: dead load = 35 psf; live load = 75 psf.
Answer: 5-in pipe at 14.6 lb/ft.

6-22 Redo Example 6-11 using the following data: dead load = 1.75 kPa; live load = 3.75 kPa; column contributory area = 6.1×7.1 m, column length = 4.98 m, and $K_x = K_y = 1.0$. Use either a round pipe or a square structural tube for the column and $F_y = 250$ MPa steel.

6-23 Design member 6 of Example 6-6 (refer to Fig. E6-6a) if the dead load bar force = 283.8 kips, the maximum live load, including impact = 109.4 kips, and the minimum live load = 0.0 kip. Compare the section to that obtained in Example 6-6. Take $P_u = \psi(\beta_d \text{ DL} + \beta_L \text{ LL})$, where $\psi = 1.3$, $\beta_d = 1.0$, and $\beta_L = 1.67$ (latest AASHTO). Also, $P_u = 0.85 A F_{cr}$, where $F_{cr} = F_a$ from Eq. (6-9) or Eq. (6-10), without using the SF = 2.12.
Answer: Use the same section as in Prob. 6-16.

Figure VII-1 Beam–columns and beams making up an industrial frame. Note alternate orientation of strong axis of beam–columns along sides of frame. A closeup of selected joints is shown in Fig. 9-15.

CHAPTER
SEVEN

BEAM–COLUMN DESIGN

7-1 INTRODUCTION

When a structural member is loaded in a manner to produce more than one stress mode, some adjustments must be made in the allowable stresses. Where the stresses are produced as a combination of bending about the X and Y axes as in Sec. 4-8, the final stresses used for design are obtained by superposition.

$$\pm f_{bx} \pm f_{by} \leq F_b \qquad (a)$$

Since F_{bx} may not be equal to F_{by} (particularly in the case of W shapes from flange geometry) the beam design in Chap. 4 was obtained by iteration. Accumulation of compressive (or tension) stresses at one edge of one of the flanges was used in the following form of Eq. (a):

$$\frac{f_{bx}}{F_{bx}} + \frac{f_{by}}{F_{by}} \leq 1 \qquad (b)$$

This equation was obtained by dividing Eq. (a) by F_b.

A problem similar to this is often involved when the structural member is loaded in a combination of bending and axial load. These situations are always produced in rigid frame building construction (i.e., the columns carry the building load axially as well as end moments from the girders that frame into them). In industrial buildings column brackets may be used to carry crane runway girders and, ultimately, the crane load. The resulting bracket eccentricity produces a bending moment in addition to the axial loads in the column. In this case the column moment is not at the column ends. Similarly, wind pressure on long vertical members may produce bending moments, since a large distance between floors (or ground to roof) may disallow the concept of wind being carried, analogous to one-way slab action. In Examples 2-5 and 2-6 the framing of the side sheds to the columns of the main bay produces large column moments which must be allowed for in their design (to be considered in a later section).

Other design conditions produce bending in addition to axial forces. For example, the top chords of roof and bridge trusses are normally "pin-ended" compression members, but the weight of the member will produce bending as well. Purlins placed between panel joints of a roof truss or rafter as a means of reducing both the purlin member size and roofing span will produce bending in the chord or rafter.

In general, compression members are loaded with axial forces and moments. The moment(s) may be at the ends of the member, as in rigid framed buildings, or developed at an interior point from a bracket, local beam, cable attachment, or other loading. When the moment effect produces single curvature (see Fig. 7-1) a much more critical design condition is developed than when the moment(s) produce reverse curvature.

Bending may also be produced in tension members such as the bottom chords of bridge trusses where floor beams may frame into them. Bottom chords of building trusses may be used to attach hoisting devices; other temporary loads attached to the bottom chords of building trusses will produce bending in addition to the axial load present.

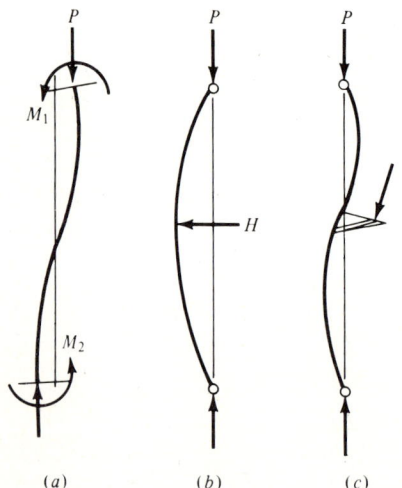

Figure 7-1 Column loading curvature resulting. The single curvature of (*b*) is often the most critical. (*a*) Reversed curvature in building frame. (*b*) Single curvature interior loading. (*c*) Reversed curvature interior loading.

In many of these situations, particularly with truss members, the bending stresses are neglected. This may be a reasonable procedure where the bending stress results from the member weight, or even from purlins if they are relatively small (and light weight), and the resulting effects are possibly less than about 10 percent of the analyzed stresses. There are undoubtedly small bending stresses in the truss chord elements due to continuity across one or more panel points (a technique used to reduce fabrication costs), and additionally there is usually some overdesign, since it is a common practice to use a constant-size top (and bottom) chord member (again to reduce fabrication costs). Exceptions to this may be obtained when it is safe to use only one bolt or rivet at the ends of the truss members, so that rotation is less inhibited—it is, of course, not practical to fabricate "pinned" joints for the usual bridge or roof truss. The effects of actual joint bending can be minimized by keeping the joint length with respect to the member lengths into the joint as short as practical and, coupled with the fact that truss members usually have small EI/L ratios (i.e., long member with small moment of inertia), the moment gradient is sharp and much of the member length is essentially moment-free.

The moments that are produced in trusses due to end fixity (and commonly ignored in analysis by using "pinned" joints) are collectively termed *secondary effects*. It is not necessary to ignore secondary effects, since modern computational methods (such as the analysis computer program in the Appendix) can analyze a "rigid" (three-degrees-of-freedom) truss almost as easily as a pinned (two-degrees-of-freedom) truss. The early computers had only limited cpu/core storage, and before efficient methods of matrix solution, the "pinned" truss geometry was necessitated, since the difference in matrix size for a truss with 100 nodes or joints was:

Number of equations: $100 \times 2 = 200$ or rigid: $100 \times 3 = 300$

CPU requirements: $200^2 = 40,000$ words $\qquad 300^2 = 90,000$ words

Since each number (or word) requires approximately 4 bytes the requirements become:

Pinned: $40,000 \times 4 = 160,000 = 160K$ bytes

$90,000 \times 4 = 360,000 = 360K$ bytes

A capacity of 360K + program requirements would tax the capacity of some of the larger computers currently available had not efficient methods of matrix solution been developed so that with clever coding (reduce the bandwidth of the matrix to a minimum) and/or use of boundary conditions, the three-degree-of-freedom truss might only require some 30 to 50K, which is well within the capacity of all but the smaller desktop minicomputers.

7-2 GENERAL CONSIDERATIONS OF AXIAL LOAD WITH BENDING

When tension axial load and bending occur simultaneously, the principal of superposition may be safely applied. This is because (see Fig. 7-2a) the tension load tends to reduce the bending effects, so that the value of Δ is reduced somewhat, and, of course, the actual stress is also somewhat less so that the maximum allowable stress conditions computed as

$$\frac{f_t}{F_t} + \frac{f_b}{F_b} \leq 1$$

will be safe. This safety is partially obtained by neglecting the effects of $P-\Delta$ on f_b, which could be computed (to be strictly correct) as

$$f_b = \pm \frac{Mc}{I} + P_t \Delta \frac{c}{I}$$

When P_b is at midspan, we have (with the limitation $\Delta \geq 0$, see Fig. 7-2)

$$\Delta = + \frac{P_b L^3}{48 EI} - \frac{P_t \Delta L^2}{8 EI} \qquad (a)$$

and the resulting bending stress is

$$f_b = \pm \frac{P_b L}{4 S_x} + \frac{P_t}{EI S_x} \left(\frac{P_b L^3}{48} - \frac{P_t \Delta L^2}{8} \right) \qquad (b)$$

The actual bending stress can only be obtained by iteration of Eq. (a) until the Δ value used on the right side is sufficiently close to the Δ obtained on the left side. This iterative solution may be reasonably practical on a computer, but with hand computations, in only a few design situations (where the number of members is limited) is this approach economically justified. Neglecting the $P-\Delta$ effect in Eq. (b) is an error, but on the conservative side. Inspection of Eq. (a) indicates that the tension stress reduces the deflection and also reduces the compression stress due to bending. With the allowable tension stress F_t a constant value and recalling that F_b may depend on the unbraced length of the compression flange (and possibly reducing the allowable compressive stress), we see that use of

$$\frac{f_b}{F_b} + \frac{f_t}{F_t} \leq 1$$

provides a satisfactory solution.

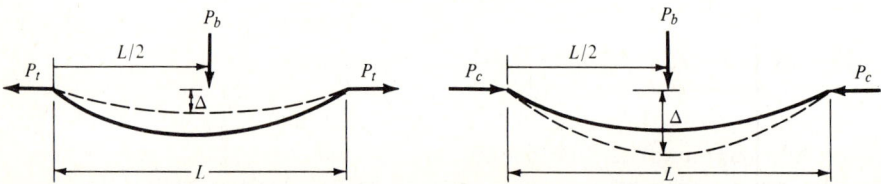

Figure 7-2 $P-\Delta$ effects on tension and compression members.

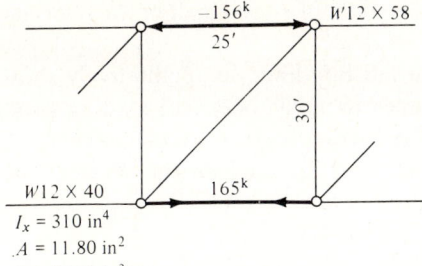

Figure E7-1

Example 7-1 Given the portion of a highway bridge truss with loads and members as in Fig. E7-1, what is the maximum tension stress in the lower chord?

SOLUTION (neglect the P–Δ effect)

$$f_b = \frac{wL^2}{8S_x} = \frac{0.040(25^2)(12)}{8(51.9)} = 0.72 \text{ ksi}$$

$$f_t = \frac{P}{A} \pm f_b = \frac{165}{11.8} \pm 0.72 \text{ ksi}$$

$$f_t = 13.98 + 0.72 = 14.70 \text{ ksi} \quad \text{(maximum)}$$

$$f_t = 13.98 - 0.72 = 13.26 \text{ ksi} \quad \text{(minimum)} \quad ///$$

When a compressive axial load acts together with a bending moment, the deflection is amplified and the compressive stress increases. Since the allowable compression stresses take into account possible buckling (lateral deflections), member design is more sensitive to this loading mode than to one producing tension stresses.

Referring to Fig. 7-2b, we note that the axial load (assumed to be applied last as being easier to visualize) increases the deflection. The order of load application does not affect the outcome, however, as long as yielding is not produced. The important concern is that there is an increase in the deflection due to the P–Δ effect, with a corresponding increase in the bending stresses. The value of the deflection is (with P_b at midspan) for this case

$$\Delta = \frac{P_b L^3}{48 EI} + \frac{P_c \Delta L^2}{8 EI} \tag{c}$$

and the resulting bending stresses with $P_c = (-)$ are

$$f_b = \pm \frac{P_b L}{4 S_x} - \frac{P_c}{EI S_x} \left(\frac{P_b L^3}{48} + \frac{P_c \Delta L^2}{8} \right) \tag{d}$$

A critical evaluation of Eq. (c) indicates that an iterative solution is required as for the tension mode, but that the deflection "feeds" upon itself (deflection causes more deflection), and for members with an I too small or an L too large a buckling failure can occur.

The P-Δ effect can also develop in tall buildings, as qualitatively shown in Fig. 7-3, when wind or earthquake forces or unsymmetrical loading produces lateral displacements of the upper floors with respect to the lower building elements. A computer analysis can be made to analyze the P-Δ effect but requires iteration. The steps include:

1. Make a conventional computer analysis using the lateral forces.
2. Obtain ith-story lateral displacements X_i and for X_{i+1} (next story above).
3. Compute an additional P-matrix moment as

$$M_i = P_{i+1}(X_{i+1} - X_i)$$

and a shear as indicated in Fig. 7-3b using

$$V_i = \frac{M_i}{L}$$

which is applied at the top and bottom of the story with due regard to signs.

P_{i+1} = axial force in column between i and $i + 1$ floor levels

4. Compute new X_i and compare to previous values used in step 3. Iterate until satisfactory convergence is obtained, such as, say, 0.02 ft or 0.006 m (approximately $\frac{1}{4}$ in).

The P-Δ effects are also called "secondary effects" and have been largely ignored until more recently. Some designers have arbitrarily increased the design

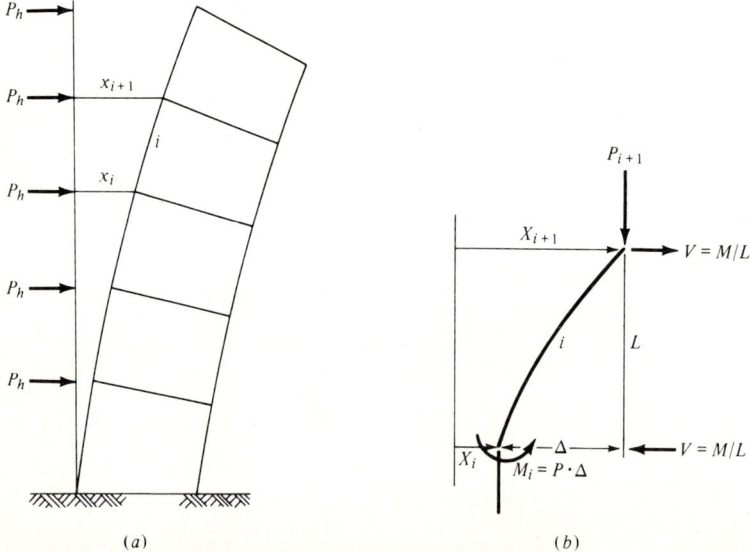

Figure 7-3 P-Δ effect for tall buildings. (*a*) Structure with lateral loads. (*b*) ith story with deflections greatly exaggerated.

stresses (from loads) by a factor such as 10 percent to allow for P-Δ. Another factor that tends to mitigate the P-Δ effect is that it is usually developed from wind or earthquake analysis where the designer can use allowable stresses that are increased by one-third. The P-Δ effect would only exceed this in rare cases. The analysis computer program in the Appendix can be easily modified to automatically scan the deflection matrix for the appropriate X values, recompute the P matrix using these values, then storing them for comparison with the new values from the current cycle until convergence and exit.

7-3 EFFECTIVE LENGTHS OF COLUMNS IN BUILDING FRAMES

The concept of effective column length $L' = KL$ was introduced in Sec. 6-3 and a value of K was obtained for several common cases. It was observed that when the column ends were laterally restrained so that the P-Δ effect (as in Fig. 7-3) could not develop, the K factor was $K \leq 1.0$. With the "flagpole" of Fig. 6-3e or the pinned base of Fig. 6-3f, the K factor was 2.0 or more.

We shall find that K in multistory buildings, which can translate, may be considerably more than 2.0, as illustrated in Fig. 7-4a. For the portion of the elastic frame shown in Fig. 7-4b and using Eq. (6-1):

$$x = A \sin ky \tag{6-1}$$

With $k = (P/EI)^{1/2}$ and using the effective length $L' = KL$, we obtain

$$x = A \sin \frac{\pi y}{KL} \tag{7-1}$$

Taking the origin of coordinates at a point of inflection as shown in Fig. 7-4b, we have at the top of the column:

$$y = y_1 \qquad x = x_1$$

$$x_1 = A \sin \frac{\pi y_1}{KL} \tag{a}$$

At the bottom of the column:

$$y = y_1 - L \qquad x = -x_2$$

Noting that $\sin(\alpha - \beta) = \sin \alpha \cos \beta - \cos \alpha \sin \beta$,

$$x_2 = -A\left(\sin \frac{\pi y_1}{KL} \cos \frac{\pi}{K} - \cos \frac{\pi y_1}{KL} \sin \frac{\pi}{K}\right) \tag{b}$$

At the top and base of the column, the slope is dx/dy, and we obtain, from Eqs. (a) and (b):

$$\theta_1 = \frac{\pi A}{KL} \cos \frac{\pi y_1}{KL} \tag{c}$$

$$\theta_2 = \frac{\pi A}{KL}\left(\cos \frac{\pi y_1}{KL} \cos \frac{\pi}{K} + \sin \frac{\pi y_1}{KL} \sin \frac{\pi}{K}\right) \tag{d}$$

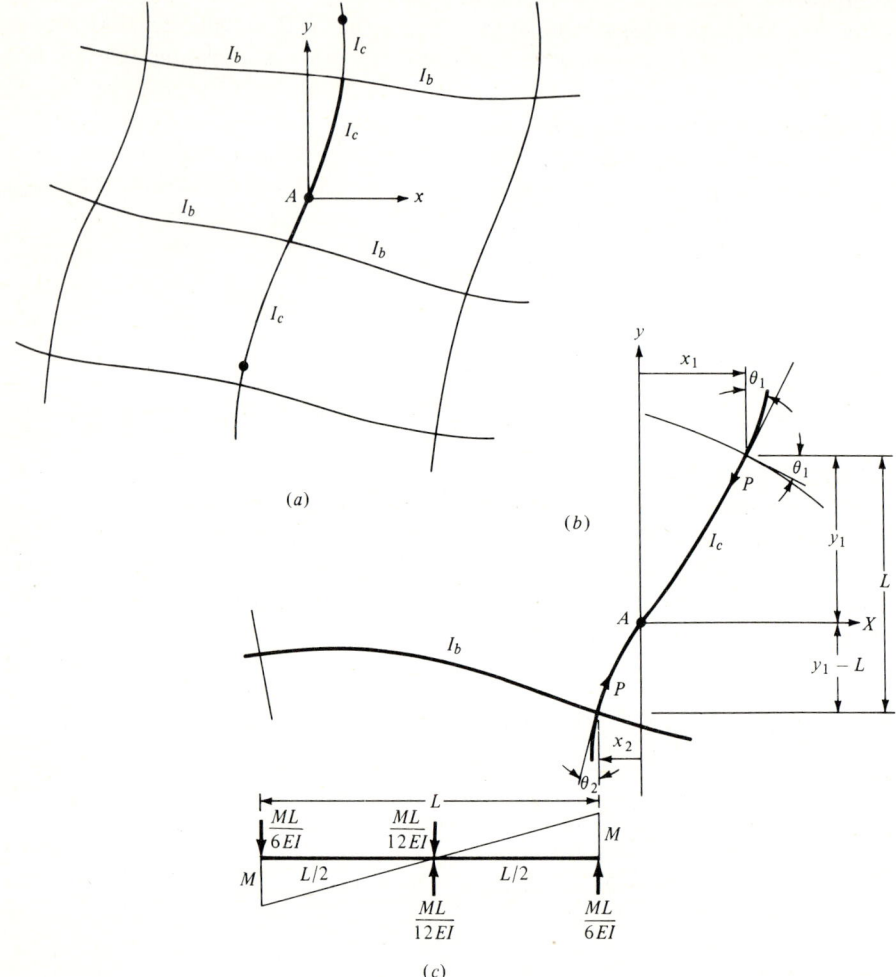

Figure 7-4 Elastic frame for derivation of G_a and G_b terms to obtain effective column length KL. (a) Part of an elastic frame. KL defined as distance between points of inflection. (b) Column element isolated from (a) with terms used in derivation identified. (c) Conjugate beam and moment variation (assumed).

From the bending moment diagram for the assumed moment distribution along the beams linearly varying as shown in Fig. 7-4c, the slope of the beam at the juncture with the column (and using conjugate beam principles) gives

$$\theta_1 = \frac{\Sigma P x_1 L_b}{\Sigma 6 E I_b} \qquad \theta_2 = \frac{\Sigma P x_2 L_b}{\Sigma 6 E I_b}$$

where the summation (Σ) is taken because load and moment are coming from both directions. From the earlier derivation of the Euler equation and summing

P, we obtain

$$\Sigma P = \Sigma \left(\frac{\pi}{K}\right)^2 \frac{EI_c}{L_c}$$

Substituting, we obtain

$$\theta_1 = \frac{\Sigma(\pi/K)^2 EI_c/L_c x_1}{\Sigma 6 EI_b/L_b} = \frac{1}{6}\left(\frac{\pi}{K}\right)^2 G_a x_1 \quad \text{where } G_a = \frac{\Sigma EI_c/L_c}{\Sigma EI_b/L_b} \quad (e)$$

Similarly,

$$\theta_2 = \frac{1}{6}\left(\frac{\pi}{K}\right)^2 G_b x_2 \quad (f)$$

Substituting for x_1 and x_2 [using Eqs. (a) and (b)], we obtain

$$\theta_1 = \left(\frac{\pi}{K}\right)^2 \frac{A}{6L_c} G_a \sin\frac{\pi y_1}{KL} \quad (g)$$

$$\theta_2 = -\left(\frac{\pi}{K}\right)^2 \frac{A}{6L_c} G_b \left(\sin\frac{\pi y_1}{KL}\cos\frac{\pi}{K} - \cos\frac{\pi y_1}{KL}\sin\frac{\pi}{K}\right) \quad (h)$$

At the column–beam intersection of rigid frames, the rotation of the column equals the rotation of the beam, so equating the two values of Θ_1, we obtain

$$\frac{\pi}{K}\frac{G_a}{6}\tan\frac{\pi y_1}{KL} = 1 \quad (i)$$

Similarly equating Eqs. (h) and (d), we obtain

$$-\frac{\pi}{K}\frac{G_b}{6}\left(\tan\frac{\pi y_1}{KL} - \tan\frac{\pi}{K}\right) = 1 + \tan\frac{\pi y_1}{KL}\tan\frac{\pi}{K} \quad (j)$$

Substituting Eq. (i) for $\tan \pi y_1/KL$, we obtain

$$\frac{G_a G_b (\pi/K)^2 - 36}{6(G_a + G_b)} = \frac{\pi/K}{\tan \pi/K} \quad (7\text{-}2)$$

We may program Eq. (7-2) for increments of G_a and G_b and iterate until a value of $F = \pi/K$ is obtained to satisfy the equality. The value of π/K thus obtained is used to obtain K as

$$K = \frac{\pi}{F}$$

A plot of G_a and G_b vs. K can be made as shown in Fig. 7-5a. This nomograph was first developed by Julian and Lawrence in unpublished lecture notes and as cited by several references, including the Structural Stability Research Council, *Guide to Stability Design Criteria for Metal Structures*, 3rd ed., edited by Johnston and published by John Wiley & Sons, Inc., New York.

In a somewhat similar manner, since the boundary conditions are different, one may develop equations for G_a, G_b, and K for frames restrained against lateral translation as

$$\frac{G_a G_b}{4}\left(\frac{\pi}{K}\right)^2 + \frac{G_a + G_b}{2}\left(1 - \frac{\pi/K}{\tan \pi/K}\right) + \frac{2 \tan \pi/K}{\pi/K} = 1 \quad (7\text{-}3)$$

306 STRUCTURAL STEEL DESIGN

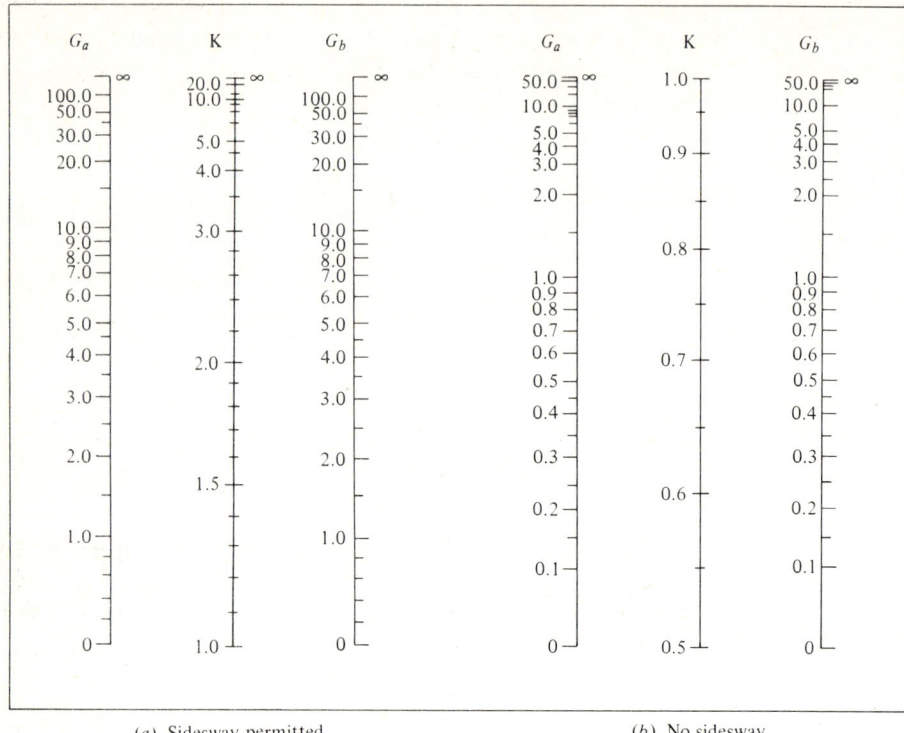

(a) Sidesway permitted   (b) No sidesway

Figure 7-5 Nomographs for the effective length of columns in continuous frames for lateral restraint conditions indicated.

This equation may also be programmed for values of G_a and G_b and to find the corresponding value of $F = \pi/K$ to satisfy the equality. A plot of this is shown in the nomograph in Fig. 7-5b.

The use of both nomographs shown in Fig. 7-5 involves computing values of

$$G_a = \frac{\Sigma EI_c/L_c}{\Sigma EI_b/L_b} \quad \text{(at far end of column)} \quad (7\text{-}4)$$

and

$$G_b = \frac{\Sigma EI_c/L_c}{\Sigma EI_b/L_b} \quad \text{(at near end of column)} \quad (7\text{-}5)$$

From the derivation involving G_a and G_b, it is evident that if we call one of the values G_a, the other end produces G_b (i.e., the values can be used interchangeably).

When E = constant it may cancel from Eqs. (7-4) and (7-5); however, when inelastic buckling is developed, E_t should be used for E in the EI_c/L_c ratio. Other considerations include:

1. When the column is pinned to the base, the EI_b/L_b ratio is zero, since the theoretical value of $I \to 0$ that results is $G \to \infty$. For this situation it is

suggested that one use $G = 10$ since there is some difficulty in producing a true pinned connection.
2. When the column is (rigidly) fixed to an infinitely rigid base, the EI_b/L_g ratio $\to \infty$ and the corresponding $G \to 0$. For this situation it is suggested to use $G = 1.0$ since there is some difficulty in producing a truly rigid connection.
3. When a beam or girder is used with adequate attachment to the column of interest but:
   a. The far end is hinged: multiply the EI_g/L_g ratio $\times$ 1.5.
   b. The far end is fixed: multiply the EI_g/L_g ratio $\times$ 2.0.
4. If beams are simply framed to columns, use Fig. 6-3 for K.

The use of K factors obtained in the manner just described has been required by AISC since 1963 and by AASHTO since 1974. The K factors tend to be considerably larger than those of Fig. 6-3, which were used by AISC prior to 1963. Since smaller K values had been used in structures that had an adequate service history, a new look was taken of the derivation for G_a, G_b and the resulting K. Yura (see "The Effective Length of Columns in Unbraced Frames," *AISC Engineering Journal*, April 1971) correctly pointed out that where the KL/r ratio was less than C_c, inelastic buckling should be considered and E_t should be used in Eqs. (7-4) and (7-5). The use of E_t is equivalent to

$$G_{\text{inelastic}} = \frac{E_t}{E} G_{\text{elastic}} \qquad (7\text{-}6)$$

Since $E_t = \lambda E$ and $\lambda \leq 1$, it follows that use of $G_{\text{inelastic}}$ gives a $K_{\text{inelastic}} < K_{\text{elastic}}$. Since E_t is somewhat awkward to obtain and recalling in the derivation of the AISC equations for F_a in the inelastic region,

$$0 < \frac{KL}{r} \leq C_c$$

we used essentially a SF on F_{cr} where

$$F_{cr} = \frac{\pi^2 E_t}{(KL/r)^2} \qquad F_a = \frac{F_{cr}}{\text{SF}}$$

and in the elastic region

$$F_a = \frac{\pi^2 E}{\text{SF}(KL/r)^2}$$

From these equations it follows (using i = inelastic, e = elastic) that

$$G_i = \frac{F_{ai}}{F_{ae}} G_e \qquad (7\text{-}7)$$

This computation neglects the variable SF, which ranges from 1.67 at $KL/r = 0$ to 23/12 (use 1.92 for hand computations) at $KL/r = C_c$. For most columns in the range of $KL/r = 40$ to 60, the variation in SF is essentially negligible. The use of Eq. (7-7) requires values of F_{ae} for the same KL/r value as F_{ai}, so it is necessary to compute a table of values such as Table II-7 or VI-7 (SSDD) and using $F_{ae} = F'_e$ to correspond to the AISC Specifications

$$F'_e = \frac{12\pi^2 E}{23(KL/r)^2} \qquad (7\text{-}8)$$

Note that this value is independent of F_y. The values shown in the table of $F'_e > F_y$ are for the purpose of using Eq. (7-7) and are not intended to be real stresses that might be used in a design.

The use of Eq. (7-7) in an actual design requires iteration; that is, assume a column; compute G_a, G_b, and find K; compute KL/r and F_a, F'_e; revise G_a, G_b; and find a new K for

1. Several cycles, or
2. To convergence, or
3. To an arbitrary limiting K, such as 1.2 or 1.5.

Of course, if the section chosen is not adequate, a new section must be selected and the computations repeated. Disque (see *AISC Engineering Journal*, No. 2, 1973) proposed that the iterations for K could be eliminated by using $f_a = P/A$ instead of F_a to obtain

$$G_i = G_e \frac{f_a}{F'_e} \qquad (7\text{-}9)$$

In lieu of using f_a, which might reduce G_e excessively, one may elect to use $F_a = 0.6 F_y$. Again carefully note that the reduced G is used only when $KL/r < C_c$. It should be self-evident that values of $K \leq 1.0$ are not to be adjusted.

Example 7-2 Given a portion of a high-rise frame shown in Fig. E7-2 with sidesway permitted, assume adequate bracing perpendicular to bent so that K_y does not have to be considered (e.g., close contact with interior masonry walls) and $F_y = 36$. Using AISC specifications, design columns CD.

SOLUTION Assume that $KL \cong 14$ ft and use Table II-4 to obtain a tentative section W14 × 99:

$$A = 29.10 \text{ in}^2$$
$$I_x = 1110.0 \text{ in}^4$$
$$r_y = 3.71$$
$$\frac{r_x}{r_y} = 1.66$$

Compute

$$\frac{KL}{r_x} = \frac{14(12)}{3.71(1.66)} = 27.2 < C_c$$

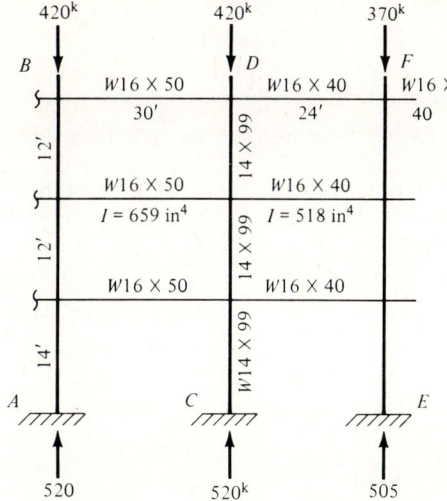

Figure E7-2

Find K: use W14 × 99 through three floors:

$$G_a = G_{top} = \frac{\Sigma I_c/L_c}{\Sigma I_b/L_b} = \frac{1110/14 + 1110/12}{659/30 + 518/24}$$

$$= 3.94$$

Reduce for inelastic effects:

$$f_a = \frac{520}{29.1} = 17.86 \text{ ksi}$$

From Table II-5 for $f_a = 17.9$ ksi, obtain $KL/r = 55$. For $KL/r = 55$ compute

$$F'_e = \frac{\pi^2 E}{23/12 \times 55^2} = 49.4 \text{ ksi} \quad \text{(calculated also in Table II-7)}$$

The revised

$$G = G_a\left(\frac{f_a}{F'_e}\right) = 3.94\left(\frac{17.9}{49.4}\right) = 1.43$$

For G_b we note that the column is rigidly attached to the base so that $I_b \to \infty$:

$$G_b = \frac{\Sigma I_c/L_c}{\Sigma I_b/L_b} = \frac{1110/14}{\infty} = 0 \quad \text{use 1.0}$$

This value is not reduced for inelastic behavior. From Fig. 7-5a, we obtain $K = 1.38$. With $K = 1.38$,

$$\frac{KL}{r_x} = \frac{1.38(14)(12)}{3.71(1.66)} = 37.6$$

310 STRUCTURAL STEEL DESIGN

From Table II-5 with this value of KL/r, we obtain

$$F_a = 19.37 \text{ ksi}$$

$$P_{\text{allow}} = 29.1(19.37) = 563.6 > 520 \text{ kips} \quad \text{O.K.}$$

Since this value is about 40 kips larger than needed, try a W14 × 90:

$$I_x = 999.0 \text{ in}^4 \quad r_y = 3.70 \text{ in} \quad \frac{r_x}{r_y} = 1.66 \quad A = 26.50 \text{ in}^2$$

$$G_a = \frac{999/14 + 999/12}{659/30 + 518/24} = 3.55$$

$$f_a = \frac{520}{26.50} = 19.62 \text{ ksi} \quad \frac{KL}{r} = 34 \quad \text{(Table II-5)}$$

From Table II-7 obtain (note that this table is computer-generated and uses SF = 23/12):

$$F'_e = 129.18 \text{ ksi}$$

$$G_a = 3.55\left(\frac{19.62}{129.18}\right) = 0.54$$

$G_b = 1$ as before, and from Fig. 7-5a obtain $K = 1.25$.

$$\frac{KL}{r_x} = 1.25(14)(12)/(3.70 \times 1.66) = 34.2$$

$$F_a = 19.64 \text{ ksi} \quad \text{(Table II-5)}$$

$$P_a = 26.50(19.64) = 520.5 > 520 \text{ kips} \quad \text{O.K.}$$

Use a W14 × 90 for column 1. ///

Example 7-3 Given the frame shown in Fig. E7-3, with sidesway possible, use the AISC specifications and $F_y = 345$ MPa to find the required column size. Use the same section for both.

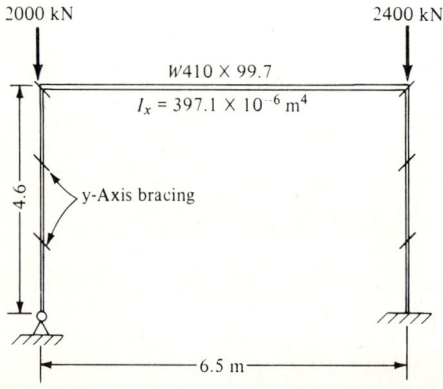

Figure E7-3

SOLUTION Make an initial column-size estimate that $KL/r = 40$, for which $F_a = 178.2$ MPa. For $F_a = 178.2$ MPa, the tentative column area is

$$A = \frac{2400}{178.2} = 13.5 \times 10^{-3} \text{ m}^2$$

Try W250 × 114.6:

$$A = 14.58 \times 10^{-3} \text{ m}^2$$
$$r_x = 114.05 \text{ mm}$$
$$I_x = 189.4 \times 10^{-6} \text{ m}^4$$

$$G_{a(\text{top})} = \frac{189.4/4.6}{397.1/6.5} = 0.673$$

$$f_a = \frac{2400}{14.58} = 164.6 \text{ MPa} \qquad \frac{KL}{r_x} = 53 \quad \text{(Table VI-6)}$$

$$F'_e = \frac{\pi^2(200\ 000)}{23/12 \times 53^2} = 366.6 \text{ MPa} \quad \text{(also Table VI-7)}$$

The adjusted $G_a = 0.673(164.6/366.6) = 0.302$.
Right column:

$$G'_b = \frac{189.4/4.6}{\infty} = 0 \qquad \text{use 1.0 (AISC recommendation)}$$

Left column:

$$G_b = \frac{189.4/4.6}{0} = \infty \qquad \text{use 10.0 (AISC recommendation)}$$

Obtain from Fig. 7-5a:

$$K = 1.2 \qquad \text{for right column}$$
$$K = 1.75 \qquad \text{for left column}$$

Check the left column first, since K is much larger than the right column:

$$\frac{KL}{r} = \frac{1.75(4.6 \times 1000)}{114.05} = 70.6$$

$$F_a = 143.7 \text{ MPa} \quad \text{(Table VI-6)}$$
$$P_a = AF_a = 14.58 \times 143.7 = 2095 \text{ kN} > 2000 \qquad \text{O.K.}$$

Now check the right column:

$$\frac{KL}{r_x} = \frac{1.2(4.6 \times 1000)}{114.05} = 48.4$$

$$F_a = 169.7 \text{ MPa}$$
$$P_a = 14.58(169.7) = 2474.2 > 2400 \text{ kN} \qquad \text{O.K.}$$

A solution: use W250 × 114.6 sections for the columns. ///

Example 7-4 Redo Example 7-3 with sidesway somehow restricted.

SOLUTION One solution is to use $K = 0.80$ for the left column according to Fig. 6-3b; correspondingly, one would obtain $K = 0.65$ using Fig. 6-3a for the right column and in both cases, using the "recommended" design values. Alternatively, use Fig. 7-5b.

Since the K must be less than 1.0, let us use the "experience" of the previous problem to check a somewhat lighter section. Try a W250 × 101.2:

$$A = 12.9 \times 10^{-3} \text{ m}^2 \qquad r_x = 112.78 \text{ mm} \qquad I_x = 164.0 \times 10^{-6} \text{ m}^4$$

$$G_a = \frac{164/4.6}{397.1/6.5} = 0.583$$

$$f_a = \frac{2400}{12.9} = 186 \text{ MPa} \qquad \frac{KL}{r} = 31.6$$

$$F_e' = 1032 \text{ MPa} \quad \text{(interpolating Table VI-7)}$$

$$G_a = 0.583 \left(\frac{186}{1032} \right) = 0.105$$

$$G_b = 10 \qquad \text{for left column}$$

$$G_b = 1.0 \qquad \text{for right column}$$

Using Fig. 7-5b, we obtain

$$K = 0.72 \qquad \text{for left column}$$
$$K = 0.65 \qquad \text{for right column}$$

Check the right column:

$$\frac{KL}{r_x} = \frac{0.65(4.6 \times 1000)}{112.78} = 26.5 \qquad F_a = 190.3 \text{ MPa}$$

$$P_a = 12.9(190.3) = 2455 > 2400 \text{ kN} \qquad \text{O.K.}$$

Check the left column:

$$\frac{KL}{r_x} = \frac{0.75(4.6 \times 1000)}{112.78} = 30.6 \qquad F_a = 186.8 \text{ MPa}$$

$$P_a = 12.9 \times 186.8 = 2410 > 2000 \text{ kN} \qquad \text{O.K.}$$

Use a W250 × 101.2. ///

7-4 DEVELOPING THE BEAM–COLUMN DESIGN FORMULAS

Prior to the sixth edition of the AISC *Manual of Steel Construction* in 1963, the design of compression members subjected to bending was obtained as

$$f_a + f_b \leq F_{\text{allow}}$$

Dividing this equation by $F_{\text{allow}} = F_a$, one obtained the widely used (AISC,

AASHTO, and AREA) interaction equation with bending about both axes f_{bx}, f_{by} as well as axial load as

$$\frac{f_a}{F_a} + \frac{f_{bx}}{F_b} + \frac{f_{by}}{F_b} \leq 1 \tag{7-10}$$

Prior to 1963, the value of $F_{bx} = F_{by} = F_b$. Currently, we recall that F_b depends on several factors, including unbraced length and compact section criteria, and in general AISC allows

$\qquad F_{bx} = 0.66 F_y$

or $\qquad F_{bx} = 0.60 F_y$

or $\qquad F_{bx}$ = values from AISC Eq. (1.5-6a), (1.5-6b), or (1.5-7)

$\qquad F_{by} = 0.75 F_y$ for W shapes due to having solid rectangular flanges

Currently, Eq. (7-10) is used only in certain limited stress conditions. For the remaining stress cases, other more complicated formulas based on research, plastic design, and elastic stability concepts are used. These will be partially developed in the next several paragraphs to indicate some of the limitations so that the practitioner will have some idea of how to follow through should the design vary from routine.

Refer to Fig. 7-6 for a short ($L/r \rightarrow 0$) rectangular section of dimensions $b \times d$ that is stressed with both an axial force and a moment sufficient to develop a plastic hinge. The plastic moment in the presence of a compressive force P is

$$M_{pc} = \frac{\sigma_y b}{4}(d^2 - 4y_0^2) \tag{a}$$

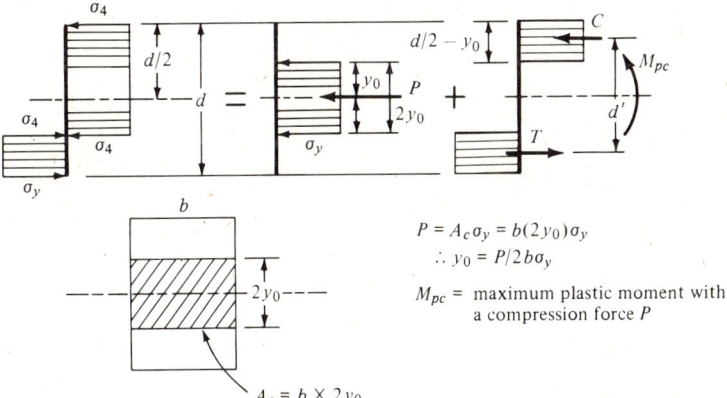

Figure 7-6 Plastic hinge formation in a very short member subjected to both an axial force and moment.

Substituting for y_0 (shown on Fig. 7-6), we obtain

$$M_{pc} = \frac{\sigma_y b}{4}\left(d^2 - \frac{P^2}{b^2\sigma_y^2}\right) \qquad (b)$$

Multiplying the P ratio by d^2/d^2 and noting that $b^2 d^2 \sigma_y^2 = P_y^2$, we obtain

$$M_{pc} = \frac{\sigma_y b d^2}{4}\left(1 - \frac{P^2}{P_y^2}\right) \qquad (c)$$

However, from Sec. 3. Example 3-3, $M_p = \sigma_y bd^2/4$; thus we obtain

$$\frac{M_{pc}}{M_p} = 1 - \left(\frac{P}{P_y}\right)^2 \qquad (7\text{-}11)$$

Although the development above has been made for a rectangular cross section, it is also valid for all (including W, S, and M) shapes. A plot of Eq. (7-11) is shown in Fig. 7-7. Also shown is the plot of a linear equation of the form

$$\frac{M_{pc}}{M_p} = 1.18\left(1.0 - \frac{P}{P_y}\right) \qquad (d)$$

If one were to plot

$$\frac{P}{P_y} + \frac{M}{M_p} = 1.0 \qquad (e)$$

for $kL/r = 40$ and for $KL/r = 120$, the straight lines shown also on Fig. 7-7 would be obtained. These curves will be somewhat in error, since the P–Δ effect has been neglected. We could use an iterative approach to include the P–Δ

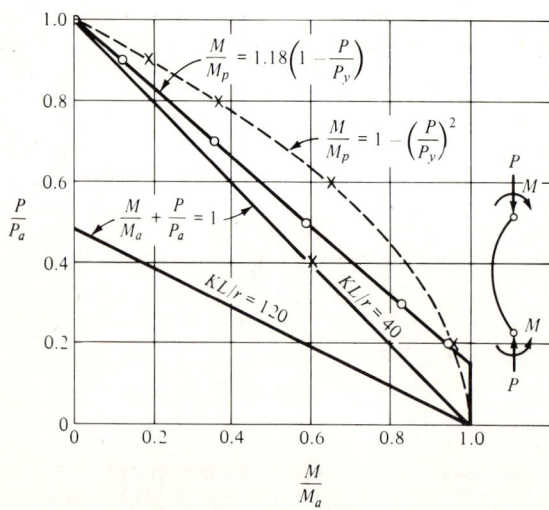

Figure 7-7 Plot of interaction equations as shown. P_a and M_a = allowable values.

effect as Eq. (*b*) of Sec. 7-2. However, it can be shown [see, for example, Timoshenko and Gere, *Theory of Elastic Stability*, 2nd ed. (New York: McGraw-Hill Book Company), Sec. 1-11] that it is sufficiently accurate (order of 1 to 2 percent error) to amplify the moment for P–Δ:

$$M' = \frac{M_0}{1 - (P/P_e)} \rightarrow \frac{M_0}{1 - (f_a/F'_e)} \qquad (f)$$

This factor may be called an *amplification factor*, since its effect is to amplify or increase M_0. This value has been used in the curves shown in Figs. 7-8 to 7-10. The P/P_e ratio is the ratio of the actual column load to the Euler column load and f_a/F'_e is simply dividing both loads by the column area. With this adjustment in bending moment, we may rewrite Eq. (*e*) to obtain

$$\frac{P}{P_y} + \frac{M}{M_p(1 - f_a/F'_e)} = 1 \qquad (g)$$

Shown in Fig. 7-8 is a plot of the loading situation where $M_1 = M_2 = M$ and in Fig. 7-9 the case where $M_1 = M$ and $M_2 = 0$. These two plots represent the extreme range of cases where a column is loaded with end moments, as in a building frame. The curves shown in Figs. 7-8 and 7-9 have been made using a modification of Eq. (*e*) proposed by Galambos and Ketter (*Transactions*, ASCE, Vol. 126, pp 1–25, 1961), which gives, for equal end moments:

$$\frac{M}{M_p} = 1 - C\left(\frac{P}{P_y}\right) - D\left(\frac{P}{P_y}\right)^2 \qquad (h)$$

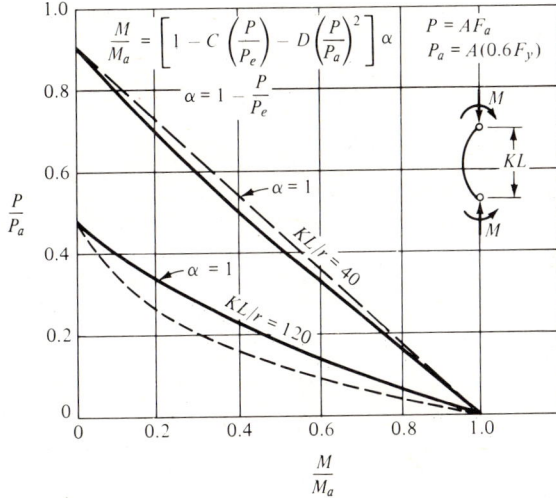

Figure 7-8 Influence of KL/r and amplification factor on column interaction.

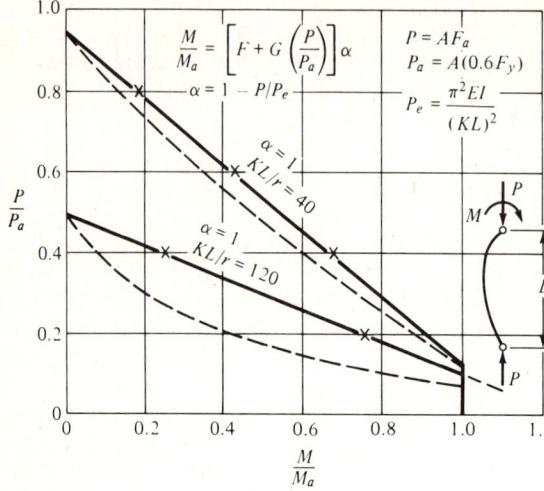

Figure 7-9 Plot of KL/r for 40 and 120 for a column with moment on one end.

and for unequal end moments a linear equation of the form

$$\frac{M}{M_p} = F - G\left(\frac{P}{P_y}\right) \tag{i}$$

The coefficients are KL/r-dependent, and several values given by Galambos and Ketter are as follows:

KL/r	C	D	F	G
0	0.42	0.77	1.13	1.11
20	0.70	0.46	1.14	1.18
40	0.99	0.17	1.16	1.23
80	1.81	−0.72	1.19	1.52
120	3.16	−2.51	1.25	2.53

The plots using Eqs. (h) and (i) are reasonably satisfactory for all of the cases with equal end moment, but it is rather conservative for those cases of unequal end moment, in particular for the extremity of $M_2 = 0$. This may be accounted for by a reduction factor C_m. Two of the several values proposed by a number of researchers are as follows:

$$C_m = 0.60 - \frac{0.4 M_1}{M_2} \geq 0.4 \tag{7-12a}$$

$$C_m = \frac{1}{1.75 - 1.05 M_1/M_2 + 0.3(M_1/M_2)^2} \geq 0.4 \tag{7-12b}$$

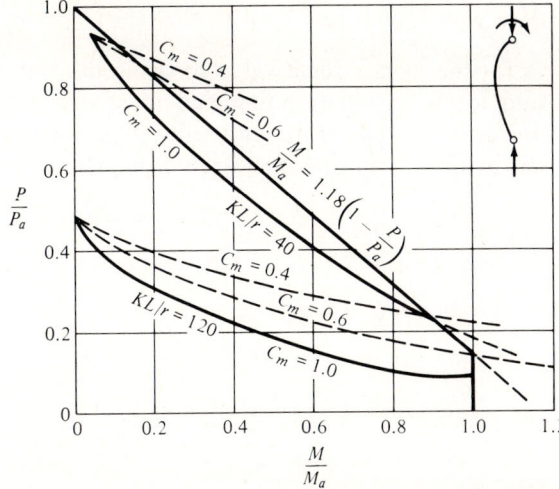

Figure 7-10 Plot of column interaction for end moment on one end and with effect of using C_m (dashed lines) for comparison.

We may note that the reciprocal of Eq. (7-12b) is used as the C_b amplification factor for laterally unbraced beams. Equation (7-12a) is used as C_m in the AISC and AASHTO design specifications. We must consider signs when using the ratio M_1/M_2, with M_1 being the *smaller* of the two end moment values, as follows:

$M_1/M_2 = (-)$ $\qquad\qquad M_1/M_2 = (+)$

This is in agreement with the definition for C_b (i.e., single curvature bending is more critical for buckling instability than reversed curvature, and similarly the P–Δ effects will be reduced).

The effect of using C_m is to generally decrease the effect of the amplification factor and is illustrated in Fig. 7-10. The use of C_m produces an intersection of the actual column interaction curve with the curve of

$$\frac{M}{M_p} + \frac{1.18P}{P_u} = 1.18 \qquad [\text{rearranging Eq. }(d)]$$

at values other than at $M_1/M_2 = 1.0$ and with a rapidity determined by the M_1/M_2 ratio. To the right of the intersection of the two curves, Eq. (d) controls. This requires use of two equations in design and using the most critical of

$$\frac{P}{P_y} + \frac{M}{M_p}\frac{1}{1 - P/P_e}C_m = 1.0 \qquad (h)$$

or
$$\frac{M}{M_p} + 1.18\frac{P}{P_y} = 1.18 \qquad (d)$$

This is because it is easier to solve for the most critical value by using the two equations than to make a plot and locate the intersection and then use the governing equation. These equations are adjusted for design use by substitution of P_{cr} for P_y and M_{allow} for M_p and with section area and section modulus to obtain stresses. This gives

$$F_a = \frac{P_{cr}}{A \times \text{SF}} \qquad f_a = \frac{P}{A}$$

F_b = allowable bending stress

$$f_b = \frac{M}{S}$$

Combining terms and extending the case to biaxial bending, we obtain

AISC Eq. 1.6-1a

$$\frac{f_a}{F_a} + \frac{C_{mx}f_{bx}}{(1 - f_a/F'_{ex})F_{bx}} + \frac{C_{my}f_{by}}{(1 - f_a/F'_{ey})F_{by}} \leq 1.0 \qquad (7\text{-}13)$$

and

AISC Eq. (1.6-1b)

$$\frac{f_a}{0.6F_y} + \frac{f_{bx}}{F_{bx}} + \frac{f_{by}}{F_{by}} \leq 1 \qquad (7\text{-}14)$$

Referring again to Fig. 7-7, we note no reduction in moment capacity until $P/P_y > 0.18$, so rounding 0.18 to 0.15 for convenience and to be conservative, we obtain:

Limitation of $f_a/F_a \leq 0.15$: AISC Eq. (1.6-2)

$$\frac{f_a}{F_a} + \frac{f_{bx}}{F_{bx}} + \frac{f_{by}}{F_{by}} \leq 1 \qquad (7\text{-}15)$$

Since Eqs. (7-13) to (7-15) are somewhat awkward to use except on a computer, let us multiply through by AF_a to [using Eq. (7-13) without f_{by} for a particular illustration] obtain

$$f_a A + \frac{C_m f_b A F_a}{(1 - f_a/F'_e)F_b} = AF_a$$

but $f_a A$ = the actual column load P and AF_a = maximum allowable column load P_a (and is not the same value shown on Figs. 7-8 and 7-9 and used to develop the curves shown). Also noting that $f_b = M/S$, define $A/S = B$, multiply f_a/F'_e by A/A, and take $F'_e A = P_e = 0.149 \times 10^6/(KL/r)^2$ ksi (in fps).

We obtain for bending about the X axis the following:

$$P + C_{mx} B_x M_x \left(\frac{F_a}{F_{bx}}\right) \frac{a_x}{a_x - P(KL)^2} \le P_{allow}$$

where $\quad a_x = 0.149 A r^2 \times 10^6$ ksi $\quad$ (fps units)

$P(KL)^2 =$ ksi

By analogy, Eq. (7-13) becomes

$$P + B_x C_{mx} M_x \frac{F_a}{F_{bx}} \frac{a_x}{a_x - P(KL_x)^2} + B_y C_{my} M_y \frac{F_a}{F_{by}} \frac{a_y}{a_y - P(KL_y)^2} \le P_{allow} \tag{7-13a}$$

For Eq. (7-14), we obtain in a similar manner:

$$P \frac{F_a}{0.6 F_y} + B_x M_x \frac{F_a}{F_{bx}} + B_y M_y \frac{F_a}{F_{by}} \le P_{allow} \tag{7-14a}$$

When f_a/F_a (or P/P_a) ≤ 0.15, we can obtain

$$P + B_x M_x \frac{F_a}{F_{bx}} + B_y M_y \frac{F_a}{F_{by}} \le P_{allow} \tag{7-15a}$$

A careful analysis of these equations shows that F_a is based on P_{cr}, which depends on the *critical value of KL/r*. Bending resistance and any amplification/reduction of moment effects is with respect to the bending axis with subscripting shown. Thus in many problems the allowable axial stress F_a will be based on KL/r_y, but the $P(KL)^2$ term will depend on the axis resisting bending. Since the X axis is usually used, it is the $P(KL_x)^2$ product that would usually be required.

The modified form of these interaction equations is generally considered to be easier to use, since the right-hand side (P_{allow}) can be tabulated for a number of column sections and for several assumed column lengths based on KL/r_y. Similarly, the terms a_x, a_y, B_x, and B_y can be computed and tabulated. These are shown in Tables II-4 ($F_y = 36$ ksi) and VI-4 ($F_y = 250$ MPa) for the W shapes commonly used as columns. The AISC manual also has these tabulations for $F_y = 50$ ksi steel and includes use of S shapes and tube and pipe sections.

7-5 DETERMINATION OF THE INTERACTION REDUCTION COEFFICIENT C_m

When a column in a structural frame is restrained against lateral translation with end moments, as illustrated in Figs. 7-8 to 7-10 and 7-11a, the value of C_m

320 STRUCTURAL STEEL DESIGN

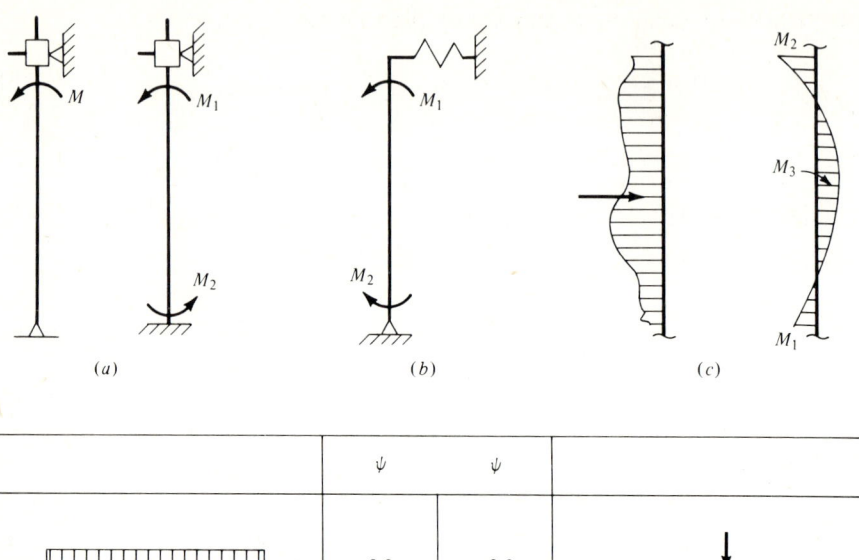

Figure 7-11 C_m reduction factor for beam–column interaction equations. (a) No sidesway: $C_m = 0.6 - 0.4 M_1/M_2$. (b) Sidesway: $C_m = 0.85$. (c) Column with transverse loading: $C_m = 1 + \psi f_a/F_e'$. (d) Several cases of transverse loading and ψ factors shown.

is computed using Eq. (7-12a):

$$C_m = 0.6 - \frac{0.4 M_1}{M_2} \geq 0.4$$

with attention to signs (single curvature $= -M_1/M_2$). Note also that M_1 is defined as the smaller of the two end moment values. When sidesway (Fig. 7-11b) is possible, AISC specification allows:

$$C_m = 0.85 \quad \text{(sidesway present)}$$

When the column has transverse loadings as in Fig. 7-11c, use

$$C_m = 1.0 + \frac{\psi f_a}{F'_e} \qquad (7\text{-}16)$$

where f_a = actual column stress
F'_e = Euler stress as previously defined (including SF = 23/12)
ψ = factor determined from Fig. 7-11d, which depends on end restraint and transverse loading

7-6 AASHTO AND AREA BEAM–COLUMN DESIGN FORMULAS

AASHTO working stress design uses essentially the same equations as AISC, with some additional adjustment to the amplification factor. The C_m factor is defined similar to AISC.

$$\frac{f_a}{F_a} + \frac{C_{mx}f_{bx}}{(1 - f_a/F''_{ex})F_{bx}} + \frac{C_{my}f_{by}}{(1 - f_a/F''_{ey})F_{by}} \leq 1 \qquad (7\text{-}17)$$

and

$$\frac{f_a}{0.472F_y} + \frac{f_{bx}}{F_{bx}} + \frac{f_{by}}{F_{by}} \leq 1 \qquad (7\text{-}18)$$

$$F''_e = \frac{\pi^2 E}{2.12(KL/r)^2}$$

Generally use $C_m = 0.85$ for end conditions of Fig. 7-11b and c; use $C_m = 1.0$ when the interior moment is greater than end values or with an interior moment and zero end moment.

The AREA equations are similar to the AISC equations. For $f_a/F_a > 0.15$:

$$\frac{f_a}{F_a} + \frac{f_{bx}}{(1 - f_a/F'''_{ex})F_{bx}} + \frac{f_{by}}{(1 - f_a/F'''_{ey})F_{by}} \leq 1 \qquad (7\text{-}19)$$

where

$$F'''_e = \frac{\pi^2 E}{1.431(KL/r)^2}$$

Also, at points braced in the plane of bending:

$$\frac{f_a}{0.55F_y} + \frac{f_{bx}}{F_{bx}} + \frac{f_{by}}{F_{by}} \leq 1 \qquad (7\text{-}20)$$

For $f_a/F_a \leq 0.15$, use

$$\frac{f_a}{F_a} + \frac{f_{bx}}{f_{by}} + \frac{f_{by}}{F_{by}} \leq 1 \qquad (7\text{-}21)$$

7-7 BEAM-COLUMN DESIGN USING INTERACTION EQUATIONS

The design of beam columns using the interaction equations is essentially a trial (iterative) process. A section is tentatively selected and analyzed and if the section is too small, a new section is selected and the process repeated until a satisfactory section (both strength and weight) is obtained. The steps may be outlined as follows:

1. Determine the axial force and column moments. We note that this step is also iterative, since indeterminate frame values are not found until a tentative section is used in the analysis.
2. Compute K to obtain KL. It may be necessary to determine both K_x and K_y depending on column end conditions and lateral bracing.
3. Estimate the moment contribution (and we will use bending about a single axis for illustration) as an equivalent axial load ΔP:

$$\Delta P = C_m BM \frac{F_a}{F_b} \beta \quad [\text{using a part of Eq. (7-13a)}]$$

or
$$\Delta P = BM \frac{F_a}{F_b} \quad [\text{using a part of Eq. (7-14a)}]$$

Inspection of Table II-4 (or VI-4) indicates that

$$C_{mx} B_x \frac{F_a}{f_{bx}} \beta = 0.2 \text{ to } 0.3 \quad (6.5 \text{ to } 9.0 \text{ in SI})$$

and
$$B_x \frac{F_a}{F_{bx}} = 0.1 \text{ to } 0.2 \quad (3.3 \text{ to } 6.5 \text{ in SI})$$

$$\beta = \text{estimation for} \quad \frac{a_x}{a_x - P(KL_x)^2}$$

The values for bending about the Y axis are considerably larger, but a ΔP estimate can be made similarly. An estimate for ΔP can be made in a somewhat similar manner for using the AASHTO and AREA equations if desired.

4. Compute the equivalent allowable column load as

$$P_{eq} = P + \Delta P = P + (0.2 \text{ to } 0.3) M \quad (M \text{ in in} \cdot \text{kips})$$
$$= P + (6.5 \text{ to } 9.0) M \quad (M \text{ in kN} \cdot \text{m and SI})$$

5. With this value of P, enter a Table such as II-4 or VI-4 (or in AISC Manual or tables which have been prepared for design use using a computer) with the KL_y value and obtain P_{table}. If $K_x/K_y \leq r_x/r_y$, the KL/r_y values control; if $K_x/K_y > r_x/r_y$, the KL_x/r_x ratio controls and one must use an adjusted

value of KL' to enter these tables, obtained as

$$KL' = \frac{KL_x}{r_x/r_y}$$

It is necessary to use r_x/r_y, since the ratio is fixed for a section but is not known until a section is tentatively selected.

6. Record P_{table}, A, r_y, r_x/r_y, L_c, L_u, B_i, and a_i. L_c and L_u are needed, so a rapid determination of F_b can be made:

$F_b = 0.66F_y$ if $L_{unbraced} \leq L_c$

$F_b = 0.6F_y$ if $L_{unbraced} \leq L_u$

$F_b =$ computed from Eq. (4-23), (4-26), or (4-27)   if $L_{unbraced} > L_u$

The unbraced length is the *actual* and not the KL value and is taken with respect to the bending axis.

7. Compute KL/r critical, obtain or compute F_a, and compute $f_a = P/A$. Compare $f_a/F_a \leq 0.15$. If the ratio is less, use Eq. (7-15) to see if the section is adequate. If $f_a/F_a > 0.15$, it will be necessary to check *both* Eqs. (7-13a) and (7-14a). In this case compute:

$P(KL)^2$ use KL with respect to the bending axis, which may be different from the critical KL/r used to compute the allowable axial stress F_a

8. Check both Eqs. (7-13a) and (7-14a) to satisfy

$$P + P_{eq} \leq P_{table}$$

One should attempt to achieve an equality as closely as possible for one of the equations (it cannot be done simultaneously for both equations) and if:

$P + P_{eq} \ll P_{table}$ check for lighter section

$P + P_{eq} > P_{table}$ use a larger section to satisfy design

With care and some preliminary scrutiny of Table II-4 (or VI-4), an adequate section can be obtained in one to three trials. This is possible, since the tables show little change in r_x/r_y and B for the two or three sections on either side of the selected section.

Example 7-5 Given column and bending moments shown in Fig. E7-5 as part of building frame in which sidesway is possible, we will limit the column section to not over a W12. Use the AISC specifications and A-36 steel. Select a tentative section and use the basic equations (7-13), (7-14), or (7-15) as applicable.

SOLUTION To keep the design of a beam column using the interaction equations in perspective, we will assume that $K_x = 1.3$ and $K_y = 1.0$ and

324 STRUCTURAL STEEL DESIGN

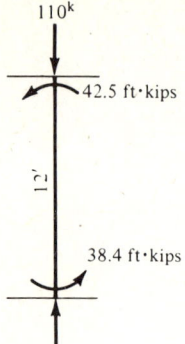

Figure E7-5

that the K values do not change with section size.

$$C_m = 0.6 - 0.4\left(+\frac{38.4}{42.5}\right) = 0.24 < 0.4$$

Use $C_m = 0.85$ with *sidesway*. After some study of Table II-4 (with $P > 110$), let us try W12 × 50:

$$A = 14.70 \text{ in}^2 \quad \text{(use Table II-4)} \quad P_{\text{allow}} = 236 \text{ kips}$$

$$r_y = 1.96 \text{ in} \quad \frac{KL}{r_y} = \frac{144}{1.96} = 73.5$$

$$\frac{r_x}{r_y} = 2.64 \quad F_a = 16.06 \text{ ksi} \quad \text{(Table II-5)}$$

$$L_c = 8.5 < 12 \quad L_u = 19.6 > 12 \rightarrow F_b = 0.6F_y = 22 \text{ ksi}$$

$$S_x = 64.7 \text{ in}^3 \quad \text{(Table I-3)}$$

$$f_a = \frac{P}{A} = \frac{110}{14.70} = 7.48 \text{ ksi}$$

$$\frac{f_a}{F_a} = \frac{7.48}{16.06} = 0.47 > 0.15$$

We must use both Eqs. (7-13) and (7-14):

$$F'_e = \frac{\pi^2 E}{1.92(KL_x/r_x)^2} = \frac{\pi^2 E}{1.92(36.2)^2} = 113.8 \text{ ksi}$$

$$\beta = 1 - \frac{f_a}{F'_e} = 1 - \frac{7.48}{113.8} = 0.934$$

$$f_{bx} = \frac{M}{S_x} = \frac{42.5(12)}{64.7} = 7.88 \text{ ksi}$$

By Eq. (7-13):

$$\frac{f_a}{F_a} + \frac{C_{mx}f_{bx}}{\beta F_{bx}} \leq 1$$

$$0.47 + \frac{0.85(7.88)}{0.934(22)} = 0.796 \ll 1.0$$

By Eq. (7-14):

$$\frac{f_a}{0.6F_y} + \frac{f_{bx}}{F_{bx}} \leq 1$$

$$\frac{7.48}{22} + \frac{7.88}{22} = 0.70 \ll 1.0$$

Try a smaller section—try W12 × 40:

$$A = 11.80 \text{ in}^2 \qquad f_a = \frac{110}{11.80} = 9.32 \text{ ksi}$$

$$r_y = 1.94 \text{ in} \rightarrow \frac{KL}{r_y} = 74.2 \qquad F_a = 15.98 \text{ ksi}$$

$$\frac{r_x}{r_y} = 2.64 \qquad \frac{KL_x}{r_x} = \frac{1.3(12)(12)}{1.94(2.64)} = 36.6$$

$$L_u = 16 > 12 \text{ ft} \rightarrow F_b = 0.6F_y = 22 \text{ ksi}$$

$$S_x = 51.9 \text{ in}^3$$

$$\frac{f_a}{F_a} = \frac{9.32}{15.98} = 0.58 > 0.15 \qquad f_b = \frac{42.5(12)}{51.9} = 9.83 \text{ ksi}$$

$$F'_{ex} = \frac{\pi^2 E}{1.92(36.6^2)} = 111.3 \text{ ksi}$$

$$\beta = 1 - \frac{9.32}{111.3} = 0.92$$

By Eq. (7-14):

$$0.58 + \frac{0.85(9.83)}{0.92(22)} = 0.991 < 1.0 \qquad \text{O.K.}$$

By Eq. (7-14):

$$\frac{f_a}{0.6F_y} + \frac{f_a}{F_b} \leq 1.0$$

$$\frac{9.32}{22} + \frac{9.83}{22} = 0.87 < 1.0 \qquad \text{O.K.}$$

Use a W12 × 40 column. ///

Example 7-6 Given the column and bending moments shown for a building frame, with sidesway for K_y restricted by use of bracing and shear walls, use the AISC specifications and $F_y = 250$ MPa steel to select a tentative column section.

SOLUTION Refer to Fig. E7-6 and assume that
$$K_x = 1.25 \quad K_y = 1.0$$
$$C_m = 0.6 - \frac{0.4 M_1}{M_2} \geq 0.4$$
$$= 0.6 - (0.4)\left(-\tfrac{53}{61}\right) = 0.95 > 0.4$$

Use $C_m = 0.95$ (AISC actually allows $C_m = 0.85$ if desired).
Tentatively:

$$P \cong P_{\text{given}} + P_{\text{equiv}}$$
$$\cong 445 + 7(61) \quad [\text{estimate factor as 7 (between 6.5 and 9)}]$$
$$\cong 872 \text{ kN}$$

Scan Table VI-4 and select W310 × 59.5:

$$P_{\text{table}} = 860.5 \text{ kN}$$

$$A = 7.61 \times 10^{-3} \text{ m}^2 \qquad \frac{r_x}{r_y} = 2.64 > \frac{1.25}{1} \quad \text{O.K.}$$

$$r_y = 49.28 \text{ mm} \qquad \frac{KL}{r_y} = \frac{3.45(1000)}{49.28} = 70.0$$

$F_a = 114.1$ MPa (Table VI-5)
$L_c = 2.58$ m $L_u = 4.87$ m > 3.45
$F_b = 0.6 F_y = 150$ MPa
$B_x = 8.94 \qquad a_x = 132.8 \times 10^3$ kN · m^2

$$P(KL_x)^2 = 445(1.25 \times 3.45)^2 = 8.27 \times 10^3 \text{ kN} \cdot \text{m}^2 \quad (\text{in same form as } a_x)$$

$$f_a = \frac{445}{7.61} = 58.5 \text{ MPa} \qquad \frac{f_a}{F_a} = \frac{58.5}{114.1} = 0.51 > 0.15$$

Actual $P_{\text{allow}} = A F_a = 7.61(114.1) = 868.3$ (vs. 860.6 interpolated)
Using Eq. (7-13a):

$$P + C_{mx} B_x M_x \frac{F_a}{F_{bx}} \frac{a_x}{a_x - P(KL_x)} \leq P_{\text{allow}}$$

$$445 + 0.95(8.94)(61)\frac{114.1}{150}\frac{132.8}{132.8 - 8.27} = 865.3 < 868.3 \text{ kN} \quad \text{O.K.}$$

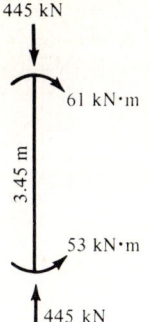

Figure E7-6

Check Eq. (7-14a):

$$P\frac{F_a}{0.6F_y} + B_x M_x \frac{F_a}{F_b} \le P_{\text{allow}}$$

$$(445)\frac{114.1}{150} + 8.94(61)\frac{114.1}{150} = 753 < 868.3 \text{ kN} \qquad \text{O.K.}$$

Using Eqs. (7-13) and (7-14) would give:
Eq. (7-13):

$$\frac{865.3}{868.3} = 0.997 < 1.0$$

Eq. (7-14):

$$\frac{753}{868.3} = 0.867 < 1.0$$

Use as tentative section, W310 × 59.5. ///

Example 7-7 The top chord member (No. 6) of the truss used in Examples 6-4 and 6-5 will be designed to include the member weight and a temporary concentrated force of 2.2 kips that will be applied to the center of the chord during maintenance operations (see Fig. E7-7a). The bridge may be temporarily closed to traffic if the maintenance load is too large to be carried safely with traffic (live loading). Other data:

$$\text{Dead load} = -283.75 \text{ kips}$$
$$\text{Live load} = -109.44 \text{ kips}$$
$$P = -393.19 \text{ kips}$$

SOLUTION Since a pair of C15 × 40 channels was used in Example 6-4 for a somewhat smaller load, we will try a pair of C15 × 50 channels (Fig. E7-7b):

$$A = 14.70 \text{ in}^2 \qquad I_y = 11.00 \text{ in}^3$$
$$r_x = 5.24 \text{ in} \qquad \bar{x} = 0.799 \text{ in} \qquad S_x = 53.80 \text{ in}^3/\text{channel}$$

328 STRUCTURAL STEEL DESIGN

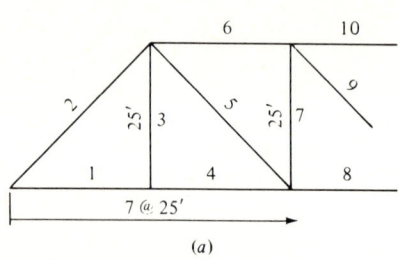

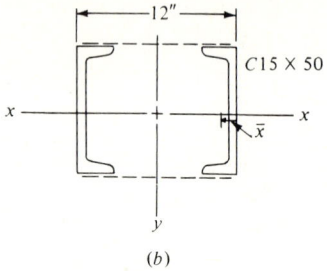

Figure E7-7a   **Figure E7-7b**

Compute the new I_y and r_y:

$$I_y = 2(11.0) + 2(14.70)(6 - 0.799)^2 = 817.3 \text{ in}^4$$

$$r_y = \left(\frac{817.3}{2(14.70)}\right)^{1/2} = 5.27 \text{ in} > 5.24 \to r_x \text{ controls}$$

$$\frac{KL}{r_x} = \frac{25 \times 12}{5.27} = 56.9$$

$$F_a = \frac{F_y}{2.12}\left[1 - 0.5\left(\frac{KL/r}{C_c}\right)^2\right]$$

$$F_a = \frac{36}{2.12}\left[1 - 0.5\left(\frac{56.9}{126.1}\right)^2\right] = 15.25 \text{ ksi}$$

$$f_a = \frac{393.19}{29.4} = 13.37 < 15.25 \text{ in}^2 \quad \text{O.K.}$$

The lacing will be the same as used in Example 6-6 and will not be again designed. Now check column weight using interaction equations.

$$M_{\text{beam}} = \frac{wL^2}{8} = \frac{2(0.05)(25)^2(12)}{8} = 93.75 \text{ in} \cdot \text{kips}$$

$$f_b = \frac{M}{S_x} = \frac{93.75}{2(53.8)} = 0.87 \text{ ksi} \quad \text{(with two channels in parallel)}$$

Compute F_e'' using Eq. (7-19):

$$F_e'' = \frac{\pi^2 E}{2.12(KL/r_x)^2} = \frac{\pi^2(29\,000)}{2.12(57.3)^2} = 41.1 \text{ ksi}$$

Using Eq. (7-17) with $C_m = 1$, since the beam weight is interior loading:

$$\frac{f_a}{F_a} + \frac{C_m f_b}{(1 - f_a/F_e'')F_b} \leq 1$$

Compute

$$F_b = 0.55 F_y \left[1 - 0.5 \left(\frac{KL/r}{C_c} \right)^2 \right] = 20 \left[1 - 0.5 \left(\frac{56.9}{126.1} \right)^2 \right] = 17.96 \text{ ksi}$$

$$\frac{13.37}{15.23} + \frac{1(0.87)}{(1 - 13.37/41.1)(17.96)} = 0.878 + 0.071 = 0.949 < 1 \quad \text{O.K.}$$

Check interaction with maintenance load added:

$$M_b = \frac{PL}{4} = \frac{2.2(25)(12)}{4} = 165 \text{ in} \cdot \text{kips}$$

$$f_b = \frac{165}{2(53.8)} = 1.53 \text{ ksi}$$

$$\frac{13.37}{15.23} + \frac{1(0.87 + 1.53)}{0.674(17.96)} = 0.878 + 0.198 = 1.08 > 1.0 \quad \text{N.G.}$$

Check if the bridge is closed to traffic:

$$f_a = \frac{P_d}{A} = \frac{283.75}{29.4} = 9.65 \text{ ksi}$$

$$\frac{9.65}{15.23} + \frac{1(0.87 + 1.53)}{(1 - 9.65/41.1)(17.96)} = 0.634 + 0.175 = 0.809 < 1 \quad \text{O.K.}$$

Now check Eq. (7-18). If adequate for bridge loads and the bridge is closed to traffic, it is O.K.

For bridge loads:

$$\frac{f_a}{0.472 F_y} + \frac{f_a}{F_b} \leq 1$$

$$\frac{13.37}{0.472(36)} + \frac{0.87}{17.96} = 0.787 + 0.048 = 0.835 < 1 \quad \text{O.K.}$$

For maintenance and no traffic:

$$\frac{9.65}{16.992} + \frac{0.87 + 1.53}{17.96} = 0.568 + 0.134 = 0.702 < 1 \quad \text{O.K.}$$

Use two C15 × 50 channel sections. Close the bridge for maintenance. ///

7-8 STEPPED COLUMNS AND COLUMNS WITH INTERMEDIATE AXIAL LOAD

The analysis of stepped columns frequently arises in the design of industrial buildings, particularly when the columns support crane runway girders. The general conditions for this are illustrated in Fig. 7-12.

Figures 4-1a and 10-1a illustrate actual crane runways being supported by built-up stepped columns. The design is heavily influenced by column end conditions. The column base can usually be assumed to be fixed with respect to

330 STRUCTURAL STEEL DESIGN

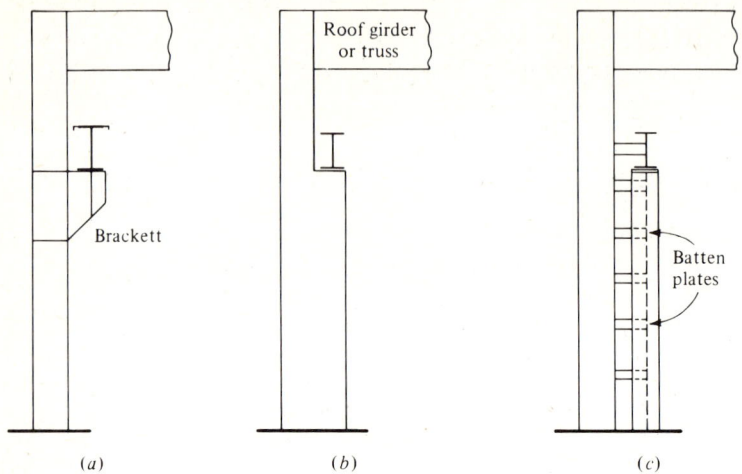

Figure 7-12 Crane columns in an industrial building. (*a*) Column with bracket. (*b*) Stepped column. (*c*) Built-up stepped column.

both the X and Y axes. Now if we take the X axis of the main column member oriented for bending in the plane of the bent and the Y axis for bending out of plane (with respect to building length), we have the following considerations:

X Axis

1. *Fixed or pinned against rotation at the roof truss level.* The roof truss of the industrial bents of Examples 2-5 and 2-6 provides rotation fixity (at least nearly so), but translation may take place.
2. *Sidesway control.* The side sheds in Examples 2-5 and 2-6 plus any bracing in the plane of the first and last bents will act to control sidesway. If it is still excessive, knee bracing may be required from the column to the roof truss.
3. *There will likely be moments in the column at the roof truss.* There will possibly be moments in the column at the crane girder level due to lateral thrust of the crane trolley against the rails making the track. There will be a column moment at the base due to assumed base fixity. These moments produce beam–column interaction for which Eqs. (7-13) and (7-14) must be used.

Y Axis

1. *Fixed or pinned at roof level.* If we put some cross bracing in the plane of the lower truss chord and vertical cross bracing in one or more of the bays on opposite sides, the major amount of sidesway can be controlled.
2. *Lateral intermediate bracing.* This is necessary to reduce KL/r_y of the main column and of the column segment above the crane girder point. Girts may

be used to support the siding and for lateral bracing. Caution is necessary, however, that girts should be continuous and, should building repair require removal of a girt, that it be done in only one bay at a time so that the lateral support is not lost.

3. *There will be a column moment at the crane runway level due to the longitudinal thrust of the crane starting or stopping suddenly.* This will also produce a moment at the base plate even if the analysis is made in such a manner as to ignore the moment at the crane level. This force will also produce a column shear that must be resisted by the anchor bolts at the base.

The general design of a column of this type proceeds as follows:

1. As in Examples 2-5 and 2-6, tentatively analyze the structure and revise until reasonable member sizes and deflections are obtained. Different member sizes can be used for any member. This was not done initially in the two examples, as the work is still highly preliminary.
2. Based on the computer output, begin to redesign the members. Increase or decrease sizes, depending on forces and deflections.
3. Reprogram and check output for forces and deflections.
4. Repeat as necessary.

7-8.1 Modification of K for Stepped Columns

A pin-ended column free to buckle and with a load P_0 and an interior load P_0' as in Fig. 7-13 is in a state of unstable equilibrium if the loads are sufficiently large. If we use the differential equation

$$EI\frac{dx}{dy} = -Px$$

and allow for boundary conditions of different loads in the lower segment, no lateral displacement at the column ends, change in I for the lower segment and a common slope at the junction of the upper and lower segments, we obtain

$$\frac{kLJ}{(P_0/P_t)^{1/2}}\cot\left[\left(\frac{P_0}{P_t}\right)^{1/2}(kL - akL)\right] + \frac{kLJ}{(1 + P_0/P_t)^{1/2}}\cot\left[akL\left(1 + \frac{P_0}{P_t}\right)^{1/2}\right]$$

$$= \frac{1}{P_0/P_t(1 + P_0/P_t)}$$

(7-22)

where $J = 1 + P_0/P_t - a$
$a = L_1/(L_0 + L_1)$

This is a solution as given by Sandhu, "Effective Length of Columns with Intermediate and Axial Load," *AISC Engineering Journal*, October 1972. The kL

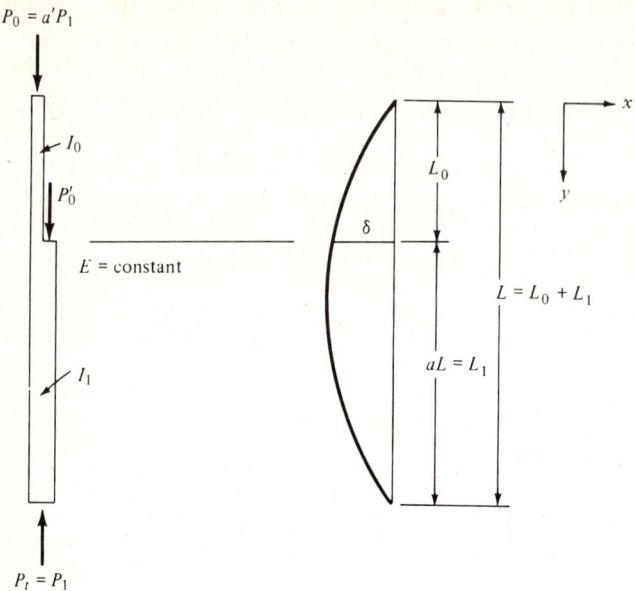

Figure 7-13 Figure for derivation of effective length coefficient for a stepped column.

value obtained as the solution of Eq. (7-22) is as used in Chap. 6:

$$kL = \left(\frac{P}{EI}\right)^{1/2}$$

Rearranging yields

$$P = \frac{k^2\pi^2 EI}{L^2} = \frac{\pi^2 EI}{(K_e L)^2}$$

but

$$P_{\text{total}} = P_0' + a'P_0' = P_0' + P_0$$

and equating this load to the Euler load, we obtain

$$P_e = \frac{\pi^2 EI}{L^2}$$

Solving for the equivalent length factor K_e yields

$$K_e = \left[\frac{\pi^2}{(1+a')k^2}\right]^{1/2}$$

This value of K_e is for a pin-ended column, and it is necessary to multiply this by the AISC value of K as obtained from Fig. 7-5 or 6-3, where the actual end conditions are taken into account. For general design office use, Eq. (7-22) should be programmed on a computer, so a plot of $a' = P_0/P_0'$ versus K_e can be made for selected ratios of a.

BEAM–COLUMN DESIGN

Example 7-8 Make a tentative design for the crane column for the industrial building in Example 2-6. Refer to Fig. E7-8a and assume the following:

1. Adequate bracing with respect to the Y axis of main column (along bays).
2. Sidesway in plane of bent with no top rotation and base fixed.
3. For initial tentative design, use only dead and live loads from computer output + crane loads.
4. Crane girder will be placed at a level with the bottom of the side shed truss.
5. Use a stepped (built-up) section.

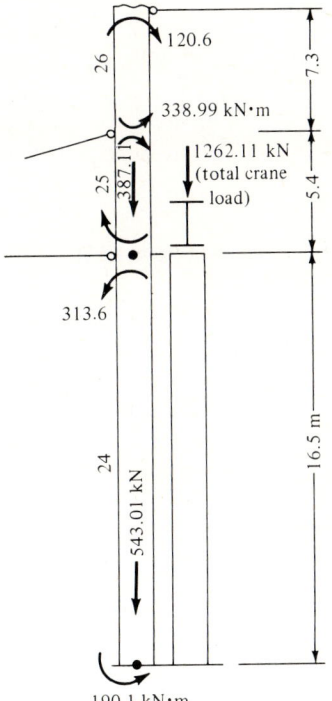

Figure E7-8a

SOLUTION A single W690 × 264.9 was used to obtain preliminary output shown in Example 2-6. This size section (area and moment of inertia) is necessary to reduce the lateral deflections at main truss roof level and at side shed-to-column intersection to tolerable values. The built-up section will require this or larger values to provide satisfactory lateral displacements. From the computer output (and for a tentative initial built-up section iteration) for $LC = 1$, the axial loads and moments (for left column) are:

Member 26 (uppermost):

$$P = 318.24 \text{ kN} \qquad \text{moment} = 120.59 \text{ kN} \cdot \text{m}$$

334 STRUCTURAL STEEL DESIGN

Member 25 (intermediate):
$$P = 387.11 \text{ kN} \qquad \text{moment} = 338.99 \text{ kN} \cdot \text{m}$$
Member 24 (lower section):
$$P = 543.01 \text{ kN} \qquad \text{moment} = -313.6 \text{ kN} \cdot \text{m}$$
$$\text{base moment} = -190.1 \text{ kN} \cdot \text{m}$$

Let us somewhat arbitrarily try a section made up of one W360 × 314 main column and one W360 × 101.2 crane column, as shown in Fig. E7-8b.

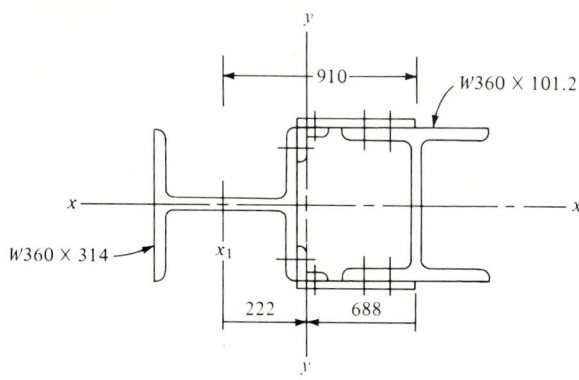

Figure E7-8b

W360 × 314 data:
$$I_x = 1107.2 \times 10^{-6} \text{ m}^4 \qquad I_y = 428.7 \times 10^{-6} \text{ m}^4$$
$$A = 40.0 \times 10^{-3} \text{ m}^2 \qquad d = 399 \text{ mm} \qquad b_f = 401 \text{ mm}$$

W360 × 101.2 data:
$$I_x = 300.9 \times 10^{-6} \text{ m}^4 \qquad I_y = 50.4 \times 10^{-6} \text{ m}^4$$
$$A = 12.9 \times 10^{-3} \text{ m}^2 \qquad d = 357 \text{ mm} \qquad b_f = 255 \text{ mm}$$

Compute r_x and r_y. Use $\Sigma M_{x_1} = 0$ to locate the new Y-Y axis.
$$(40 + 12.9)x = 910(12.90)$$
$$x = \frac{910(12.9)}{52.9} = 221.9 \text{ mm} \qquad (\text{use } 222 \text{ mm})$$
$$I_{yy} = 1107.2 + 50.4 + 40(0.222)^2(10^3) + 12.9(0.688)^2(10^3)$$
$$= 9235.09 \times 10^{-6} \text{ m}^4$$
$$r_y = \left(\frac{0.009235}{0.0529}\right)^{1/2} = 0.418 \text{ m} \rightarrow 418 \text{ mm}$$
$$I_{xx} = 428.7 + 300.9 = 729.6 \times 10^{-6} \text{ m}^4$$
$$r_x = \left(\frac{0.0007296}{0.0529}\right)^{1/2} = 0.117 \text{ m} \rightarrow 117 \text{ mm}$$

Check allowable stress F_a based on KL/r_y (since the other direction is considered stable by adequate use of girts and siding). Use the largest force in the upper column, 387.11 kN.

$$a' = \frac{P_0}{P_1} = \frac{387.11}{543.01 - 387.11 + 1262.11} = 0.273$$

$$a = \frac{16.5}{16.5 + 5.4 + 7.3} = 0.565$$

Substituting values into Eq. (7-22) yields

$$J = 1 + 0.273 - 0.565 = 0.708$$

$$\frac{kL(0.708)}{\sqrt{0.273}} \cot\left[\sqrt{0.273}\, kL(1 - a)\right]$$

$$+ \frac{kL(0.708)}{(1 + 0.273)^{1/2}} \cot\left[(1 + 0.273)^{1/2}(0.565kL)\right] = 2.877$$

$$1.356 kL \cot(0.227 kL) + 0.627 kL \cot(0.637 kL) = 2.877$$

Noting that the term to be evaluated as cotangent must be identified in radians, the problem can be quickly programmed on a pocket programmable calculator to obtain $kL = 3.525$.

$$(kL)^2 = 3.525^2 = 12.426$$

$$P_t = 543.01 + 1262.11 = 1805.1 \text{ kN (actual)}$$

$$P_t = \frac{12.426(1 + a')EI}{L^2} = \frac{15.82 EI}{L^2}$$

Also,

$$\frac{15.82 EI}{L^2} = \frac{\pi^2 EI}{(K_e L)^2}$$

$$K_e = \left(\frac{\pi^2}{15.82}\right)^{1/2} = 0.790$$

The actual $K = K_e \times K_{\text{end conditions}}$. From Fig. 6-3c obtain $K_{\text{end conditions}} = 1.2$.

$$K = 0.79(1.2) = 0.95$$

$$\frac{KL}{r_y} = \frac{0.95(16.5)}{0.418} = 37.5$$

From Table VI-5, obtain $F_a = 134.6$ MPa. From Table VI-7, obtain $F'_e = 732.8$ MPa. Check the interaction equation [Eq. (7-13)].

$$f_a = \frac{P_t}{A} = \frac{1805.1}{52.9} = 34.12 \text{ MPa}$$

$$\text{ratio} = \frac{34.12}{134.6} = 0.25 > 0.15$$

Use $C_{mx} = 0.85$. (We will not check for bending about the X axis at this time, since computations are very preliminary. After the next computer run, if this section is still satisfactory, we would check bending about both axes.)

$$f_b = \frac{Mc}{I} = \frac{313.6(0.816)}{9.235} = 27.7 \text{ MPa} \qquad \text{(note decimal shifted for } I\text{)}$$

Assume that $F_b = 0.6F_y$. [We will have to compute this later using Eqs. 1.5-6 and 1.5-7 (AISC numbers) as appropriate.]

$$\frac{f_a}{F_a} + \frac{C_m f_b}{F_b}\left(\frac{1}{1 - f_a/F'_e}\right) \leq 1$$

$$0.25 + 0.85\left(\frac{27.7}{150}\right)\left(\frac{1}{1 - 34/733}\right) = 0.25 + 0.16 = 0.41 \ll 1.0$$

By inspection, Eq. (7.14) will be satisfied for the lower column segment.

The upper column segment should be checked for interaction to make sure that the W360 × 314 is adequate as a column for the full height of the bent.

We can now reprogram this example with the new column sections and any revised sections for the truss members and see if the lateral deflections for the several load conditions is satisfactory and that the bending moments and axial forces are compatible with the section being analyzed. This should be done prior to refining the final design, to keep the engineering calculations to a minimum and maximize use of the computer. ///

7-9 CONTROL OF SIDESWAY

It is evident that the most efficient column design results when the frame is adequately braced against sidesway. With no sidesway:

1. C_m can be less than 0.85 (but may be more in selected cases).
2. The effective length factor K is not greater than 1.0.

A rigidly framed structure can translate laterally sufficiently to undergo "sidesway." Note also that the development of the equations for the G factors is based on a common slope at a joint that can only be obtained for a "rigid" joint. It is therefore necessary to provide specific resistance to sidesway to obtain the most efficient columns. This may be obtained (see Fig. 7-14) via:

1. Shear walls (use rigid vertical walls of brick, tile, or concrete block to contain the lateral movement). If masonry walls are used, a close contact with the column should be provided so that the column cannot translate in the construction void between materials.

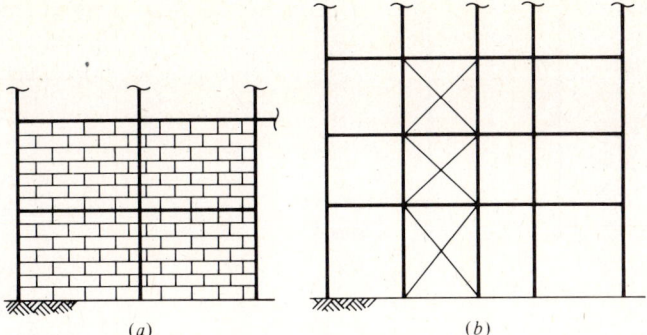

Figure 7-14 Sidesway control. (*a*) Masonry shear wall. (*b*) Diagonal bracing in selected bay.

2. Diagonal bracing (essentially produce a vertically oriented truss) in one or more bays.

It may not be necessary to use rigid joints with shear walls or diagonal bracing. However, some material economy may be obtained via use of rigid joints as well as providing some additional room for uncertainties.

Diagonal bracing requirements are usually small. The controlling criteria may be the L/r ratio rather than the cross-sectional area requirements. Galambos ("Lateral Support for Tier Building Frames," *AISC Engineering Journal*, January 1964), using essentially the same method proposed by Winter for lateral bracing of columns and beams, develops an expression for the area of the diagonal brace member:

$$A_b = \frac{2(1 + \alpha^2)^{3/2}}{(\alpha)^2 E} \Sigma P \qquad (7\text{-}23)$$

where A_b = area of bracing, in² or m²
  ΣP = sum of all column loads at a story level in plane of bracing
  α = horizontal run of diagonal brace/length of column
  E = modulus of elasticity assuming column and brace of same material, ksi or MPa

Diagonal bracing is usually designed only for tension. It is assumed that the bracing member is so small and flexible that it will buckle (with stresses well below F_y) under a very small compression load. For opposite-direction loading, the buckled member straightens with no damage due to small buckling stresses and prevents sidesway from occurring by carrying the necessary tension load.

Example 7-9 Given the story of a tiered building shown in Fig. E7-9, design the diagonal bracing for the interior bay. Assume that this will be placed in alternate bents in the out-of-plane direction.

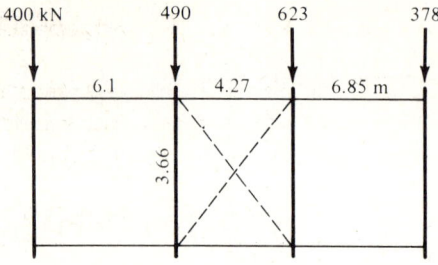

Figure E7-9

SOLUTION

$$L_b = 4.27 \text{ m} \qquad L_c = 3.66 \text{ m}$$

$$\alpha = \frac{4.27}{3.66} = 1.167$$

Use Eq. (7-23):

$$A_b = \frac{2(1 + 1.167^2)^{1.5}}{1.167^2(200\,000)}\Sigma(400 + 490 + 623 + 378) = 0.0504 \times 10^{-3} \text{ m}^2$$

The diagonal brace has a length of

$$L = (4.27^2 + 3.66^2)^{1/2} = 5.62 \text{ m}$$

The maximum L/r for a tension member is 300.

$$r_{min} = \frac{5.62 \times 1000}{300} = 18.7 \text{ mm}$$

Let us arbitrarily select a channel shape for the brace. From Table V-7 select a MC150 × 22.47:

$r_y = 22.6$ mm   (no standard C with adequate r_y and reasonable weight)

$$A = 2.86 \times 10^{-3} \text{ m}^2 \gg 0.0504 \qquad \text{required}$$

No channel section lighter and controlled by r_y. ///

7-10 BEAM–COLUMN DESIGN USING LRFD

Beam–column design using LRFD requires the separate components of load effects. The current interaction form of the interaction equations that have been proposed by the research group working on this design procedure are as follows:

$$\frac{P_{ud}}{\phi P_u} + \frac{C_{mx} M_{udx}}{\phi M_p(1 - P_{ud}/\phi P_{Ex})} + \frac{C_{my} M_{udy}}{\phi M_p(1 - P_{ud}/\phi P_{Ey})} \leq 1.0 \qquad (7\text{-}24)$$

and

$$\frac{P_{ud}}{\phi P_y} + \frac{M_{udx}}{\phi M_{px}} + \frac{M_{udy}}{\phi M_{py}} \leq 1.0 \qquad (7\text{-}25)$$

If $P_{ud}/\phi P_u \leq 0.152$,

$$\phi M_p \geq M_{ud} \qquad (7\text{-}26)$$

Additionally, we must always satisfy the following:

$$P_{ud} \leq \phi_c P_u \qquad (7\text{-}27)$$

When there is bending about only one axis we have the following:

$$\frac{P_{ud}}{\phi P_u} + \frac{M_{udx}}{1.18\phi M_p} \leq 1.0 \qquad \text{(about the } X \text{ axis)} \qquad (7\text{-}28)$$

and

$$\left(\frac{P_{ud}}{\phi P_u}\right)^2 + \frac{M_{udy}}{1.19\phi M_p} \leq 1.0 \qquad \text{(about the } Y \text{ axis)} \qquad (7\text{-}29)$$

In these equations $\phi = 0.86$ (for beams) and ϕ_c = value given in Table 3-1 for columns and noting ϕ_c varies from 0.65 to 0.86 depending on KL/r. The C_m terms are as in AISC. The values of P_u are

$$P_u = AF_y(1 - 0.25\eta^2) \qquad \eta \leq \sqrt{2} \qquad (7\text{-}30)$$

$$P_u = \frac{AF_y}{\eta^2} \qquad \eta > \sqrt{2} \qquad (7\text{-}31)$$

$$P_E = \frac{AF_y}{\eta^2} \qquad \text{(but use appropriate bending axis)} \qquad (7\text{-}32)$$

The design forces and moments are as computed in Chaps. 6 and 4.

The use of these LRFD equations will be illustrated by the following example.

Example 7-10 Given the beam column and loading shown in Fig. E7-10, select the lightest W310 shape using $F_y = 250$ MPa and LRFD.

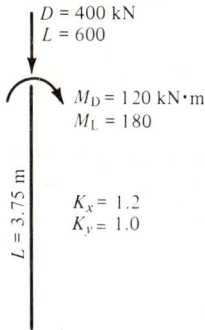

Figure E7-10

SOLUTION Use $F_L = 1.5$ (author's choice).

$$M_{ud} = 1.1[1.1(120) + 1.5(180)] = 442.2 \text{ kN} \cdot \text{m}$$

$$P_{ud} = 1.1[1.1(400) + 1.5(600)] = 1474 \text{ kN}$$

With bending only about the X axis, Eq. (7-28) applies, so that

$$\frac{P_{ud}}{\phi P_u} + \frac{M_{ud}}{1.18\phi M_p} \quad 1.0$$

Inspection of this equation indicates that

$$1.18\phi M_p > M_{ud}$$

$$1.18(0.86)ZF_y > 442 \qquad Z > \frac{442}{1.18(0.86)(250)} = 1.7422 \times 10^{-3} \text{ m}^2$$

$$AF_y > P_{ud}$$

$$A > \frac{1474}{0.86(250)} > 6.86 \times 10^{-3} \text{ m}^3$$

Using these values of A and Z, try a W310 × 157.7 section:

$$A = 20.13 \times 10^{-3} \text{ m}^2$$
$$r_y = 79.0 \text{ mm} \qquad r_x = 138.9 \qquad r_y \text{ controls}$$
$$Z_x = 2.687 \times 10^{-3} \text{ m}^3$$

$$\eta = \frac{KL}{\pi r}(F_y/E)^{1/2} = \frac{1(3750)}{\pi r}(250/200\ 000)^{1/2}$$

$$\eta = \frac{42.2}{r_y} = 0.534 > 0.16$$

from which

$$\phi_c = 0.90 - 0.25(0.534) = 0.77$$
$$P_u = AF_y[1 - 0.25(0.534)^2] = 20.13(250)(0.9287) = 4674 \text{ kN}$$
$$M_p = ZF_y = 2.687(250) = 672 \text{ kN} \cdot \text{m}$$

Substituting values into Eq. (7-28), we obtain

$$\frac{1474}{0.86(4674)} + \frac{442}{1.18(0.86)(672)} = 0.367 + 0.648 = 1.015 > 1.0 \qquad \text{N.G.}$$

Check:

$$\phi_c P_u \geq P_{ud}$$
$$0.77(4674) = 3599 \gg 1474 \qquad \text{O.K.}$$

Since this column is just over, the next largest W310 section should work. Tabulating data so that a comparison can be made, we obtain

$$\text{W310} \times 178.6: \quad A = 22.77 \times 10^{-3} \text{ m}^2$$
$$r_y = 79.5 \text{ mm} \quad \text{(and controls, as before)}$$
$$Z_x = 3.055 \times 10^{-3} \text{ m}^3$$

By inspection of this data it is evident that the section is adequate. It is not necessary to check that $P_{ud} \leq \phi_c P_u$, since this ratio is almost that obtained from the first term of Eq. (7.28). Use a W310 × 178.6 section. ///

PROBLEMS

7-1 Using the general solution for column buckling,

$$x = A \sin ky + B \cos ky + Cy + D \quad \text{and} \quad k = \sqrt{\frac{P}{EI}}$$

show that $K = 0.7$ for the column of Fig. 6-3b. Note the boundary conditions:

$$x = 0 \text{ at } y = 0 \text{ and } y = L$$

$$M = 0 \text{ at } y = 0 \qquad \frac{dx}{dy} = 0 \text{ at } y = L$$

7-2 Determine the effective length coefficients K for the columns in the frame shown in Fig. P7-2. Note that the far end of the girder of column A is pinned.
  Answer: AB: 1.75; FG: 0.78.

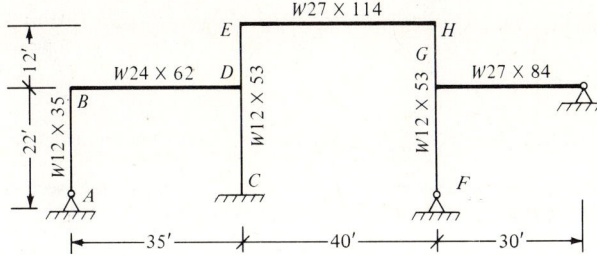

Figure P7-2

7-3 Determine the effective length coefficients K for the columns in the frame shown in Fig. P7-3.
  Answer: $BC = 0.65$; $EF = 0.63$.

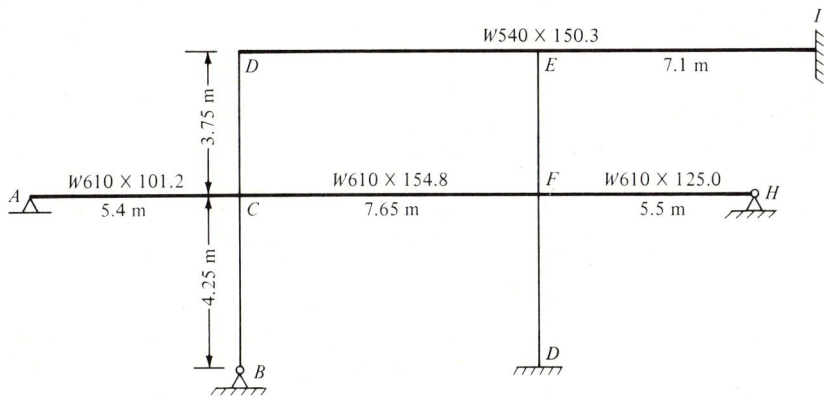

Figure P7-3

7-4 Select the lightest column section for the column loading and geometry of Fig. P7-4. Use the AISC specifications and A-36 steel. Also use no column larger than the nominal depth of the W12 sections. Take $K_x = K_y = 1.0$.

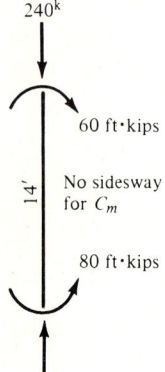

Figure P7-4

7-5 Select the lightest column section for the column shown in Fig. P7-5, using no column section larger than W310. Take $K_x = 1.2$, $K_y = 0.8$. Use the AISC specifications and $F_y = 250$ MPa.
  Answer: W310 × 142.9.

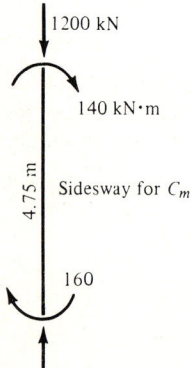

Figure P7-5

7-6 Given the frame in Fig. P7-6, use the computer program to obtain the axial forces and moments. Check the columns and beams for adequacy and reprogram as required to obtain reasonably light beams and columns. Limit the midspan deflection to $L/360$. Do not consider any wind loading. Use the AISC specifications, A-36 steel, and $K_y = 1.0$.

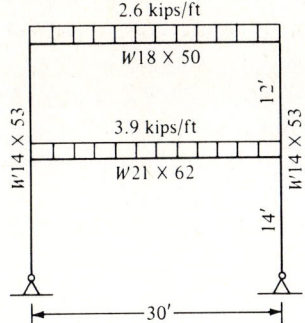

Figure P7-6

7-7 Given the frame in Fig. P7-7, use the computer program and for the tentative sections given, obtain the axial forces and moments. Check the sections for adequacy and reprogram until reasonably light beams and columns are obtained. Limit the roof beam deflection to $L/360$ at midspan and the lateral deflection at D to not over 0.040 m. Use the AISC specifications and $F_y = 250$ MPa steel.

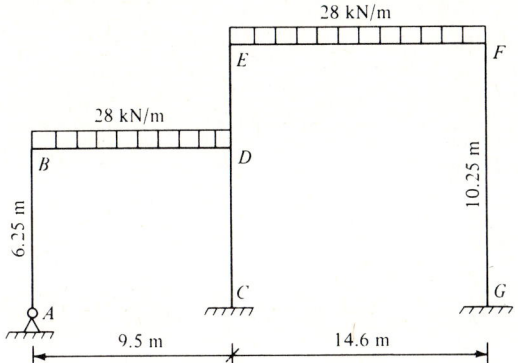

Figure P7-7

7-8 Design the lightest column (use a built-up shape with standard rolled W shapes, if required) for the loading and geometry shown in Fig. P7-8. Use $F_y = 345$ MPa (or higher if required) and the AISC specifications. Take $K_x = 1.2$ and $K_y = 1.0$.
  Answer: Use three W920 × 446.4 with two 1969 × 40 mm cover plates.

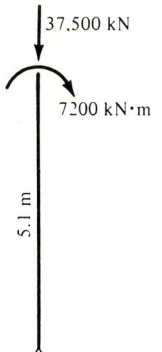

Figure P7-8

7-9 Design the top chord of a railroad truss with a panel length of 27.583 ft for the following conditions:

$$\text{Live load} = 867.2 \text{ kips (E-80 loading)}$$
$$\text{Impact} = 473.6 \text{ kips}$$
$$\text{Dead load} = 402 \text{ kips (estimated)}$$

Use a built-up section somewhat as shown in Fig. P7-9. Use the AREA specifications and A-36 steel. The chord member ends will be either riveted or bolted.

Answer: Two S24 × 100 and one S24 × 79.9 with a 30 × $\frac{1}{2}$ cover plate.

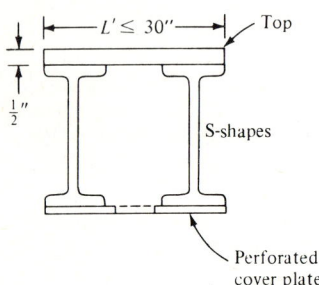

Figure P7-9

7-10 Check the section of the column above the crane runway girder of Example 7-8 and redesign as required.

7-11 Using Example 7-8 as a guide, make a tentative redesign of the main column of Example 2-5. Use A-36 steel and the AISC specifications.

7-12 Redo Example 7-2 with the column size limited to W12.
Answer: W12 × 96.

7-13 Design the diagonal bracing for the bent shown in Fig. E7-3 to inhibit sidesway. Use the lightest pair of angles with a 12-mm gusset plate.
Answer: Two L63 × 51 × 4.8.

7-14 Check the exterior columns of Example 2-3 using the computer output and resize the columns if necessary. Note that both exterior columns are to be the same size. Use a single column (no splices) for full building height. Use A-36 steel and the AISC specifications.

7-15 Check the interior columns of Example 2-3 using the computer output and resize the columns as necessary. Note that basement columns are not necessarily the same size as the upper column which is to be used for full building height. Use A-36 steel, the AISC specifications, and not over W10 columns.
Answer: W8 × 48 bottom; W8 × 40 upper.

7-16 Use the computer output of Example 2-4 for the exterior columns as outlined in Prob. 7-14. Limit column size to W250.
Answer: W250 × 67.0.

7-17 Use the computer output of Example 2-4 for the interior columns as outlined in Prob. 7-15. Limit column size to W250.

7-18 Verify with computations that the W310 × 178.6 column section of Example 7-10 is adequate.

7-19 Redo Example 7-10 if $M_L = 210$ kN · m and $D = 425$ kN. All other data are the same.
Answer: W310 × 178.6.

7-20 Redo Example 7-10 if M_y moments are present: $M_{dy} = 50$ kN · m; $M_{Ly} = 75$ kN · m.
Answer: W310 × 282.8.

Figure VIII-1 High-strength bolted joints. (*a*) Splicing smaller-to-larger column using filler plates. (*b*) Splicing same-size column. (*c*) Diagonal bracing.

CHAPTER
EIGHT

BOLTED AND RIVETED CONNECTIONS

8-1 INTRODUCTION

A steel structure is produced as an assemblage of the structural members making up the framework. Connections are required where the various member ends must be attached to other members sufficiently to allow the load to continue an orderly flow to the foundation. Since the connection serves to carry load from or to adjoining members, it must be adequately designed. The connection design involves producing a joint that is safe, economical of materials, and capable of being built (it must be practical). The more practical connections are usually more economical, since fabrication costs greatly affect the economy of both connections (or joints) and the members themselves, as illustrated earlier, particularly concerning built-up tension and compression members. Several structural connections are illustrated in Fig. 8-1.

Connections (or structural joints) may be classified according to:

1. Method of fastening, such as rivets (hardly ever), bolts, or welding. Connections using bolts are further classified as *bearing* or *friction*-type connections.
2. Connection rigidity, which may be simple, rigid (as produced by an indeterminate structure analysis), or of intermediate rigidity. The AISC, in Sec. 1.2 of the specifications, classifies joints based on connection rigidity as:
   Type 1: rigid connections that develop the full moment capacity of the connecting members and retain a constant relative angle between the connected parts under any joint rotation.

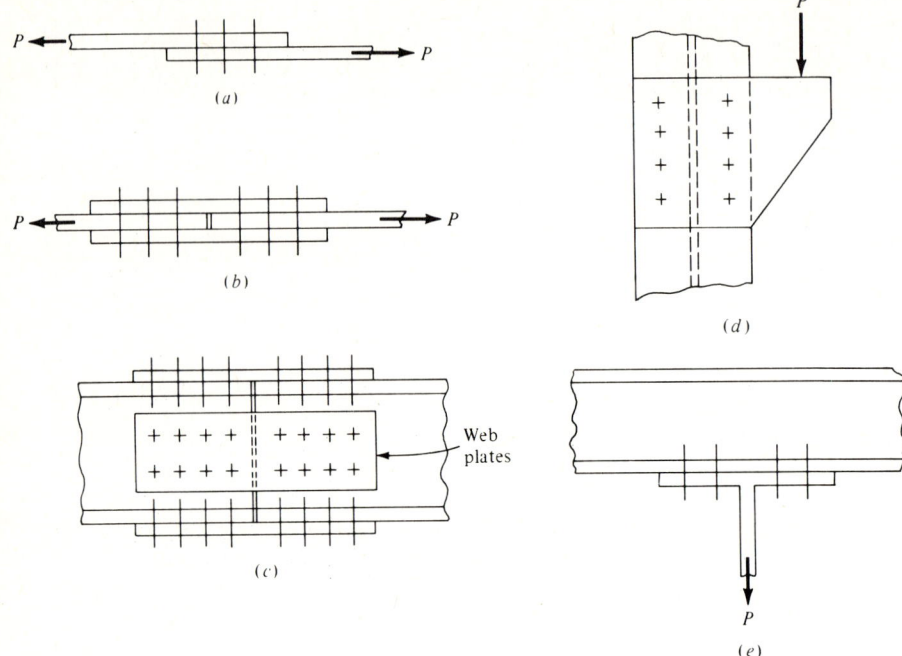

Figure 8-1 Several structural connections using bolts (or rivets). (*a*) Lap joint single shear. (*b*) Butt joint double shear. (*c*) Bridge beam splice. Flange plates single shear; web plates double shear if on both sides. (*d*) Bracket connection with single shear but torsion must be considered. (*e*) T section used as tension hanger.

Type 2: simple framing with no moment transfer assumed between the connected parts. Actually, a small amount of moment will be developed, but it is ignored in the design. Any joint eccentricity less than about $2\frac{1}{2}$ in (63 mm) is neglected.

Type 3: semirigid connections with less than the full moment capacity of the connected members being transferred. Design of these connections requires assuming (with adequate documentation) an arbitrary amount of moment capacity (e.g., 20, 30, or 75 percent of member capacity).

3. Type of forces transferred across the structural connection:
   a. Shear forces: common for floor beams and joists.
   b. Moment: either bending or torsion.
   c. Shear and moment: as in type 1 or 3 connections.
   d. Tension or compression: as for column splices and for "pinned" truss members.
   e. Tension or compression with shear: as for diagonal bracing.
4. Connection geometry:
   a. Framing angles used to connect floor joists and stringers to beams and columns.

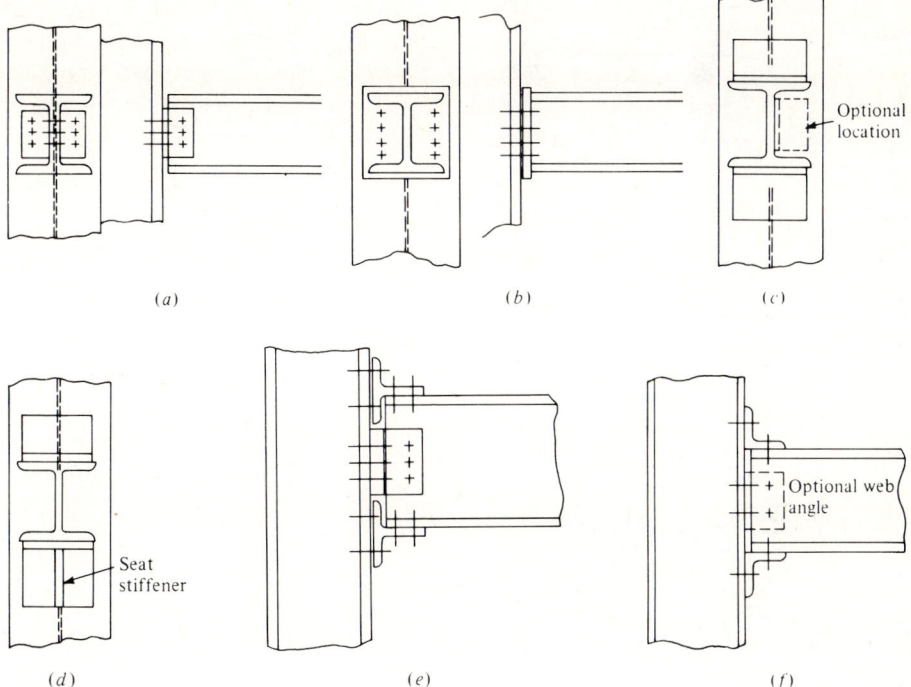

Figure 8-2 Several frame connections. Inspection of mode of carrying load determines if bolts are in shear or tension. Welding is commonly used in lieu of bolts for all of above connections except (*f*). (*a*) Framed using a pair of angles. (*b*) Framed using a shop-welded end plate and field bolting. (*c*) Using seat angle and top or side angle. (*d*) Stiffened seat angle. (*e*) Seat and top clip angle with additional optional web angle. (*f*) Rigid to semirigid connection using T-shapes to carry moment and a web angle for shear. Web angle is optional.

    b. Welded connections using plates and angles.
    c. End plates on beams or rafters.
    d. Plates or angles used on one side of a floor joist or beam.
    e. Seat angles with or without stiffeners.
   Several of these connections are illustrated in Fig. 8-2.
5. Fabrication location:
    a. Shop connections: produced in the fabrication shop.
    b. Field connections: joint parts fabricated in the shop but assembled on the job site.
6. Joint resistance. When considering joint (or connection) resistance, we have:
    a. Friction connections. Connections designed as friction connections have their primary resistance assumed to be developed as shear on the connectors (bolts or rivets) on the plane of potential slip between the connected parts. No relative movement between the connected parts develops until the design load is substantially exceeded. Actually, the strength of friction

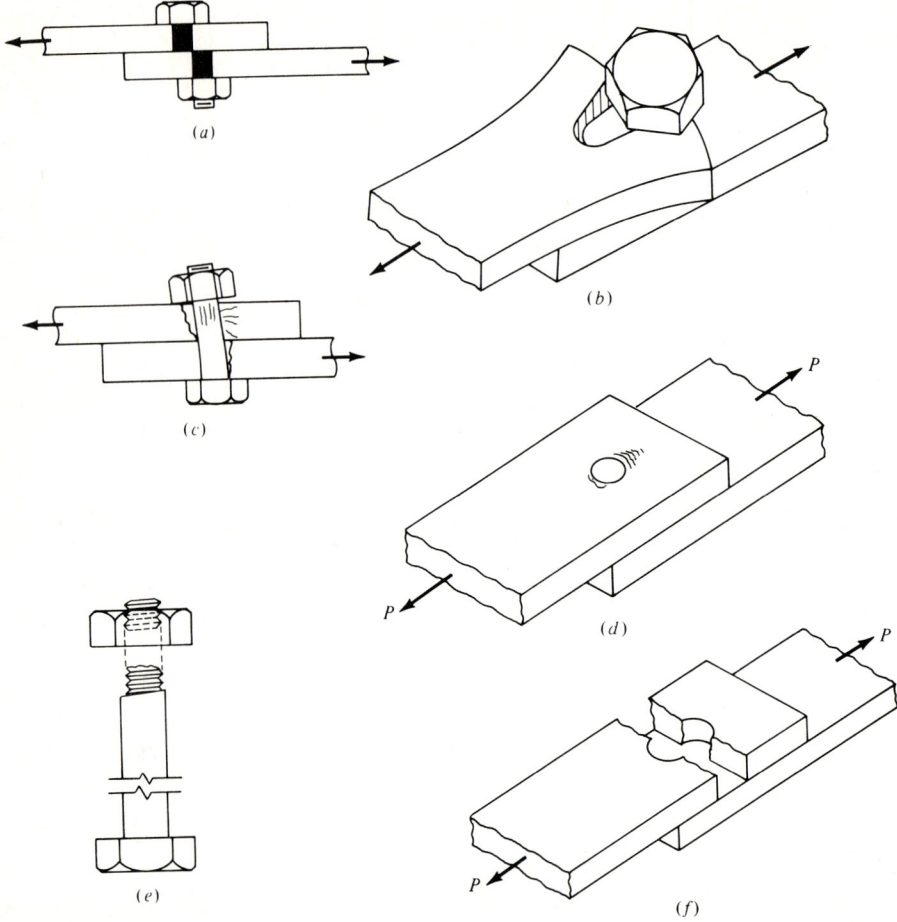

Figure 8-3 Several modes of joint resistance. (*a*) Bolt shear. (*b*) Plate shear or tear-out. (*c*) Bolt bearing. (*d*) Plate bearing. (*e*) Bolt tension failure. (*f*) Tension on net section.

connections is not developed as the shear resistance of the connectors; rather, it is developed as the product of the clamping force produced by tightening the bolts (or driving the rivets) and the coefficient of friction between the clamped parts. It is expected that in load factor resistance design this will be directly used as the design parameter, producing an equation of the general form used in several other design codes (outside the United States) as

$$P = \phi\mu\Sigma A_b F_y$$

where ϕ = performance factor (0.67 to 0.70)
μ = coefficient of friction × number of slip surfaces
$\Sigma A_b F_y$ = total developed clamping force as the sum from all the bolts used in the connection

b. **Bearing connections.** Connections where the joint resistance is taken as a combination of connector shear resistance and bearing of the connected material against the connector. This mode of behavior develops as sufficient slip occurs to bring the connected material into contact with the back projection of the connector near the working or design load. Since connector shear is a portion of the resistance in bearing connection analysis, the reduced shear area available for threaded connectors when the threads are in any slip plane requires a reduction in the design load. In actual practice, threads in the shear plane result in a lower allowable design shear stress for the fastener.

The design of both friction and bearing connections involves use of an allowable shear stress. The value is much lower for friction connections, since we do not want any joint slippage under working loads. The value is considerably larger for bearing connections, since some small amount of relative movement between the parts making up the joint can be tolerated. Both types of joints, in addition to being designed for "shear," are routinely checked for tension on the net section and for bearing of the connected material against the connector.

Current fabrication practice tends to use of slotted over-sized holes in friction connections. The slot allows easier field erection since more alignment tolerance is available for temporary erection bolts.

8-2 RIVETS AND RIVETED CONNECTIONS

For many years rivets were the sole practical means of producing safe and serviceable metal connections. The process required punching or drilling holes approximately 1/16 in (1.5 mm) oversize, assembling the parts using drift pins to align the holes, and using one or more bolts to hold the parts together temporarily. Rivets were heated in a furnace (portable for onsite use) to a cherry red color (approximately 980°C) and inserted into the aligned hole through the several parts to be connected. One member of the riveting crew then applied a bucking bar with a head die to the manufactured rivet head to hold the rivet in place and to shape. Another crew member used a pneumatic driver with a head die to forge the protruding rivet shank to produce the other head. The forging operation simultaneously reworked the rivet metal and caused a shank enlargement to very nearly fill the oversized hole. This reworking and shank enlargement, together with the shrinking of the hot rivet, produced a substantial joint most of the time. The rivet contraction during cooling is resisted by the joint material and develops tension in the rivet so that a riveted joint is intermediate between a friction- and a bearing-type connection (a bearing type is commonly assumed). This joint transmits the design load primarily by friction between the clamped plates making up the joint. The riveted joint has had a long history of success under fatigue stresses as in railroad bridges. Only recently has AREA allowed use of high-strength bolts and welds in joints for railroad bridges.

The combination of driving the rivet tight plus the additional rivet contraction the rivet attempts as it cools produces tension stresses which in well-driven rivets approach F_y of the rivet metal. Some problems occur, however, which include the problem of occasional overheating of rivets, which has a deleterious effect on the strength. If a rivet is cooled much below 540°C before the installation is complete, the metal is not as strong, since the forging is more of a "cold-forming" process, and cold working tends to make the metal brittle. If a rivet is not tightly driven, it will be loose after it cools and can be detected by tapping it with a hammer. A loose rivet cannot be redriven to make it tight and must be dug out by hand (chip off the head). Digging out a rivet is slow work and careful inspection is required to ensure that the rivet crew does not take a hammer and chisel and caulk the edge of the rivet head to make it appear to be tight when the inspector taps it with his hammer.

A riveting crew requires at least four persons (iron workers), consisting of one who heats and tosses the rivets to the driving crew; one to catch the hot rivet in a bucket and insert it in the hole; one to handle the bucking (or backup) bar, and one to drive the rivet with a pneumatic hammer.

It is possible, with pneumatic hammers, to drive cold rivets (cold-formed rivets), but if considerable care is not taken, the rivets may not be very tight and there is, of course, no later cooling to affect a final contraction of length to draw the parts firmly together. In operations such as aircraft fabrication and handles for cutlery, the rivets are routinely driven cold. In many fabrication shops, rivets can be driven cold, but here large hydraulic presses are used in the driving process. In fact, with the large pressures developed, the connected parts are deformed and upon removal of the pressure tend to restore their shape and develop tension stresses in the cold-formed rivets.

At present almost no riveting is used in engineered structures, either in fabrication shops or in the field, for several reasons:

1. Labor costs associated with large riveting crews.
2. Careful inspection required for riveted joints and the large costs involved in digging out poorly installed rivets.
3. Development and the high reliability of high-strength bolts.
4. Development and high reliability of welding.
5. High level of noise associated with driving rivets, which would be unacceptable under present environmental standards in most urban areas.

Riveted construction for railroad bridges was among the last to survive since rivets had a higher fatigue resistance than did the early bolts. The forging process tended to fill the rivet holes and, coupled with the high tension stresses, produced a high level of reliability compared to the early bolts (similar to the current grade A-307 to be considered later), which could not develop high tensile stresses; further, the nuts tended to loosen under continued dynamic loadings. Shop fabrication using rivets continued for some time after field use was discontinued in favor of high-strength bolts or welding. However, at present,

Table 8-1 Allowable bolt and rivet stresses for building construction[a]

Material	Tensile stress, F_t		Shear stress, F_v			
			Friction connection[b]		Bearing connection	
	ksi	MPa	ksi	MPa	ksi	MPa
Rivets						
A-502 grade 1	23	160	—	—	17.5	120
grade 2	29	200	—	—	22	150
Bolts						
A-307 (unfinished)	20	138	—	—	10	70
A-325N[c]	44	305	17.5	120	22	150
A-325X[d]	44	305	17.5	120	30	205
A-449N	$0.33F_u$	$0.33F_u$	—	—	$0.17F_u$	$0.17F_u$
A-449X	$0.33F_u$	$0.33F_u$	—	—	$0.22F_u$	$0.22F_u$
A-490N	54	370	22	150	28	193
A-490X	54	370	22	150	40	275

[a] All stresses in this table to be used with bolt cross-sectional area based on the *nominal* fastener diameter.

[b] See AISC *Manual for Steel Construction* for connection stresses using slotted holes.

Bolt	Ultimate strength, F_u, ksi (MPa)		
	$\frac{1}{2}$ to 1 in diam.	$1\frac{1}{8}$ to $1\frac{1}{2}$ in diam.	$1\frac{9}{16}$ to 3 in diam.
A-325	120 (825)	105 (725)	—
A-449	120 (825)	105 (725)	90 (620)
A-490	150 (1035)	150 (1035)	—

[c] N, threads *included* in shear plane (may neglect thread runout).

[d] X, threads *excluded* from shear plane.

shop welding and bolting has almost totally displaced rivet use, particularly with the advent of automatic and semiautomatic welding methods.

We note specifically in referring to tables such as Table 8-1 that the allowable rivet shear stress in bearing-type connections is the same as allowed for A-325 and A-490 high-strength bolts in friction connections where the bolt threads may be either included or excluded from the shear planes. Thus one may substitute a high-strength bolt for a rivet of the same diameter on a one-for-one basis. The labor requirements are two persons, at most, for the bolt installation.

Since rivets are no longer used (at least in American practice), the remainder of this chapter will focus on bolted connections, but the reader should be aware

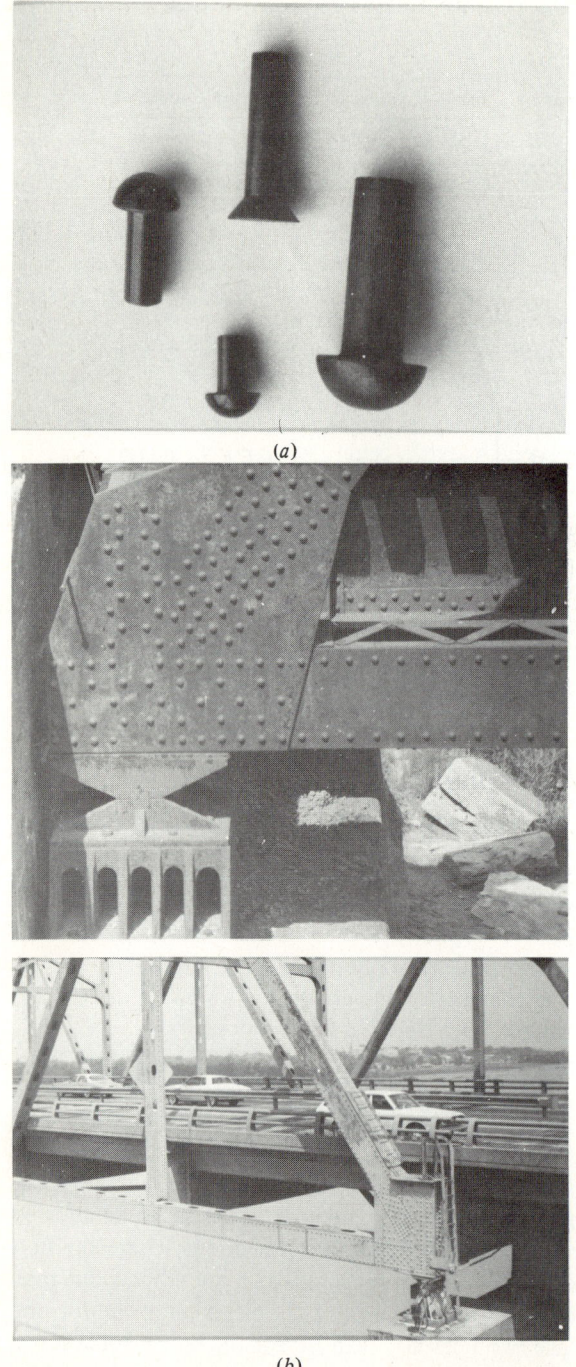

Figure 8-4 (*a*) Several sizes of undriven structural rivets. (*b*) Rivets in structural applications.

that the procedure for the design of a riveted connection is exactly the same as for a bolted connection. Figure 8-4 illustrates several sizes of undriven rivets and structural applications using rivets.

8-3 HIGH-STRENGTH BOLTS

There are two general classes of bolts used in structural applications. These are general-use A-307 (ASTM designation), sometimes called unfinished bolts. These bolts have a somewhat rough shank and bearing surfaces, since not as great care is taken in their manufacture. The A-307 bolts are made of steel with an ultimate tensile strength F_u on the order of 60 (grade A) to 100 (grade B) ksi (415 to 690 MPa) and available from $\frac{1}{4}$ in (6 mm) to 4 in (102 mm) in diameter and in lengths from 1 to 8 in, in increments of 1/4 in, and over 8 in, in increments of 1/2 in. A-307 bolts are available with several head and nut configurations, but the hexagonal and square head are most commonly used. Several sizes of A-307 bolts are illustrated in Fig. 8-5.

A-307 bolts are cheaper than A-325 and A-490 bolts and should be used in static load structural applications whenever possible. Applications include use in small structures, locations where the bolt installation is visible for regular serviceability checks, and in service loads which are relatively small.

High-strength bolts are available in the ASTM classifications, sizes, and ultimate tensile strength shown in the lower inset of Table 8-1.

The general length, head, and nut configurations are the same as for A-307 bolts except that larger diameters may not be available. The A-325 bolts can be obtained with metallurgy for special purposes, such as high resistance to corrosion. A-325 bolts may also be obtained with a galvanizing coating.

When high-strength bolts were first introduced into structural applications, washers were required to spread the bolt load to a larger area of the softer metal of the fastened parts. This requirement was partially caused by the nut and head tending to dig (called galling) into the A-33 and A-7 (F_y = 33 ksi) steel available at that time. Current high-strength bolting applications require that a hardened washer be used under the turned element as follows:

	Method of tightening	
	Turn-of-nut	Specified torque
A-325[a]	No	Yes
A-490[b]	Yes	Yes

[a] Washers are required when using oversized bolt holes. Washers are required when flange slope is greater than 1 : 20 (S and C shapes both have a slope of inner flange face of approximately 1 : 6).

[b] Use washers on sloping flange faces as above. Use two washers when the fastened material has F_y < 40 ksi.

356 STRUCTURAL STEEL DESIGN

Figure 8-5 A-307 general-purpose bolts are available with either hexagonal or square heads. May be used with washers.

Bevel washers are required on the slope surface between the bolt head or nut when the slope of the connected part exceeds 1 : 20, as for C and S shapes. Several A-325 and A-490 bolts are illustrated in Fig. 8-6.

Proprietary bolts and accessories (see Fig. 8-7) are available and include interference bolts, which must be driven into the bolt holes and provide a tight fit similar to the driven rivet. Slotted bolts are also available so that a single

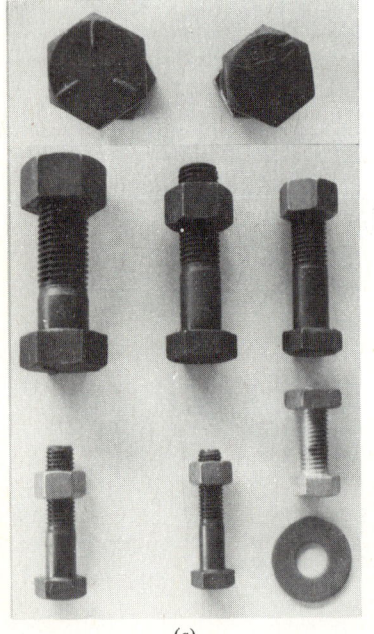

(a)   (b)

Figure 8-6 High-strength bolts. (*a*) A-325 bolts. Note three marks at 180° always indicates A-325 bolts. Some manufacturers also stamp "A-325." (*b*) A-490 bolts. Note grade is always marked on bolt head as shown.

Figure 8-7 Bolts and accessories for special applications. Shown are two types of interference bolts, locking nuts and a slotted end bolt.

installer using a power driver can torque the bolt without need of a helper to hold the bolt head against turning. Also available are ribbed washers, which flatten at the specified torque so that the installer can visually observe adequate bolt tension. Self-locking nuts are also available for installation in connections subjected to dynamic loadings, so that the nut does not loosen during service.

Bolts in structural applications range from $\frac{1}{2}$ to $1\frac{1}{2}$ in (occasionally larger diameters are used, particularly as column anchor bolts in moment resisting connections to the concrete foundation). Since a balance between joint size net section and number of fasteners must be made, the most common bolt and rivet diameters are 3/4 and 7/8 in (20 and 22 mm).

High-strength bolts are installed with a developed tension in the bolt shank of approximately 70 percent of the ASTM specified minimum tensile strength $(0.7F_u)$ by either:

1. *Turn-of-nut method.* Nut is initially tightened to a snug fit (snug being the point at which an impact wrench begins to impact, or about 1/2 turn from the time some nut resistance is developed using a wrench). From this point the nut is turned with respect to the bolt shank an additional 1/2 turn (3/4 turn when $L > 8D$ or 200 mm).
2. *Torque control.* Either calibrated torque or impact wrenches are used. This method requires use of hardened washers under the turned element (either nut or bolt head) to prevent galling and to provide a more uniform friction.

Tests on large numbers of bolted joints indicate that either of these two methods puts sufficient strain into the bolt shank to produce the desired tension. The reader should note that this procedure of installation proof-tests the bolt. If the bolt is overstrained, it simply pulls apart and a new bolt can be installed. A

Table 8-2 Minimum prescribed bolt tension for proper installation (these values are also called "bolt proof load")

Bolt diameter, in	Bolt tension, kips			Bolt diameter, mm	Bolt tension, kN		
	A-325	A-449	A-490		A-325	A-449	A-490
$\frac{1}{2}$	12	↑	17	12.5	52	↑	73
$\frac{5}{8}$	19		27	15	75		105
$\frac{3}{4}$	28		40	20	141		200
$\frac{7}{8}$	39	Use A-325 or A-490 bolts	55	22	170	Use A-325 or A-490 bolts	240
1	51		73	25	220		310
$1\frac{1}{8}$	56		91	28	275		388
$1\frac{1}{4}$	71		116	30	322		454
$1\frac{3}{8}$	85		138	35	438		618
$1\frac{1}{2}$	104	↓	168	40	591	↓	834
Over $1\frac{1}{2}$		$0.7F_u$ [a]		Over 40		$0.7F_u$ [a]	
2		137.5 ksi		50		628 MPa	
$2\frac{1}{4}$		178.8		60		886	
$2\frac{1}{2}$		220		65		1057	
$2\frac{3}{4}$		271		70		1215	
3		328		75		1440	

[a] F_u = ultimate tensile strength (see Table 8-1).

poorly fabricated bolt would be readily detected. What cannot be detected is that adequate strain has been developed. Careful job inspection is therefore necessary with frequent checking of the installation equipment to ensure joint adequacy.

A bolt tension of approximately $0.7F_u$ gives adequate reserve strength should the bolt be somewhat overstressed (say, 3/4 turn instead of 1/2 turn). The bolt tension acts as a massive spring in tension to hold the fastened parts in relative position. This clamping effect also tends to hold the joint against nut loosening in fatigue load situations, so that most of the time a locking nut is not required. If A-325 bolts have not been excessively overstressed (not more than 1/2 to 3/4 turn of nut) they may be reused one or more times. Tests on reuse indicate that A-490 bolts should not be reused in any situations.

The minimum bolt tension based on $0.7F_u$ produces the minimum bolt proof load or the installation tension as shown in Table 8-2. With some number of bolts in a joint, each developing tensions shown in Table 8-2, the load on the joint that is necessary to produce a relative slip between the connected parts is readily computed:

$$P_{\text{slip}} = m\mu NT \qquad (8\text{-}1)$$

where μ = slip coefficient (usually can use 0.35 for clean mill scale; most other surfaces are less than this value, and it may be necessary to determine the value by test)

m = number of slip surfaces
N = number of fasteners
T = proof load of each fastener (as in Table 8-2)

From consideration of Eq. (8-1) we can readily see that regardless of the type of joint designed (*friction*, or no slip, or *bearing* with some slip acceptable), the applied joint load must exceed P_{slip} before either bolt shear or bolt (or material) bearing is developed. We may now see the rationale for $P_t = A_g F_t$, since the clamping effect is distributed away from the holes along the critical net section. After P_{slip} the joint actually develops resistance as a combination of bolt shear and bearing.

Example 8-1 What is the nominal safety factor against relative slip in a friction-type joint using 20-mm A-325 bolts? Take $\mu = 0.35$.

SOLUTION The bolt proof load is 141 kN (Table 8-2). The allowable shear stress using AISC specifications (Table 8-1) is 120 MPa with threads in shear plane. From Eq. (8-1):

$$P_{\text{slip}} = m\mu NT = \mu 1(141)$$

since it is only necessary to consider one bolt and we will consider only one slip plane, as in a lap joint. The safety factor is always defined as

$$SF = \frac{P_{\text{resisting}}}{P_{\text{allowable}}}$$

The value of P_{allow} is

$$P_{\text{allow}} = F_v A_{b(\text{nominal})}$$

Combining, we obtain the safety factor:

$$SF = \frac{0.35(141)}{120(0.7854 \times 0.020^2)10^3} = 1.31$$

The reader should note that this safety factor is against slip and is not the safety factor of the joint, which is on the order of 1.67 (for a tension joint and may depend on tension on net section). ///

8-4 FACTORS AFFECTING JOINT DESIGN

The preceding discussion has considered several factors involved in connection design, particularly as it pertains to fasteners. We will now consider several others factors.

8-4.1 Joint Length

One factor of considerable importance is joint size. Obviously, smaller joints are economical of material. However, since an assumption is made that each fastener in a joint carries a prorated share (equal for constant-size fasteners), a problem arises for long joints. Referring to Fig. 8-8, we see that the distribution of strain is unequal from the frontmost bolt to the rear bolt. If the joint is too long (with "too" not being specifically defined here), it is evident that the first bolt will carry more than P/N of the load and the last bolt will carry nothing to almost nothing. With the base metal or plate designed to be adequate for tension in the net section, the plate does not pull apart but does stretch based on PL/AE, so the forward bolts (or rivets) will either undergo compatible shear strains or will shear off if the strain and resulting forward displacement of the hole is too great. The loss of the forward bolt will transfer the load to the next bolt(s) in line, and the next bolt may shear, and so on—producing a progressive joint failure (a process called "unbuttoning"). Note, however, that with the large loads involved, this process is very nearly instantaneous. If the joint is short enough that all the bolts carry load, the first bolt strains with the plate. When strains corresponding to yield stress (shear) develop, the bolts continue to strain with no increase in load and the next bolt(s) in line will pick up the transferred load. The ultimate joint load is reached when all the bolts have yielded. Strain compatibility analyses are seldom made, since the factors of safety used together with the property of steel ductility are such that except for long joints, only the first bolts (if any) in a connection are yielded or are close to yield.

We should note that the safety factor for the connection (particularly the fasteners) should be higher than for the members being connected. This is so that a member failure will always occur before a joint failure. A joint failure will generally be catastrophic, whereas a member failure is likely to allow time for safety measures to be undertaken.

Recalling that no joint (with holes and in tension) is more than 85 percent efficient, and based on the work of Bendigo, Hansen, and Rumpf ("Long Bolted Joints," *Proceedings, ASCE*, Vol. 89, ST6, December 1963), the efficiency of a

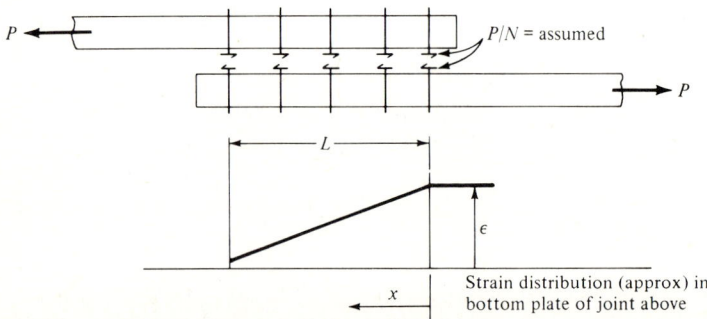

Figure 8-8 Load and strain distribution in bolted joints.

joint assuming reasonable bolt spacing on the order of 3 × diameter may be computed:

$$E = 0.85 - C_1(L - C_2) \qquad L \geq C_2 \tag{8-2}$$

where

fps: $C_1 = 0.007$ \qquad SI: $C_1 = 0.00275$

$C_2 = 16$ in \qquad $C_2 = 406$ mm

This equation indicates that connections with joint lengths up to 406 mm have an efficiency of 85 percent (i.e., no reduction in connection capacity for joint length). For greater lengths there is nearly a linear loss of joint capacity, reaching a capacity of about 60 percent of the short joint when the length is on the order of 50 in (1250 mm).

The current AISC specifications indirectly allow for long joints by use of the 85 percent efficiency factor and adjusting the allowable fastener stress. Values for fastener stress are considered valid (for bearing connections) up to a joint length of 50 in (1250 mm). Above this length the allowable shear stress is to be reduced 20 percent. Recalling that the specifications provide minimum requirements, the structural designer has the option of using Eq. (8-2) for intermediate joint lengths, between 16 and 50 in.

8-4.2 Edge Distance

If bolts in the line of stress are located too close to the edge, it may be possible to tear out the plate as shown in Fig. 8-3b and in the actual joint shown in Fig. 8-9. This may be avoided by using an edge distance obtained by equating shear forces using F_v = constant for both bolt and base metal, to obtain

$$A_b F_v = dt F_v$$

from which the edge distance d is

$$d = \frac{A_b}{t}$$

This edge distance is required by AISC in Sec. 1-16.4 and using an edge distance $d' = 2d$ when the bolt is in double shear.

When punching holes, it is necessary to have an adequate edge and end distance to avoid warping damage to the material. The AISC specifications (Sec. 1-16.4) give distances for this based on the nominal bolt diameter. For bolt diameters of $d \leq 1\frac{1}{4}$ in (30 mm) nominal dimension measured from center of hole to edge:

Sheared edge: \quad $D_e = 1.7 \times$ diameter (rounded to nearest $\frac{1}{8}$ in or 3 mm)

Rolled edge: \quad $D_e = 1.4 \times$ diameter (rounded to nearest $\frac{1}{8}$ in or 3 mm)

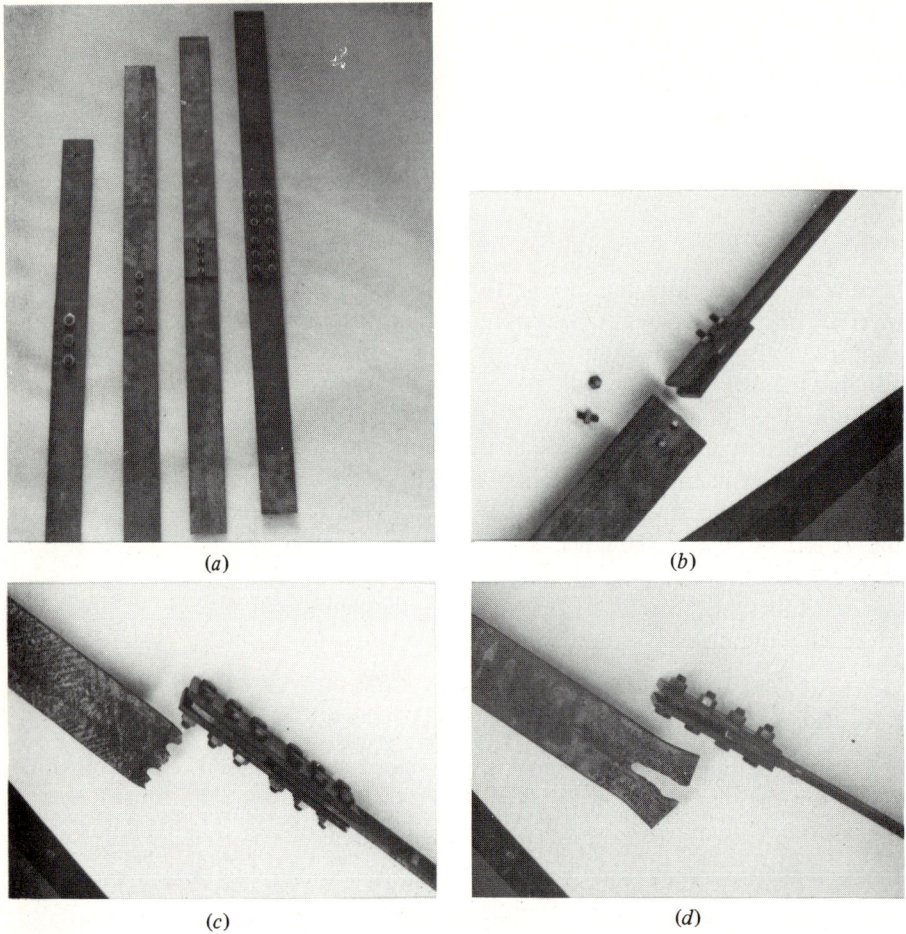

Figure 8-9 Several modes of joint failure. (*a*) Several joints. (*b*) Bolt shear failure. (*c*) Tension on net section failure. (*d*) Tear-out failure due to bolts being too close to end in the direction of the stress.

For bolt diameters larger than $1\frac{1}{4}$ in (or 30 mm) use:

Sheared edge: $D_e = 1.75 \times$ diameter

Rolled edge: $D_e = 1.25 \times$ diameter

8-4.3 Bolt Distribution and Gage Distances

It is necessary to ensure a reasonably compact joint and one where the connected material is in reasonably good contact so that the developed friction from clamping is uniform between the parts. If the bolts are too close together, interference is obtained, since the maximum coefficient of friction is $\mu \simeq 0.35$, and superposition of effects eventually produces a limiting friction resistance (i.e., μN does not increase without bound with increasing N). A spacing that is too close can cause difficulty in installing the fasteners, since the wrench head requires a minimum working space. These problems are resolved by using the minimum spacing requirements:

AISC (Sec 1-16.4): $s_{min} = 2.67 \times$ diameter ($3 \times$ diameter preferred)

AASHTO (Sec 1-7.22C): $s_{min} = 3 \times$ diameter

AREA (Sec. 1-9.3): $s_{min} = 3 \times$ diameter

Maximum spacing of a single line of fasteners in the direction of stress should generally be limited to $12t$, where $t =$ thickness of thinnest part being clamped. This spacing can be used for AISC, AASHTO, and AREA.

Fabrication practice coupled with wide usage of the $\frac{3}{4}$- through 1-in-diameter fasteners has led to certain standard gage distances. These values are shown in the tables for rolled shapes (see Tables I-3 to I-7 and V-3 to V-7 of SSDD) and depend on width of angle leg for angles, as in Tables I-13 and V-13 and in the AISC manual.

The user must also check the minimum edge distance for angles so that the fastener size does not result in a hole too close to the edge to satisfy the specifications.

The rolled shapes, which have very wide flanges, may have a second gage line. Values for the very wide flanges (e.g., W14 larger than 142 lb/ft) are shown in the AISC manual, as well as in tables available from the steel producers, but are not shown in the SSDD tables.

The reader should note that the "standard" gage distances will generally result in a more economical fabrication cost, but these distances are not the only ones that can be used. The designer should, however, consult the fabricator if other than standard gages are contemplated so that an economical joint is produced.

8-4.4 Minimum Joint Design

AISC specifications require that all connections, except those in trusses, carrying calculated stresses be designed for the design load but not less than 6 kips (or 27 kN).

AISC specifications require that truss joints in either tension or compression be designed for the design load but not less than 50 percent of the effective strength of the member based on the type of design stress.

The AASHTO specifications require that connections be designed on the basis of the average of the design load and the effective strength of the member but not less than 75 percent of the effective strength of the member. This is because many of the structural members in AASHTO design are controlled by factors other than stress, such as L/r. At least two fasteners are required in any AASHTO-designed connection.

The AREA requires connections for main members to be designed for the full effective strength of the member. For secondary members and bracing, the connection is designed for the average of the member strength and the design stress. A minimum of three fasteners is required in any AREA-designed connection.

8-4.5 Shear Lag

Long joints are undesirable from the standpoint of reduced efficiency (below 85 percent when $L > 406$ mm), but in cases where W, S, or C shapes are used with gusset plates on the flanges (see Fig. 8-10), it is necessary to produce a joint sufficiently long that the stress in the section at $A-A$ can be transferred to and through the flanges and into the gusset plates. The axial force in the tension member shown lags the transfer to the gusset plate, because of its distribution across the tension member. If the joint length is too large, the tension member may fail from a large concentration of stresses in the web, producing a tear and a resulting progressive failure across the section. A measure of shear lag efficiency is based on the distance from the gravity axis of the member to the fastener (or gusset plate) plane. Munse and Chesson ("Riveted and Bolted Joints: Net Section Design," *Proceedings, ASCE*, STI, February 1963) give an equation for shear lag efficiency:

$$E_{sl} = \left(1 - \frac{\bar{x}}{L}\right)100 \qquad (8\text{-}3)$$

where $\bar{x}$ = distance from gravity axis to fastener plane ($= d/2$ for W shapes and simply $\bar{x}$ given in tables for angles)
L = joint length

For most well-designed sections, Eq. (8-3) should give a value of e_{sl} from 85 to 90 percent. The AISC reductions for shear lag were presented in Sec. 5-3.2.

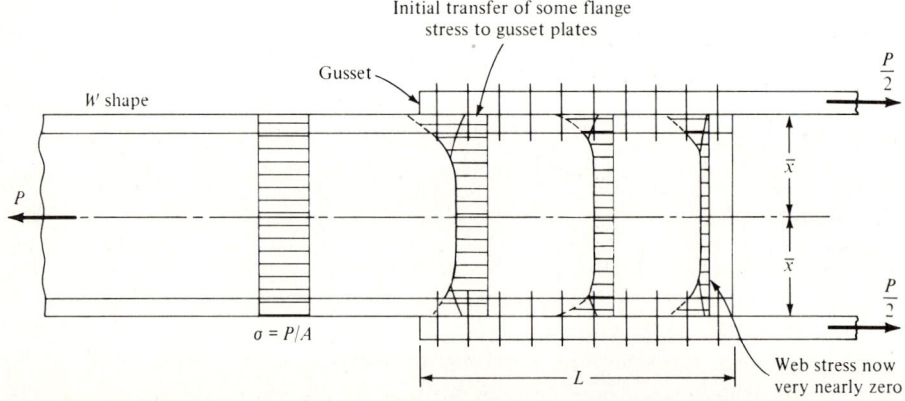

Figure 8-10 Shear lag in a W shape connected to a pair of gusset plates.

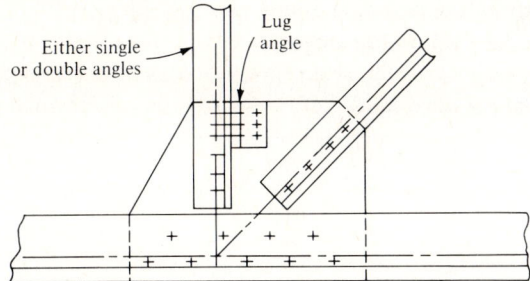

Figure 8-11 Lug angle. May use lug angles on both sides of gusset plate when a pair of angles are used for strut.

8-4.6 Long Bolts

Early tests on rivets indicated a loss of efficiency due to bending when the L/d ratio was greater than about 5. High-strength bolt tests for L/d up to about 9 indicated no significant loss of efficiency. Based on these considerations, an increase in rivets and A-307 bolts is required as follows:

	Increase required
AISC (rivets and A-307 bolts)	1 percent for each $\frac{1}{16}$ in (3 mm) in excess of $L = 5D$
AASHTO	1 percent for each $\frac{1}{16}$ in (3 mm) in excess of $L = 4.5D$
AREA	Same as AASHTO

8-4.7 Lug Angles

Lug angles may be used to reduce the joint length and shear lag of single or double angle members to a gusset plate, as illustrated in Fig. 8-11. Use of a lug angle at the forward end of the joint tends to reduce the term $\bar{x}/L$ in Eq. (8-3) to zero. Some ingenuity may be required, however, so that the net section is not reduced excessively from the extra bolt holes for the lug angle.

8-4.8 Shear and Bearing Distribution among Fasteners in a Group

A basic premise in connection design is that each fastener carry a prorated share of the connection load. For a joint with constant-size fasteners in a symmetrical pattern and loaded such that the load passes through the center of the pattern, the load is

$$P_{\text{fastener}} = \frac{P_{\text{total}}}{\text{number of fasteners}}$$

This assumption is made for either fastener shear or bearing. Bearing is considered in some specifications as bearing of fastener on base metal and of base

Table 8-3 Table to determine required bolt length based on grip and thread length (use table to determine if threads will fall in shear plane; see example below)

L_{reqd} = grip + L; round L_{reqd} to next larger $\frac{1}{4}$ in or 6 mm

Bolt size, in	Thread L	L, in	Bolt size, mm	Thread L[a]	L, mm[a]
$\frac{1}{2}$	1	$\frac{11}{16}$	12.5	25	18
$\frac{5}{8}$	$1\frac{1}{4}$	$\frac{7}{8}$	15	30	22
$\frac{3}{4}$	$1\frac{3}{8}$	1	20	35	25
$\frac{7}{8}$	$1\frac{1}{2}$	$1\frac{1}{8}$	22	38	28
1	$1\frac{3}{4}$	$1\frac{1}{4}$	25	45	30
$1\frac{1}{8}$	2	$1\frac{1}{2}$	28	50	38
$1\frac{1}{4}$	2	$1\frac{5}{8}$	30	50	42
$1\frac{3}{8}$	$2\frac{1}{4}$	$1\frac{3}{4}$	35	55	45
$1\frac{1}{2}$	$2\frac{1}{4}$	$1\frac{7}{8}$	40	55	50

[a] Approximate and based on soft conversion.

Example: Given the connection shown in sketch and using a $\frac{3}{4}$-in-diameter bolt. Are threads in the shear plane (defined by the plate junction)?

$$\text{Grip} = \tfrac{3}{8} + \tfrac{7}{16} + \tfrac{3}{8} = 1\tfrac{3}{16} \text{ in}$$

From the table, L = 1 in:

$$L_{reqd} = 1\tfrac{3}{16} + 1 = 2\tfrac{3}{16}$$

Round to nearest larger $\frac{1}{4}$:

$$L_{reqd} = 2\tfrac{1}{4} \text{ in}$$

Thread length = $1\frac{3}{8}$ (from table):

$$2.25 - 1.375 = 0.875 \text{ in}$$

Distance for two plates = $\frac{3}{8} + \frac{7}{16}$ in = 0.8125 in
Therefore, threads are out of the shear plane.

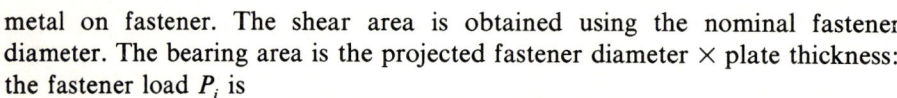

metal on fastener. The shear area is obtained using the nominal fastener diameter. The bearing area is the projected fastener diameter × plate thickness: the fastener load P_i is

$$P_s = A_b \times F_v \quad \text{(shear)}$$
$$P_b = D \times t \times F_b \quad \text{(bearing)}$$

No allowance is made for the hole size being $\frac{1}{16}$ in (1.5 mm) larger than the bolt shank in bearing.

When the fastener group is unsymmetrical or the load does not pass through the centroid of the fastener group, the fasteners are not equally stressed. This situation is considered in detail in the next section.

We note that the allowable shear stress F_v depends on the design assumption that the joint is either a *friction* or a *bearing* type, and whether the bolt

Table 8-4 Allowable bolt and rivet stresses for AASHTO specifications (use given stress with nominal bolt area except for A-307 bolts in tension)[a]

| | | | | | Allowable shear stress, F_v | | | |
| | Tension, F_t | | Bearing, F_b[b] | | Friction[c] | | Bearing[d] | |
Material	ksi	MPa	ksi	MPa	ksi	MPa	ksi	MPa
Rivets								
A-502 grade 1	—	—	40	275	—	—	13.5	93
grade 2	—	—	40	275	—	—	20.0	138
Bolts								
A-307	13.5	93	20	138	—	—	11.0	76
A-325N[e]	36	250	40	275	13.5	93	—	—
A-325X	36	250	40	275	—	—	20.0	138
A-490N	48	330	50	345	18.0	124	—	—
A-490X	48	330	50	345	—	—	29.0	200

[a] Data from AASHTO specifications, 12th ed., Secs. 1-7.22 and 1-7.41.
[b] F_b = allowable bearing stress on rivet or bolt from fastened material.
[c] Friction connections are required for connections subject to stress reversal.
[d] Bolts in bearing-type connections must be X-type (threads excluded from shear planes). Bearing-type connections are to be used only for compression members or secondary members. Reduce high-strength bolt values 20 percent if joint length is greater than 24 in and $F_y < 42$ ksi.
[e] N, threads in shear plane; X, threads excluded from shear plane (same as AISC).

threads are in the shear plane for the bearing-type connection. Table 8-3 may be used to determine if the threads are in the shear plane. Tables 8-1, 8-4, and 8-5 may be used to obtain the allowable stresses by AISC, AASHTO, and AREA specifications, respectively. Table 8-6 may be used to obtain the allowable bearing stress for base metal-to-bolt or bolt-to-metal. The AISC values shown in this table are based on the ultimate strength of the fastened material, computed as

$$F_b = 1.5 F_u$$

No distinction is made for fasteners in either single or double shear. The AASHTO bearing stress based on metal to fastener is

$$F_b = 1.22 F_y$$

and will generally be limited by fastener-to-metal stresses, as shown in Table 8-4 under the column headed F_b.

The AREA bearing values are stipulated for the fastener type except that bearing is not considered in the design for connections using high-strength bolts. These values are shown in Table 8-6.

Table 8-5 AREA fastener stresses (all connections are "friction" type)

Fastener	F_t		F_v	
	ksi	MPa	ksi	MPa
Rivets[a]				
Hand-driven	—	—	11.0	76
Power-driven	—	—	13.5	93
Bolts[b]				
A-325	36	248	20	138
A-490	36	248	27	186

[a] Use bearing stress on rivets: single shear, 27 ksi or 185 MPa; double shear, 36 ksi or 250 MPa.

[b] Need not consider bearing on bolts in friction connections.

Table 8-6 Bearing values for rivets and bolts by several specifications (top part of table is metal-on-fastener, bottom part is fastener-to-metal[a]

Material		F_u		AISC		AASHTO		AREA	
ksi	MPa	ksi	MPa	ksi	MPa	ksi	MPa	ksi	MPa
A-36	250	58	400	87.0	600	44.0	305	—	—
$F_y = 46$	315	67	460	100.5	690	56.1	390	—	—
$F_y = 50$	345	65	450	97.5	675	61.0	421	—	—
$F_y = 60$	415	75	520	112.5	780	73.2	506	—	—
A-307 bolts				—	—	20	140	—	—
Rivets A-501 and A-502				—	—	40	275	—	—
Power driven: single shear				—	—	—	—	27	185
double shear				—	—	—	—	36	250
A-325 and A-490 bolts				—	—	40	275	Not required	

[a] Top part: $F_b = 1.5F_u$; AASHTO: $F_b = 1.22F_y$. Bottom part: depends on fastener; fastener bearing generally controls for bridge design.

8-4.9 Nominal Factor of Safety of Fasteners and Connections

The nominal safety factor against slip for bolted joints is on the order of 1.25 to 1.30 (as computed in Example 8-1). One might ask what the nominal safety factor of the joint and of the mechanical fasteners is against failure.

The safety factor of the mechanical fasteners can be readily estimated based on the ultimate bolt tension and shear stress values divided by the allowable values given in Tables 8-1, 8-4, and 8-5.

Shear strength tests on high-strength bolts (see Wallaert and Fisher, "Shear Strength of High Strength Bolts," *Journal of Structural Division, ASCE*, ST5,

October 1965) give average values of ultimate shear in terms of ultimate tensile strength F_u (see Table 8-1) of

$$\tau_u = 0.62 F_u \quad \text{(plates in tension)}$$
$$\tau_u = 0.68 F_u \quad \text{(plates in compression)}$$

These tests indicated that the initial bolt tension (plate clamping effect) has no effect on the ultimate bolt shear strength.

If we designate the safety factor F as

$$F = \frac{\tau_u}{F_v}$$

we obtain for A-325 bolts in friction connections with diameter ≤ 25 mm (most commonly used)

$$F = \frac{0.62(825 \text{ MPa})}{120} = 4.26$$

Joint geometry reduces this value of F to something on the order of 3.3 for compact joints (in tests) and to around 2.0 for joints whose length is in excess of 1270 mm (50 in). This value of F compares to the tension value on the cross section for A-36 steel of

$$F = \frac{F_u}{0.58 F_u} = 1.72 \quad \text{(net section after slip)}$$

or

$$F = \frac{F_u}{0.6 F_y} = \frac{400}{150} = 2.67 \quad \text{(gross section before slip)}$$

Observations of joints in long service indicate that a safety factor of $F \geq 2.0$ for the fasteners gives satisfactory service.

8-4.10 Splices in Beams

Beams are often spliced to produce continuous spans. Splices are usually placed close to the location of zero shear in the span. Any use of mechanical fasteners in the tension flange will reduce the effective area somewhat. Based on tests, AISC and recent AASHTO specifications allow the designer to use the gross flange area for stress calculations as long as the holes (in the tension flange) are less than 15 percent of the tension flange area. When the area of holes exceeds 15 percent, the flange area is reduced by that portion of holes in excess of 15 percent.

For example, if A_f (of tension flange) = 20, A_{holes} = 4, the percent holes = 4 × 100/20 = 20 percent. The excess hole area = 20 − 15 = 5 percent and the tension flange area is 20(1 − 0.05) = 19, not 16, as would be obtained by taking out all the hole area (20 − 4 = 16).

These computations should also be used for the flange splice plates.

370 STRUCTURAL STEEL DESIGN

Example 8-2 How many A-325 bolts are required for the tension splice for the C310 shown in Fig. E8-2a if the metal is $F_y = 250$ MPa and the connection is assumed to be friction type, or bearing type.

SOLUTION We will use 22-mm bolts, so the hole diameter = 22 + 3.0 = 25 mm. Also, since each bolt must be sheared twice for the beam web to slip out from between the two splice plates, the load per bolt is

$$P_{bolt} = 2 \times A_b \times F_v$$

For a friction-type connection:

$$P_{bolt} = 2(0.7854 \times 0.022^2)(120) = 91.2 \text{ kN/bolt} \quad \text{(double shear)}$$

The number of bolts required is $N = 442.5/91.2 = 4.85$. Use five bolts. Use the bolt pattern shown in Fig. E8-2b, so that the splice plate width will be a maximum but the maximum net section is obtained for the channel.

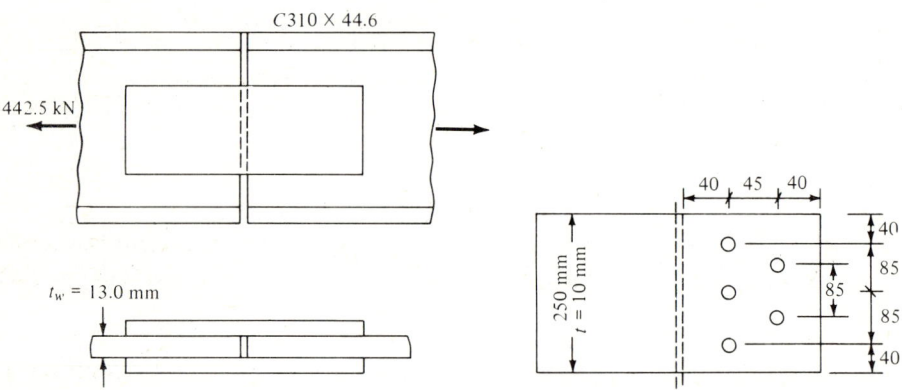

Figure E8-2a   **Figure E8-2b**

Check the net section of the splice plates. With three bolts out, the area requirements are:

Gross: $$A_p = \frac{442.5}{2 \times 0.6 F_y} = 1.475 \times 10^{-3} \text{ m}^2$$

Net: $$A_p = \frac{442.5}{2 \times 0.5 F_u} = 1.106 \times 10^{-3} \text{ m}^2$$

$$A_g \geq \frac{1.106}{0.85} = 1.301 \times 10^{-3} \text{ m}^2$$

Try two plates 250 × 5 mm with three holes in the critical section.

$$A_{net} = [250 - 3(25.0)]0.01 = 1.75 \times 10^{-3} > 1.106 \quad \text{O.K.}$$

BOLTED AND RIVETED CONNECTIONS 371

For a three bolt line and sheared edges, the minimum edge distance is
$$1.7(22) = 37.4 \text{ mm} \quad \text{use 40 mm}$$
$$\text{Minimum bolt spacing} = 3D = 3(22) = 66 \text{ mm}$$
$$\text{Use spacing} = \frac{250 - 2(40)}{2} = 85 > 66 \text{ mm} \quad \text{O.K.}$$

For the forward two bolts:
$$\text{Center bolts as } 40 + \frac{85}{2} = 82.5 \text{ mm from edges}$$

Spacing to edge is less than maximum allowed of $12t = 12 \times 10 = 120$ mm. Use the distance from the front bolt to the edge of splice plate at 40 mm (1.75D for sheared edges) and similarly from the back bolts to the edge of the W410. Set the gage distance so that only two holes are deducted from the critical net section.

$$w_{net} = 250 - 3(25) = 250 - 5(25) + \frac{4s^2}{4g}$$

$$\frac{4s^2}{4(85/2)} = 50$$

$$s = \sqrt{2125} = 46.1 \text{ mm} \quad \text{use } s = 50 \text{ mm} \quad \text{(arbitrary choice)}$$

Check the bearing:

On splice plates: $P_b = N \times A_b \times F_b = 5(0.01)(22)(1.50 \times 400)$
$$= 660 > 442.5 \quad \text{O.K.}$$
On web of C310: $P_b = 5(0.013)(22)(1.50 \times 400)$
$$= 858 > 442.5 \text{ kN} \quad \text{O.K.}$$

Use five bolts for the friction connection.

For a bearing-type connection:

$$\text{Grip} = 2 \times 5 \times 13.0 = 23.0 \text{ mm}$$
$$L_{reqd} = 23.0 + 28 = 51 \quad \text{use 55 mm}$$
$$L_{thread} = 38 \text{ mm} \quad \text{(Table 8-3)}$$

Width of one splice plate + beam web = $5 + 13 = 18.0$ mm

Distance from bolt head to end of thread = $55 - 38 = 17$ mm < 18 mm
This computation shows that the threads are in the shear plane, so $F_v = 150$ MPa (instead of 205) and P_{bolt} is

$$P_{bolt} = 2 \times A_b \times F_v = 2(0.7854 \times 0.022^2)(150)10^3 = 114 \text{ kN}$$

$$N_{bolts} = \frac{P}{P_{bolt}} = \frac{452.5}{114} = 3.97 \quad \text{say 4 bolts}$$

$$N_{brg} = \frac{452.5}{0.013 \times 22 \times 1.50 \times 400} = 2.6 \quad \text{(also use 4)}$$

Use two columns of two bolts for bearing-type splice. ///

Example 8-3 Design the connection for the vertical compression member (No. 7) for the highway truss of Examples 5-7, 6-4, and 6-6. Refer to Fig. E6-6 of Example 6-6 and to Fig. E8-3 (refer also to the bridge shown in Fig. 8-4b). The previous examples were used to design the end post and web members 7 and 9. Let us use the same cross section for the top chord as was used for the end post (all in compression). The remaining web members may use the same sections as for 7 and 9 (since L/r controlled rather than stress). In any case, let us design the connection for the vertical web member (No. 7), which is in compression. Data for W12 × 58:

$$A = 17.00 \text{ in}^2 \qquad d = 12.19 \text{ in} \qquad t_f = 0.640 \text{ in}$$

Use A-36 steel and the AASHTO specifications. Use A-325 high-strength bolts. $P_{max} = -99.3$ kips ($P_{min} = -40.6$ kips). The gusset plate $t = 5/16$ in (minimum t allowed by AASHTO for a plate).

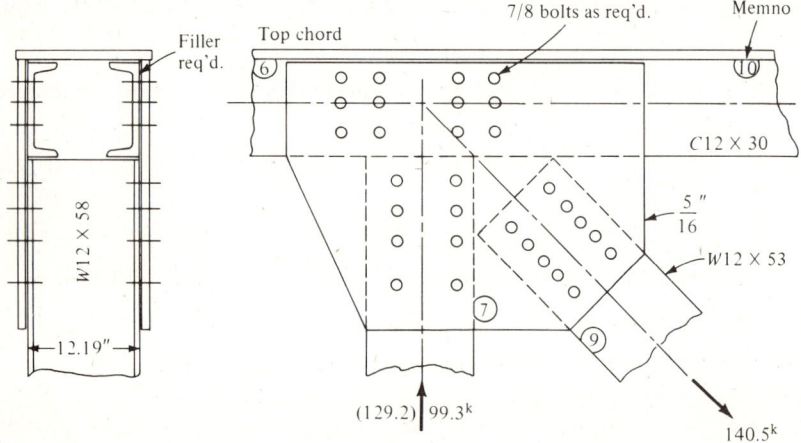

Figure E8-3

SOLUTION The fastener design will be based on a friction-type connection. AASHTO does not allow a bearing-type connection in a main member. Stress range does not have to be considered for connection design.

We will try four 7/8-in-diameter bolts at a section as shown in Fig. E8-3, the same as assumed in the design of member 9 for tension in Example 5-7. A deduction for net area does not have to be made for compression members unless a complete stress reversal occurs (this does not occur here).

AASHTO (Sec. 1-7.16) requires that a connection be designed for the average of the design load and the full effective member strength but not less than 75 percent of the effective strength of the member. From Example

6-4, the allowable axial compressive stress was 9.35 ksi.
$$P_{str} = AF_a = 17.00(9.35) = 159.0 \text{ kips}$$
$$P_{av} = \frac{159.0 + 99.3}{2} = 129.2 \text{ kips}$$

The 75 percent member strength criteria give
$$P_{0.75} = 159.0(0.75) = 119.3 \text{ kips} < 129.2$$

From Table 8-4, the allowable bolt shear stress $F_v = 13.5$ ksi. The number of bolts required in the connection to transfer 129.2 kips is
$$N = \frac{P}{A_b F_v} = \frac{129.2}{0.7854 \times 0.875^2 \times 13.5} = 15.9 \text{ bolts}$$

Use $N = 16$ bolts for symmetry and since 0.9 bolt is not possible. The use of 16 bolts requires four rows. Use a bolt spacing of $3D$:
$$s = 3(\tfrac{7}{8}) = 2\tfrac{5}{8} \text{ in}$$

Use a minimum edge distance in line of stress = $1\tfrac{1}{2}$ in (1.75D for cut end):
$$\text{Total length of joint} = 4(2.625) + 2(1.5) = 13.5 \text{ in}$$

With a joint length of $13\tfrac{1}{2}$ in (nominal), shear lag is not a factor. The minimum required transverse edge distance = $1.25D$ rounded to the nearest larger 1/8 in or 3 mm for flanges of beams and channels (but not for plates and other elements). This gives
$$d_e = 1.25(\tfrac{7}{8}) = 1\tfrac{1}{8} \text{ in } (1.125 \text{ in})$$

The distance furnished and based on the standard gage distance = 5.5 in (see Table I-3) is computed as
$$d_{furn} = \frac{b_f - 5.5}{2} = \frac{10.01 - 5.5}{2} = 2.26 \text{ in} > 1.125 \quad \text{O.K.}$$

Check the bolt bearing on the gusset plate, since $t_p = 0.313 < 0.640$ of flange of a W12.
$$P_{brg} = 16(0.875 \times 0.313)(40) = 175.3 \text{ kips} > 129.2 \quad \text{O.K.}$$

The gusset plate can be made sufficiently wide that tension on a net section is not a factor. There will be some filler plates needed between the gusset plate and the channels, since the W12 sections are deeper than 12 in.   ///

Example 8-4 Design the connection for the vertical members of the main roof truss of Example 2-6 that were designed in Example 5-6. $F_y = 250$ MPa. Other data include:
$$P = 70.18 \text{ kN (tension)}$$
$$2 \text{ L}127 \times 89 \times 6.3 \ (L/r\text{-controlled})$$
$$A_{furn} = 2.66 \times 10^{-3} \text{ m}^2 \qquad A_e = 1.695 \times 10^{-3} \text{ m}^2$$

This design was based on using a 12-mm gusset plate and 25-mm high-strength bolts.

SOLUTION AISC requires that connections for truss members be designed for either the design load or 50 percent of the effective strength of the member.

$$P_{50} = 0.5(0.6F_y)A_g = 0.5(150)(2.66) = 199.5 \text{ kN} \quad \text{controls}$$
$$P_{50} = 0.5(0.5F_u)A_e = 0.5(200)(1.695) = 169.5 \text{ kN}$$

The bolts through the angles and gusset plate will be in double shear, as illustrated in Fig. E8-4b. Assume that a bearing-type connection (slip can be tolerated) is satisfactory (a designer's prerogative). Check the bolt length using Table 8-3 to see if the threads are in the shear plane.

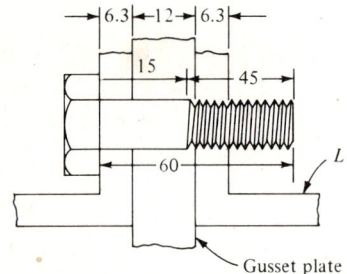

Figure E8-4a

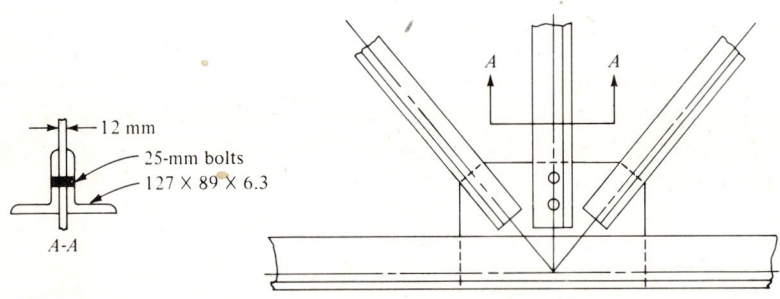

Figure E8-4b

$$\text{Bolt grip} = 6.3 \times 2 + 12 = 24.6 \text{ mm}$$
$$L = 24.6 + 30 = 54.6 \quad \text{use 60 mm}$$
$$L_{\text{threads}} = 45 \text{ mm} \quad \text{(Table 8-3)}$$
$$\text{One angle} + \text{gusset} = 6.3 + 12 = 18.3 \text{ mm}$$
$$\text{Thread runout location} = 60 - 45 = 15 \text{ mm}$$

15 mm < 18.3 mm. ∴ threads in the shear plane (see Fig. E8-4a)

With the threads in the shear plane, $F_v = 150$ MPa (Table 8-1):

$$P_{\text{bolt}} = 2(0.7854 \times 0.025^2)(150)10^3 = 147 \text{ kN}$$

$$N = \frac{199.5}{147} = 1.36$$

Use two bolts for shear, since a fraction is not possible.
Check the bearing:

$$F_b = 1.5 F_u = 1.5(400) = 600 \text{ MPa} \ (F_u \text{ from Table 8-6})$$

On angles: $P_b = 2(2 \times 0.0063)(25)(600) = 378.0 > 175.9$ kN O.K.
On gusset plate: $P_b = 2(0.012 \times 25)(600) = 360.0 > 175.9$ kN O.K.
Use two 25-mm A-325 bolts and a 12-mm gusset plate. ///

8-5 RIVETS AND BOLTS SUBJECTED TO ECCENTRIC LOADING

Generally, when the eccentricity of the load on a bolt group is less than about $2\frac{1}{2}$ in (60 mm), it is neglected. Joints such as the simple frame connection of Fig. 8-2a, which is widely used, are in this category. The bracket connection of Fig. 8-12a is loaded with an eccentricity that is obviously too large to be neglected. The framed beam connection may be large enough that the resulting eccentricity is also too large to neglect. One may note that the standard framed connection angles in the AISC design manual neglect the eccentricity for values up to about 3.7 in (one of the standard framing angle connections with nearly this value of maximum eccentricity is shown in Fig. 8-12b).

A load to be resisted by a bolt group that is eccentric with respect to the centroid of the group pattern can be replaced with a force that has a line of action through the pattern centroid and a moment with the magnitude $M = Pe$, where e is the eccentricity of the load. This is illustrated in Fig. 8-13a and b. Again considering that each bolt in a pattern that is centrally loaded carries its prorated share of the total load, we have, for equal-sized bolts,

$$P_{si} = \frac{P}{N}$$

where P_{si} is the shear force on the ith bolt with a vector to resist the applied force P.

An additional bolt force is developed by the eccentric moment $M = Pe$. Assuming a group of bolts acting as an elastic unit, we have a concept similar to that of beam resistance being developed and as related to the beam moment of inertia. Referring to Fig. 8-14, we have a bolt pattern with an applied moment M which produces a resisting moment for rotational equilibrium that is equal to

$$M_r = \sum_{i=1}^{i=n} R_i d_i = M_{\text{applied}} \qquad (a)$$

If we assume that the value of R_i is proportional to the distance from the

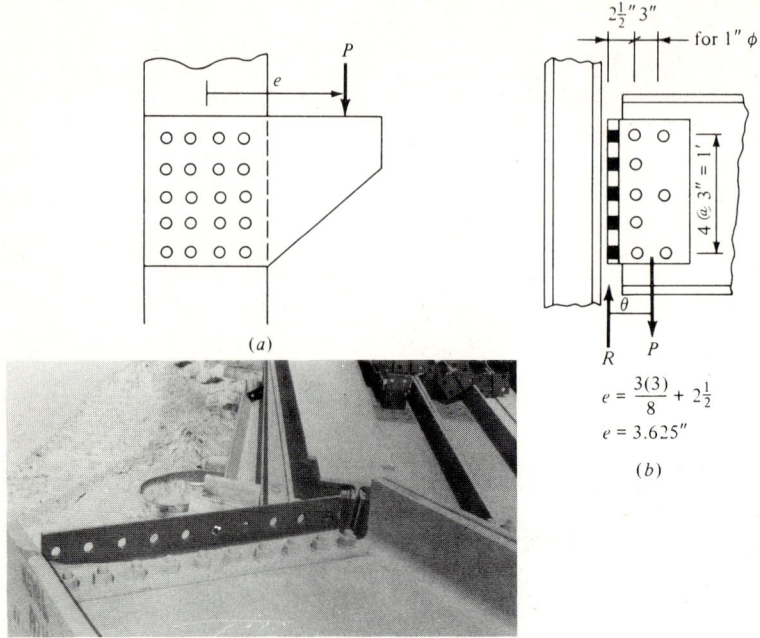

Figure 8-12 Connections with varying amounts of eccentricity. Refer also to Fig. 8-15a, b, and c. (a) Column bracket. (b) One of several standard beam-to-column framing connections as in the AISC design manual. (c) Photograph of beam with framing angles as in (b). Note small beams with framing angles with two holes in outstanding legs in upper right background.

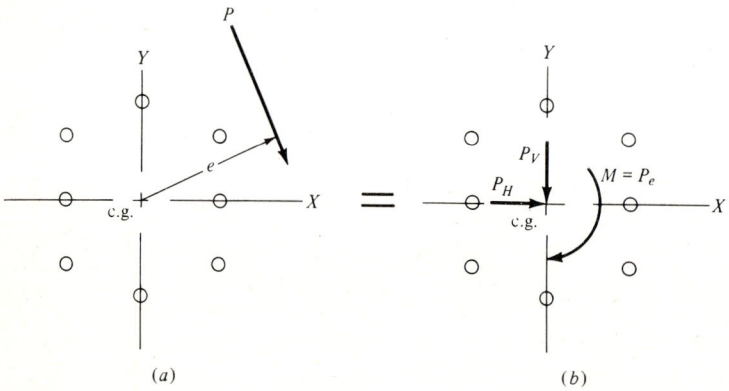

Figure 8-13 Eccentrically loaded bolt pattern. (a) Actual bolt pattern and loading. (b) Equivalent system for analysis.

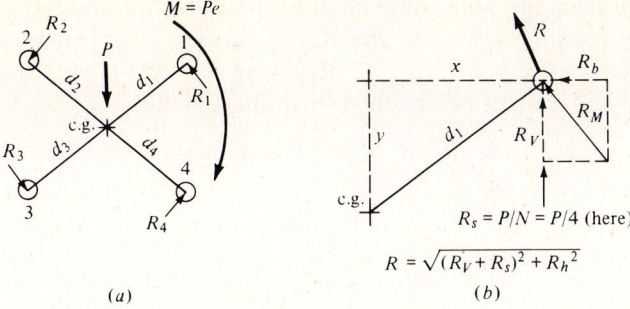

Figure 8-14 Eccentric bolt pattern to obtain resisting bolt force or stress. (a) Bolt pattern. (b) Forces on bolt No. 1 (but similar for any bolt).

centroid of the fastener pattern, we have

$$\frac{R_1}{d_1} = \frac{R_2}{d_2} = \frac{R_3}{d_3} = \cdots = \frac{R_n}{d_n} \qquad (b)$$

In terms of R_1, we have

$$R_1 = R_1\frac{d_1}{d_1} \qquad R_2 = R_1\frac{d_2}{d_1} \qquad R_3 = R_1\frac{d_3}{d_1} \qquad R_n = R_1\frac{d_n}{d_1} \qquad (c)$$

Substituting of Eq. (c) into Eq. (a), we obtain

$$M_r = Pe = R_1\frac{d_1^2}{d_1} + R_1\frac{d_2^2}{d_1} + R_1\frac{d_3^2}{d_1} + \cdots + R_n\frac{d_n^2}{d_1} \qquad (d)$$

Collecting terms and simplifying yields

$$M_r = \frac{R_1}{d_1}(d_1^2 + d_2^2 + d_3^2 + \cdots + d_n^2)$$

This can be simplified to

$$M_r = \frac{R_1}{d_1}(\Sigma d_i^2)$$

and solving for R_1 yields

$$R_1 = \frac{Pe(d_1)}{\Sigma d_i^2} = \frac{M(d_1)}{\Sigma d_i^2} \qquad (e)$$

From Eq. (c),

$$R_2 = \frac{Md_1}{\Sigma d_i^2}\frac{d_2}{d_1} = \frac{Md_2}{\Sigma d_i^2}$$

Similarly,

$$R_3 = \frac{Md_3}{\Sigma d_i^2} \qquad \text{etc.}$$

378 STRUCTURAL STEEL DESIGN

Since the action of P through the bolt group centroid produces an additional bolt resistance R_s which must be added to the R_{moment} vector, it is better to obtain the H and V components of R_{moment}. Referring to Fig. 8-14b, the components of R_i are R_h and R_v can be obtained by proportion as:

$$\frac{R_h}{y} = \frac{R_i}{d_i} \qquad \frac{R_v}{x} = \frac{R_i}{d_i}$$

$$R_h = R_i \frac{y}{d_i} = \frac{Md_i y}{\Sigma d_i^2 d_i} = \frac{My}{\Sigma d_i^2} \qquad (f)$$

Also, for R_v, we obtain

$$R_v = \frac{Mx}{\Sigma d_i^2} \qquad (g)$$

Similarly, from Fig. 8-14b, it is evident that

$$\Sigma d_i^2 = \Sigma(x^2 + y^2)$$

and can be interpreted as the polar moment of inertia of a group of unit areas. Note that if we use the area A with the denominator of either Eq. (f) or (g), R_h and R_v are obtained as stresses. Multiplying the numerator by the area A of the ith bolt produces the force R_v or R_h. With A in both the numerator and denominator, it cancels, giving R as a force. For general design, the equations are

$$R_h = \frac{My}{\Sigma(x^2 + y^2)} \qquad (8\text{-}4)$$

$$R_v = \frac{Mx}{\Sigma(x^2 + y^2)} \qquad (8\text{-}5)$$

Example 8-5 Given the bracket connection shown in Fig. E8-5a and that the fasteners are 25-mm A-325 bolts and the plate is $F_y = 250$ MPa steel, is

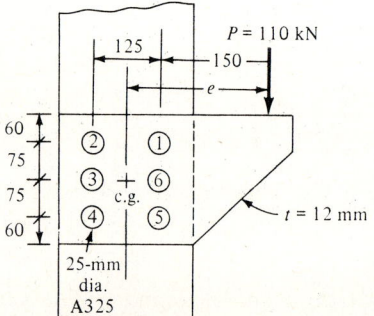

Figure E8-5a

the bolt pattern and plate adequate for the given load in a bearing-type connection assuming threads in the shear plane?

SOLUTION Since the bolt pattern is symmetrical (as in most practical problems), the centroid of the pattern is readily located and marked as c.g., as shown on sketch.

Compute $\Sigma(x^2 + y^2)$:

$$x = \frac{125}{2} = 62.5 \text{ mm}$$

$$\Sigma x^2 = 6 \times 62.5^2 = 23\ 437.5 \text{ mm}^2$$

$$\Sigma y^2 = 2(2 \times 75^2) = 22\ 500.0 \text{ mm}^2$$

$$\Sigma = 45\ 937.5$$

The number of bolts is shown in Fig. E8-5b. Draw rays from c.g. and place the R_i vector perpendicular to each ray d_i and in a direction to develop a resisting moment to oppose the applied moment.

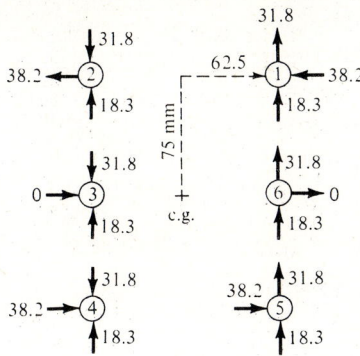

Figure E8-5b

Compute

$$R_s = \frac{P}{N} = \frac{110}{6} = 18.33 \text{ kN}$$

shown on Fig. E8-5b to resist applied P.

Compute $e = 150 + 125/2 = 212.5$ mm:

$$R_{i(h)} = \frac{212.5(110)x}{45\ 937.5} = 0.5088x$$

$$R_{i(v)} = 0.5088y$$

Set up a table as follows and omit the signs of x and y (use vectors previously drawn on Fig. E8-5b to determine the direction of the h and v

vectors):

Bolt	x, mm	y, mm	R_v, kN	R_h, kN
1	62.5	75	31.8	38.2
2	62.5	75	31.8	38.2
3	62.5	0	31.8	0
4	62.5	0	31.8	0
5	62.5	75	31.8	38.2
6	62.5	75	31.8	38.2

Placing these values on Fig. E8-5b, it is easy to see that bolts 1 and 5 are the most highly stressed (critical). Bolt 3 is loaded the least amount (13.47 kN↓). The resisting force on bolt 1 is computed as

$$R = \sqrt{(31.8 + 18.33)^2 + 38.2^2} = \sqrt{3972.2} = 63.03 \text{ kN}$$

$$f_v = \frac{63.03}{0.7854 \times 0.025^2 \times 1000} = 128.4 < 150 \text{ MPa} \quad \text{O.K.}$$

Check the plate bearing:

$$f_b = \frac{63.03}{0.025(12)} = 210.1 < 1.5(400) \quad \text{O.K.}$$

Check the possible tension rupture of the plate along the forward bolt line:

$$\text{Moment of inertia, } I = \frac{0.012(0.270)^3}{12} - 2(0.012 \times 0.025)(0.075)^2$$

$$I = 16.308 \times 10^{-6} \text{ m}^4$$

$$\text{Section modulus } S = \frac{I}{c} = \frac{16.308(2)}{270} = 0.1208 \times 10^{-3} \text{ m}^3$$

Moment at forward bolt line = $Pe' = 110(0.15) = 16.5$ kN · m

$$f_b = \frac{M}{S} = \frac{16.5}{0.1208} = 136.6 \text{ MPa} < 0.6F_y \quad \text{O.K.}$$

Check plate buckling:

$$\frac{b}{t} = \frac{150}{12} = 12.5 < 250/\sqrt{F_y} \quad \text{O.K.}$$

The joint is adequate for bolt shear, plate bearing on bolt, and bending.

///

Example 8-6 Given a crane runway bracket that carries a load as shown in Fig. E8-6. Use A-36 steel and either 7/8- or 1-in-diameter A-325 bolts. Assume a friction-type connection. Find the number of bolts and the bracket plate thickness.

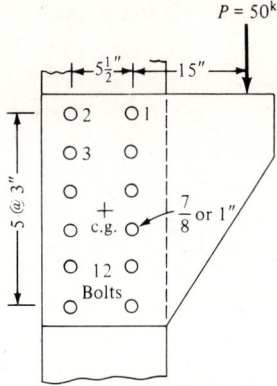

Figure E8-6

SOLUTION Compute

$$e = 15 + \frac{5.5}{2} = 17.75 \text{ in}$$

$$M = 17.75(50) = 887.5 \text{ in} \cdot \text{kips}$$

$$x = 2.75 \text{ in}$$

y_i depends on the number of bolts. Try 12 bolts. Note that bolt 1 is always the most stressed for a bolt pattern such as this.

$$\Sigma(x^2 + y^2) = 12(2.75)^2 + 4(7.5^2 + 4.5^2 + 1.5^2)$$
$$= 405.75$$

$$R_s = \frac{50}{12} = 4.17 \text{ kips}$$

$$R_h = \frac{887.5}{405.75}(7.5) = 16.40 \text{ kips}$$

$$R_v = \frac{887.5}{405.75}(2.75) = 6.01 \text{ kips}$$

Since $16.40/0.7854 = 20.88 > 17.5$ ksi, the number of bolts at a 1-in diameter is too small without computing R.

Try 16 bolts (bypassing 14, since 12 bolts were so highly stressed):

$$\Sigma(x^2 + y^2) = 16(2.75)^2 + 4(10.5^2 + 7.5^2 + 4.5^2 + 1.5^2) = 877.0$$

$$R_s = \frac{50}{16} = 3.12 \text{ kips}$$

$$R_h = \frac{887.5(10.5)}{877} = 10.62 \text{ kips}$$

$$R_v = 1.012(2.75) = 2.78 \text{ kips}$$

$$R = \sqrt{(2.78 + 3.12)^2 + 10.62^2} = 12.15 \text{ kips} < 0.7854(17.5) \quad \text{O.K.}$$

Use sixteen 1-in diameter A-325 bolts.

Find the plate thickness for the bearing (assume that the column flange is adequate):

$$t_p(1)(1.5 \times 58) = 12.15 \text{ kips}$$

$$t_p = \frac{12.15}{87(1)} = 0.139 \text{ in}$$

Check the plate bearing along the forward fastener line and neglect bolt holes 1/16 in larger than bolt:

$$h = 7 \times 3 + 3 = 24 \text{ in} \quad M = 50(15) = 750 \text{ in} \cdot \text{kips} \quad F_b = 22 \text{ ksi}$$

$$I_p = \frac{t_p h^3}{12} - 2t_p(10.5^2 + 7.5^2 + 4.5^2 + 1.5^2) = 774 t_p$$

$$f_b = F_b = 22 + \frac{Mc}{I} = \frac{750(12)}{774 t_p}$$

$$t_p = \frac{750(12)}{774(22)} = 0.53 \text{ in}$$

Check the AISC, Sec. 1-9.1.2, for unstiffened elements under compression $b/t \leq 95/\sqrt{F_y}$.

$$\frac{b}{t} \leq \frac{95}{\sqrt{36}} = 15.8$$

$$t = \frac{15.0}{15.8} = 0.95 \text{ in} \quad \text{use } t_p = 1.0 \text{ in}$$

Summary: use sixteen 1-in-diameter A-325 bolts and a bracket plate $t_p = 1.0$ in. ///

8-5.1 AISC Reduced Eccentricity for Connections

The preceding method of joint analysis with eccentricity is widely used. Based on a series of tests which displayed that this method of joint analysis is considerably conservative, AISC uses (in tables for design of eccentrically loaded fastener groups in the design manual) a reduced eccentricity which is computed, for a single line of n fasteners:

$$e_{\text{eff}} = e - \frac{1 + 2n}{4} \tag{8-6}$$

For two or more lines of fasteners with n fasteners in any line:

$$e_{\text{eff}} = e - \frac{1 + n}{2} \tag{8-7}$$

We might note that use of Eq. 8-7 for the fastener eccentricity shown in Fig. 8-12c gives

$$e_{\text{eff}} = 3.625 - \frac{1+5}{2} = 0.625 \text{ in}$$

With this small apparent eccentricity (for joint performance), it is easy to see that the long-established practice of neglecting the eccentricity for most framed beam connections was a reasonable design procedure.

Example 8-7 Redo Example 8-6 taking into account the reduced eccentricity.

SOLUTION Referring to Fig. E8-7, try twelve 1-in-diameter bolts.

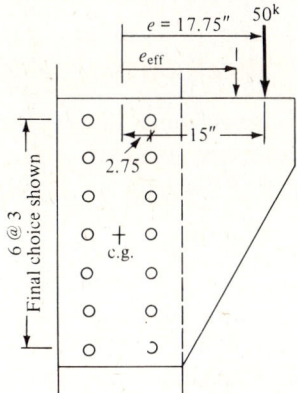

Figure E8-7

$$e_{\text{eff}} = 17.75 - \frac{1+6}{2} = 14.25 \text{ in}$$

$$M = 50(14.25) = 712.5 \text{ in} \cdot \text{kips}$$

$$\Sigma(x^2 + y^2) = 405.75 \quad (\text{Example 8-6})$$

$$R_s = \frac{50}{12} = 4.17 \text{ kips}$$

$$R_v = \frac{712.5}{405.75}(2.75) = 4.83 \text{ kips}$$

$$R_h = 1.756(7.5) = 13.17 \text{ kips}$$

$$R = \sqrt{(4.17 + 4.83)^2 + 13.17^2} = 15.95 \text{ kips}$$

$$f_v = \frac{15.95}{0.7854} = 20.3 > 17.5 \quad \text{N.G.}$$

Try fourteen 1-in-diameter bolts:

$$e_{\text{eff}} = 17.75 - \frac{8}{2} = 13.75 \text{ in}$$

$$M = 50(13.75) = 687.5 \text{ in} \cdot \text{kips}$$

$$\Sigma(x^2 + y^2) = 14(2.75)^2 + 4(9^2 + 6^2 + 3^2) = 609.9$$

$$R_s = \frac{50}{14} = 3.57 \text{ kips}$$

$$R_v = \frac{687.5}{609.9}(2.75) = 3.10 \text{ kips}$$

$$R_h = 1.127(9) = 10.14 \text{ kips}$$

$$R = \sqrt{(3.57 + 3.10)^2 + 10.14^2} = 12.14 \text{ kips}$$

$$f_v = \frac{12.14}{1.0(0.7854)} = 15.45 \text{ ksi} < 17.5 \quad \text{O.K.}$$

Using the reduced eccentricity requires two fewer bolts in the connection. The plate thickness of 1.0 in is still necessary to satisfy b/t, and bearing is not a problem. ///

The AISC Design Manual gives tables based on one, two, and four vertical columns of fasteners which may be used to design eccentric connections. By assuming the number of fasteners in a row, a computation for e_{eff} is made, and with n and e_{eff} we obtain a coefficient that is multiplied by the allowable fastener load to give the total group eccentric load. One may readily derive an equation for the allowable eccentric load for a single vertical (or horizontal) line of fasteners. Equations are given in several textbooks which may be used in an attempt to reduce the computational effort in finding the number of fasteners for an eccentrically loaded connection. The author suggests that with the increased computational efficiency available to the designer with the pocket calculators, it is as easy to "punch it out" as to try to use an equation developed by someone else. This is so particularly because no simple equation exists and most equations require some iteration anyway.

8-6 BEAM FRAMING CONNECTIONS

Figures 8-2a and 8-15 illustrates the most common methods of framing steel structures for small buildings of five or fewer stories where the connections are AISC Type 2 (simple shear connections). The eccentricity of the beam shear is neglected with these connections, and no moments are assumed to be transferred across the connection. Frame stability is provided by use of wind bracing or masonry walls closely fitted to the columns as shear walls.

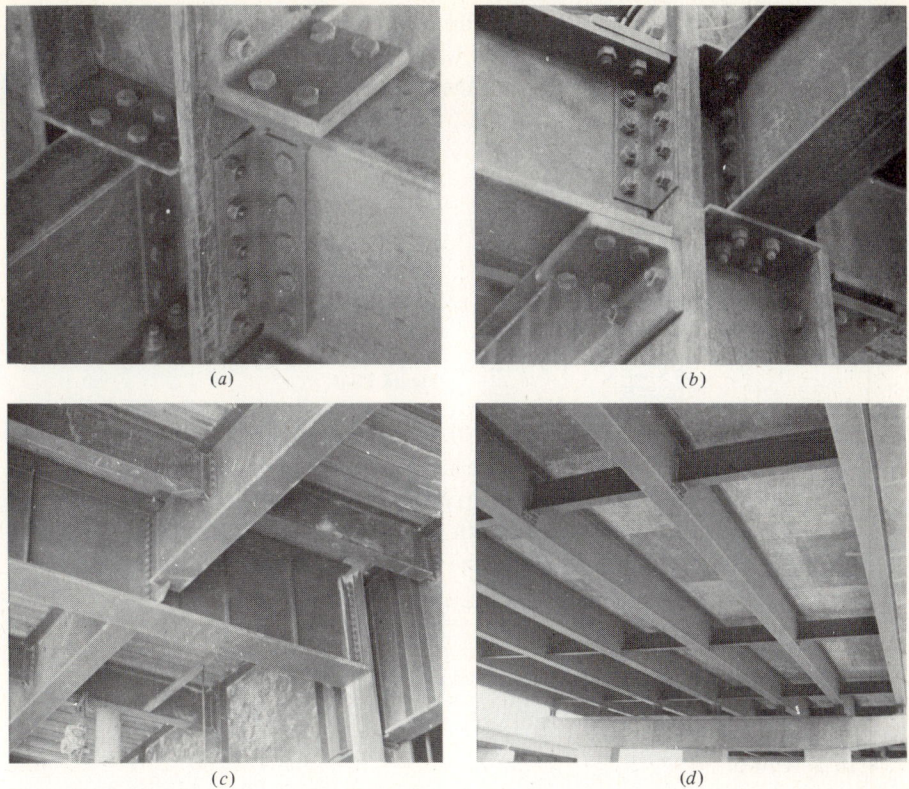

Figure 8-15 Framing connections. (*a*) Moment-resisting connection using top and seat angles bolted to column flange. Web angles carry shear. View looking down. (*b*) Moment-resisting connection into column web. Note that top and bottom plates are welded into column web and flange to act as column stiffeners. (*c*) Framing floor system in power station. Note shear stiffeners in web of main girder. Coping is shown for small floor beams in near foreground. (*d*) Framing for bridge stringers.

A number of the joints shown in Fig. 8-2a are "standardized" as to bolt pattern and with an angle length that depends on the beam size and as given in the AISC design manual. The angle is selected based on bolt bearing and with leg dimensions that depend on producing adequate edge distance and without interference between bolts and wrench during installation if the bolt holes are aligned both vertically and horizontally. Use of these tables often produces a connection that is overdesigned. However, the cost of overdesign is generally more than offset by reduced fabrication costs from using standard dimensions.

The simple framing connection is used to connect stringers to floor beams and floor beams to girders in bridge fabrication. In bridge design the connections should be standardized for the given bridge to reduce fabrication costs; general standardization for bridges is not as easily done.

386 STRUCTURAL STEEL DESIGN

Example 8-8 Design a beam framing connection for a W14 × 30 floor beam to frame into a W18 × 50 girder. The W14 × 30 carries a uniform load (live + dead) = 2.5 kips/ft and the span is $L = 18.0$ ft. Use the AISC specifications, A-36 steel, and A-325 high-strength bolts.

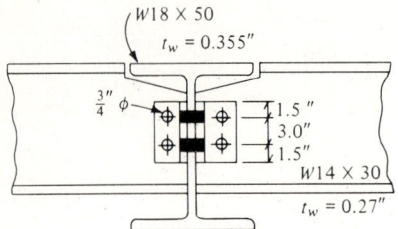

Figure E8-8

SOLUTION The top of the W14 × 30 should be coped as shown in Fig. E8-8 and with a pair of angles (one on each side) as shown, but we may note that in some cases one may use a single angle on only one side of the web.

With beams framing from both sides it is evident that the bolts in the webs of both the floor girder and the floor beam will be in double shear.

$$V = \frac{wL}{2} = \frac{(2.5 + 0.03)(18)}{2} = 22.77 \text{ kips}$$

Use 3/4-in A-325 bolts in a friction connection.

$$P_{\text{bolt}} = 15.46 \text{ kips double shear} \quad \text{(Table II-7, SSDD)}$$

In a W14 × 30:

$$N_{\text{shear}} = \frac{22.77}{15.46} = 1.47 \quad \text{use two bolts}$$

$$N_{\text{brg}} = \frac{22.77}{1.5 \times 58 \times 0.75 \times 0.27} = 1.29 \quad \text{use two bolts}$$

In the web of a W18 × 50:

$$N_{\text{brg}} = \frac{22.77(2)}{1.5 \times 58 \times 0.75 \times 0.355} = 1.96 \quad \text{requires two bolts}$$

$$N_{\text{shear}} = \frac{22.77(2)}{15.46} = 2.95 \quad \text{use four bolts for symmetry}$$

Use a web angle with a length of 6 in and t to be determined:
For bearing:

$$2t(0.75)(1.5 \times 58) = 22.77$$

$$t = 0.174 \text{ in}$$

For shear:

$$2t(6 - 2(0.75)(0.4F_y)) = 22.77$$

$$t = 0.176 \quad \text{use } t = \tfrac{1}{4} \text{ in}$$

Use an angle with a length of 6 in. Use two L4 × 4 × 1/4 × 6 in long with two 0.82-in holes punched (3/4 + 1/16) as in Fig. E8-8—this gives adequate clearance to install the bolts. ///

Example 8-9 Design the simple floor beam framing connections for the AREA bridge which was partly designed in Examples 4-15 and 10-5. $F_y = 250$ MPa. Use A-325 high-strength bolts. $F_v = 138$ MPa (Table 8-5). Note that AREA only considers friction-type connections.

SOLUTION From Example 4-15, obtain
$$V_{dead} = 28.3 \text{ kN}$$
$$V_{live} = 140.5 \text{ kN}$$
$$V_{imp} = 178.0 \times 0.795 = 111.7 \text{ kN}$$
$$\text{Total shear} = 280.5 \text{ kN}$$

The floor beam is a W760 × 160.7 and frames into a plate girder (Example 10-5):
$$t_{wg} = 16.0 \text{ mm}$$
$$t_{wb} = 13.8 \text{ mm}$$

Let us use 22-mm-diameter bolts:
$$P_{bolt} = 0.7854(0.022^2)(138)(10^3) = 52.3 \text{ kN}$$

For the web of a W760:
$$N_{shear} = \frac{280.5}{2 \times 52.3} = 2.7 \quad \text{use three bolts (double shear)}$$

For bearing with bolts in double shear, $F_b = 250$ MPa (Table 8-5):
$$N_{brg} = \frac{280.5}{0.022 \times 250 \times 13.8} = 3.7 \quad \text{use bolts as required}$$

For the web of a plate girder [bolts in single shear and $F_b = 185$ MPa (rivet value); Table 8-5]:
$$N_{brg} = \frac{280.7}{0.022 \times 185 \times 16} = 4.31 \quad \text{use six bolts}$$
$$N_{shear} = \frac{280.7}{52.3} = 5.36 \quad \text{use six bolts}$$

Place bolts as shown in Fig. E8-9.

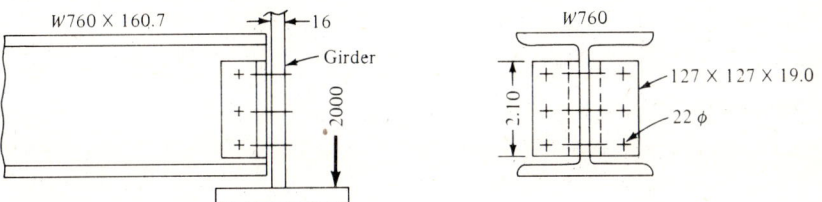

Figure E8-9

Design the framing angle:

$$L = 5 \times 3D + 2 \times 1.75D = 18.5D = 18.5(22) = 407 \text{ mm} \qquad \text{use 410 mm}$$

The angle thickness will be controlled by the bearing of the three bolts in the outstanding leg (o.s.l.) fastening the beam to the girder; therefore, $t \geq 16$ mm.

$$2t(0.022)(3)(185) = 280.5$$
$$t \geq 11.5 \text{ mm}$$

Checking the tables, try: L127 × 127 × 11.1 (Table V-9):

$$g = 75 \text{ mm} \quad \text{(Table V-13)}$$

Effective edge distance $d_e = 127 - 75 = 52$ mm

$$d_{e(\text{reqd})} \geq 1.5D = 33 \text{ mm}$$

A L102 × 102 × 11.1 will not give sufficient clearance between the W760 bolts and the bolts through the girder web in the o.s.l. The small amount of overdesign is rather negligible in any case. The design is summarized in Fig. E8-9. Note that AREA does not require a "bearing" check using A-325 bolts. This check established the approximate angle t and is recommended whether required or not. ///

8-7 FASTENERS SUBJECTED TO TENSION

Figure 8-16 illustrates the usual conditions for bolts in tension. When the bolt is tightened to develop the proof load, the shank elongates. Simultaneously the clamped plates are compressed. When we apply a load to the connection, we have the free body of Fig. 8-16c, which gives

$$NT - \sigma A - P_{\text{applied}} = 0 \qquad (a)$$

or
$$P = NT - \sigma A \qquad (b)$$

Obviously, as P becomes larger and larger, two events occur simultaneously:

1. The bolt tension T increases slightly, producing a slight shank elongation.
2. Shank elongation reduces the plate clamping pressure σ, since the plate compression $e_p = 2Tt_p/A_p E$ is small and caused by t_p being relatively small and A_p relatively large. On the other hand, $e_{\text{bolt}} = \Delta TL/AE$ is nearly of constant magnitude as long as the plates remain in contact.

In equation form and now considering a single bolt as in Fig. 8-16d, we have

$$-\Delta e_p = \Delta e_{\text{bolt}}$$

since the two clamped plates expand, and taking P' = prorated part of total P

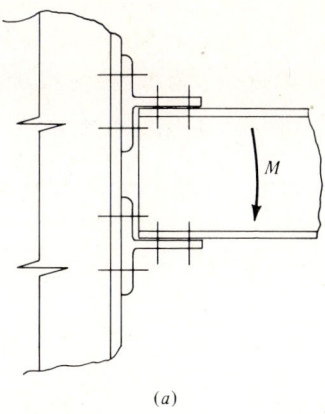
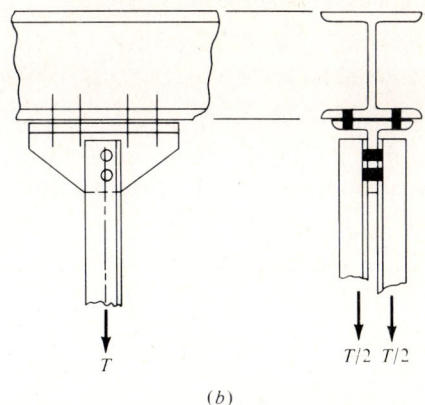
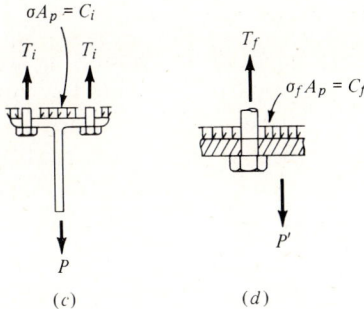

Figure 8-16 Bolts in tension. (*a*) Beam moment connection; may include shear as well as tension. (*b*) Hanger connection with bolts in pure tension. (*c*) Initial conditions. (*d*) Final tension in bolt from P'.

applied for one bolt:

$$-e_p = \frac{(C_i - P')2t_p}{A_p E} \qquad e_b = \frac{(T_f - T_i)L_b}{A_b E} \qquad (c)$$

$$(d)$$

Also, $C_i = T_i$, from statics:

$$P' + C_f = T_f \qquad \text{[also from statics (Fig. 8-16}d\text{)]} \qquad (e)$$

With $2t_p \simeq L_b$ and for E = constant,

$$\frac{C_i - C_f}{A_p} = \frac{T_f - T_i}{A_b} \qquad (f)$$

Substituting the values for C_i and C_f from Eqs. (*d*) and (*e*) into Eq. (*f*):

$$T_f - T_i = \frac{(T_i - T_f + P')A_b}{A_p}$$

with some rearranging, we obtain

$$T_f = T_i + \frac{P'}{1 + A_p/A_b} \tag{8-8}$$

For common spacings of $g = 5.5$ in and 1.5 in edge distances, the contributory area for a bolt in a connection of this type is on the order of 10 to 20 in^2. For nominal bolts of 7/8- to 1-in-diameter, the bolt area is on the order of 0.7, taking A_p/A_b as approximately 10 (a reasonable and conservative estimate), we obtain

$$T_f = T_i + \frac{P'}{1 + 10} = T_i + 0.1 P' \tag{8-8a}$$

Note, however, that if P' is applied directly to the fastener, $T_f = T_i + P'$, and not as given in Eq. (8-8a). When the plates separate, there is no C_f in Eq. (e) and a simple free-body analysis of Fig. 8-16d gives

$$T_f = T_i + P'$$

Referring to Fig. 8-17a, we see that the external edges of the connection are somewhat like cantilevered beams. It may be necessary to account for this "prying action" since the Q force shown decreases the connection capacity. Tests indicate that Q should be considered if the connection parts are very stiff. Tests also show that unless the connection is stiffened adequately, the outer fasteners in a connection with more than two fastener lines (Fig. 8-17b) is not very effective. From a large number of tests performed at the University of Illinois and reported in *Structural Research Series Bulletin No. 353*, the following empirical equations have been proposed:

For A-325 bolts:

$$Q = F\left[\frac{100b(D^2) - 18L(t_f^2)}{70a(D^2) + 21L(t_f^2)}\right] \tag{8-9}$$

For A-490 bolts:

$$Q = F\left[\frac{100b(D^2) - 14L(t_f^2)}{62a(D^2) + 21L(t_f^2)}\right] \tag{8-10}$$

where $F =$ load for one fastener $= P/N$
$\quad a =$ distance from fastener line to edge of flange (but $a \leq 2t_f$)
$\quad b =$ contributory width of flange, as in Fig. 8-17c
$\quad D =$ nominal fastener diameter
$\quad L =$ contributory length of flange for each fastener $= L/2$ for two fasteners as in a four-bolt connection.

Beam flanges supporting hangers (either the type in Example 8-10 or hanger rods as for stairs) must be adequate for the load. If the load is too large, the flange will deform unless it is stiffened. A solution proposed by Dranger ("Yield Line Analysis of Bolted Hanging Connections," *AISC, Engineering Journal*, Vol.

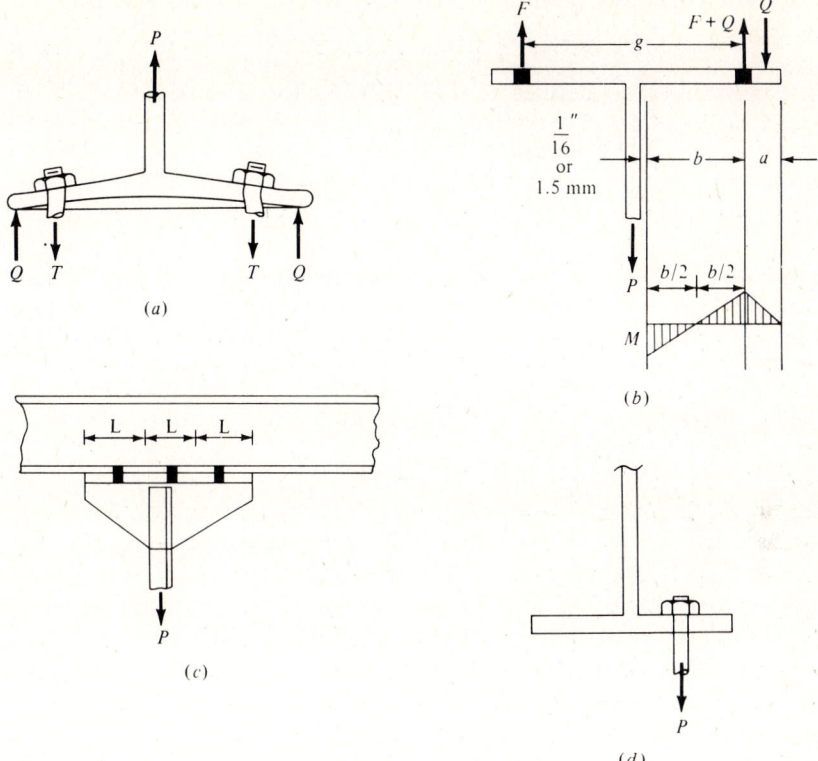

Figure 8-17 Hanger connections. (*a*) Fastener with prying. (*b*) General assumptions for bending and critical section. (*c*) Contributory length to a bolt for prying. (*d*) Flange stability with a hanger loading.

14, No. 3, 1977) gives some solutions in terms of F_y (and requires using the LF = 1.70 to obtain the working loads).

For clamped hangers (as Example 8-10):

Large clamped length:

$$P = \frac{F_y t_b^2 L}{2b(\text{LF})} \tag{8-11}$$

Short clamped length:

$$P = F_y t_b^2 (2r)^{\frac{1}{2}} \left(1 + \frac{a}{b}\right) \frac{1}{\text{LF}} \tag{8-12}$$

For hanger rods (see Fig. 8-17*d* and Example 8-11), the smaller of

$$P = F_y t_b^2 (2r)^{1/2} \left(1 + \frac{a}{b}\right) \frac{1}{\text{LF}} \tag{8-13}$$

$$P = F_y t_b^2 \left[r\left(1 + \frac{a}{b}\right)\right]^{1/2} \frac{1}{\text{LF}} \tag{8-14}$$

392 STRUCTURAL STEEL DESIGN

where terms not previously defined in Eqs. (8-9) and (8-10) and Fig. 8-17 are:

P = total force carried by flange on one side of beam web

r = stress ratio, defined as $(F_y - F_b)/F_y$; for A-36 steel,

$$r = (36 - 22)/36 = 0.3889$$

t_b = thickness of beam flange supporting load

Example 8-10 Design a hanger using a WT section for a load to be carried by a pair of angles. Use A-325 bolts and A-36 steel; the load is 62.4 kips. The load is carried from the bottom of a W33 × 221 beam as shown in Fig. E8-10a.

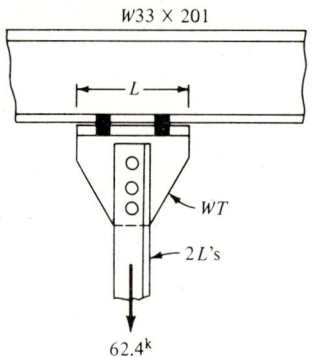

Figure E8-10a

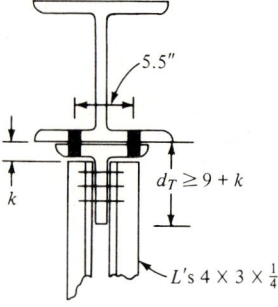

Figure E8-10b

SOLUTION The hanger design will require design of a large-enough WT to attach the angles for the load and to select enough bolts of the proper size to carry the load in tension.

Step 1. Design the angles.

Assume that L/r is not critical and choose 7/8-in bolts:

$$A_g = \frac{62.4}{22} = 2.84 \text{ in}^2$$

$$A_n = \frac{62.4}{29} = 2.15 \text{ in}^2$$

$$A_g \geq \frac{2.15}{0.85} = 2.53 \text{ in}^2$$

Select angles at least 1/4 in thick, so that bearing is not a problem. Try two L4 × 3 × 1/4 (long legs back to back):

$$A_{\text{holes}} = 2[1/4 \times (7/8 + 1/8)] = 0.50 \text{ in}^2$$

$$A_{\text{furn}} = (3.38 - 0.50)0.90 = 2.60 > 2.15 \text{ required} \qquad \text{O.K.}$$

Step 2. Number of bolts for angles:

$$A_{bolt} = 0.7854(0.875)^2 = 0.601 \text{ in}^2$$

$F_v = 22$ ksi (bearing-type connection and threads in shear plane)

$$N = \frac{62.4}{0.601 \times 22 \times 2} = 2.36 \quad \text{use three bolts}$$

Check the bearing:

$$P_b = 3(2 \times 0.25 \times 1.50 \times 58) = 131 < 62.4 \text{ kips} \quad \text{O.K.}$$

Note if thickness of the WT web $\geq 2 \times t_{angle}$ bearing and shear not a problem.

Step 3. Design WT.

This step simply involves studying tables after making a computation for approximate depth based on edge distances and bolt spacing in angles:

$$d_T \geq 1.5 + 2 \times 3 + 1.5 + k$$
$$= 9 + k \text{ in}$$

Try a WT12 × 47:

$$d = 12.15 \text{ in} \quad b_f = 9.065$$
$$t_f = 0.875 \quad t_w = 0.515 \quad k = 1.53 \text{ in}$$
$$d_{edge} = \frac{9.065 - 5.5}{2} = 1.78 > 1\tfrac{1}{4} \quad \text{(Table 1.16.4, SSDD)}$$

Required depth $= 9 + 1.53 = 10.53 > 12.15$ in   O.K.

Check the bending moment at the toe of the fillet in web:

$$b = \frac{5.5}{2} - \frac{1}{16} = 2.6875 \text{ in}$$

$$T = \frac{62.4}{2} = 31.2 \text{ kips}$$

$$M = 31.2\left(\frac{2.6875}{2}\right) + 41.92 \text{ in} \cdot \text{kips}$$

$$F_b = 0.75 F_y = 27 \text{ ksi}$$

$$S = \frac{Lt^2}{6} = \frac{L(0.875)^2}{6} = 0.1276 L$$

$$S = \frac{M}{F_b}$$

$$L = \frac{41.92}{0.1276(27)} = 12.16 \text{ in} \quad \text{use } L = 12.5 \text{ in}$$

Estimate the number of bolts to carry the hanger force. Use 7/8-in-diameter bolts for the hanger to connect the WT to the beam:

$$N = \frac{62.4}{0.601(44)} = 2.35$$

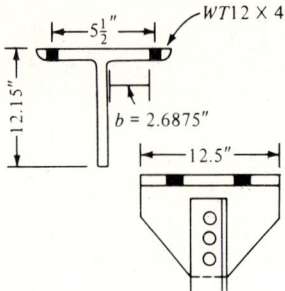

Figure E8-10c

Use four bolts for symmetry. Also, data for a W33 × 201:

$$b_f = 15.745 \text{ in} \qquad t_f = 1.15 \text{ in} \qquad g = 5.50 \text{ in}$$

Step 4. Check the "prying action" using Eq. (8-9):

$$F = \frac{62.4}{4} = 15.6 \text{ kips}$$

$$a = d_{\text{edge}} = 1.78 > 2(0.875) \qquad \text{use } a = 2(0.875) = 1.75 \text{ in}$$

$$Q = 15.6 \left\{ \frac{100(2.69)(0.875)^2 - 14(6.25)(0.875)^2}{70(1.75)(0.875)^2 + 21(6.25)(0.875)^2} \right\}$$

$$= 11.1 \text{ kips}$$

The possible moment at the bolt is

$$M_1 = 11.1 \left(\frac{1.75}{2} \right) = 9.71 < 41.92 \text{ in} \cdot \text{kips} \qquad \text{O.K.}$$

$$F + Q = 15.6 + 11.1 = 26.7 > 0.601(44) \, (= 26.44 \text{ kips})$$

Since this value is only 300 lb/bolt over the allowable, take it as O.K. Use a WT12 × 47 with four 7/8-in-diameter A-325 bolts to the flange of the W33 × 201 beam.

$$L = 12.5 \text{ in}$$

2L4 × 3 × 1/4 with long legs back to back

Step 5. Check the beam flange for adequacy without using stiffeners.

$$t_b = 1.15 \text{ in } (t_{\text{flange}} \text{ of W33} \times 201) \qquad t_w = 0.715 \text{ in } g = 5.5 \text{ in}$$

From which

$$a = \frac{b_f - 5.5}{2} = \frac{15.745 - 5.5}{2} = 5.12 \text{ in}$$

$$b = \frac{g - t_w - 1/8}{2} = \frac{5.5 - 0.715 - 0.125}{2} = 2.33 \text{ in}$$

Substitute into Eq. (8-12), since L is only 12.5 in and the yield load for each

side of the flange is
$$P = 36(1.15)^2(2 \times 0.3889)^{\frac{1}{2}}\left(1 + \frac{2.33}{5.12}\right) = 61.1 \text{ kips}$$
$$P' = \frac{P}{1.70} = \frac{61.1}{1.7} = 35.9 \text{ kips} > \frac{62.4}{2} \quad \text{O.K.}$$

The flange does not require stiffening for this load. ///

Example 8-11 A stairway hanger rod is attached to the lower flange of a W410 × 59.5 floor beam. The load to be carried by the hanger rod is estimated to be 38.75 kN. F_y = 250 MPa for both rod and beam. Design a hanger rod and check if the beam can carry this load without a web-to-flange stiffener.

SOLUTION Design the rod:
$$A = \frac{P}{F_t} = \frac{38.75}{150} = 0.2583 \times 10^{-3} \text{ m}^2$$
$$A_e = 0.7854\left(D - \frac{0.9743}{n}\right)^2$$

By trial, obtain D = 22 mm (at 10 threads/25 mm), or
$$A_e = 0.7854\left(22 - \frac{24.747}{10}\right)^2 \times 10^{-3} = 0.299 \times 10^{-3} \text{ m}^2 > 0.2583 \quad \text{O.K.}$$

Check the beam for the hanger force. Neglect the torsion produced by this hanger rod on only one side of the beam flange, since it is located very near the end of the beam. Take the hanger rod hole at the standard gage distance of 88.9 mm for a W410 × 59.5 section. Also,
$$b_f = 178 \text{ mm} \qquad t_w = 7.7 \text{ mm} \qquad t_f = 12.8 \text{ mm} = t_b$$
$$b = \frac{89 - 7.7 - 1.5}{2} = 39.9 \text{ mm}$$
$$a = \frac{178 - 89}{2} = 44.5 \text{ mm}$$

By Eq. (8-13) and directly incorporating LF = 1.7, we obtain
$$P = \frac{250(12.8)^2(2 \times 0.389)^{\frac{1}{2}}(1 + 44.5/39.9) \times 10^{-3}}{1.7} = 44.95 \text{ kN}$$

By Eq. (8-14):
$$P = \frac{250(12.8)^2[0.389(1 + 44.5/39.9)]^{\frac{1}{2}}(10^{-3})}{1.7} = 19.8 \text{ kN} < 38.75 \quad \text{N.G.}$$

Since 19.8 < 38.75 kN, the flange of the W410 × 59.5 is too thin and we must either use a section with a thicker flange or reinforce the flange of this section. It will be about as economical to use a section with a thicker flange.

///

8-8 CONNECTIONS SUBJECTED TO COMBINED SHEAR AND TENSION

Figure 8-18 illustrates several cases of connections subjected to combined shear and tension. We note that a connection in a combination of shear and compression is adequately conservative if designed as either a friction or a bearing connection. It is evident that the compressive force on the connection increases the slip resistance, whereas a tension force tends to decrease it.

Generally, a fastener subjected to combined shear and tension can be treated somewhat similarly to the beam–column condition for combined stresses. That is, the actual shear f_v and tension f_t stresses computed on the basis of the *nominal* fastener area are

$$f_v + f_t \leq f_{\text{allow}} \qquad (a)$$

A better fit is obtained using a quadrant of a stress ellipse, since F_{allow} has separate values for tension F_t and shear F_v, which give

$$\left(\frac{f_v}{F_v}\right)^2 + \left(\frac{f_t}{F_t}\right)^2 \leq 1 \qquad (b)$$

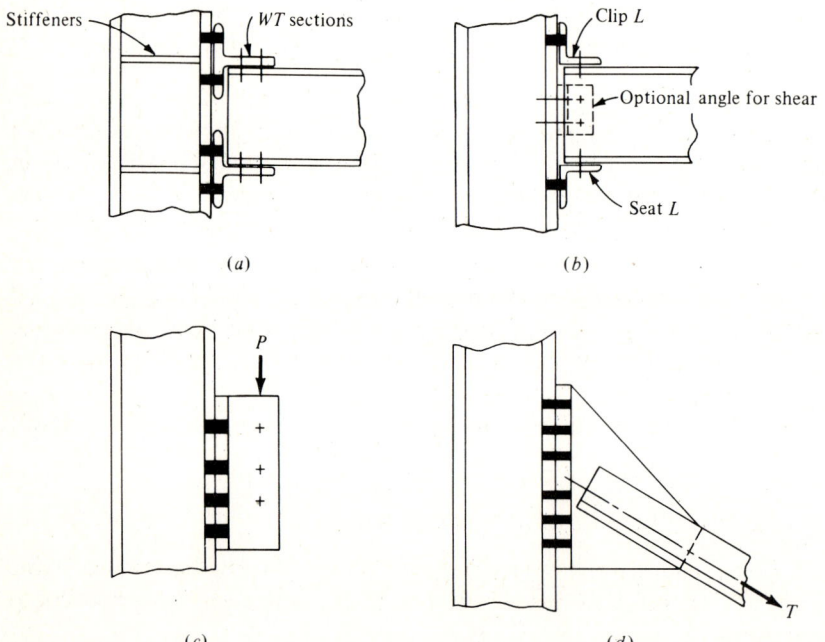

Figure 8-18 Connections where fasteners are subjected to combined shear and tension. (*a*) Moment-resisting connection using WT and column stiffeners. (*b*) Type 2 (simple) or type 3 (semirigid) connection. If angle is used with clip angles, assume that it carries all of shear. (*c*) Bracket connection using a pair of angles back to back or a WT. (*d*) Common type connection used for wind bracing. One may use a cable instead of the angle shown.

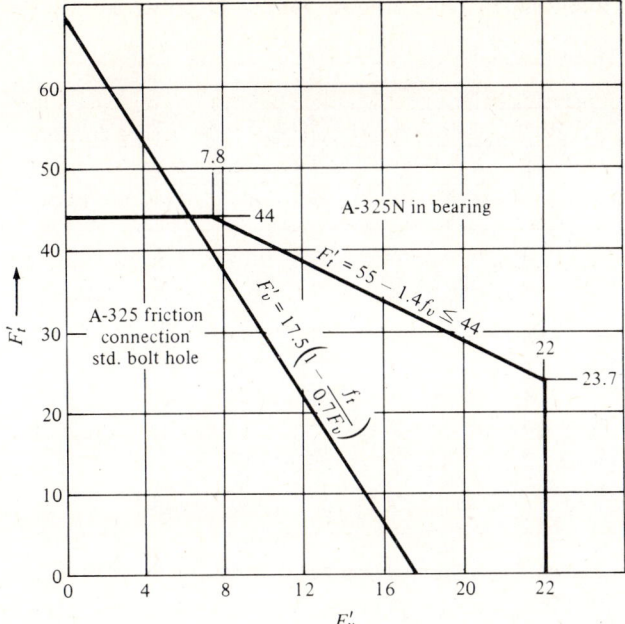

Figure 8-19 Bolt shear and tension interaction diagrams for allowable stresses.

The AISC replaces this stress ellipse with three straight lines for bearing connections and a single straight line for friction connections, as illustrated in Fig. 8-19. The general AISC equation for allowable tensile stress on a *bearing* connection with shear is

$$F'_t = C_1 - C_2 f_v \le C_3 \tag{8-15}$$

Table 8-7 gives the values for the C factors. The AISC equation for combined shear and tension in a friction-type connection is obtained from Eq. (b) using $F_f = T_b/A_b$, where the bolt proof load T_b is used and taking $F'_v = f_v$ to obtain

$$F'_v \le C_4\left(1 - \frac{f_t A_b}{T_b}\right) \tag{8-16}$$

where T_b = bolt proof load as obtained from Table 8-2
$f_t A_b$ = nominal bolt tension force ($f_t = T/A_b$)

The AASHTO equation for the combined stress cases are the same as AISC but with some additional conservatism and simplification, giving, for friction-type connections,

$$F'_v \le 13.5 - 0.22 f_t \quad \text{ksi} \tag{8-17}$$

$$F'_v \le 93 - 0.22 f_t \quad \text{MPa} \tag{8-17a}$$

and is obtained using an average $A_b/T_i = 0.017$ from considering the nominal area of $\frac{3}{4}$- and $\frac{7}{8}$-in bolts and the proof loads given in Table 8-2.

Table 8-7 Coefficients for allowable fastener stresses for combined shear and tension connections

	AISC							AASHTO
	Bearing					Friction		
	C_1		C_2	C_3		$C_4{}^a$		C_5
Fastener	fps	SI		fps	SI	fps	SI	
A-502 grade 1	30	207	1.3	23	158	—		0.75
grade 2	38	260	1.3	29	200	—		0.75
Bolts								
A-307	26	180	1.8	20	138	—		—
A-325N[b]	55	380	1.8	44	303	17.5	121	—
A-325X[c]	55	380	1.4	44	303	17.5	121	0.555
A-449N	38.7[d]	270[d]	1.8	30	207	—		—
A-449X	38.7[d]	270[d]	1.4	30	207	—		—
A-490N	68	470	1.8	54	372	22	152	—
A-490X	68	470	1.4	54	372	22	152	0.604[e]

[a] Standard holes 1/16 in (or 1/5 mm) larger than nominal bolt diameter; see the AISC manual (Sec. 1-6.3) for slotted or oversize holes.
[b] N, threads in shear plane.
[c] X, threads excluded from any shear plane.
[d] For bolt diameters larger than $1\frac{1}{2}$ in (38 mm); $F_u = 90$ ksi (620 MPa).
[e] Only static tensile loads allowed.

For bearing-type connections AASHTO uses:

$$F'_v \leq \sqrt{f_v^2 + (C_5 f_t)^2} \tag{8-18}$$

where f_v and f_t are the actual computed stresses and C_5 is a coefficient from Table 8-7.

The AREA currently has no provisions for fasteners in combined shear and tension.

Example 8-12 Given the tension-shear connection of Fig. E8-12, what is the allowable load P for the WT to column connection using the AISC specifications, $F_y = 250$ MPa steel, and 25-mm-diameter A-325 bolts?

SOLUTION Assume a bearing-type joint with threads included in shear plane.

$$F'_t = C_1 - C_2 f_v \leq C_3$$

From Table 8-7 (A-325N), obtain

$$C_1 = 380 \quad C_2 = 1.4 \quad C_3 = 303$$

$$A_b = 0.7854(0.025)(10^3) = 0.4908 \times 10^{-3} \text{ m}^2$$

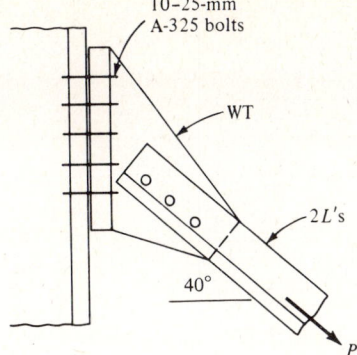

Figure E8-12

Set $F_t = 303$ MPa and find P based on shear limitations.

$$f_v = \frac{P \sin 40}{10(0.4908)} = 0.13094P$$

$$303 = 380 - 1.4(0.13094P)$$

$$P = 420 \text{ kN}$$

Find P based on maximum tension stress.

$$F_t = 303 = \frac{P \cos 40°}{10(0.4908)}$$

$$P = 1941 \gg 420 \text{ kN} \quad \text{use } P = 420 \text{ kN}$$

Check $P = 420$ kN.

$$f_v = \frac{420}{10(0.4908)} = 85 \text{ MPa} < 150 \text{ MPa of Table 8-1} \quad \text{O.K.}$$

$$F_t' = 380 - 1.4(85) = 260 < 303 \text{ MPa} \quad \text{O.K.} \qquad ///$$

8-9 MOMENT (TYPE 1) CONNECTIONS

Figures 8-16a and 8-17a to c are commonly used building connections, with those of Figs. 8-16a and 8-18a and b used for rigid (or AISC type 1) connections using mechanical fasteners. Figure 8-18b (see also Fig. 8-15a and b) is commonly used for both simple (type 2) and semirigid (type 3) connections. The designations "simple" and "semirigid" for this connection are determined to some degree by the thickness of the top clip angle. For simple connections this angle thickness t is often limited to 1/4 in (or 6.3 mm), so that the angle can deform in such a way that a moment is not developed. The WT for the connection of Fig. 8-16a may be designed similar to Example 8-10. The clip angle for the connection of Fig. 8-18b requires a design for bending that is somewhat similar to the WT of Fig. 8-18a. Critical sections and assumptions for the clip and seat angle design are shown in Fig. 8-20.

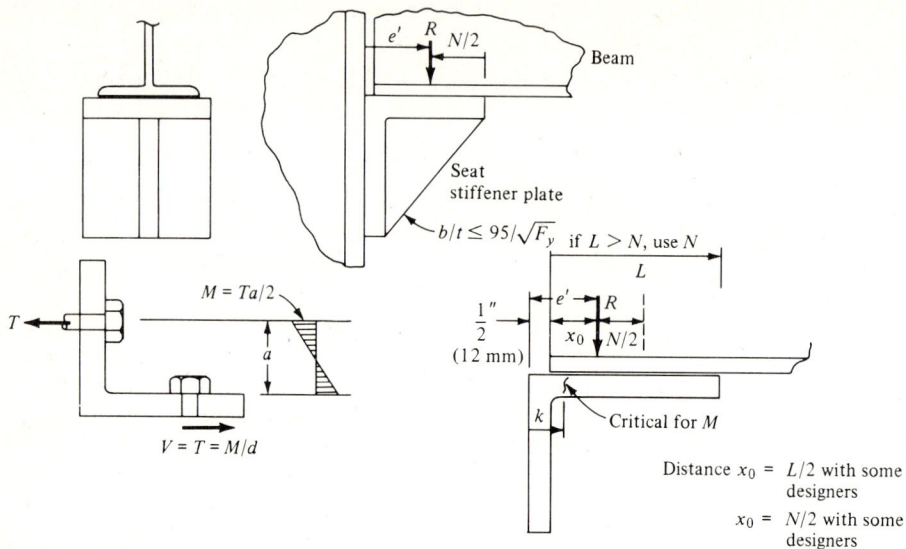

Figure 8-20 Critical sections and dimensions for clip (top) and seat angle design for use in framing beams to columns.

Example 8-13 Design a semirigid (type 3) connection to resist one-half the capacity of a W460 × 74.4 section, as shown in Fig. E8-13a. Use 25-mm-diameter A-325 bolts, F_y = 250 MPa steel, and the AISC specifications.

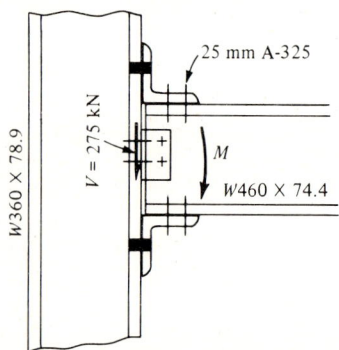

Figure E8-13a

SOLUTION W460 × 74.4 data:

$$d = 457 \text{ mm} \qquad b_f = 190 \text{ mm} < 205 \qquad \text{O.K.}$$
$$t_f = 14.5 \text{ mm} \qquad t_w = 9.0 \text{ mm} \qquad S_x = 1.46 \times 10^{-3} \text{ m}^3$$

Assume that web angles will be used to carry shear so that the seat angle carries only compression due to moment and the clip angle carries only tension.

Find the tension force to be resisted by the clip angle (assume that $F_b = 0.6F_y = 150$ MPa):

$$M = 0.5(0.6F_y) \times 1.46 = 109.5 \text{ kN} \cdot \text{m}$$

$$T = \frac{M}{d} = \frac{109.5}{0.457} = 239.6 \text{ kN}$$

The number of bolts in tension in the clip angle is based on $F_t = 305$ MPa and $A_b = 0.4908 \times 10^{-3}$ m².

$$N = \frac{239.6}{305(0.4908)} = 1.6 \qquad \text{use two bolts}$$

We note that it is good to not have to use more than two bolts, since the additional rows would have a considerably reduced efficiency. Referring to Fig. E8-13b, the number of bolts needed to carry the value of T in shear using $F_v = 150$ MPa is

$$N = \frac{239.6}{150(0.4908)} = 3.25 \qquad \text{use four for symmetry}$$

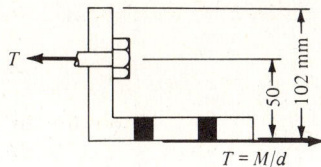

Figure E8-13b

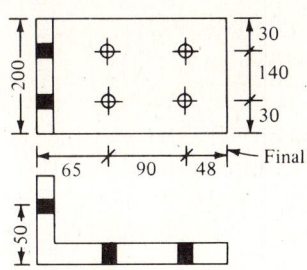

Figure E8-13c

Design the angle. Take $L \le b_f = 190$ to 205 mm → use 200 mm
Use an unequal leg angle with the long leg outstanding. The long leg will have to be long enough to place two bolts at $3D$ + edge distance + value of g_1 from Table I-13 of SSDD.

$$L \ge 28 + 75 + g_1 = 28 + 75 + 65 = 168 \text{ mm}$$

Try L178 × 102 × 22.2 (refer to Fig. E8-13c):

$a = 50 - 22.2 = 27.8$ mm

$$M = \frac{Ta}{2} = (239.6)\frac{0.0278}{2} = 3.330 \text{ kN} \cdot \text{m}$$

$$S = \frac{Lt^2}{6} = \frac{200(0.0222)^2}{6} = 0.01643 \times 10^{-3} \text{ m}^2$$

$F_b = 0.75 F_y \qquad$ due to type of connection, location, and due to rectangular shape of cross section

$F_b = 0.75(250) = 187.5$ MPa

$$f_b = \frac{3.350}{0.01643} = 202.7 > 187.5 \qquad \text{N.G.}$$

Since this is thickest angle in this group, go to the top of the next group, since thickness controls. We would not make $L > 205$ mm, because it would extend outside the column flange and also because the bending analysis may become questionable with a very long width and only two fasteners.

Try L203 × 102 × 25.4 mm.

$$a = 50 - 25.4 = 24.6 \text{ mm}$$

$$M = (239.6)\frac{0.0246}{2} = 2.947 \text{ kN} \cdot \text{m}$$

$$S = \frac{200(0.0254)^2}{6} = 0.0215 \times 10^{-3} \text{ m}^3$$

$$f_b = \frac{2.947}{0.0215} = 137 < 187 \text{ MPa} \quad \text{O.K. (and no angle thinner will work)}$$

A routine check for tension shows the section to be adequate. Let us check to see if it is necessary to use a web angle for shear, since this angle is rather thick. Refer to Fig. E8-13d for critical dimensions and other data for the analysis.

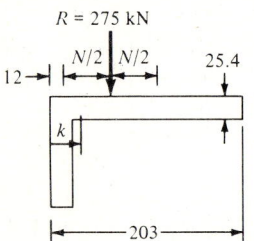

Figure E8-13d

With 12-mm standard beam end clearance, $R = V$ acts at an eccentricity of

$$e = \frac{203 - 12}{2} = 95.5 \text{ mm}$$

Since this value is so large and because we are using the long leg outstanding to provide adequate length for two rows of bolts, let us compute the N distance required for the beam and use that value to determine the bending moment in the seat angle. Assume that the reaction is concentrated at $N/2$ from the end of the beam for computing the moment.

$$(N + k)t_w F_b = R \quad [\text{Eq. (4-5)}]$$

$$k = 27.8 \text{ mm} \quad t_w = 9.0 \text{ mm}$$

$$(N + 27.8)(0.009)(187.5) = 275$$

$$N = 135.2 \text{ mm}$$

$$e = \frac{135.2}{2} = 67.6 \text{ mm}$$

$$e' = e + 12 - k_{\text{angle}} = 67.6 + 12 - 38.1 = 41.5 \text{ mm}$$

$$M = Re' = 275(0.0415) = 11.412 \gg 2.947 \text{ kN} \cdot \text{m as used for top angle}$$

This is not satisfactory even if we assumed the top angle carries one-half of shear; therefore, web shear angles are required.   ///

The moment connection shown in Fig. 8-18c may be treated in one of two ways:

1. Assuming initial tension in the bolts (which is always developed with high-strength bolt connections).
2. Assuming no initial tension.

These two assumptions are illustrated in Figs. 8-21 and 8-22. With high-strength bolts the assumption of initial bolt tension allows the connection to act as an elastic unit and the stresses can be calculated using the usual bending moment equation, $f_b = Mc/I$, which is valid up until the connection plates separate. Since the joint is never designed for a moment large enough to separate the plates, the method shown in Fig. 8-21 is adequately conservative for design use. The method shown in Fig. 8-22 is necessary to design or analyze a joint using rivets where no initial tension is commonly assumed.

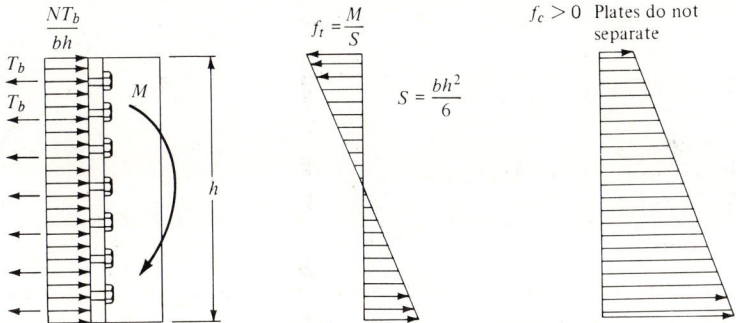

Figure 8-21 Moment connection with initial bolt tension assumed.

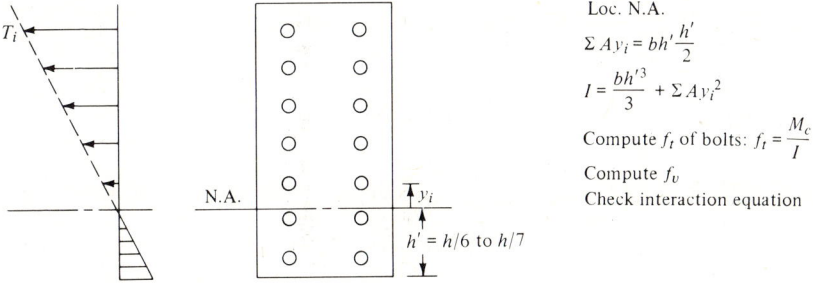

Figure 8-22 Moment connection with initial bolt tension neglected.

Example 8-14 The bracket connection shown in Fig. E8-14a uses a piece of WT and two pieces of angle to make a stiffened beam seat. The fasteners are A-325 high-strength bolts and A-36 steel. Is the connection adequate for the shear and moment to be resisted? Use the AISC specifications.

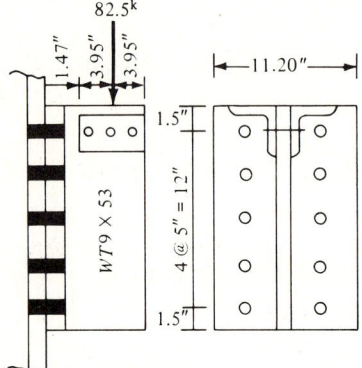

Figure E8-14a

SOLUTION Assume a friction-type joint. For 7/8-in-diameter bolts, $T_b = 39$ kips/bolt (Table 8-2).

$$\frac{P}{A} = \frac{10(39)}{15(11.20)} = 2.32 \text{ ksi}$$

$$f_v = \frac{82.5}{10(0.601)} = 13.72 \text{ ksi}$$

$$S = \frac{bh^2}{6} = \frac{11.20(15)^2}{6} = 420.0 \text{ in}^3$$

$$M = 82.5(1.47 + 3.95) = 447.15 \text{ in} \cdot \text{kips}$$

$$f_t = f_c = \frac{M}{S} = \frac{447.15}{420} = 1.06 \text{ ksi}$$

The resulting stress diagram is shown in Fig. E8-14b. The hatched part of the M/S diagram is tension due to moment, which must be carried by the

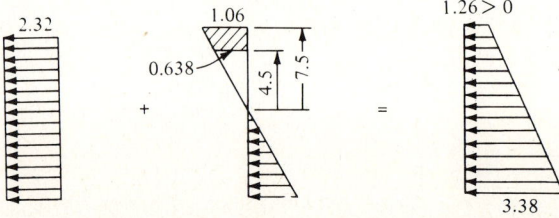

Figure E8-14b

top two bolts:

$$T_m = \frac{1.06 + 0.638}{2} \times \frac{3(11.20)}{2} = 14.3 \text{ kips/bolt}$$

$$F'_v = 17.5\left(1 - \frac{A_b f_t}{T_b}\right) = 17.5\left(1 - \frac{14.3}{39}\right) = 11.08 < 13.7 \text{ ksi} \qquad \text{N.G.}$$

It will be necessary to redesign the connection using either larger bolts or more bolts. ///

8-10 LOAD RESISTANCE FACTOR DESIGN (LRFD) FOR CONNECTIONS

The current form of the LRFD equation for connection design using A-325 and A-490 high-strength bolts (and with size limited to diameter ≤ 1.5 in) is

$$R_n = 1.1(1.1D + 1.4L)$$

This value is compared to the fastener resistance or plate-to-bolt bearing as follows:

Bearing-type connection:
$$R_n = 0.625 A_b F_u \qquad \phi = 0.70$$

Friction-type connection:
$$R_n = 0.7 m \mu A_t F_u \qquad \phi = 1.00$$

Combined shear and tension:
$$(P_{u(\text{shear})})^2 + (0.6 P_{u(\text{tension})})^2 \leq \phi(0.6 A_b F_u)^2 \qquad \phi = 0.75$$

Plate bearing on bolt:
$$R_n = 3 t d F_{u(\text{plate})} \qquad \phi = 0.64$$

where A_b = nominal area of bolt
 A_t = bolt tension area (in thread zone) [see Eq. (5-4)]
 F_u = ultimate tensile strength (nominal) of bolt
 m = number of slip surfaces (= 1 for single shear lap joints)
 μ = coefficient of friction (commonly 0.35 for clean mill scale)

Example 8-15 Given the connection shown in Fig. E8-15, determine if the plate thickness shown is adequate and find the number of 20-mm A-325 bolts using a friction-type connection, and a bearing-type connection.

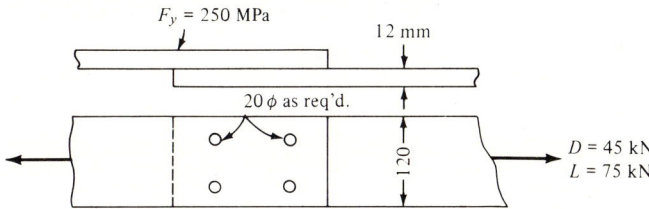

Figure E8-15

SOLUTION
$$R_u = 1.1(1.1D + 1.4L)$$
$$= 1.1[1.1(45) + 1.4(75)] = 170 \text{ kN}$$

Checking the plate dimensions.

$A_n \phi F_y = 170$ (see Sec. 5-10)

$$A_n = \frac{170}{0.88(250)} = 0.77272 \times 10^{-3} \text{ m}^2$$

$A_n \phi F_u = 170$ kN

$$A_n = \frac{170}{0.74(400)} = 0.5743 \times 10^{-3} \text{ m}^2$$

$$A_g = \frac{0.5743}{0.85} = 0.6756 \times 10^{-3} \text{ m}^2 < 0.77272 \quad \text{does not control}$$

The A_g furnished is

$$A_g = [120 - 2(23)]0.012 = 0.888 \times 10^{-3} \text{ m}^2 > 0.77272 \quad \text{O.K.}$$

Find the number of bolts required in a friction-type connection (no slip tolerated, but note that after plates slip, the bolts must still shear for a connection failure). Assume that $\mu = 0.35$ (clean mill scale on faying surface).

$$A_t = 0.7854\left(D - \frac{0.9743}{n}\right)^2 \text{ (fps)} \quad [\text{Eq. (5-4)}]$$

$$A_t = 0.7854\left(20 - \frac{24.747}{10}\right)^2 = 0.24122 \times 10^{-3} \text{ m}$$

Assume 10 threads/25.4 mm for the 20-mm-diameter bolt.

$$\phi R_n = 0.7 m \mu A_t F_u = 170 \text{ kN}$$

$m = 1$ for single shear

$$0.70 \times 1 \times 0.35 \times 0.24122 \times 725 \times N = 170 \text{ kN}$$
$$N = 3.97 \quad \text{use four bolts}$$

Find the number of bolts required in a bearing-type connection (slip tolerated and joint failure by bolt shear):

$$\phi R_n = \phi(0.625 A_b F_u) N = 170 \text{ kN}$$

$$A_b = \frac{\pi d^2}{4} = \frac{\pi (0.020)^2 \times 10^3}{4} = 0.31416 \times 10^{-3} \text{ m}^2$$

$$N = \frac{170}{0.70 \times 0.625 \times 0.31416 \times 725} = 1.71 \quad \text{use two bolts}$$

Check the plate bearing using two bolts:

$$R_n = 3tdF_u \quad \phi = 0.64 \quad F_u = 400 \text{ MPa}$$
$$\phi R_n = 0.64(3)(0.012 \times 20)(400)(2) = 368.6 \text{ kN} \gg 170 \text{ kN} \quad \text{O.K.}$$

The connection is "safe" with either two or four bolts. Choose the number based on the final decision as to connection type, "friction" or "bearing."
///

PROBLEMS

8-1 Determine the tensile capacity of the connection shown in Fig. P8-1 using 3/4-in-diameter A-325 bolts. Set the gage so that the critical net section is a minimum. Also determine the minimum thickness of cover plates to the nearest multiple of 1/16 in. Use the AISC specifications and A-36 steel. Show a neat final sketch with all critical items required. Use a bearing connection.
Answer: P = 159 kips.

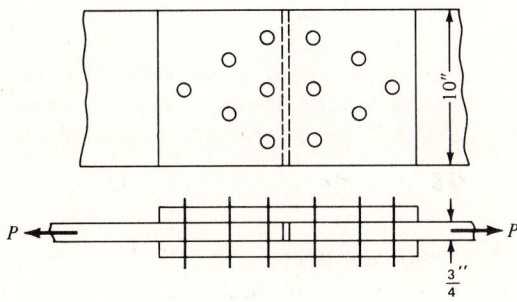

Figure P8-1

8-2 Determine the number of 22-mm-diameter A-490 bolts needed to develop the full effective capacity of the tension connection shown using a pair of angles and the gusset plate shown in Fig. P8-2. Use a bolt spacing of at least $3D$. Make a neat sketch of the final joint design showing the bolts required, length, and any other critical design information. Use the AISC specifications, $F_y = 345$ MPa steel, and a friction connection.

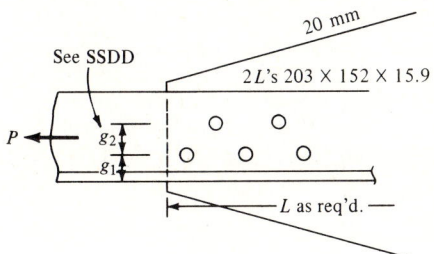

Figure P8-2

8-3 Redo Rob. 8-2 using $F_y = 250$ MPa and the AASHTO specifications for the full effective capacity of the angles. Use A-325 bolts and a friction connection.
Answer: 18 bolts, $L = 610$ mm just under L for no reduction.

8-4 Do Prob. 8-2 for the fps equivalent of the pair of angles and a 3/4-in gusset plate for an axial load of 240 kips, A-36 steel, and A-325 bolts in a bearing connection.
Answer: Eight 7/8-in-diameter bolts.

8-5 Given the beam splice shown in Fig. P8-5, use the AASHTO specifications to (a) Design bolts and cover plates for the full moment capacity of the beam. (b) Design bolts and web plates for the full web shear capacity of the beam. Use 7/8-in-diameter A-325 bolts for all splice parts, and A-36 steel.

Partial answer: $M = 1950$ in · kips, $T = 107$ kips.

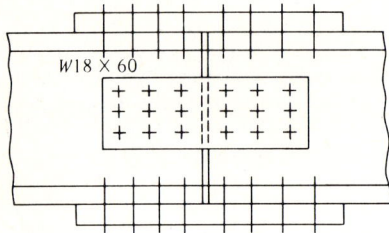

Figure P8-5

8-6 Design the eccentrically loaded bracket connection for the load shown in Fig. P8-6. Use 25-mm-diameter A-325 bolts in a friction connection. Determine the plate thickness for both bearing and tear along the forward row of bolts.

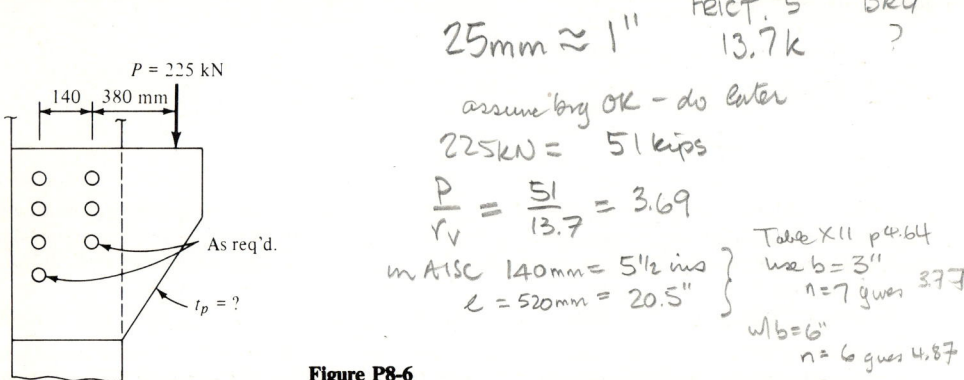

Figure P8-6

8-7 Determine the number and placing of 22-mm-diameter A-325 bolts for a cable connection to a W360 × 314 column as shown in Fig. P8-7. Assume any needed T dimension to produce L that is available. Check the tension in the stem of the T so that a large enough section is used. Assume that the hole for the cable attachment will be reinforced so that capacity is not limited at that point. Use the AISC specifications and $F_y = 250$ MPa for the T section.

Answer: 10 bolts.

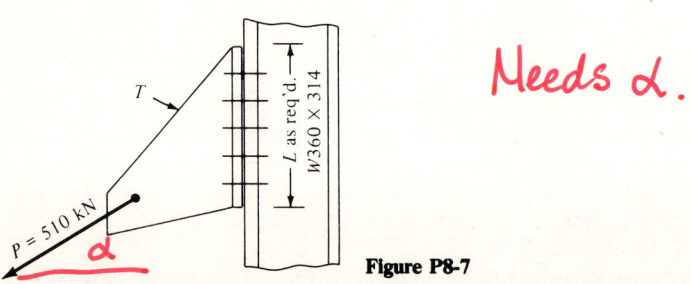

Figure P8-7

8-8 Determine the maximum P allowed for the stiffened seat connection shown in Fig. P8-8 using the AISC specifications and A-36 steel with 7/8-in-diameter A-325 bolts. Assume that the angles and T are adequate for tearing but that the bearing should be checked as appropriate. All bolt lines in the angles are on standard gage distances, as in Table I-13 of SSDD.

Answer: 52.2 kips.

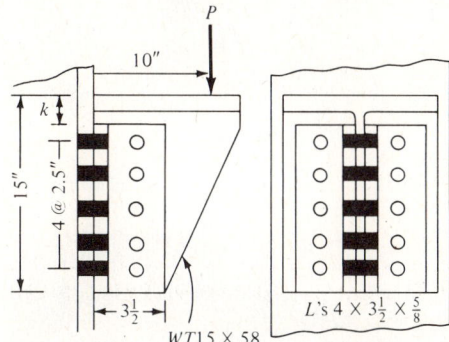

Figure P8-8

8-9 Redo Example 8-14 for twelve 7/8-in-diameter A-325 bolts.
Answer: 12.

8-10 Redo Example 8-14 for ten 1-in-diameter A-325 bolts.

8-11 Redesign the connection of Example 8-8 if the floor beam on the left does not frame into the W18 × 50 section.

8-12 Check the connection of Example 8-7 if the two bolts at the c.g. of the bolt pattern are omitted. Is the pattern adequate?
Answer: Yes.

8-13 Design the fasteners for member 9, which frames into the joint designed for the highway bridge in Example 8-3.

8-14 Determine the W410 section that can safely carry the hanger rod stairway load of Example 8-11.

8-15 Design the bottom connection for vertical member 7 of the highway bridge using the bottom chord section you designed in Prob. 5-28.

8-16 Do Prob. 8-1 using LRFD.
Answer: P_u = 217 kips, N = six $\frac{3}{4}$-in bolts.

8-17 Do Prob. 8-2 using LRFD.

8-18 Do Prob. 8-5 using LRFD. Assume that the 225-kN load is made up of D = 100 kN and L = 125 kN. Check only for number of bolts.
Answer: 16 bolts with 1 percent overstress.

(a)

(b)

(c)

Figure IX-1 Welds. Most structural welds (80 to 90 percent) are fillet welds. (*a*) Rare use of butt weld in a column splice. Note lower column end plate butt welded to column but upper plate is fillet-welded. (*b*) Column base plate fillet-welded to column. (*c*) Fillet welds. Split tee used as beam girder (open-web joist) seat. Note lower plate providing lateral stability for lower flange is also fillet-welded to column. Beam in left foreground has shopfillet-welded framing angles (field bolted to column web). An erection bolt (also for structural stability) can be seen in flange of WT seat element.

CHAPTER NINE

WELDED CONNECTIONS

9-1 GENERAL CONSIDERATIONS

Welding is the joining of pieces of metal by heating the contact points to a fluid, or nearly fluid, state and with or without the application of pressure. The earliest welding (some 3000 years ago) involved heating the pieces to a plastic state and then hammering them together (i.e., using pressure). Currently, very little of this type of welding is used except in some small local blacksmith operations involving repair of farm equipment and shoeing of animals. Most farm equipment is, at present, repaired using more modern methods of gas and electric welding.

Structural welding is nearly all electric, and numerous processes are available. Some gas welding is used ("gas" is used here to denote use of a gas/oxygen mixture to produce a very hot flame to heat the parts and the weld filler material), but gas is used primarily for cutting pieces to shape. Mechanically controlled gas cutting equipment in fabrication shops can make cuts that approach sawed cuts in smoothness.

As stated earlier, most welding uses an electric current. The current is used to heat an electrode to a liquid state, which is then deposited as a filler along the interface of the two or more pieces of metal being joined. The process simultaneously melts a portion of the base metal (metal being joined) at the interface so that the electrode intermixes with the base metal and develops continuity of material at the joint when cooling takes place. If the quantity of deposited electrode is small relative to the thickness of the joined parts, the process tends to be unreliable (i.e., insufficient melting of the base metal occurs so that the weld may pop off or not fully join). This event can be avoided by either preheating the base metal or limiting the minimum size of the weld. When the welding operation takes place in a very cold environment, it may be necessary to preheat the parts, particularly where the parts are thick so that a very large temperature differential does not develop in a short distance such that the resulting thermal stresses are so high that the weld zone fails.

Electric welding involves passing either dc or ac current through an electrode. By holding the electrode a very short distance from the base metal which is connected to one side of the circuit, an arc forms as the circuit is essentially "shorted." With this "shorting" of the circuit, a very large current flow takes place which melts the electrode tip (at the arc) and the base metal in the vicinity of the arc. The electron flow making the circuit "carries" the molten electrode metal to the base metal to build up the joint. Careful control of electrode size and current are necessary to produce a quality weld with sufficient heat to define an adequate melt zone and while keeping electrode splatter to a minimum.

The electrode may be either the anode (+) side of the circuit or the cathode (−) side. Most commonly, the electrode is the anode and the resulting operation is conducted using "reversed polarity." When the weld electrode is the cathode (−), the circuit uses straight polarity. Most welding is done using dc current; if ac is used as the power supply, it is first transformed to dc.

Of the numerous welding processes available, the following are most likely to be used for structural applications:

1. *Shielded metal arc welding (SMAW)*. This is the most common welding method using stick electrodes. The electrodes are available in lengths of 9 to 18 in and are coated with a material that produces an inert gas and slag when the welding current melts the metal. This gas surrounds the weld zone to prevent oxidation (see Fig. 9-1a), which is a critical factor if more than one weld pass is necessary to build the weld to the required size. The slag, being lighter than the metal, floats to the top of the weld and can be brushed away. On subsequent passes it is necessary to brush the earlier passes to remove any slag, dirt, or other foreign material whose presence might cause a flaw in the weld. This type of welding is the most commonly used field method using dc welding equipment. The maximum size of weld produced in one pass is about 5/16 in or 8 mm.

2. *Gas-shielded metal arc welding (GMAW)*. This method of welding is most often used in shop welding, where uncoated electrodes are used in a mechanical welding unit. The unit controls electrode spacing and welding speed and has an inert gas source to shield the weld against the surrounding atmosphere.

3. *Submerged arc welding (SAW)*. This method of welding is also used in fabrication shops. The joint is aligned and covered with a blanket of granular fusible material containing alloying and fluxing agents as well as inert gas producers. The electrode is inserted into the granular material, the arc produced, and the melting of electrode and base metal takes place. The heat fuses the

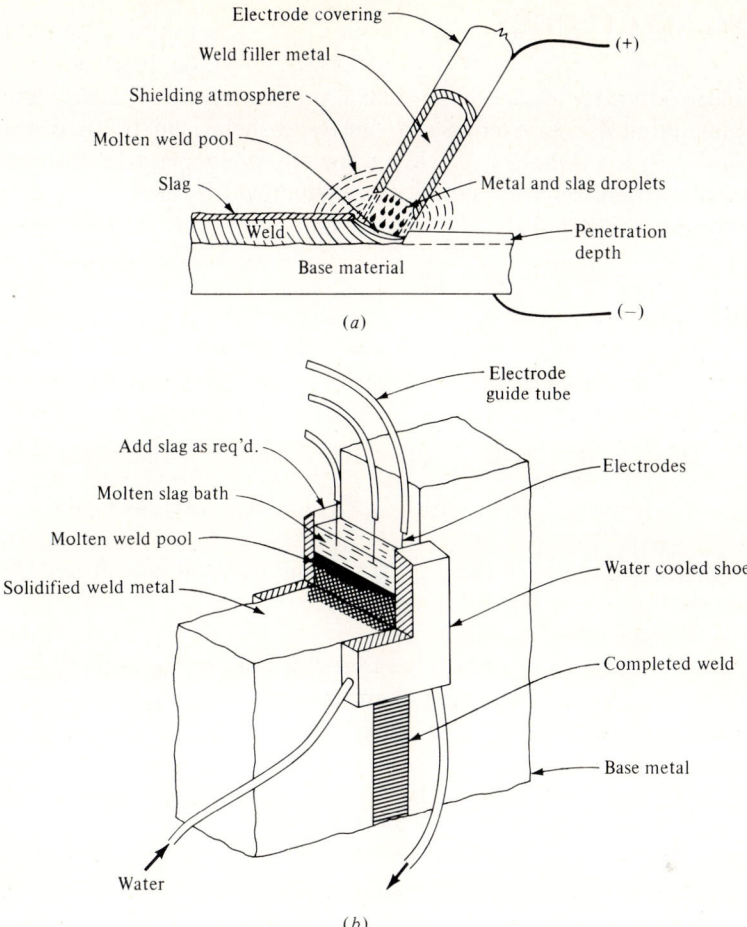

Figure 9-1 The two very common welding methods. (*a*) Shielded metal arc. (*b*) Electroslag.

granular blanket to develop the gas shield and to obtain any other effects desired from the material. The slag is later brushed away to expose the weld.

   4. *Electroslag welding.* This welding process is very similar to the submerged arc process but uses an electroconductive slag which is held in position between the two pieces of metal to be joined by water-cooled retaining plates (see Fig. 9-1*b*). The slag material is melted and current passed through it to maintain the molten state of slag and filler metal. The filler is obtained from the weld electrode, which is passed into the slag. The process is generally done in a vertical jig so that as the filler melts the retaining plates are slowly raised, leaving behind the completed, partly cooled weld, which has a thin slag covering that must be brushed away.

   Electroslag welding is used to shop-weld thick plates together. It has been rather popular in bridge work to weld girder plates and floor plates. Plates on the order of 20 to 450 mm can be welded by this process in one pass.

9-2 WELDING ELECTRODES

A variety of electrodes are available so that a proper match of base metal strength and metallurgical characteristics to the weld metal can be made. In structural applications, the American Welding Society, in cooperation with the ASTM, has established an electrode numbering system which classifies welding electrodes (or rods) as follows:

$$Eaaabc$$

where E = electrode
     aaa = two- or three-digit number establishing the ultimate tensile strength of the weld metal. Currently, the following values are available:

        60,   70,   80,   90,   100,   110,   and 120 ksi

        415, 485, 550, 620, 690, 760, and 825 MPa

     b = digit for indicating suitability of welding position, which may be flat, horizontal, vertical, or overhead.
      1 = suitable for all positions
      2 = suitable for horizontal fillets and flat positioning of work
     c = digit indicating current supply and welding technique: ac, dc, dc straight polarity, dc reversed polarity, etc.

For example, an E7013 is an electrode with $F_u = 70$ ksi, usable in all positions, with either ac or dc current and either straight or reversed polarity.

For structural design the information of interest is whether the electrode is E60, E70, E80, or whatever. Hardware stores commonly carry E7014 and E6011 electrodes (all-position welding and for use with either ac or dc current; the 4 indicates that an iron powder has been added to the electrode covering so that the arc is easier to maintain). Currently, the E70 electrodes are the most widely used for structural work and are compatible with all grades of steel to $F_y = 60$ ksi.

9-3 TYPES OF JOINTS AND WELDS

A number of different types of welds may be used in structural applications, including groove, fillet, plug, seam, slot, and spot. Most welds in structural applications are groove (about 15 percent of all welds) and fillet (about 80 percent of all structural welds) welds. The basic structural joints shown in Fig. 9-2 can be produced by using one or more of the welds listed above. The butt joint is obtained from a groove weld, as shown in Fig. 9-3a. This joint is taken to have 100 percent efficiency if constructed so that full penetration of the weld is obtained.

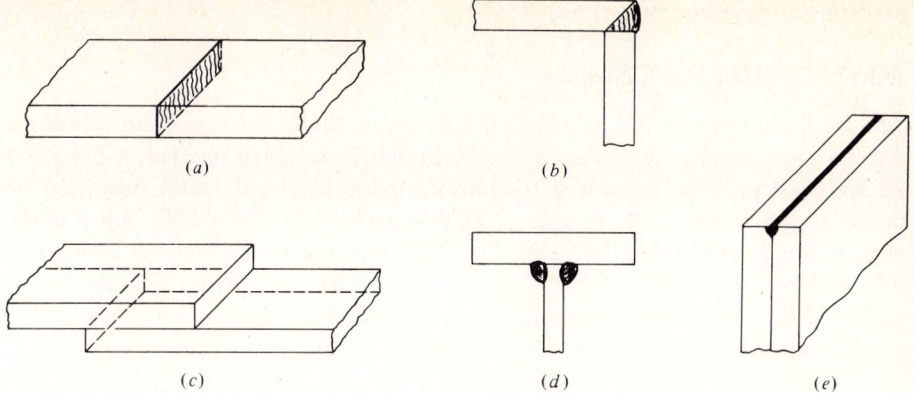

Figure 9-2 Basic structural joints using welds. (*a*) Butt joint. (*b*) Corner joint. (*c*) Lap joint. (*d*) Tee joint. (*e*) Edge joint.

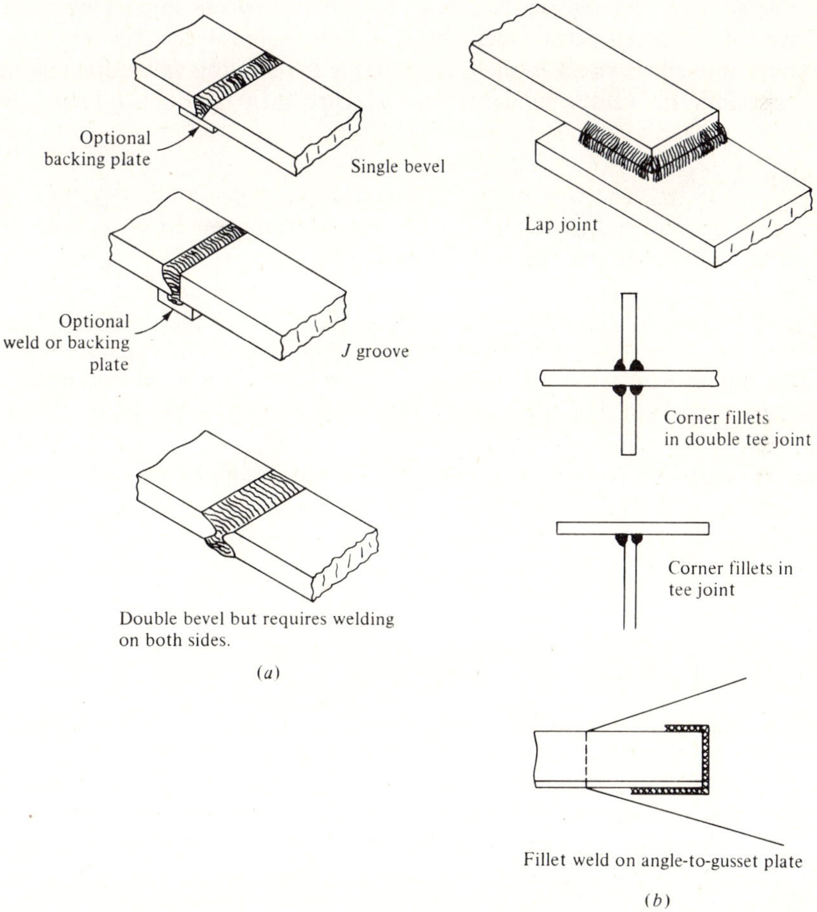

Figure 9-3 Groove and fillet welds in structural applications. (*a*) Groove welds used in butt joints. (*b*) Fillet welds used as shown.

415

9-3.1 Allowable Weld Stresses

The butt joint is the only joint likely to be in direct tension. The allowable tension or compression stresses for weld metal is given in Table 9-1. The allowable stresses in tension or compression for the weld metal may also be taken as $0.6F_y$, but with $F_{u(\text{electrode})}$ at 60 (E60) or 70 ksi (E80), the limiting stresses are those of the base metal. The AWS has introduced some conservatism into this by further limiting the F_y of the base metal to 42 ksi for E60 electrodes and to 55 ksi for E70 electrodes in structural grade steel.

The allowable shear stress for fillet welds is limited to

$$F_v = 0.3 F_{u(\text{electrode})}$$

in the AISC specifications, but it is always necessary to check that there is sufficient base metal to resist the same shear stresses. In general, the weld shear stress of $0.3F_{u(\text{electrode})}$ will produce larger allowable shear stresses than $0.4F_y$ for the base metal. For this reason the base metal shear stress must always be checked against the specifications being used.

The maximum shear stress in the weld metal is the limiting value of stress in connections where the weld is subjected to combined shear and tension stresses.

9-3.2 Fillet Welds

The *fillet weld* shown in Fig. 9-4 is approximately triangular in cross section. Care must be taken that the throat dimension shown in Fig. 9-4c is built out adequately. In most cases the legs of the weld D are made equal, but this is not necessary and unequal legs may be used. If equal legs are used for the fillet weld, the throat dimension represents the minimum area for shear and is computed as

$$T = D \times \cos 45° = D \times 0.70711$$

where T = throat dimension
D = nominal leg dimension

The leg dimension for fillet welds D should be taken to the nearest 1/16 in or 1 mm.

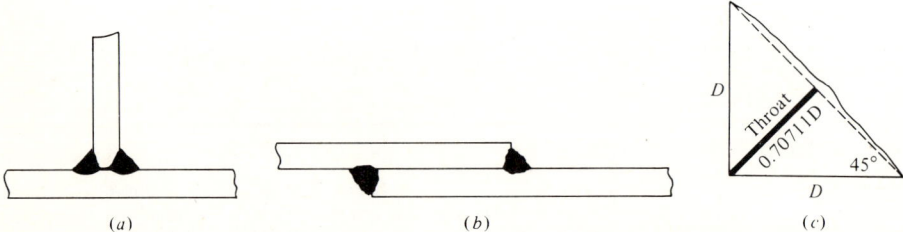

Figure 9-4 Critical shear area for fillet welds. (*a*) Fillet weld for tee joint. (*b*) Fillet weld for lap joint. (*c*) Throat dimension for minimum shear area.

Table 9-1 Allowable weld stresses by several specifications

Type of weld	Type of stress	Allowable stress		
		AISC	AASHTO	AREA
Full-penetration groove	Tension or compression parallel or normal to weld axis	Same as base metal[a]	Same as base metal[a]	Same as base metal[a]
Partial-penetration groove	Tension or compression parallel or normal to weld axis	Same as base metal	Same as base metal	Same as base metal
All groove welds	Shear	$0.30 F_{u(\text{electrode})}$	$0.27 F_{u(\text{electrode})}$	$0.35 F_{u(\text{electrode})}$
Fillet welds[b]	Shear	$0.30 F_{u(\text{electrode})}$	$0.27 F_{u(\text{electrode})}$	$0.35 F_{u(\text{electrode})}$
Plug and slot welds	Shear	$0.30 F_{u(\text{electrode})}$	$0.27 F_{u(\text{electrode})}$	$0.35 F_{u(\text{electrode})}$

[a] The base metal must be compatible with the electrode: for example, E60 electrodes are limited to base metal with F_y not more than 42 ksi (290 MPa); E70 electrodes for base metal with F_y not more than 55 ksi (380 MPa); E80 electrodes for F_y not more than 65 ksi (415 MPa).

[b] Shear stress may be limited by maximum allowable shear stress in the base metal ($F_v = 0.4 F_y$ in the AISC specifications; $F_v = 0.33 F_y$ in the AASHTO specifications; $F_v = 0.35 F_y$ in the AREA specifications).

9-3.3 Plug and Slot Welds

A *slot weld* (Fig. 9-5) is a special type of fillet weld sometimes used where it is difficult to obtain the necessary length to develop the required shear resistance in a lap joint. *Plug welds* (Fig. 9-5b) may be used to prevent buckling of long lap joints in compression or to connect top plates to the lower plates in built-up members. They can be used in either round holes or slots with rounded ends. The AISC specifications cover slot and plug weld dimensions as shown on the figure. These dimensions ensure adequate holding and transfer of shear after the substantial weld shrinkage (contraction upon cooling) has occurred. If a fillet weld is used in a slot (slot cavity not filled, as with slot or plug welds) the design and dimensions are based on strength requirements as for other fillet welds. This is because the quantity of molten weld metal is much smaller and the resulting shrinkage is much less.

9-3.4 Minimum Weld Size

Welds must be of some minimum size based on the thickness of the base metal. This is to ensure adequate heat and reasonably slow cooling of the weld area so that the weld and base metal do not crystallize and tend to crack or pop off due to insufficient fusing of weld into the base metal. The AWS code values are used as a guide by AISC and AASHTO and are given in Table 9-2.

9-3.5 Maximum Weld Size

The maximum size of fillet welds along the edges of connected parts is:

Part		
fps, in	SI, mm	AISC and AASHTO
$t \leq 1/4$	$t \leq 6$	Use D = thickness of part t
$t > 1/4$	$t > 6$	Use $D = t - 1/16$ in or $t - 1$ mm[a]

[a] Unless the throat dimension is specifically built out to use the full value of t.

It is possible in the fabrication shop when using the submerged arc welding process to specifically produce a fillet throat thickness by rounding the weld out so that for

$D \leq 3/8$ in   use throat = $D \times 0.70711$ (in)

$D > 3/8$ in   use throat = $D \times 0.70711 + 0.11$ (in)

If this is contemplated, care must be taken to differentiate between the shop

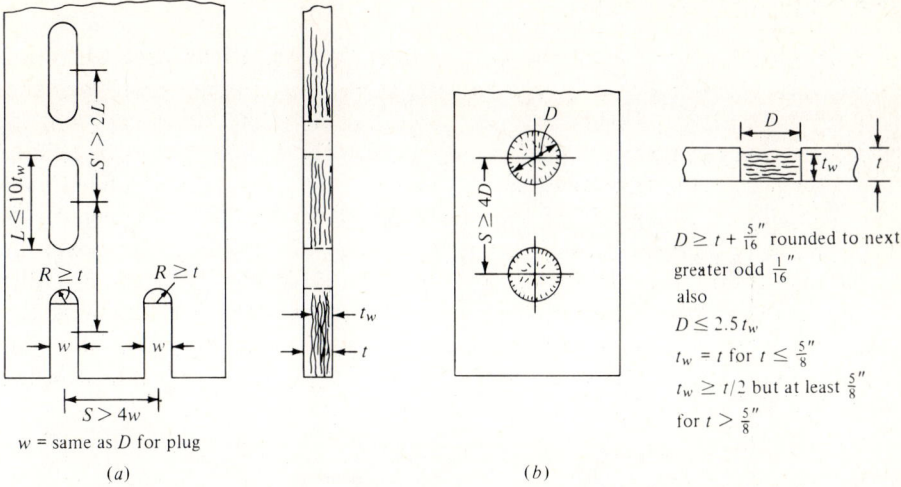

Figure 9-5 AISC specifications for plug and slot welds. (*a*) Slot welds. (*b*) Plug welds.

welding where the throat dimension can be controlled and the field welding where the nominal dimension is obtained.

The maximum weld size may also be limited by the shear resistance of the base metal. This is particularly true where fillet welds extend to the edge of the gusset plate or in matching D to the plate thickness. In general, for evaluation of the maximum effective D, one has the following:

$$0.70711 D(\beta_1 F_u) \leq t_{\text{part}}(\beta_2 F_y)$$

where $\beta_1 = 0.3, 0.27$, etc. from the appropriate code for weld rod
$\beta_2 = 0.33, 0.4$, etc. from appropriate code for the base metal

Table 9-2 Minimum weld

Thickness of base metal[a]		Minimum size of fillet weld[b]			
		AISC		AASHTO	
fps, in	SI, mm	fps, in	SI, mm	fps, in	SI, mm
$t \leq 1/4$	$t \leq 6$	1/8	3	3/16	5
$1/4 < t \leq 1/2$	$6 < t \leq 12.5$	3/16	5	3/16	5
$1/2 < t \leq 3/4$	$12.5 < t \leq 22$	1/4	6	1/4	6
$t > 3/4$	$t > 22$	5/16	8	5/16[c]	8

[a] Base metal thickness is that of the thicker part being joined.
[b] Note that the AREA has no minimum weld requirements.
[c] See the AASHTO specifications for $t > 1\frac{1}{2}$ in or 38 mm.

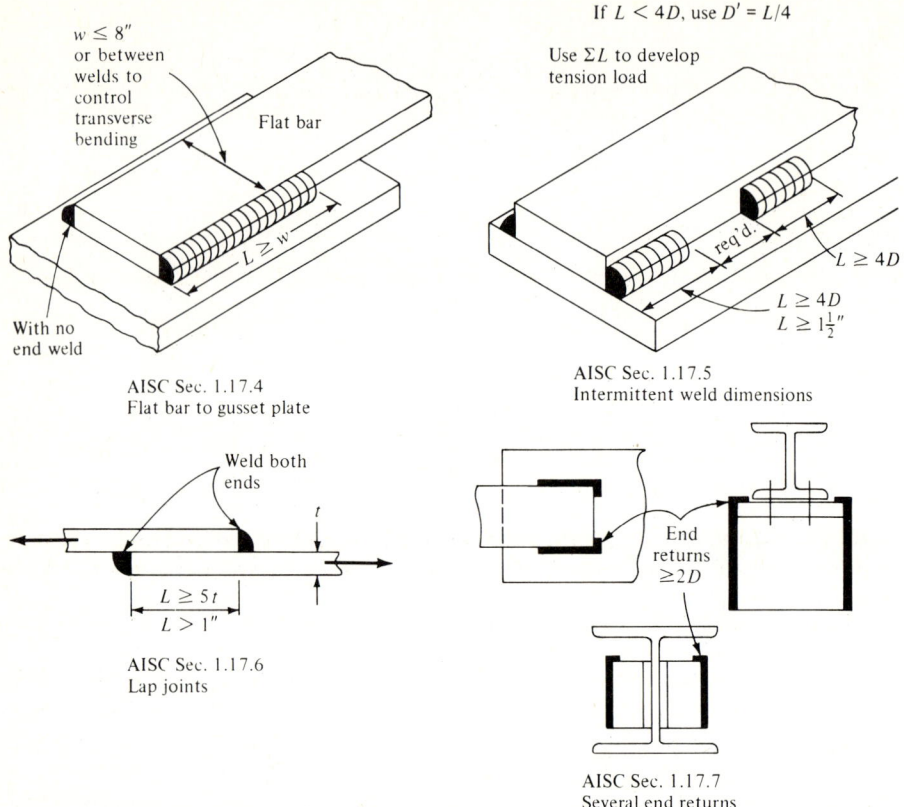

Figure 9-6 Several AISC welded connection specifications. Since these specifications are in the AWS *Structural Welding Code*, they should be generally followed for AASHTO and AREA.

Several other AISC specifications concerning welded connection design for building construction are shown in Fig. 9-6.

9-4 LAMELLA TEARING

Lamella tearing (see Fig. 9-7) is a phenomenon that may occur in certain welded joints. It is not a common condition because it involves several factors:

1. There must be large relative strains in the base metal (lamella tearing does not occur in the weld metal). These strains occur where large localized stresses occur.
2. Loading is generally perpendicular to the mill rolling direction that produced the member being welded. Beams welded to column flanges produce this type of loading in the column flanges (but not in the beam flange).
3. There must be strain restraint in the base metal.

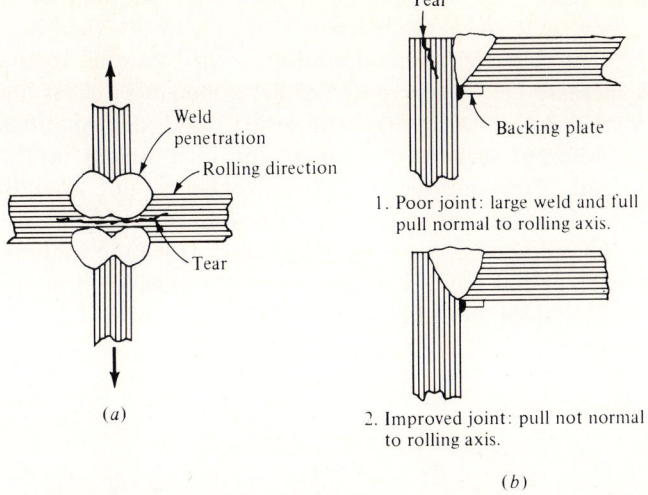

Figure 9-7 Lamella tearing in base metal at a welded joint. Also shown is one solution to reduce the likelihood of a lamella tear occurring. (*a*) Lamella tear in base metal. (*b*) Joint geometry to reduce lamella tear.

If a large weld (or welds from both sides) is imposed on a thick piece of base metal, a lamella tear can occur. The tear can occur because the shrinkage strains from the welding operation will be large and restrained. The restraint may be developed from a weld on the far side or from the member thickness or a combination. A large overmatch of electrode and base metal in a full-penetration butt weld tends to increase the possibility of tearing (i.e., from using an E80 instead of an E70 electrode with A-36 base metal). A thin, stiffened column flange is also susceptible to lamella tearing, since the flange stiffeners being welded to the column flange produce restraint.

The use of fillet welds, a joint design that allows strain relief or loading other than normal to the rolling direction, and the order of welding to minimize shrinkage strains are practical methods used to avoid lamella tearing. The AISC paper "Commentary on Highly Restrained Welded Connections" (*AISC Engineering Journal*, vol. 3, 1973) discusses lamella tearing in some detail and gives a number of joint alternatives that may be used to reduce lamella tearing.

9-5 ORIENTATION OF WELDS

Laboratory tests on small to medium-size joints show that butt welds do not limit joint capacity where the electrode has been "matched" to the base metal. The orientation of the applied stresses does not have a significant effect on the butt joint strength.

The orientation of stresses for fillet welds is a significant factor in the ultimate joint strength. Tests (see Butler and Kulak, "Strength of Fillet Welds as a Function of Direction of Load," *Welding Journal*, May 1971) on a number of joints using fillet welds show that welds loaded perpendicular to the weld axis are approximately 44 percent stronger than fillet welds loaded parallel to the weld axis. This strength increase can be attributed to development of shear lag in the longitudinally loaded welds with resultant larger weld deformations ("deformation" being a preferred term to "yielding" for fillet welds) in the connection nearest the load. This increase in weld strength is not directly considered in design specifications because the direction of load is not sufficiently reliable. It is indirectly considered by AISC (Sec. 1-17.7) in limiting the longitudinal weld length for flat bars. It is also indirectly considered in lap joints by requiring fillet welds along both bar ends (AISC Sec. 1-17.9).

9-6 WELDED CONNECTIONS

Welded connections are often easier to fabricate than are bolted connections. However, in framed connections the joint will usually require erection bolts in either temporary (or permanent) erection plates or angles to hold the parts during alignment and fitting up for the welding.

In general, welded connection design involves proportioning the parts based on a gross section and welds. Butt welds are generally designed for the tension (or compression) strength of the base metal. Fillet and other welds in shear are designed on the basis of the minimum dimensions in the shear plane and the allowable shear strength of the weld.

It should be noted that a butt weld used to resist a moment develops stresses

$$f_b = \frac{Mc}{I}$$

where f_b is compared to the allowable bending stress of the base metal.

If a fillet weld is used in the same type of connection (see Fig. 9-8), the value of f_b is compared to the allowable weld shear stress.

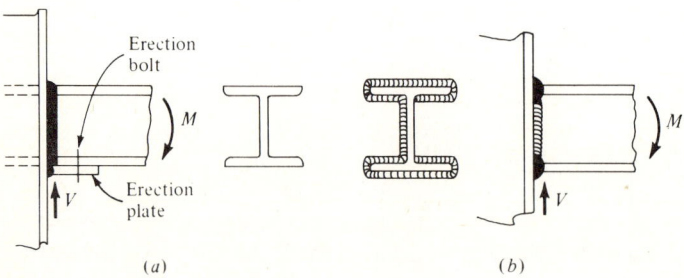

Figure 9-8 Welded moment connections. (*a*) Butt-welded moment connection. (*b*) Fillet-welded connection.

WELDED CONNECTIONS 423

Example 9-1 Design the welds for the lap joint shown in Fig. E9-1a using a flat bar and E60 electrodes, $F_y = 250$ MPa, and the AISC specifications.

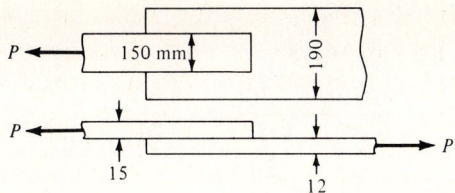

Figure E9-1a

SOLUTION

$$\text{Joint efficiency} = 100 \text{ percent}$$
$$P = 0.015(150)(0.6F_y) = 337.5 \text{ kN}$$

Use $D = 15 - 2.0 = 13$ mm.

$$F_w = 0.3 \times F_u = 0.3(415) = 124.5 \text{ MPa}$$

$$L_{\text{weld}} = \frac{337.5}{(0.013 \times 0.70711)(124.5)} = 295 \text{ mm} < 300 \text{ mm } (2w) \quad \text{O.K.}$$

Use a 150-mm weld on each side of joint as shown in Fig. E9-1b (just satisfies AISC Sec. 1-17.4). Use 26-mm end returns per AISC Sec. 1-17.7.

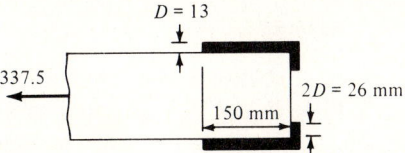

Figure E9-1b

Question: Why not 150 mm of end weld and 75 mm along each side?   ///

Example 9-2 Design the welds for connecting an L4 × 3½ × 1/4 to a 3/8-in gusset plate as shown in Fig. E9-2a. Use the AISC specifications, E70 electrodes, and A-36 steel. Design for static loading and dynamic loading.

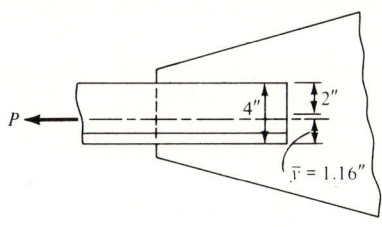

Figure E9-2a

SOLUTION For static loading

$$A_{\text{angle}} = 1.81 \text{ in}^2$$
$$P = 1.81(22) = 39.8 \text{ kips}$$

Use $D = 3/16$ in
$$F_v = 0.3 \times 70 = 21 \text{ ksi}$$
Compare the capacity of the weld and angle.
$$0.70711 D(0.3F_u) \le t_a(0.4F_y)$$
$$D \le \frac{0.25(0.4 \times 36)}{0.70711 \times 21} = \frac{3.6}{14.84}$$
$$\le 0.242 \left(\simeq \frac{3.9}{16}\right)$$

Use $D = 3/16$ in.
$$L_w = \frac{39.8}{(0.70711 \times 0.1875)(21)} = 14.3 \text{ in} \quad \text{use } L_w = 14.5 \text{ in}$$

Use the weld shown in Fig. E9-2b.

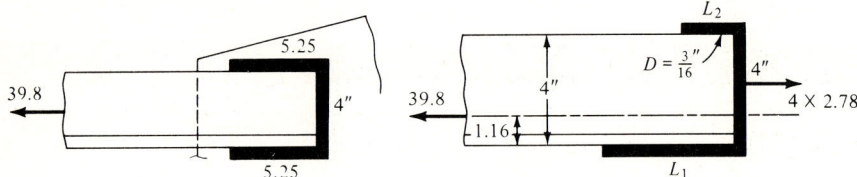

Figure E9-2b   **Figure E9-2c**

For dynamic loading it is necessary to balance the weld about the neutral axis of the angle (AISC, Sec. 1-15.3). $L_w = 14.5$ in, same as above. Referring to Fig. E9-2c and placing the weld across the back of angle to reduce the joint length, we have

$$L_1 + L_2 + 4 = 14.5$$
$$L_1 + L_2 = 10.5$$
$$L_1 = 10.5 - L_2$$

Take the sum of moments about the neutral axis of the angle so that P is eliminated; also,

$$P_w = 0.70711 \times 0.1875 \times 21 = 2.78 \text{ kips/in}$$
$$L_2(2.78)(4 - 1.16) + 4(2.78)\left(\frac{4}{2} - 1.16\right) - L_1(2.78)(1.16) = 0$$

Canceling P_w, we obtain
$$2.84L_2 - 1.16L_1 = 3.36 \quad\quad (a)$$

Substituting $L_1 = 10.5 - L_2$, we obtain
$$2.84L_2 - 1.16(10.5 - L_2) = 3.36$$
$$L_2 = 3.89 \text{ in}$$

Substituting $L_2 = 10.5 - L_1$ into Eq. (*a*) yields

$$L_1 = \frac{26.46}{4} = 6.62 \text{ in}$$

Check: $6.62 + 3.89 + 4 = 14.5$ in. ///

9-6.1 Rigid Beam–Column Connections

Rigid or AISC Type 1 connections for beams to columns can be easily made using welding. Figure 9-9 illustrates several beam-to-column connections using welding.

In most of rigid connections it is common practice to use a web weld either as a fillet or in combination with a plate or angle to carry the shear. This design proceeds as if no shear is carried by the flange welds and avoids the use of combined stress computations. With beam webs carrying the principal shear and the flanges the moment, this assumption is reasonable and works well in practice.

Figure 9-10 illustrates a line detail of a typical welded connection. Note that the use of a top plate to develop the beam moment allows a looser fitting tolerance of beam-to-column spacing. The top plate should not be welded in a zone of about $L_0 = 1.2 \times$ width of plate so that shear lag effects do not cause high local stresses and plate (or weld) failure.

Stiffeners may be required opposite the tension and/or compression flanges of the beam delivering the load to the column. The remainder of this section will

(*a*)   (*b*)

Figure 9-9 Rigid beam-to-column connections. (*a*) Beam stubs butt-welded to column. Note use of column web stiffeners opposite beam flanges. Column in supply yard for erection. (*b*) Rigid connection using seat angle welded on top (not in view). Top welds to upper moment retaining element can be seen along with welds holding framing angle to column for shear.

426 STRUCTURAL STEEL DESIGN

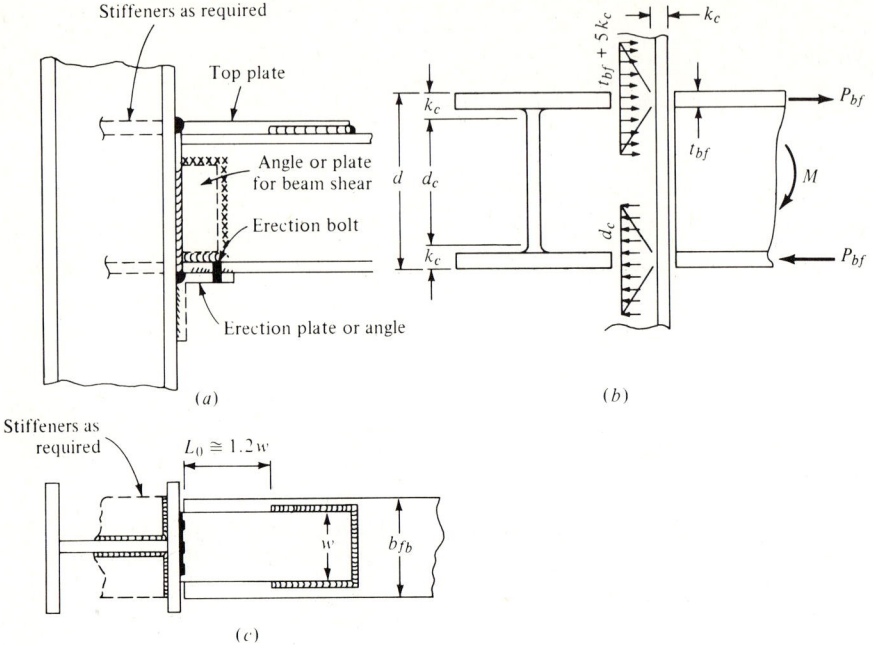

Figure 9-10 Commonly used rigid welded beam-to-column connection. (*a*) Rigid connection. (*b*) Column web stress zone; identification of terms used in Eq. (9-1). (*c*) Top plate and welding.

be concerned with determination of whether these column web and flange stiffeners are required.

Referring to Fig. 9-10*b*, the column web opposite the compression flange of the beam may be treated as an edge-loaded plate. Using Eq. (3-5) with $k_c = 1.15$ (Table 3-2, case 2, for a fixed-free edge plate), $E = 29\,000$ ksi, $\mu = 0.3$, $d_c = 0.95d$, and $F_{cr} = F_{yc} = F_y$ of column to allow for the possibility of the column using a higher F_y steel than the beam, we obtain

$$\frac{d_c}{t_w} = \frac{183}{\sqrt{F_{yc}}}$$

The earlier AISC specifications rounded 183 to 180. The current AISC requirement returns to Eq. (3-5); rearranging to obtain $P_{cr} = F_{cr}(A_e)$, where the effective area of column web in compression is taken as $d_c t_w$, we obtain by substitution of values

$$P_{cr} = \frac{33\,400}{(d_c/t_w)^2} d_c t_w \left(\frac{F_y}{36}\right)^{1/2}$$

where the additional term $(F_y/36)^{1/2}$ is used to adjust for other grades of steel. A further adjustment to the factor 33 400 to incorporate test results (see Chen and Newlin, "Column Web Strength in Beam-to-Column Connections," *Journal of Structural Division, ASCE*, ST9, September 1973) gives

$$P_{cr} = \frac{4100 t_w^3 \sqrt{F_y}}{d_c}$$

Now if we equate P_{cr} to the beam flange compressive force ($P_{bf} = A_f f_f$) and look at the dimension d_c, we obtain the current AISC equation [Eq. (1.15-2)]:

$$d_c \leq \frac{4100 t_w^3 \sqrt{F_y}}{P_{bf}} \quad \text{(fps)} \tag{9-1}$$

$$d_c \leq \frac{10.73 t_w^3 \sqrt{F_y}}{P_{bf}} \quad \text{(SI)} \tag{9-1m}$$

where P_{bf} = beam flange force (compressive) × F
 F = 5/3 for dead and live and 4/3 for dead + live + wind, kips or kN
 t_w = column web thickness, in or mm
 d_c = required column web thickness as $d - 2k$, in or mm
 F_y = yield stress of column steel, ksi or MPa

A stiffener is required opposite the compression flange if the actual column d_c as previously defined is greater than that given by the right side of Eq. (9-1).

Stiffeners (for the column web) are required opposite the beam tension flange (see Figs. 9-10b and 9-15a) as follows:

$$P_{bf} \leq F_{yc} t_w (t_{bf} + 5k_c)$$

If this inequality is satisfied as shown (right side equal or larger), no column stiffener is required. A more convenient method to determine the stiffener requirements is to equate

$$P_{st} + P_{cw} = P_{bf}$$

With $P_{st} = A_{st} F_{yst}$, we obtain

$$A_{st} = \frac{P_{bf} - F_{yc} t_w (t_{bf} + 5k_c)}{F_{yst}} \tag{9-2}$$

This is the current AISC equation for column stiffeners opposite the beam tension flange. Note that only a positive stiffener area is valid.

The column flange must be of sufficient thickness to resist the beam flange tension force without excessive deformation. A yield line analysis has given the following equation:

$$t_{cf} < 0.4 \left(\frac{P_{bf}}{F_{yc}} \right)^{1/2} \quad \text{(fps)} \tag{9-3}$$

$$t_{cf} < 12.7 \left(\frac{P_{bf}}{F_{yc}} \right)^{1/2} \quad \text{(SI)} \tag{9-3m}$$

428 STRUCTURAL STEEL DESIGN

If the column flange thickness t_{cf} is less than that given on the right sides of the equations above, flange stiffeners are required. The AISC specifications require that any column web or flange stiffeners meet the following criteria:

1. $A_{st} \geq$ Eq. (9-1) if this equation is applicable.
2. Width of both stiffeners + $t_{wc} \geq 0.67 b_{fb}$.
3. Stiffener thickness $t_s \geq t_{fb}/2$ (also, the b/t ratio must be satisfied).
4. For beams on one side of the column, the stiffeners may extend only one-half of the column depth.
5. The weld joining the stiffeners to the column web must be sized to carry the unbalanced moments on each side of the column.
6. Stiffeners for tension requirements must be welded to the column flange sufficient to carry $A_{st}F_{yst}$ (i.e., use full penetration butt welds).
7. Compression stiffeners must be welded or accurately fitted to the column flange opposite the beam flange delivering the compression load.

Example 9-3 Design the moment connection shown in Fig. E9-3a using E70 electrodes, the AISC specifications, and A-36 steel.

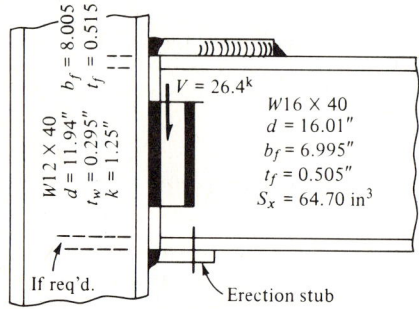

Figure E9-3a

SOLUTION Designing for full moment capacity:

$$M = F_b S_x = 24 \times 64.7 = 1552.8 \text{ in} \cdot \text{kips}$$

$$T = C = \frac{M}{d} = \frac{1552.8}{16.01} = 97 \text{ kips}$$

Use a top plate that is 8 in wide at the column weld and tapered to 3 in, as shown in Fig. E9-3b.

$$t_p = \frac{97}{22 \times 6} = 0.73 \quad \text{use 3/4-in plate}$$

Take D for fillet welds at 1/2 in.

$$P_w = 0.70711(0.5)(21) = 7.42 \text{ kips/in}$$

$$L_w = \frac{97}{7.42} = 13 \text{ in and distribute as shown}$$

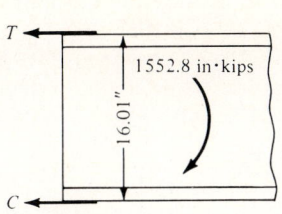
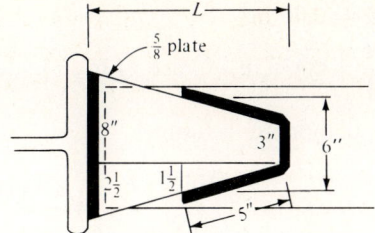

Figure E9-3b

Compute the plate length. Note that the plate will have to be long enough to allow placing 5 in of weld at $D = 0.5$ in. It will also have to have some length between the end of the beam welds and the butt weld to the column in order to develop adequate strain. By proportion:

$$\frac{L}{2.5} = \frac{5}{1.5} \qquad L = 8.33 \text{ in}$$

$$1.2w_{av} + 5 = 1.2(5.5) + 5 = 11.6 \qquad \text{use } L = 12 \text{ in}$$

For a compression of 97 kips, use a full-penetration butt weld using the erection stub for a backing plate.

For shear (Fig. E9-3c), use a plate 12 in deep on one side.

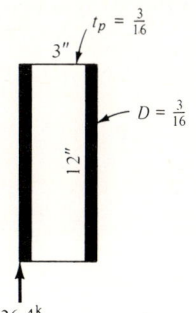

Figure E9-3c

$$t_p = \frac{26.4}{12 \times 0.4F_y} = 0.15 \qquad \text{use 3/16-in plate}$$

$$\text{Weld } D = \frac{26.4}{0.70711 \times 12 \times 21} = 0.148 \text{ in} \qquad \text{use 3/16 in}$$

Check if the column needs reinforcing (AISC Sec. 1-15.5):

$$A_{st} \geq \frac{P_{bf} - F_{yc}t_w(t_b + 5k_c)}{F_{yst}}$$

$$= \frac{97(5/3) - 36(0.295)[0.505 + 5(1.25)]}{36} = \frac{161.7 - 71.74}{36} = 2.49 \text{ in}^2$$

Therefore, a pair of stiffener plates are required opposite the tension flange. Use plates 3 in wide ($2 \times 3 > 0.67 \times 8$ O.K.).

$3 \times t \times 2 = 2.49 \text{ in}^2$

$t = \dfrac{2.49}{2(3)} = 0.415 \text{ in}$ use 7/16-in plate $\left(t = 7/16 > \dfrac{0.505}{2} \text{ O.K.}\right)$

Check the compression flange:

$$d_{c(w)} > \dfrac{4100 t_w^3 \sqrt{F_y}}{P_{bf}} = \dfrac{4100(0.295)^3 \sqrt{36}}{161.7} = 3.91$$

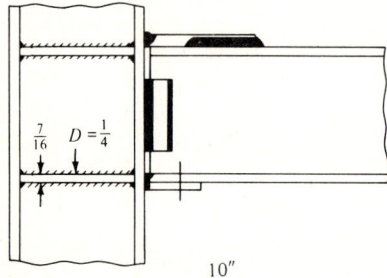

Figure E9-3d

Since d_c is greater than 3.91, compression flange stiffeners are also required. Use the same stiffeners as for a tension flange. Weld stiffeners to the column to carry 97 kips and across the ends to the flange adjacent to loads. The stiffener may be only one-half of the column depth, since moment is only on one side of column. Due to t_w of column, use a full-depth stiffener plate welded on three sides to column giving $L \simeq 10.91 + 2(3) = 16.91$ in.

$$D_{\text{weld}} = \dfrac{97}{(16.91 \times 2)(0.70711 \times 21)} = 0.193 \text{ in}$$

Use a $\tfrac{1}{4}$-in weld on both top and bottom of each stiffener plate.   ///

9-7 ECCENTRICALLY LOADED WELDED CONNECTIONS

Many framed connections produce welds that are eccentrically loaded as shown in Figs. 9-11 and 9-12. The design of these two types of connections have some similarity of design. The stress analysis for the weld involves using the concept that the maximum weld stress is

$$R_w = \left(R_{\text{moment}}^2 + R_{\text{shear}}^2\right)^{1/2}$$

where R = force or stress/unit of length (or area). The value of R_w is limited to

$$R_w \leq R_{\text{allowable}}$$

The concept of using the moment of inertia (or polar moment of inertia) of a line is used for weld stress analysis.

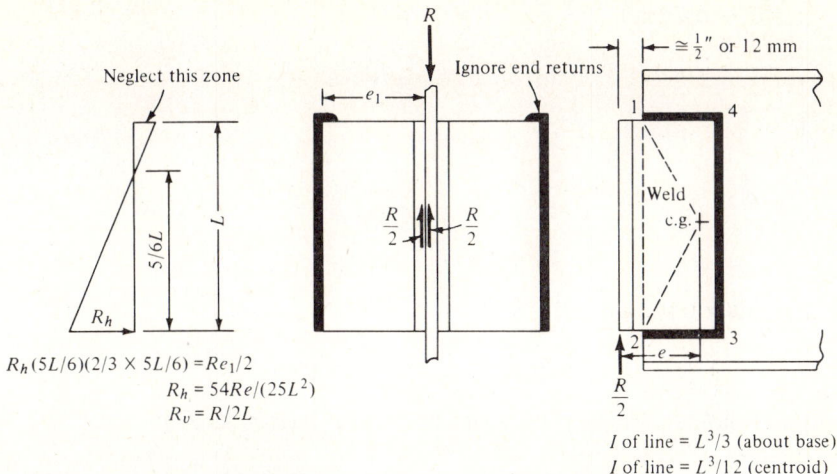

$R_h(5L/6)(2/3 \times 5L/6) = Re_1/2$
$R_h = 54Re/(25L^2)$
$R_v = R/2L$

I of line $= L^3/3$ (about base)
I of line $= L^3/12$ (centroid)

Figure 9-11 Eccentrically loaded beam framing angles (see also Fig. 9-12d).

9-7.1 Welded Beam Framing Angles

The design of beam framing angles consists in selecting an angle (or pair) to satisfy shear. The maximum weld force for the angle to beam web is obtained from computing the polar moment of inertia of the weld. Using the polar moment of inertia, it is possible to compute horizontal and vertical components of weld shear resistance to balance the moment from $R/2$ and the eccentricity shown. An additional vertical shear is computed from the force R, and the resultant weld resistance is the vector sum of these several forces. The application of these several concepts is illustrated on Fig. 9-11 and used in the following example.

Generally, 4×3 or 3×3 angles are used for web framing. If all welded construction is used with 4×3 angles, the 4-in leg should be used on the beam web to reduce the eccentricity of the weld along the o.s.l. on the column flange. If the angle is welded to the beam web and field-bolted to the column flange (or girder web), the 3-in leg should be welded to the beam and bolts in the 4-in o.s.l. so that there will be adequate erection clearance and edge distance.

Example 9-4 Design the web angle connections for a W18 × 55 as shown in Fig. E9-4a for a simply framed beam-to-column flange connection carrying a 55-kip end reaction. Use two L3 × 3 × t in × 12 in long. Use the AISC specifications, E70 electrodes, and A-36 steel.

SOLUTION For the beam web weld, locate neutral axis as

$$(12 + 2 \times 2.5)\bar{x} = 2(2.5)\frac{2.5}{2}$$

$$\bar{x} = 0.367 \text{ in}$$

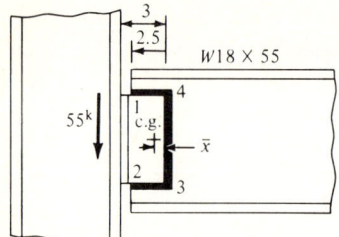

Figure E9-4a

Consider only one side, so
$$R = \frac{P}{2} = \frac{55}{2} = 27.5 \text{ kips}$$

The polar moment of inertia is (I of line about base = $L^3/3$):
$$I_p = I_x + I_y = \frac{12^3}{12} + 2(2.5)(6)^2 + 12(0.367)^2 + \frac{2(0.367^3 + 2.133^3)}{3}$$
$$= 144 + 180 + 1.62 + 6.49 = 332.1 \text{ in}^4/\text{width of weld}$$

The weld shear resistance
$$R_s = \frac{R}{L} = \frac{27.5}{17} = 1.617 \text{ kips/in}$$

Point 1 of the weld is critical by inspection (or drawing rays from c.g.)
$$M = Re = 27.5(3.0 - 0.367) = 72.4 \text{ in} \cdot \text{kips}$$
$$R_h = \frac{My}{I_p} = \frac{72.4(6)}{332.1} = 1.308 \text{ kips/in}$$
$$R_v = 0.218x = 0.218(2.5 - 0.367) = 0.465 \text{ kips/in}$$
$$R_w = ((0.465 + 1.617)^2 + 1.308^2)^{1/2} = (6.045)^{1/2} = 2.46 \text{ kips/in}$$
$$D = \frac{2.46}{0.70711 \times 0.3 \times 70} = 0.165 \text{ in} \quad \text{use 3/16-in weld}$$

Check the web shear capacity:
$$2(3/16 \times 0.70711)(21) \le 0.39(0.4 \times 36)$$

For the column flange weld (assume $1/4 \le t_f \le 3/4$: why??)
$$R_h = \frac{54Pe_1}{25L^2} \quad \text{(from Fig. 9-11, refer also to Fig E9-4b)}$$
$$R_h = \frac{54 \times 55 \times 3}{25 \times 12^2} = 2.48 \text{ kips/in}$$
$$R_v = \frac{R}{L} = \frac{27.5}{12} = 2.29 \text{ kips/in} \quad \text{(and neglecting end returns of } 2D\text{)}$$
$$R_w = (2.48^2 + 2.29^2)^{1/2} = (11.4)^{1/2} = 3.38 \text{ kips/in}$$
$$D = \frac{3.38}{0.70711 \times 21} = 0.227 \text{ in} \quad \text{use 1/4 in weld}$$

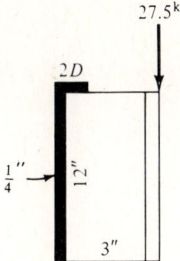

Figure E9-4b

Check the angle shear:

$$0.70711(0.25 \times 21) \le t_a(0.4F_y)$$

$$t_a = \frac{3.71}{14.4} = 0.257 \text{ in} \cong D \quad \text{O.K.}$$

Use L3 × 3 × 5/16 × 12 in long. ///

9-7.2 Welded Beam Seat Angles

The beam seat angle shown in Fig. 9-12 must be designed for bending stability and must be of sufficient thickness and leg length that an adequate fillet weld can be placed along the vertical legs to carry the shear and moment due to the eccentricity of the reaction. The angle is checked for bending at the fillet runout (k distance from tables) as in Fig. 9-12a. The allowable bending stress is taken as

$$F_b = 0.75F_y$$

since the bending section is rectangular. Computing the actual bending stress as $f_b = M/S = F_b$ and with the section modulus $S = bt/6$, the required angle thickness is computed as

$$t \ge \left(\frac{6M}{0.75F_y b}\right)^{1/2}$$

where b = seat angle width. The value of t is obtained by trial. We may initially estimate the eccentricity (assuming the k distance for any angle is $t + 0.375$ in or $t + 9$ mm) to obtain

$$e_{\text{initial}} = \frac{N}{2} + 0.5 - (t + 0.375) \quad \text{in}$$

$$e_{\text{initial}} = \frac{N}{2} + 12 - (t + 9) \quad \text{mm}$$

The actual required angle thickness t is very sensitive to the k value, so the tentative angle should always be accurately checked.

Example 9-5 Design a beam seat angle and weld for the conditions shown in Fig. E9-5a. Use $F_y = 250$ MPa and E70 electrodes. Data for a W410 ×

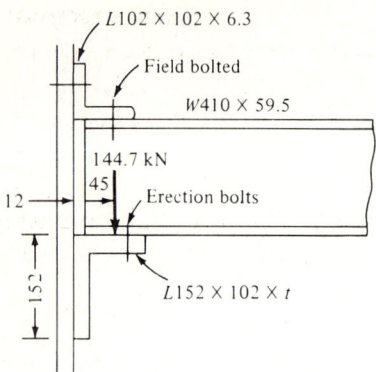

Figure E9-5a

59.5 section:

$$d = 407 \text{ mm} \qquad b_f = 178 \text{ mm}$$
$$t_f = 12.8 \text{ mm} \qquad t_w = 7.7 \text{ mm} \qquad k = 26.2 \text{ mm}$$

SOLUTION Minimum N is

$$(N + k)t_w F_b = R$$
$$N = \frac{R}{t_w \times F_b} - k$$
$$N = \frac{144.7}{0.0077 \times 0.75 \times 250} - 26.2 = 74 \text{ mm}$$

Try an L152 × 102 × 12.7 mm: $k = 25.4$ mm. Check the thickness at the toe of the fillet:

$$e = \frac{102 - 12}{2} + 12 - k = 31.6 \text{ mm} \qquad \text{(assume } R \text{ at leg/2 not } N/2\text{)}$$
$$M = Re = 144.7(0.0316) = 4.57 \text{ kN} \cdot \text{m}$$
$$F_b = 0.75 F_y = 187.5 \text{ MPa}$$

Use angle seat width = 210 mm ($b_f = 178$ mm).

$$t = \left(\frac{6M}{bF_b}\right)^{1/2} = \left(\frac{6 \times 4.57}{0.210 \times 187\,500}\right)^{1/2} = 26.3 \text{ mm} \gg 12.7 \qquad \text{N.G.}$$

Try an L152 × 102 × 22.2: $k = 34.9$ mm.
$$e = 45 + 12 - 34.9 = 22.1 \text{ mm}$$
$$t = \left(\frac{6 \times 0.1447 \times 0.0221}{0.210 \times 187.5}\right)^{1/2} = 22.1 \simeq 22.2 \text{ mm} \qquad \text{O.K.}$$

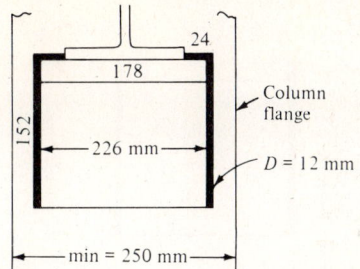

Figure E9-5b

The reader should note the necessity of conversion factors inside a square-root radical so that the dimensions are correct. We cannot just shift the decimal, since the square root of 1000 is involved. Also note that the method of computing the eccentricity of the reaction is somewhat conservative compared to using the N distance and measuring from the beam end. There is not complete agreement among designers as to just how to apply the reaction distance N to this problem.

Use a seat angle L152 × 102 × 22.2 mm. Design the welds (and neglect the $2D$ end returns, as is commonly done). The shear stress due to a load directly applied is

$$f_v = \frac{P}{2L} = \frac{144.7}{2 \times 152 \times D_e} = 0.476 D_e \text{ MPa}$$

The shear stress due to moment is

$$f_{vm} = \frac{M}{S} \qquad S = \frac{2D_e L^2}{6} \qquad M = Re$$

$$f_{vm} = \frac{6(0.1447)(45 + 12)}{2D_e \times 0.152^2 \times 1000} = \frac{1.071}{D_e} \text{ MPa}$$

$$f_w = \left[\left(\frac{1.071}{D_e}\right)^2 + \left(\frac{0.476}{D_e}\right)^2\right]^{1/2}$$

Allowable weld stress = 0.3 × 485 = 145.5 MPa

$$145.5 = \frac{1.172}{D_e}$$

$$D_e = 0.00805 \text{ m}$$

$$D = \frac{D_e}{0.70711} = \frac{8.05}{0.70711} = 11.38 \text{ mm}$$

Use weld $D = 12$ mm. Revise the seat width to $W = 178 + 4 \times 12 = 226$ mm. The column flange width must be at least 250 mm (Fig. E9-5b). ///

436 STRUCTURAL STEEL DESIGN

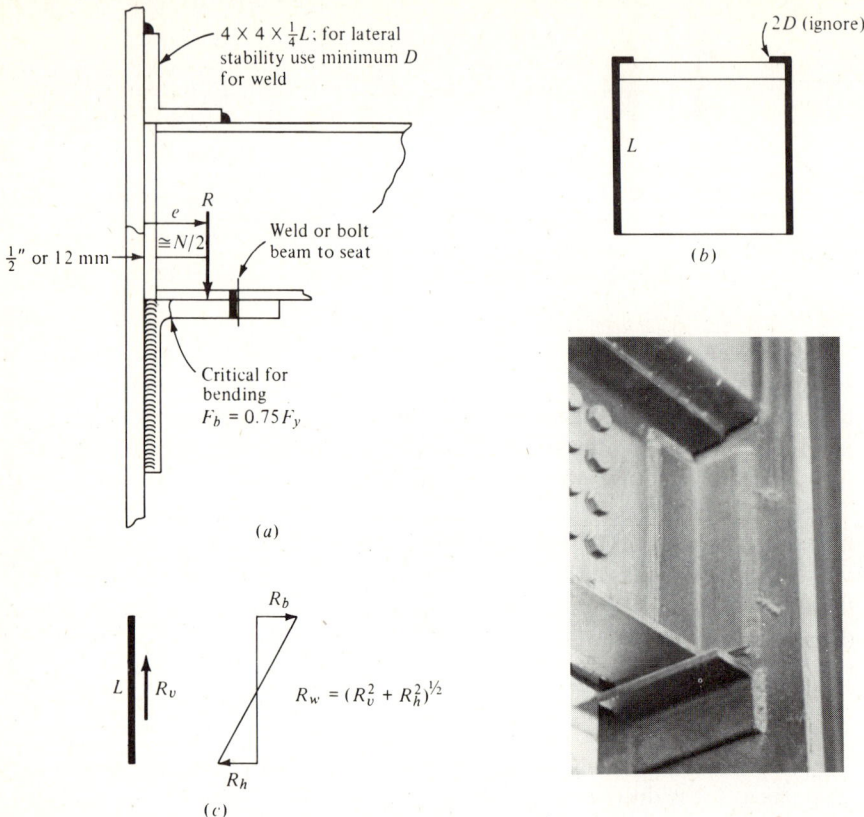

Figure 9-12 Eccentrically loaded seat angle. Top clip angle may be used for lateral stability of beam. (a) Critical location for bending in a seat angle. (b) Seat angle weld position. (c) Weld stresses covered by eccentric seat angle loading. (d) Using framing angles for shear with seat angle used for erection. Beam so heavy no erection bolts are needed. Seat angle also used to resist shear.

9-7.2 Stiffened Seat Angles

When the required seat thickness or N distance is too large for any rolled angles, one must use a stiffened seat. This can be done using a single plate as in Fig. 9-13, but more often it is a pair of angles or a WT so that erection bolts may be used to temporarily hold the beam in place on the seat. Since the seat stiffener is a plate in compression, the b/t of the unsupported edge should satisfy AISC Sec. 1-9.1. Additionally, the stiffener bending stresses should be low enough that the plate does not tear. Since the stiffener plate is supported along two edges, column stresses should not be critical if the maximum bending stresses do not exceed the allowable value. As a matter of course, the stiffener thickness should be at least as thick as the supported beam web.

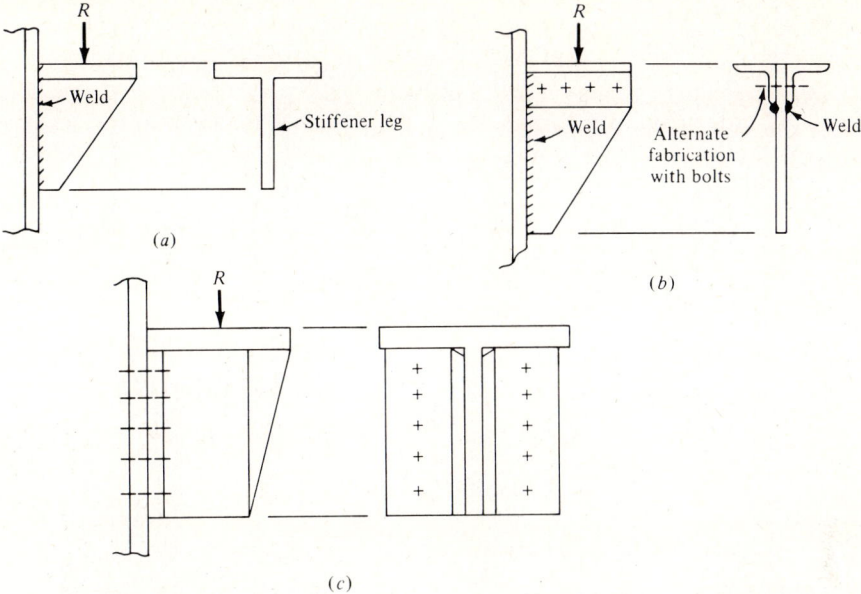

Figure 9-13 Stiffened beam seats (may use either welds or bolts [for (b) and (c)]. (a) WT for seat. (b) Two angles and plate. (c) Two angles and WT for larger reactions. Note alternate fastening using bolts (or part bolt and part weld).

Example 9-6 Redesign Example 9-5 if the beam reaction R increases to 210.5 kN.

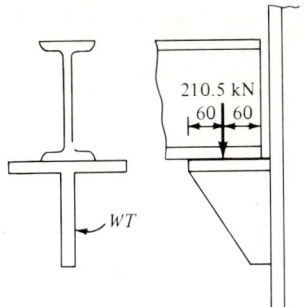

Figure E9-6a

SOLUTION Since very nearly the thickest angle section was used for Example 9-5, consider use of a stiffened beam seat made from a WT as shown in Fig. E9-6a.

$$N = \frac{R}{t_w F_b} - k$$

$$= \frac{210.5}{0.077 \times 187.5} - 26.2 = 119.6 \text{ mm}$$

Use $N = 120$ mm:

$$\frac{N}{2} = \frac{120}{2} = 60 \text{ mm}$$

Use the web of the WT at least as large as for the W140 × 59.5 (7.7 mm). Also,

$$\frac{b}{t} \le \frac{127}{\sqrt{F_y}} \quad \left(\text{or } \frac{325}{\sqrt{F_y}} \text{ in SI}\right)$$

$$\frac{b}{t} \le \frac{335}{\sqrt{250}} = 21.2$$

From an inspection of Table V-18 (SSDD), try a WT180 × 118.3.

$$b_f = 395.4 \text{ mm} \quad t_f = 30.2 \text{ mm}$$

$$t_w = 18.9 \text{ mm} \quad d = 190.2 \text{ mm}$$

Find the maximum weld size D for shear in the web of the WT with a weld on each side.

$$2(0.70711D \times 0.3 \times 485) = t_w(0.4 \times 250)$$

$$D = \frac{100 t_w}{205.8} = 9.19 \text{ mm} > 8 \text{ as minimum for } t_f$$

Check the bending stress in the WT (using only the web):

$$M = 210.5\left(\frac{N}{2} + 12\right) = \frac{210.5(60 + 12)}{1000} = 15.16 \text{ kN} \cdot \text{m}$$

$$f_b = \frac{6M}{t_w d^2} = \frac{6(15.16)}{18.9(0.1902^2)} = 132.9 \text{ MPa} < 150 \quad \text{O.K.}$$

Design the welds: try a weld as shown in Fig. E9-6b.

$$2(160 + 175 + 100)y = 2(175)\left(\frac{160}{2}\right) + 2(100)(80 + 30.2)$$

$$y = \frac{50\ 040}{870} = 57.5 \text{ mm}$$

$$I_x = \left\{2\left(\frac{160^3}{12}\right) + 2(160)(57.5)^2 + 2(175)(80 - 57.5)^2 \right.$$

$$\left. + 2(100)(80 + 30.2 - 57.5)^2\right\} D_e$$

$$= 2.4733 D_e \times 10^{-6} \text{ m}^4$$

$$S_{\text{top}} = \frac{2.4733 D_e}{c_t} = \frac{2.4733}{52.7} = 0.0693 D_e \times 10^{-3} \text{ m}^3$$

$$S_{\text{bot}} = \frac{2.4733}{137.5} = 0.0180 e D_e \times 10^{-3} \text{ m}^3$$

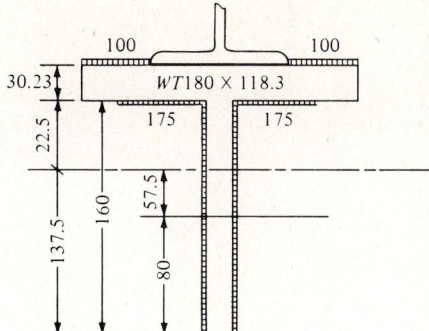

Figure E9-6b

Shear Stress due to load:
$$f_v = \frac{210.5}{870 D_e} = \frac{0.242}{D_e}$$

Shear stress due to moment:
$$f_{vm} = \frac{M}{S} = \frac{15.16}{0.180 D_e} = \frac{0.842}{D_e}$$

$$R_v = (f_v^2 + f_{vm}^2)^{1/2}$$

$$R_v = \left[\left(\frac{0.242}{D_e}\right)^2 + \left(\frac{0.842}{D_e}\right)^2\right]^{1/2} = \frac{0.876}{D_e}$$

Equating R_v to the allowable shear stress and solving for D_e:

$$\frac{0.876}{D_e} = 0.3 \times 485 = 145.5$$

$$D_e = \frac{1.006}{145.5} = 0.00602 \text{ m}$$

$$D = \frac{D_e}{0.70711} = \frac{6.02}{0.70711} = 8.51 \text{ mm} < 9.19 \quad \text{O.K.}$$

Use $D = 9$ mm. ///

9-8 WELDED COLUMN BASE PLATES

Column base plates are usually shop-welded to the column prior to shipping to the job site. Several situations using column end plates are shown in Fig. 9-14. The base plate may be either butt- or fillet-welded to the column. The decision is usually made by the steel fabricator based on economic considerations. The base plate is prepunched with bolt holes for the anchor bolts to attach the column to the foundation. Any configuration other than this (such as Fig. 9-14d and e) requires additional fabrication. The general base plate design for dimensions (width × length × thickness) is as outlined in Sec. 6-6.

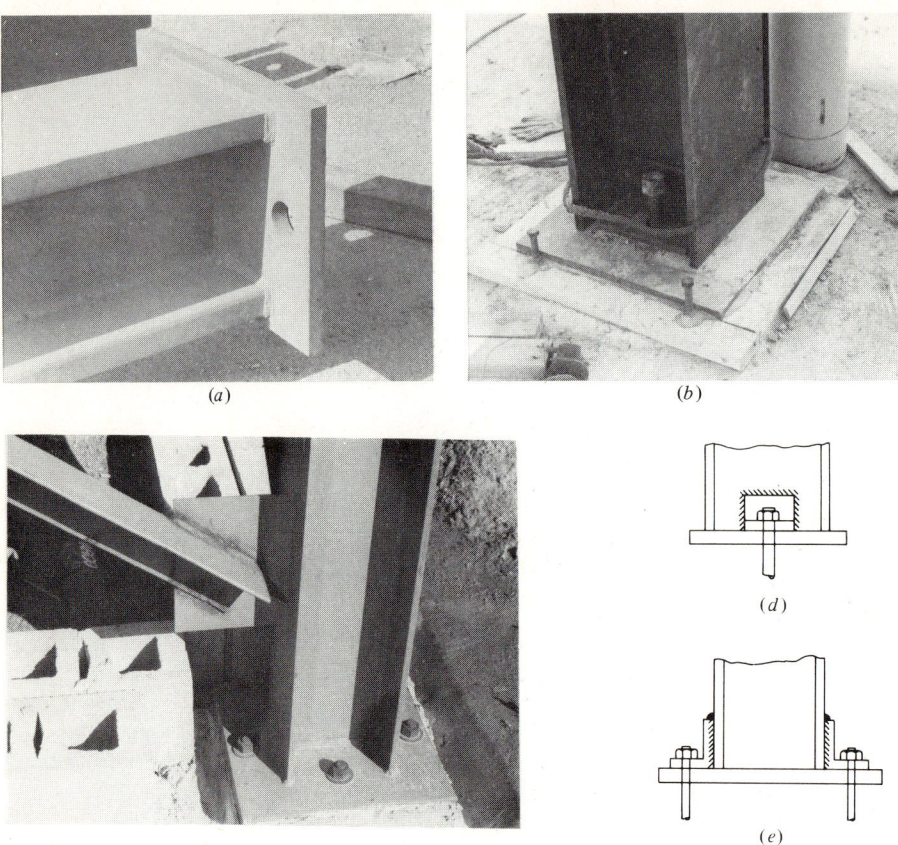

Figure 9-14 Several column base plate examples. (*a*) Column base plate on column in fabrication yard. (*b*) Column base plate using bolted angle to column. Note close fit of column on base plate. Wood frame about base plate is retaining wall for grout used to accurately align base plate to foundation. (*c*) Column base plate already grouted to grade. Note the plate welded to column for wind bracing using angle. (*d*) Line detail of photograph in (*b*) using welds. (*e*) Alternate to (*d*) for moment on column.

Example 9-7 Design the fillet welds and redesign the anchor bolts to include the axial force of the column to reduce the anchor bolt tension for the base plate designed in Example 6-9 (see Fig. E9-7a). $F_y = 250$ MPa, use E70 electrodes, and $F_w = 485$ MPa.

SOLUTION Use fillet welds for the base plate to the column so that the column end does not have to be undercut for a full-penetration weld. Design fillet welds for column stress.

$$f_b = \frac{M}{S_{x(\text{col})}} - \frac{P}{A} \quad \text{(for the tension side, which is critical with reasonable column bearing on the base plate on the compression side)}$$

$$= \frac{265}{1.84} - \frac{1670}{14.06} = 25.2 \text{ MPa} \ll 145.5 \text{ MPa of weld in shear}$$

W310 × 110.1
$S_x = 1.84$
$A = 14.06$
$t_f = 19.9$ mm
$b_f = 256$ mm

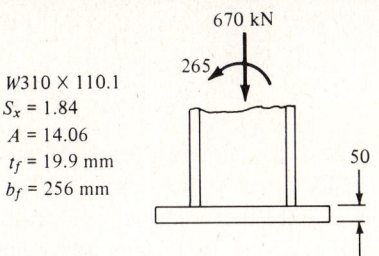
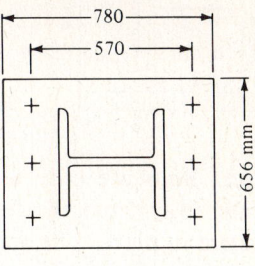

Figure E9-7a

Only a minimal fillet weld will be required for this stress. Try a minimum weld size (AISC Sec. 1-17.2) of 8 mm since the 50-mm base plate is larger than 20 mm.

The tension flange force is

$$25.2(A_f) = 25.2(19.9)(0.256) = 128.4 \text{ kN}$$

The fillet weld resistance is (using a fillet only on one side of the flange)

$$T_w = 8(0.70711)(0.3 \times 485)(0.256) = 210.7 \text{ kN} > 128.4 \quad \text{O.K.}$$

Redesign the anchor bolts using $F_y = 345$ MPa.

Let us try two 32-mm-diameter threaded ($n = 8$ threads/25 mm) anchor rods. With threads the net tension area is

$$A_e = 0.7854\left(\frac{D - 0.9743}{n}\right) \quad \text{(fps units)}$$

$$= 0.7854\left(32 - \frac{24.747}{8}\right) \times 10^{-3} = 0.6563 \times 10^{-3} \text{ m}$$

Take moments about bolt line on compression side of column.

$$M = 2(0.6 \times 345)(0.6563)(0.570) = 154.9 \text{ kN} \cdot \text{m}$$

The resulting stress obtained by superposing the resisting anchor bolt on the given stress system results in

$$q = -\frac{P}{A} \pm \frac{M_{\text{Load}}}{S} + \frac{M_{\text{bolt}}}{S}$$

$$q = \frac{1670}{0.780(656)} \pm \frac{6(265)}{0.656(0.780^2)} \pm \frac{6(154.9)}{0.656(0.780^2)}$$

Adjusting for consistent units, we obtain for the tension side of the base plate the stress

$$q = -3.264 + 3.984 - 2.329 = -1.609 \text{ MPa (compression)} \quad \text{O.K.}$$

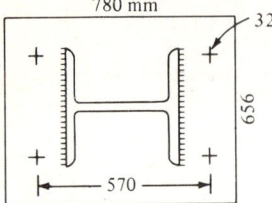

Figure E9-7b

Note: Bolts are only tightened snug. When the moment tends to pull the plate away from the foundation, the bolts develop resistance to hold the system in place. The actual bolt tension is only that necessary to produce a state of zero stress on the tension side of the base plate plus any compression residue from tightening nut during installation.

Summary: Use a 780 × 656 × 50 mm base plate. Use four 32-mm-diameter anchor bolts of F_y = 345 MPa steel. Use an 8-mm fillet weld across the column flanges as shown in Fig. E9-7b. ///

9-9 WELDED END PLATE CONNECTIONS

A common beam-to-column moment connection uses a flat plate welded to the beam and field bolted to the column. This type of connection is shown in Fig. 9-15. When a butt weld using an electrode compatible with the base metal is used, the bending stresses are limited by the allowable bending stress in the beam. When fillet welds are used, allowance must be made for the shear stresses on the weld throat, which will generally control the design. The design procedure based on recent research and proposed by Krishnamurthy ("A Fresh Look at Bolted End-Plate Behavior and Design," *AISC Engineering Journal*, Vol. 15, No. 2, 1978) is illustrated by an example.

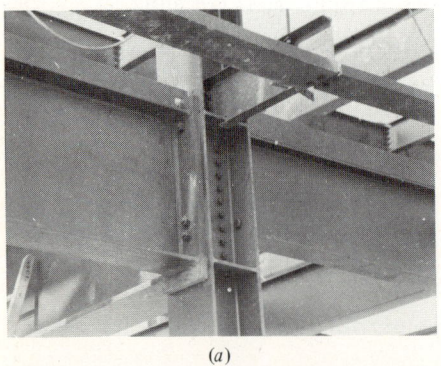
(a)

(b)

Figure 9-15 Welded beam-to-column end plates. (a) Beam end plate to column flange. Note use of column web stiffeners. (b) Use of plate welded to flange edges to provide anchorage for beam end plate.

Example 9-8 Design the interior beam-to-column connection for member 15 of Example 2-4. The computer output shows LC-1 (gravity + live) controls and the values are as shown in Fig. E9-8a. The tentative beam section used in Example 2-5 was a W410 × 59.5. Use $F_y = 250$ MPa, E70 electrodes for welding the plate, and A-325 bolts for the field erection.

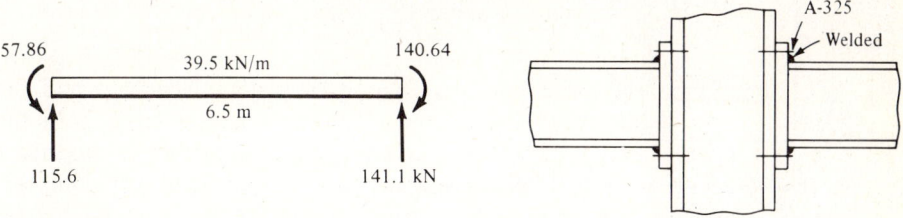

Figure E9-8a

SOLUTION Other data:

W200 × 46.1 (column): W410 × 59.5:

$d = 203$ mm $d = 407$ mm

$b_f = 203$ mm $b_f = 178$ mm

$t_f = 11.0$ mm $t_f = 12.8$ mm

$t_w = 7.2$ mm $t_w = 7.7$ mm

Step 1. The flange force (using largest moment at right end) is

$$F_f = \frac{M}{d - t_f} = \frac{140.64(1000)}{407 - 12.8} = 357 \text{ kN}$$

Step 2. The approximate bolt area/row using the equation given in the paper cited and w/8 A-325 × 22-mm-diameter bolts.

$$f_s = \frac{V}{N} = \frac{141.8}{8} = 17.7 \text{ kN/bolt} \rightarrow 46.6 \text{ MPa}$$

$$a = \frac{0.5 F_f}{F_{tb}}$$

$F_{tb} = 380 - 1.4(46.6)$ bearing connection

$\quad = 315$ MPa

$$a = \frac{0.5(357)}{315} = 0.56667 \times 10^{-3} \text{ m}^2$$

Using nominal diameter and two 22-mm-diameter bolts in either of the upper two rows, the area furnished is

$A_f = 4(0.7854)(0.022^2)(10^3) = 0.7603 \times 10^{-3}$ m$^2 \leq 0.56667$ O.K.

The edge distance $d_e \geq 1.75 \times 20 = 35$ mm. Use 55 mm, as in Fig. E9-8b, and use a 90-mm standard (approximately) gage on the column.

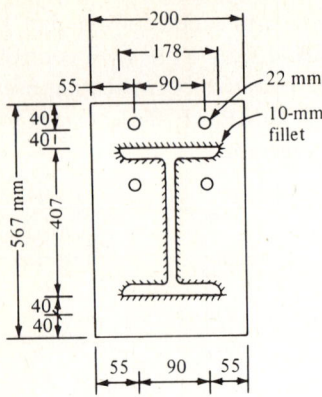

Figure E9-8b

Step 3. Design fillet welds to carry the flange force F_f.

$$0.178(2D \times 0.70711)(0.3 \times 485) = 357 \text{ kN}$$

$$D = 9.74 \quad \text{use 10 mm}$$

Place the fillet weld on the top and bottom of both flanges. Use a 10-mm web fillet to carry a shear of 141.1 kN. The effective weld capacity is

$$P_{\text{weld}} = (407 - 2t_f)(0.70711 \times 10 \times 145.5) = 392 \text{ kN} > 141.1$$

$$P_{\text{web}} = (407 - 2t_f)(0.0077)(0.4 \times 250) = 294 \text{ kN} > 141.1 \quad \text{O.K.} \leftarrow \text{controls}$$

Step 4. The effective bolt distance based on the actual distance P_f from the beam flange and the bolt diameter $d_b = 20$ mm and effective weld D:

$$P_e = P_f - \tfrac{1}{4}d_b - 0.70711D$$

$$= 40 - 0.25(20) - 0.70711(10) = 27.9 \text{ mm}$$

Step 5. The effective moment of the plate taken as a WT section is

$$M = \frac{F_f \times P_e}{4} = \frac{357(0.0279)}{4} = 2.49 \text{ kN} \cdot \text{m}$$

Step 6. A material coefficient based on the yield stresses of the plate, bolt (F_{ab}), and allowable bolt stress F_{tb} is computed as

$$C_a = 1.29 \left(\frac{F_y}{F_{ub}}\right)^{0.4} \left(\frac{F_{tb}}{0.75 F_y}\right)^{0.5}$$

$F_{ub} = 825$ MPa (bottom of Table 8-1 for bolt diameter < 25 mm)

$$C_a = 1.29 \left(\frac{250}{825}\right)^{0.4} \left(\frac{315}{0.75 \times 250}\right)^{0.5} = 1.04$$

Step 7. Compute a width correction factor for the plate, based on the beam flange width b_f and plate width b_p:

$$C_b = \left(\frac{b_f}{b_p}\right)^{0.5} = \left(\frac{178}{200}\right)^{0.5} = 0.943$$

Step 8. Find the ratio A_f/A_w of the beam:

$A_f = b_f \times t_f = 0.178(12.8) = 2.2784 \times 10^{-3}$ m²

$A_w = (d - 2t_f)t_w = [407 - 2(12.8)](0.0077) = 2.9368 \times 10^{-3}$ m²

Step 9. Find ratio $P_e/d_b = 27.9/20 = 1.395$.

Step 10. The moment modification factor α_m is

$$\alpha_m = C_a C_b \left(\frac{A_f}{A_w}\right)^{0.32} \left(\frac{P_e}{d_b}\right)^{0.25}$$

$$= 1.04(0.943)\left(\frac{2.278}{2.937}\right)^{0.32}(1.395)^{0.25}$$

$$= 0.982$$

Step 11. The design moment $M_d = Mx\alpha_m = 2.49 \times 0.982 = 2.45$ kN m. The end plate thickness t_p is (as for other rectangular bending plates):

$$t_p = \left(\frac{6M_d}{b \times 0.75 F_y}\right)^{1/2}$$

$$= \left(\frac{6 \times 0.00245}{0.200 \times 187.5}\right)^{1/2} = 0.0198 \text{ m}$$

Use end plate thickness $t_p = 20$ mm.

Step 12. Check the maximum effective plate width (must not be less than the width furnished):

$b_e = b_f + 2D + t_p$

$= 178 + 2(10) + 20 = 218 > 200$ mm furnished   O.K.

Note: If $b_e < b_p$, must repeat steps 7 through 12 using $b_p = b_e$.

Step 13. Check plate shear stress.

$$f_v = \frac{F_f}{2bt_p}$$

$$= \frac{357}{2 \times 0.200 \times 20} = 44.6 \text{ MPa} \ll 0.4 \times 250 \quad \text{O.K.}$$

Design summary: end plate 200 × 338 × 20 mm thick. Use eight 20-mm A-325 bolts to the column flange. Use a 10-mm fillet weld for beam to end plate.   ///

9-10 WELDED CORNER CONNECTIONS

Corner connections may be straight or haunched. Only straight corner connections will be considered here. The design of straight corner connections is generally controlled by shear considerations. Consider the corner connection shown in Fig. 9-16a. If the shear stress is assumed uniform over the beam web in

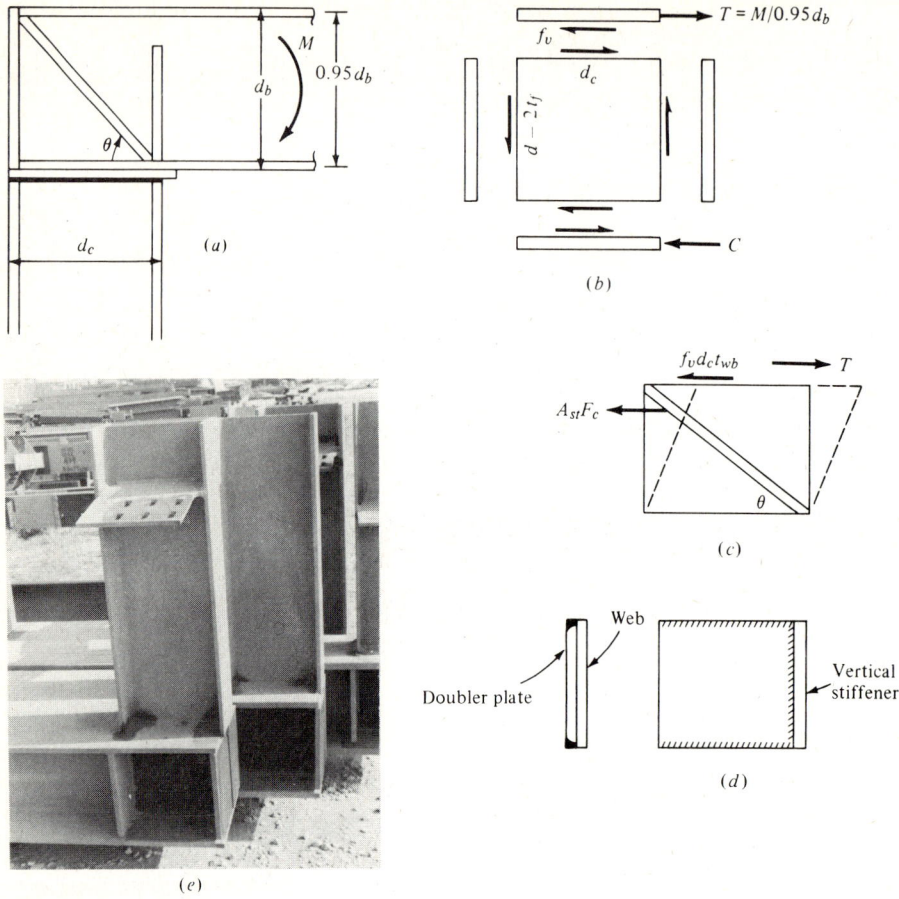

Figure 9-16 Straight column corner connections. (a) (b) (c) (d) (e) Column corner connections with beam stubs butt-welded to column.

a length d_c (Fig. 19-16b), we have

$$f_v d_c t_{wb} = T = C = \frac{M}{0.95 d_b}$$

Taking $f_v = F_v = 0.4 F_y$, we obtain, with slight rounding:

$$t_{wb} \geq \frac{32 M}{F_y d_b d_c} \quad \text{(in)} \qquad \text{for fps} \qquad (9\text{-}4)$$

$$t_{bw} \geq \frac{2.5 \times 10^6 M}{F_y d_b d_c} \quad \text{(mm)} \qquad \text{for SI} \qquad (9\text{-}4\text{m})$$

where d_c, d_b = tabulated depth of column and beam, respectively, in or mm
M = joint moment, ft · kips or kN · m

In plastic design, which is recommended by many designers, use of M_p and $f_v = F_{vy} = F_y/\sqrt{3}$ and $0.9db$ gives

$$t_{wb} \geq \frac{23 M_p}{F_y d_c d_b} \quad \text{(in)} \tag{9-5}$$

$$t_{wb} > \frac{1.8 \times 10^6 M_p}{F_y d_c d_b} \quad \text{(mm)} \tag{9-5m}$$

where F_y, d_c, and d_b are as previously defined. The moment $M_p = F_y Z$ or one may use $M_p = M \times LF$, where M = working load moment as in elastic design and LF = plastic design load factor. A value of $LF = 1.7$ is recommended. If the web thickness t_{wb} is less than given by Eq. (9-4) or (9-5), a web stiffener is required. This may be a web doubler plate (Fig. 9-16d), but a diagonal stiffener is generally preferred since it is easier to fabricate. Note that welding on all four sides of the doubler plate may result in a premature weld or plate failure due to the excessive restraint of the shrinkage stresses. A diagonal stiffener is proportioned to take the horizontal component of the web shear deficiency (Fig. 9-16c). Thus in terms of actual stresses, we have

$$T - A_{st} f_c \cos \theta = f_v d_c t_{wb}$$

Substituting allowable stresses, we have

$$A_{st} = \frac{1}{F_c \cos \theta} \left(\frac{M}{0.95 d_b} - F_v d_c t_{wb} \right) \tag{9-6}$$

The value of F_c (compression) may be taken as either $0.6 F_y$ or, preferably, $F_y/1.7$ as somewhat more conservative, since the stiffener is in compression Using plastic design (and with $F_{ys} = F_y$ of stiffener steel), we obtain

$$A_{st} = \frac{1}{F_{ys} \cos \theta} \left(\frac{M_p}{0.90 d_b} - \frac{F_y d_c t_{wb}}{\sqrt{3}} \right) \tag{9-7}$$

Consistent units should be used in both Eqs. (9-6) and (9-7). Note that it is possible to either extend the column to build the corner or, alternatively, extend the beam over the column (as shown in Fig. 9-16a). The appropriate values for d_c, d_b, and t_{wb} should be used consistent with the derivations for the given equations and the corner joint assembly.

Example 9-9 Design the corner assembly shown in Fig. E9-9a. Use E70 electrodes, the AISC specifications, and $F_y = 250$ MPa for all steel.

448 STRUCTURAL STEEL DESIGN

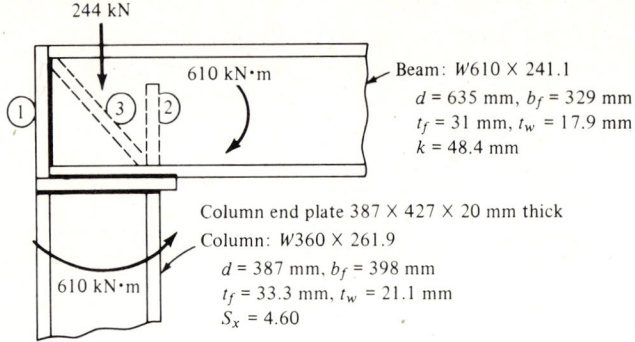

Figure E9-9a

SOLUTION
Step 1. Design vertical stiffeners 1 and 2.
For stiffener 1, opposite the tension flange of the column:

$$f_b = \frac{M}{S_x} = \frac{610}{4.60} = 132.6 \text{ MPa}$$

$$A_{fc} = b_f t_f = 0.398 \times 33.3 = 13.2534 \times 10^{-3} \text{ m}^2$$

$$P_{bf} = \tfrac{5}{3} f_b A_f = \tfrac{5}{3}(132.6 \times 13.2534) = 2929 \text{ kN}$$

$$A_{st} \geq \frac{P_{bf} - F_{yc} t_{cw}(t_{bf} = 5k)}{F_{yst}}$$

Use the beam for the "column" dimension and 2.5 instead of 5k, since the column tension flange is at the end and not centered on the 5k zone.

$$A_{st} = \frac{2929 - 250 \times 0.0179(33.3 - 2.5 \times 48.4)}{250}$$
$$= 10.146 \times 10^{-3} \text{ m}^2$$

Use end plate:

$$b_p = 398 \text{ mm (width same as column)}$$
$$t_p = 28 \text{ mm}$$
$$A_{st} = 0.028 \times 398 = 11.14 \times 10^{-3} \text{ m}^2 \quad \text{O.K.}$$

For stiffener 2, opposite the compression flange of the column:

$$d_{b(\text{web})} = d - 2k = 635 - 2(48.4) = 538.2 \text{ mm}$$

A stiffener is required if $b_b > d_{bw}$:

$$d_{bw} > \frac{10.73 t_{bw}^3 \sqrt{F_y}}{P_{bf}}$$

$$> \frac{10.73(17.9)^3 \sqrt{250}}{2929} = 332.2 \text{ mm}$$

Since d_{bw} furnished = 538.2 > 332.2 required, a web stiffener is required.

The stiffener (a pair with one on each side of the web opposite the column compression flange) only has to be one-half of beam depth, since the load is only on one side.

$$t_{stiff} > \frac{t_{cf}}{2} = \frac{33.3}{2} = 17 \text{ mm} \qquad \text{use } t_{st} = 20 \text{ mm}$$

$$b_{stiff} = \frac{(0.67 \times 329) + 17.9}{2} = 119.2 \text{ mm} \qquad \text{use } b_{st} = 120 \text{ mm}$$

Step 2. Check if a diagonal stiffener is required.
Use plastic design [Eq. (9-7)]:

$$M_p = 1.7 \times M = 1.7 \times 610 = 1037 \text{ kN} \cdot \text{m}$$

$$t_w \geq \frac{1.8 \times 10^6 M_p}{F_y d_c d_b} = \frac{1.8 \times 10^6 \times 1037}{250 \times 387 \times 635} = 30.38 \text{ mm}$$

Since the furnished $t_{wb} = 17.9 < 31.6$ mm required, it is necessary to use a diagonal stiffener.

$$\theta \simeq \tan^{-1} \frac{635}{387} = 58.64^0 \qquad \cos\theta = 0.52042$$

$$A_{st} = \frac{1}{F_{ys} \cos\theta} \left(\frac{M_p}{0.90 d_b} - \frac{F_y d_c t_w}{\sqrt{3}} \right)$$

$$= \frac{1}{250(0.52042)} \left[\frac{1037}{0.90 \times 0.635} - \frac{250(387)(0.0179)}{\sqrt{3}} \right]$$

$$= 6.2615 \times 10^{-3} \text{ m}^2$$

Use two plates 20 mm × 160 mm wide:

$$A_p = 6.4 \times 10^{-3} \text{ m}^2 \qquad \text{O.K.}$$

Step 3. Design the welds.
The weld for plate 1:

$$P_{bf} = 2929 \text{ kN} \qquad F_w = 0.3 F_u \times 1.7 = 247 \text{ MPa (plastic design)}$$

Using inside fillet welds (see Fig. E9-9b), we obtain $D_{min} = 8$ mm, since $t_p = 25 > 20$ mm. Check the effective D of the beam web for shear:

$$2 \times D_e \times 0.70711 \times 247 = \frac{17.9 \times 250}{\sqrt{3}}$$

$$D_e = 7.4 \text{ mm}$$

So only 7.4 mm of the weld on the beam web is effective, since the shear of the base metal controls. Use D along the inside flange as required and an

8-mm weld on the beam web.

$$L_{web} = [635 - 2(31)]2 = 1146 \text{ mm}$$

$$L_{flange} = (329 - 17.9)2 = 622.2 \text{ mm}$$

$$D \times 0.70711 \times 0.622 \times 247 + 1146 \times 0.0074 \times 0.70711 \times 247 = 2929 \text{ kN}$$

$$D = \frac{1447.8}{108.6} = 13.3 \text{ mm}$$

Use flange $D = 14$ mm.

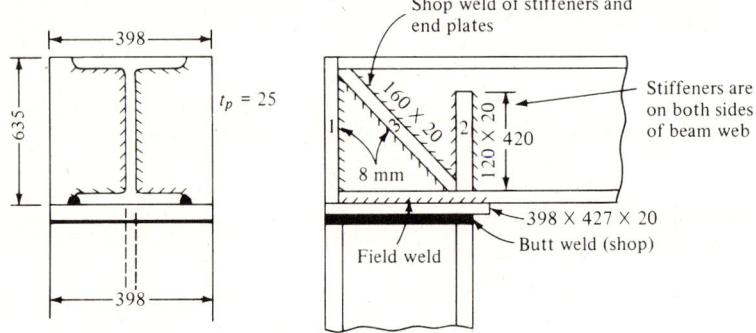

Figure E9-9b

The weld for plate 2: with maximum effective weld $D = 7.40$ mm in the web and $D = 8$ mm required for 20-mm stiffener plates and four lengths of fillet weld (refer to Fig. E9-9b), we have that the part of P_{bf} carried by the beam web is

$$P_{web} = F_y t_w (t_{cf} + 2t_{p(\text{col. base})} + 2k_b)$$

$$= 250(0.0179)[33.3 + 2(20 + 48.4)] = 761.2 \text{ kN}$$

$$P_{weld} = P_{bf} - P_{web} = 2929 - 761.2 = 2167.8 \text{ kN}$$

$$4L \times 0.0074 \times 0.70711 \times 247 = 2167.8 \text{ kN}$$

$$L = 419.3 \text{ mm}$$

$$d - 2t_f = 635 - 2(31) = 573 \text{ mm} > 419 \quad \text{O.K. (can get this in)}$$

Weld for diagonal stiffener:

$$P_{stiff} = A_{st} F_y = 0.12(20)(250) = 600 \text{ kN/stiffener}$$

Use 8-mm welds, but only 7.4 mm is effective:

$$L_{stiff} = \frac{d - 2t_f}{\cos \theta} = \frac{387 - 2(33.3)}{0.52042} = 615.7 \text{ mm}$$

Assuming that 600 mm is effective, check P:
$$P_{weld} = 2(600)(0.0074)(0.70711)(247) = 1550.9 \text{ kN} \gg 600 \quad \text{O.K.}$$

Design the weld for the column end plate: use a full-penetration butt weld. The design summary is shown in Fig. E9-9b. ///

9-11 FILLET WELD DESIGN USING LRFD

Fillet weld design using LRFD is similar to load resistance factor. Design for bolted connections. It is necessary to compute the factored load
$$R_n = 1.1(1.1D + 1.4L)$$
or the alternate factored loads using wind, earthquake, and so on. The value of R_n is compared to the allowable weld shear resistance:
$$R_n = \phi 0.6 F_{EXX} A_w \qquad (9\text{-}8)$$
where F_{EXX} = electrode F_u (E60, E70, E80, etc.)
A_w = throat area = $0.70711 D$
ϕ = 0.80 [currently (1978)]

The shear resistance of the base metal is computed as for beam webs:
$$F_{vu} = \frac{\phi F_y}{\sqrt{3}} \qquad (\phi = 0.86)$$

Since the electrode should always be matched to the base metal, butt-welded joint design uses the LRFD procedures for the type of member using the butt joint.

PROBLEMS

For all problems, be sure to summarize your connection using a neat sketch.

9-1 Design the weld for the joint shown in Fig. P9-1 using "matched" electrodes, A-36 steel, and the AISC specifications, if $t_1 = 5/8$ in, $t_2 = 1/2$ in, $w_1 = 6$ in, and $w_2 = 9$ in. Determine P based on the maximum allowable for the base metal.
  Answer: $P = 81$ kips, $L_w = 12$ in, using $D = 9/16$.

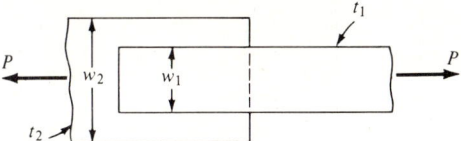

Figure P9-1

9-2 Do Prob. 9-1 if steel with $F_y = 50$ ksi is used. Plate dimensions are the same. Try to make the joint length a minimum.
  Answer: $P = 112.5$ kips.

9-3 Design the weld for the joint shown in Fig. P9-1 using "matched" electrodes, $F_y = 345$ MPa, and AASHTO allowable weld stresses. Take $t_1 = 20$ mm, $t_2 = 15$ mm, $w_1 = 175$ mm, and $w_2 = 250$ mm. Use P = AASHTO maximum allowable for base metal.

9-4 Do Prob. 9-1 using LRFD and assume that the D/L ratio $= 1.0$.
   Answer: $P_u = 118.8$ kips, $L_w = 19$ in, $D = 5/16$ in.

9-5 Find the required size of fillet welds to attach the W610 × 241.1 beam to the 387 × 427 × 20 mm column end plate in Example 9-9.

9-6 Design the welds and gusset plate for a 6 × 4 × 5/8 in angle for 50 percent of the angle capacity in compression for $L = 8$ ft. Use the AISC specifications, $F_y = 50$ ksi, and E70 electrodes, and a static load.

9-7 Do Prob. 9-6 if the load is dynamic.
   Answer: $P = 35.5$ kips, $L_1 = 2.71$, $D = 5/16$, no end weld.

9-8 Do Prob. 9-6 for a dynamic load and $D/L = 0.75$, using LRFD.

9-9 Design the welds and gusset plate for a pair of L152 × 102 × 19 mm for a load of 75 percent of the effective angle capacity in static tension. Use $F_y = 250$ MPa, E70 electrodes, and the AISC specifications. Keep the joint length to a minimum.

9-10 Do Prob. 9-9 for a dynamic load.

9-11 Design a welded framed simple beam connection to carry a beam shear of 59.8 kips from a W18 × 50 beam to a W12 × 53 column. Use A-36 steel, E70 electrodes, the AISC specifications, and 3 × 3 × t angles with a length of 12 in.
   Answer: $t = 5/16$ in; $D_{web} = 3/16$; $D_{col} = 1/4$.

9-12 Design the framing angles and weld for framing the stringers into the cross (floor) beam of a bridge truss. Use shop welding of framing angles to stringer and field bolting to web of cross beam. The stringers are W18 × 50 and assume that the reaction includes a 4-kip dead load and a 35-kip live load with impact. Use A-36 steel, E70 electrodes a 4 × 3 × t × 14 in angle, and the applicable AASHTO specifications.
   Answer: $t_a = 5/16$; $D = 1/4$ in.

9-13 Design the welded end connections for a W18 × 55 beam to develop 70 percent of the moment capacity based on $F_b = 24$ ksi. Use the AISC specifications, E70 electrodes, and A-36 steel. The column is a W12 × 65, so a moment check for stiffeners is required for the moment being developed on only one side.

9-14 Design a framed beam connection to carry a beam shear of 152 kN from a W460 × 81.8 beam. Use a pair of angles 76 × 76 × t with $L = 200$ mm. Use the AISC specifications, $F_y = 250$ MPa, and E70 electrodes.

9-15 Design the weld and plate for the loading shown in Fig. P9-15. Use A-36 steel and AISC.
   Answer: $D = 3/16$ in; $t = 1/2$ in.

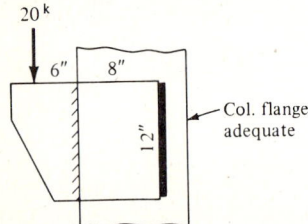

Figure P9-15

9-16 Design a seat angle (and top clip angle for lateral stability) to support A W16 × 40 on a 25-ft simple span carrying a uniform load of 1.54 kips/ft, including the beam weight. Be sure to make a neat final design sketch. Use E70 electrodes, A-36 steel, and the AISC specifications. The beam frames to a W12 × 53 column.

9-17 Design a stiffened beam seat using a WT for a W530 × 196.4 beam carrying a uniform load of 91.9 kN/m, including beam weight, on a 8.75-m simple span. Use a top clip angle for lateral stability, E70 electrodes, $F_y = 250$ MPa steel, and the AISC specifications.

9-18 Design a welded moment connection for two W16 × 40 beams which frame into the opposite flanges of a W12 × 53 column. Use top and bottom welded plates to develop the moment and a web fillet weld for shear. The beam moment is 98.9 ft · kips (both sides) and a shear of 34.5 kips. Use the AISC specifications, A-36 steel, and E60 electrodes.

9-19 Two W410 × 84.8 beams frame into a W530 × 123.5 girder as in Fig. P9-19 in a moment connection. Design the connection using E60 electrodes, $F_y = 250$ MPa steel, and the AISC specifications.

Answer: Try a top plate 24×150 mm $\times L/2 = 260$ mm and $D = 12$ mm.

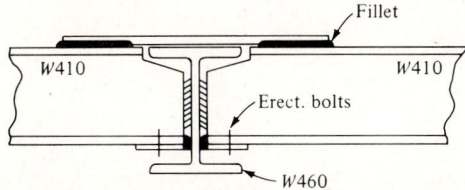

Figure P9-19

9-20 Redesign the corner connection of Example 9-9 by using an end plate welded to the beam for framing to column. Use the design data given in Example 9-9. Be sure to show a final design sketch of the solution.

9-21 Design the end plates for a bolted connection for the full moment capacity of a W16 × 50 to be bolted to the flange of a W14 × 211. The beam shear due to uniform load is 48.55 kips, including the beam weight. Use the AISC specifications, A-36 steel, A-325 bolts, and E70 electrodes. Show the final design sketch.

Figure X-1 Turbine area of power station. Note the very long continuous plate girder being used for a crane runway. Columns are stepped for the crane girder.

CHAPTER
TEN

PLATE GIRDERS

10-1 GENERAL

Plate girders, including rolled beams with cover plates **termed built-up beams**, are used when the loading and span combination is such that a standard rolled section is inadequate.

Built-up rolled sections may be used where the overall depth is limited in buildings and are sometimes used for highway and railroad bridges. In building applications where limited depth is encountered, an alternative to a built-up section is to use two lighter rolled shapes in parallel. Even where the total weight is appreciably different, the additional fabrication costs (cutting the additional plates to proper width and welding to the flanges) may produce a higher overall cost for the built-up section.

Plate girders are fabricated structural shapes formed by welding (although on occasion riveting or bolting may be used) flange plates to a web plate to form a shape intermediate between the S and W rolled shapes. Plate girders are widely used for bridges and are not uncommon for buildings. Several girders are shown in Fig. 10-1. The crane runway girder shown in the industrial building of Fig 10-1a is by far the most common use of girders in building construction.

Plate girders are commonly fabricated by welding two flange plates to a web plate as illustrated in line details in Fig. 10-2. The most common materials are A-36 steel and either E60 or E70 electrodes. In the era of cheaper labor and when the state of the art of welding had not progressed very far, girders were fabricated using angles riveted to a web plate and using one or more cover plates riveted to the flange angles, as illustrated in Fig. 10-2b. Box girders, consisting of

(a)

Enlargement of crane girder not yet erected shown in right foreground

(b)

(c)

Figure 10-1 Typical installations using plate girders. (*a*) Plate girder used in industrial building. Plate girder made of rolled W shape with a channel placed on top as shown. (*b*) Deck stringer girder bridges in interstate interchange. Note use of both simple and continuous spans. (*c*) Plate girders used for railroad bridge. Here open-deck girders are used in several spans on right and abutting an open deck through girder for the left span. The use of deck girders as shown allows the use of shorter piers. "Open-deck" refers to the cross-ties being placed directly on the girder or stringer flanges.

parallel web plates spaced some distance apart, are sometimes used for bridges but will not be considered here.

Vertical stiffeners (Fig. 10-2*c*; also Fig. 10-1*b*) are generally required when the web plate is very thin, the h/t_w ratio is large, and/or the web shear stress is relatively high. The stiffeners effectively reduce the a/b ratio discussed in Chap. 3 (Fig. 3-6 and Table 3-1) and increase the shear capacity with respect to shear buckling of the web. Stiffeners are nearly always required by specification under concentrated loads and at the reactions of plate girders, regardless of the shear intensity. Stiffener requirements along the span (intermediate transverse stiffeners) depend on the shear stress and the web proportions.

The stiffener spacing a depends on the h/t_w ratio, where h is the *clear* web distance between flanges. This distance is flange plate to flange plate for a welded girder consisting of three plates. The clear distance for girders using flange angles is as shown in Fig. 10-2*b*. It necessarily follows that an economic

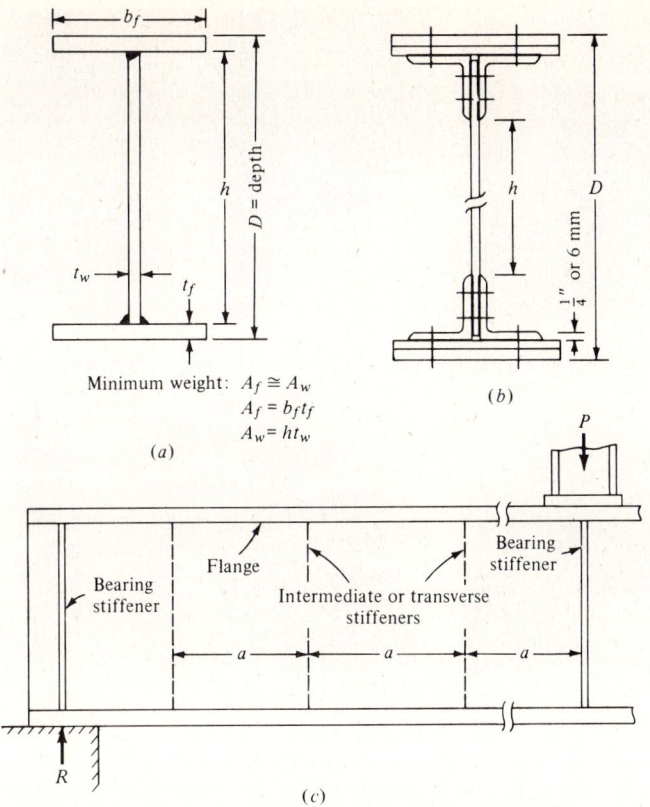

Figure 10-2 Line and other details of plate girders. Note definition of depth D and clear distance h. The ratio h/t_w is a significant parameter in plate girder design. (*a*) Welded plate girder. (*b*) Riveted or bolted plate girder. (*c*) Girder elevation (side view) illustrating other elements in design.

analysis may be made of using flange angles welded with the reduction, or elimination of stiffeners versus rectangular flange plates and larger clear web distance h.

Plate girders used in building construction may be economical for spans up to about 300 ft (100 m). Girder depths may range up to 15 ft (5 m) or more. Depths of 4 to 8 ft are common. In terms of the depth/span (D/L) ratio, the range is from $\frac{1}{10}$ to $\frac{1}{25}$, with most values being on the order of $\frac{1}{12}$ to $\frac{1}{15}$. Larger building spans may require use of a truss.

Plate girders are used for highway bridges when the span exceeds 50 to 60 ft (15 to 18 m) and are generally more economical than trusses for spans up to 300 ft (100 m) or more. Rolled beams are generally more economical for bridge spans less than 50 to 60 ft and are used in a deck stringer configuration. Although girders are generally more economical than trusses, the latter are still used in many situations for esthetics, particularly where additional lanes are required and the existing facility is a truss.

458 STRUCTURAL STEEL DESIGN

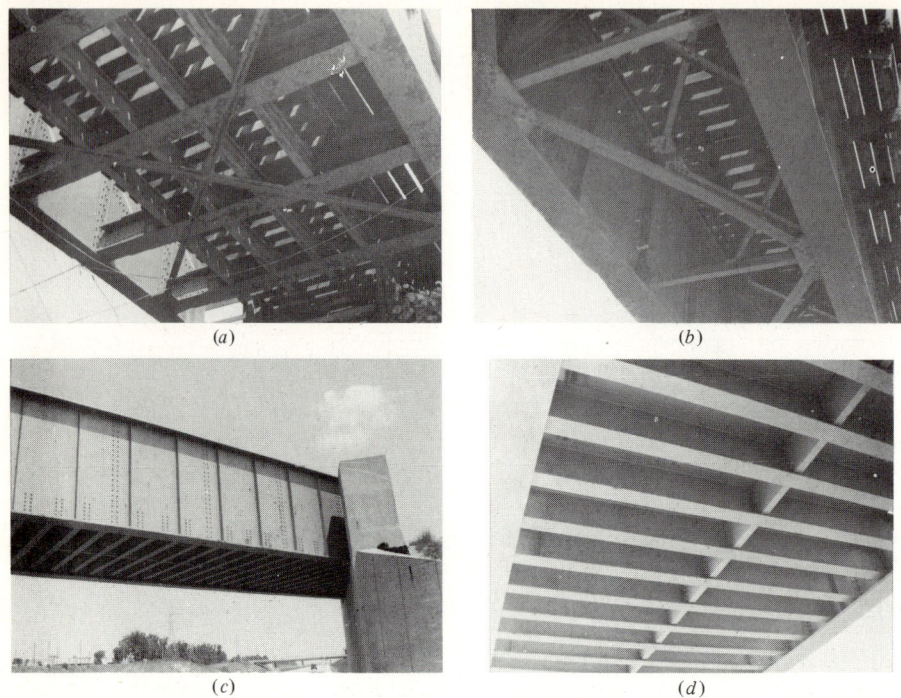

Figure 10-3 Bridge details. (*a*) View from underside of older through girder open-deck bridge using a system of cross beams and short longitudinal stringers on which the cross-ties directly rest. This is the left portion of bridge in Fig. 10-1*c*. (*b*) View from underside of older open-deck girder bridge. Note system of girder bracing both vertical and horizontal (in plane of bottom flange). This is the girder of the right portion of bridge in Fig. 10-1*c*. (*c*) Modern through-ballasted deck girder bridge. (*d*) View of underside of bridge shown in (*c*), showing the metal deck plate holding ballast and system of cross-beams to carry load to main girders. Note the single line of longitudinal diaphragms for bracing the cross-beams.

Bridges (both highway and railroad) using plate girders may be termed:

1. *Deck girder bridges*, where the bridge deck or load is placed directly on the top flange. In railroad terminology this may be an *open-deck* girder bridge, where the cross-ties rest directly on the flanges (Fig. 10-3*a* and *b*). In highway construction this type of bridge carries the floor system of stringers and cross-beams between the main load-carrying girders.
2. *Stringer deck girder bridges* (Fig. 10-1*b*), where three or more girders function as floor stringers similar to rolled beams but on longer spans. These girders must be adequately cross-braced with a system of cross-bracing or diaphragms (see Fig. 10-15). This is the most common type of highway bridge using plate girders.
3. *Through deck girder bridges*, where the load moves between and is carried to the main girders via use of a system of cross-beams or cross-beams and

stringers (Figs. 10-1c and 10-3a) located as close as practical to the bottom flange. This arrangement is widely used for railroad bridges and may be used for either an open deck or a ballasted deck. The use of a through deck is often economical in reducing pier heights and/or approach fills as compared to the deck girder and depending on the necessary underpassage clearance.

4. *Ballasted-deck bridge*, where the floor beam system is covered with a fabricated deck (usually steel plate) and overlaid with 6 to 15 in (150 to 500 mm) of track ballast into which the cross-ties and rail is placed. Figure 10-4 illustrates a ballasted-deck bridge.

The D/L ratios for bridges ranges from about $\frac{1}{10}$ to a minimum of $\frac{1}{25}$ (as in the current AASHTO specification). The depth of a highway bridge girder must be balanced against site parameters. For example, an extra 12 in (300 mm) of girder depth for an overpass structure that utilizes approach fill can require a substantial amount of extra fill and right-of-way compared to the heavier weight of the more shallow girder requiring thicker flanges and (possibly) web.

Plate girders for railroad bridges are recommended for spans from 50 to 150 ft (16 to 45 m) with rolled beams used for shorter spans. Trusses are used for longer spans. The common range of D/L for railroad girders is $\frac{1}{12}$ to $\frac{1}{20}$.

Figure 10-4 Ballasted through deck plate girder railroad bridge. (*a*) Ballasted through deck girder bridge under construction. Note that knee bracing for lateral stability also braces the compression flange at very close intervals. While not clearly visible, the deck plates are field-butt-welded and ground smooth. (*b*) Bottom view of above ballasted deck bridge. Note the maintenance walkway brackets on outside of girder with planking for floor to be added. (*c*) Completed bridge with ballast and track in place. Maintenance walkway also completed.

However, the current AREA specifications do not use D/L as a criterion; rather, the computed deflection of the girder under full dead + live load, including impact, is limited to $\Delta/\text{span} \leq 1/640$.

Girders are fabricated in segments limited by the lifting capacity of the erection/transport equipment and rolling mill capacity on plate sizes. Welded plate girders are almost exclusively used in American design practice because of the reduced fabrication costs and the fact that the welding of the flange plates to the web can be almost fully automated.

A-36 steel is most commonly used for plate girders and almost universally for webs and stiffeners. Continuous girders produce requirements of reduced sections in the positive moment region and deeper sections in the larger negative moments regions over reactions. An increase in girder depth is in many cases esthetically pleasing; however, fabrication costs are increased, since a tapered web must be cut and additional stiffeners are usually required for the larger h/t_w ratio produced. An alternative to an increase in girder depth is to use higher-strength steels in the zones of increased moment. The use of high-strength steel for the flange either throughout the span or in local zones of higher stresses produces a *hybrid* plate girder.

An inspection of Figs. 10-1 to 10-3 and observation of other girders in service readily indicates that there is no unique girder design for a given problem. Within the constraints of D/L and loading, an infinite number of possibilities exist for cross-sectional area. If we define optimization of a plate girder design as that producing the least overall cost, it is not practical, even using a computer program, to "optimize" the girder, since least weight does not necessarily (and probably will not) mean least overall cost because of fabrication considerations. What one should attempt to obtain is a balance between weight and fabrication cost to produce as economical (within reasonable computational effort, which is also a cost factor) a solution as possible.

10-2 LOADS

Plate girder loads are obtained similarly to those for beams. However, in building design the loads are often moving, as for crane runway girders in industrial buildings or warehouses. AISC specifications require use of an impact factor for these types of loadings.

Plate girder design for highway bridges involves a decision as to how many girders are to be used per lane, using the AASHTO specifications to prorate the amount of truck or lane loading to the several girders and producing an influence-line type analysis using either standard lane loads or standard truck loading to find the maximum moment and shear for design. The AASHTO specifications allow use of only one truck per lane for simple spans, in general, although as shown in Fig. 10-5, more than one truck may be on the bridge at the critical location at a given instant. When more than one truck is on a lane, the maximum moment resulting is not likely to be greater than a single truck

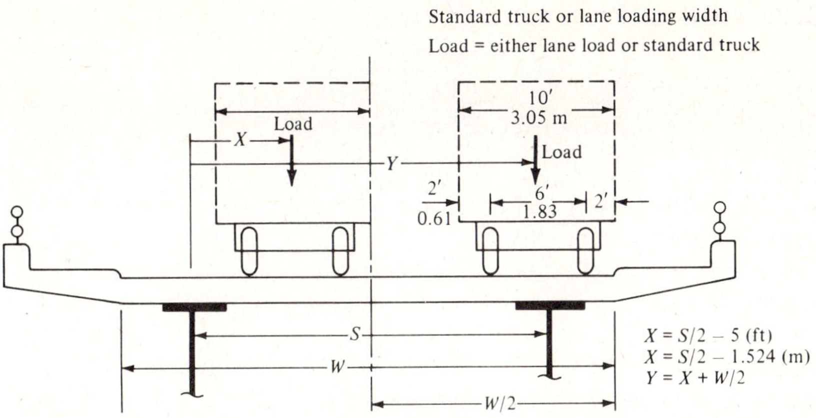

Standard truck or lane loading width
Load = either lane load or standard truck

$X = S/2 - 5$ (ft)
$X = S/2 - 1.524$ (m)
$Y = X + W/2$

Figure 10-5 Placement of standard truck loading for maximum girder loadings. Make appropriate modifications for additional lanes/girders.

because of truck length factors and positioning of wheels. When the span is sufficiently long, the lane loading with single concentrated load represents a line of trucks well enough and is more convenient to use than trying to find the number of trucks and their positioning for maximum stress effects.

A computer program that runs the standard truck across (forward and backward) the bridge with output of shear and moment at the 0.1 points is most useful for analysis of continuous girders. We may note in passing that an influence line analysis is not necessary for continuous spans when using a computer program.

In all girder design the standard lane/truck loads are increased by the impact factor, which is a function of span length.

Figure 10-5 illustrates the placing of the standard truck loading on a two-lane bridge to obtain the contribution of truck loads to either of the two girders. The designer must make a similar type of analysis for multiple lanes using more than two girders.

Example 10-1 What are the design moment and shear values for the girders of a highway bridge as shown in Fig. 10-5?

HS 20 loading, $S = 19$ ft, $W = 28$ ft, span $= 110$ ft. Dead load due to deck, sidewalk, and so on $= 3.1$ kips/ft.

SOLUTION The lane factor L_f is obtained by ΣM about the left girder of Fig. 10-5, to obtain

$$L_f = \frac{X + Y}{S}$$

where $X = S/2 - 5 = 19/2 - 5 = 4.5$ ft
$Y = X + W/2 = 4.5 + 28/2 = 18.5$ ft

Substituting values, we obtain

$$L_f = \frac{4.5 + 18.5}{19} = 1.21$$

The impact factor is computed (see Sec. 1-10) as

$$I = \frac{50}{L + 125} = \frac{50}{110 + 125} = 0.21 < 0.30 \quad \text{O.K.}$$

The moments are

$$M_d = \frac{wL^2}{8} = \frac{3.1(110)^2}{8} = 4689 \text{ ft} \cdot \text{kips}$$

Use the equations given in Sec. 1-9 for $L = 110 < 145.6$ ft.

$$M_L = \frac{W}{L}[(0.9L + 4.206)(0.5L + 2.33) - 11.2L]$$

Substituting values yields

$$M_L = \frac{40}{110}\{[0.9(110) + 4.206][0.5(110) + 2.33] - 11.2(110)\}$$

$$= \frac{40}{110}(5916.8 - 1232.0) = 1703.6 \text{ ft} \cdot \text{kips}$$

The shear values are

$$V_d = \frac{wL}{2} = \frac{3.1(110)}{2} = 170.5 \text{ kips}$$

The live-load shear (one 32 kip wheel on reaction) at reaction is

$$V_L = W\left(\frac{L - 16.8}{L} + 0.8\right)$$

$$= 40\left(\frac{110 - 16.8}{110} + 0.8\right) = 65.9 \text{ kips}$$

We note that M_d is at the centerline of the span, whereas M_L is 2.33 ft to the left of the centerline (truck moving left to right). The difference is so small that we will simply add the two values as if they both occur at the centerline to obtain a slightly conservative design M:

$$M_{\text{design}} = M_L(L_f) + \text{impact} + M_d$$

$$= 1703.6(1.21)(1 + 0.21) + 4689 = 7183 \text{ ft} \cdot \text{kips}$$

The design shear is

$$V_{\text{design}} = V_L(L_f) + \text{impact} + V_d$$

$$= 65.9(1.21)(1.21) + 170.5 = 267 \text{ kips} \quad ///$$

Railroad girders are more difficult to design, since there are 18 wheels in the Cooper E-series standard loading. Design aids are available, such as those shown in Table 1-2, which gives the approximate (to exact) maximum moments due to the moving train either at the centerline or very close to it. This aid is made based on a combination of use of $dM/dx = 0$ for short span lengths where only three or four wheels are simultaneously on the span and an influence line with apex at the centerline for a larger number of wheels on the span. Also given in Table 1-2 are the end reactions (shear) and the maximum reaction (shear) developed in a floor beam into which two stringers of lengths shown in the first column frame. [For example, for a floor beam between two 7-ft stringers, the maximum reaction is obtained when one 10-kip wheel load is on the floor beam and the other two wheels are 5 ft away, giving $R = 10 + 2(2/7)10 = 15.7$ kips; similarly, for $L = 10$ ft, the reaction $R = 10 + 2(5/10)10 = 20.0$ kips; etc.] The values in the table are shown for Cooper's E-80 loading; any other loading X is $X/80 \times$ table value.

10-3 PROPORTIONING FLANGES AND WEBS OF GIRDERS AND BUILT-UP SECTIONS

The flange cover plate of a built-up section using a rolled section as the base may be proportioned as follows (refer to Fig. 10-6). Assuming that the maximum bending stress f_b is the allowable value of F_b, the average flange value is about $0.95 F_b$. At the junction of the cover plate and the rolled section flange the stress is about $0.90 F_b$. Therefore, let

$$M_1 = f_b' S_x = 0.90 F_b S_x \qquad (a)$$

$$M_2 = a' \Delta C = (d + t_f')(t_f' b_f' \times 0.95 F_b) \qquad (b)$$

The required section capacity is

$$M_{\text{total}} = M_1 + M_2 \qquad (c)$$

Substitution of Eqs. (a) and (b) into (c) and taking $(d + C) = (d + t_f')$ defining the flange area $A_f = b_f' t_f'$, and taking the ratio $0.90/0.95$ as approximately 1, we

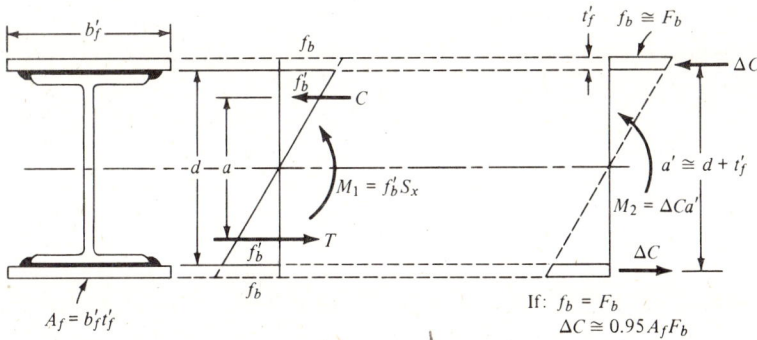

Figure 10-6 Bending stress distribution on a built-up beam cross section.

obtain

$$A_f \simeq \frac{M_{total}}{(d + C)0.95F_b} - \frac{S_x}{d + C} \qquad (10\text{-}1)$$

It should be evident that Eq. (10-1) does not give a unique solution and the final flange cover plates must be investigated using $f_b = M/S \le F_b$. A reasonable estimate of $C = 1$ in or 25 mm.

Example 10-2 Design the cover plates for a W920 × 342.3 section using $F_y = 250$ MPa for a moment of 3630 kN · m for the conditions shown in Fig. E10-2a. Use the AISC specifications.

SOLUTION Data for a W920 × 342.3:

wt = 3.36 kN/m $\qquad$ $d = 912$ mm

$b_f = 418$ mm $\qquad$ $S_x = 13.72 \times 10^{-3}$ m^3

$I_x = 6243.5 \times 10^{-6}$ m^4 $\qquad$ $t_f = 32.0$ mm

$A = 43.61 \times 10^{-3}$ m^2

The moments are:

$$M_L = \frac{wL^2}{8} = \frac{72.6(20)^2}{8} = 3630 \text{ kN} \cdot \text{m}$$

$$M_d = \frac{3.36(20)^2}{8} = 170 \text{ kN} \cdot \text{m}$$

$$M_{design} = \overline{3800 \text{ kN} \cdot \text{m}}$$

Tentatively, the area of each cover plate [using Eq. (10-1)] is

$$A_f = \frac{3800}{(0.912 + 0.025)(0.95 \times 0.6 F_y)} - \frac{13.72}{0.912 + 0.025}$$

$$A_f = 28.46 - 14.64 = 13.82 \times 10^{-3} \text{ m}^2$$

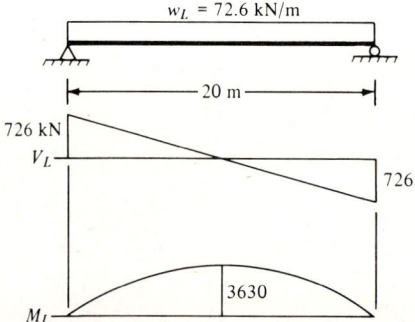

Figure E10-2a

We could use flange cover plates 25 × 548 mm; however, let us arbitrarily use a plate width 52 mm larger than the b_f of the beam, which allows a 26-mm overhang on each side for welding the plate to the flange. This gives

$$b'_f = 418 + 52 = 470 \text{ mm}$$

$$t'_f = \frac{13.82}{0.47} = 29.4 \text{ mm} \quad \text{use 30 mm}$$

Now check if cover plates 30 × 470 mm are adequate:

$$I_x = I_0 + 2Ad^2$$

$$= 6243.5 + 2(0.03)(0.470)\left(\frac{912 + 30}{2}\right)^2$$

$$= 12\ 499.4 \times 10^{-6} \text{ m}^4$$

The section modulus is

$$S_x = \frac{I}{c} = \frac{12\ 499.4}{912/2 + 30} = 25.719 \times 10^{-3} \text{ m}^3$$

The additional moment due to the flange cover plates is proportional to the flange area:

$$M_{\text{total}} = 3800 + \frac{0.030(470)}{43.61}(170) = 3855 \text{ kN} \cdot \text{m}$$

$$f_b = \frac{3855}{25.719} = 149.9 \text{ MPa} < 150 \quad \text{O.K.}$$

The cover plates will be welded as shown in Fig. E10-2b. This requires checking b''_f / t'_f for compliance with AISC Sec. 1-9.2.2 (stiffened edge element):

$$\frac{b}{t} \leq \frac{670}{\sqrt{F_y}} = 42.4$$

$$\frac{b''_f}{t'_f} = \frac{418}{30} = 13.93 \ll 42.4 \quad \text{O.K.}$$

It is not necessary to check $b/2t_f$ of beam since $t_f = 32.0$ mm. Use cover plates 30 × 470 mm.

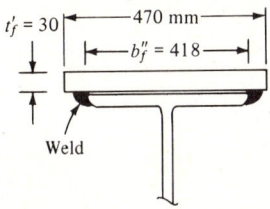

Figure E10-2b

///

10-4 PARTIAL-LENGTH COVER PLATES

From inspection of the moment diagram of Example 10-2 it is evident that the cover plate is not needed for the full length of the beam. Some economy may be obtained by providing cover plates only in the region where they are needed. However, specifications are very rigid for providing a distance beyond the theoretical cutoff point to fully develop the cover plate contribution to section bending resistance. Figure 10-7 illustrates the code requirements for this distance.

The fastener capacity to attach the cover plate to the beam can be computed as the difference between the compressive forces C and ΔC (see Fig. 10-6); alternatively, it can be computed using

$$v_s = \frac{VQ}{I} \quad \text{(kips/in or kN/m)}$$

where v_s = force/length of developed shear
Q = statical moment of cover plate with respect to neutral axis
I = moment of inertia of section, including cover plates

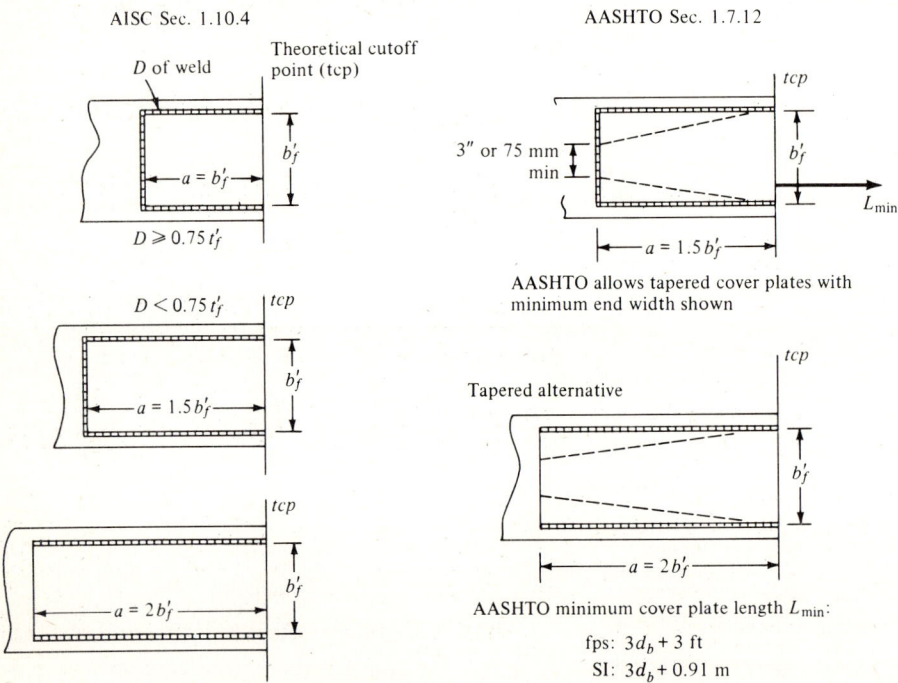

Note: AREA (Sec 1.7.2.1) merely states cover plates shall extend far enough beyond the theoretical cutoff point to develop the capacity of the plate.

Figure 10-7 Code requirements for theoretical cutoff distances for cover plates. Requirements are same for cover plates narrower (shown above) or wider than next underlying plate or beam flange.

The spacing of bolts or welds can be found by equating the weld capacity/length or bolt capacity and the spacing s as

$$s = \frac{\text{resistance capacity}}{v_s}$$

The weld capacity at the theoretical cover plate cutoff is obtained by equating the moment of the total weld capacity to the moment at the theoretical cutoff point.

$$F_w = \frac{MQ}{I} \qquad \text{(kips or kN)}$$

where F_w = total force to be carried by the weld in the cutoff length a
 M = bending moment at the theoretical cutoff point
 Q, I = terms previously defined

Example 10-3 What is the theoretical cutoff distance for the cover-plated beam of Example 10-2? Also, design the welding to fasten the plates to the beam using E60 electrodes.

SOLUTION Obtain the cutoff points (Fig. E10-3a). The capacity of the beam without cover plates is

$$M_b = S_x F_b = 13.72(150) = 2058 \text{ kN} \cdot \text{m}$$

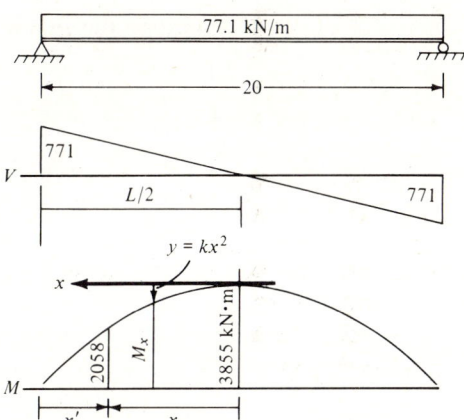

Figure E10-3a

Since the moment diagram is a parabola with slope = 0 at midspan, the equation for moment is

$$M_x = M_0 - kx^2$$

as shown in Fig. E10-2a using the origin at midspan. Since $M_x = 0$ at

$x = L/2$, we have
$$M_x = M_0 - \frac{4M_c}{L^2}x^2$$

With $M_x = M_b = 2058$ kN · m at the distance x from midspan where the beam capacity is adequate without cover plates, the theoretical cutoff point is

$$x = \left[(3855 - 2058)\frac{L^2}{4 \times 3855}\right]^{1/2} = 6.828 \text{ m}$$

The theoretical length of the cover plate $= 2x = 2(6.865) = 13.73$ m.

Design the welds (cover plate 30×470 mm; $I_x = 12\ 499.4 \times 10^{-6}$ m^4): at the cutoff point $x' = 10 - 6.83 = 3.17$ m.

Shear $V = 771 - 77.1(3.17) = 530$ kN

$$Q = A\bar{y} = 0.030(0.470)\left(\frac{912}{2} + 15\right) = 6.41 \times 10^{-3} \text{ m}^3$$

$$v_s = \frac{530(6.41)}{12499.4} = 281.6 \text{ kN/m}$$

(Noting that I is 10^{-6} and $Q = 10^{-3}$, we obtain 281.6 kN/m and not the direct calculator reading of 0.2816.)

The allowable stress using E60 electrodes:
$$F_w = 0.3(415) = 124.5 \text{ MPa}$$

We will arbitrarily use $D = 8$ mm for the weld ($t_f > 20$ mm per AISC Sec. 1-17.2) so that the theoretical spacing at the cutoff point to the interior of the span is

$$\frac{P_w}{m} = 2(0.008 \times 0.70711)(124.5) = 1.4085 \text{ MN/m}$$

Tentatively, use a 40-mm length of weld (40 mm = $5D > 2D$), which will carry

$$P'_w = 40(1.4085) = 56.34 \text{ kN}$$

This corresponds to a distance along the beam of

$$s = \frac{56.34 \times 10^3}{281.6} = 200 \text{ mm} \quad \text{O.K.}$$

Use 40-mm lengths of intermittent weld alternating on each side for each 200-mm increment of length.

The weld in the distance a ($a = 1.5b_f$ since $D < 0.75t'_f$) is continuous and the effective moment M_e is

$$M_e = [0.705(2) + 0.418](1.4085)[912 + 2(30)]$$
$$= 2503 \text{ kN} \cdot \text{m} > 2058 \quad \text{O.K.}$$

If M_e had been less than the moment at the theoretical cutoff point, it would have been necessary to either increase D or the distance a (or both). The final sketch is as shown in Fig. E10-3b.

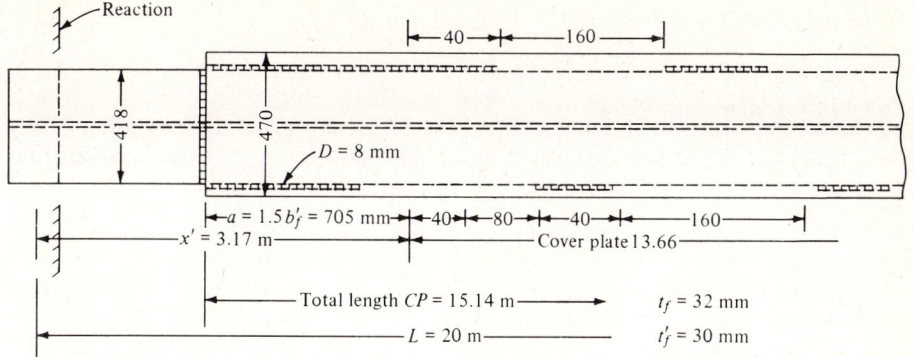

Figure E10-3b

///

10-5 GENERAL PROPORTIONS OF PLATE GIRDERS

All plate girders are proportioned by trial and revised until the bending stresses are $f_b \leq F_b$. Stiffeners are added as required until the shear stress is such that $f_v \leq F_v$. There is no unique solution (unless two persons use the same computer program with the same input data). Initial estimates of flange sizes may be made using the contributions of the web and flange in bending as (see Fig. 10-8) follows. The moment carried by the flange is

$$M_f = A_f f_b'(D - t_f)$$

The moment carried by the web is

$$M_w = f_b'' S_{xw} = f_b'' \frac{t_w h^2}{6}$$

The total moment is

$$M = M_f + M_w$$

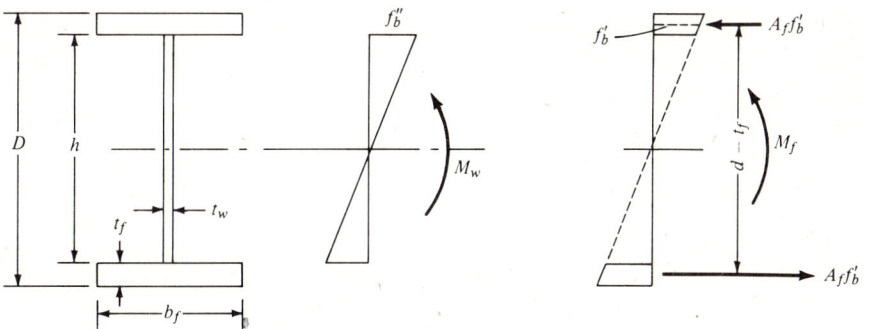

Figure 10-8 Girder section to obtain an approximate expression for the flange area.

If we let $f_b'' = f_b' \cong F_b' \cong 0.6F_y$ and take $d - t_f = h$,

$$M_f = M - M_w$$

Substitution of values yields

$$A_f F_b' h = M - F_b'\left(\frac{t_w h^2}{6}\right)$$

but $t_w h^2 = A_w h$, and we obtain the desired expression for flange area as

$$A_f = \frac{M}{F_b' h} - \frac{A_w}{6} \qquad (10\text{-}2)$$

where A_w = area of the girder web (approximately—must still check Mc/I)
F_b' = a value not to exceed $0.6F_y$; however, inspection of the derivation indicates that a value of something a little less than $0.6F_y$ should be used (e.g., 21 ksi instead of 22 ksi, 132 MPa instead of 137.5).

Note again that the overall depth is generally not exactly fixed, the web thickness can vary between reasonably wide limits, and the flange area is a combination of width × thickness, so that a number of equally valid solutions can be obtained. The only feature establishing preference for one workable design rather than another would be the final in-place cost.

10-6 PLATE GIRDER DESIGN THEORY—AISC

Plate girders are essentially truss members where the flanges are the top and bottom chords and the web constitutes the web members. Since the members are made up of an assemblage of relatively thin plates, stability theory is an essential element of the design. Two general design "theories" are used in plate girder design. One method, *buckling strength*, is used by AASHTO and AREA. The other method, *load-carrying capacity*, is used by the AISC. This latter method recognizes that after the web initially buckles locally, there is a considerable reserve load-carrying capacity until the remainder of the web buckles. This reserve capacity has been verified by several large-scale girder tests.

We will consider AISC plate girder design specifications followed by the AASHTO and AREA specifications in the next section. The several elements making up the girder section and design will be considered in following sections.

10-6.1 Girder Flanges

The compression flange of a plate girder has the usual three modes of failure shown in Fig. 10-9a. The design criteria of d/A_f and L/r_T used in developing AISC Eqs. (1.5-6a), (1.5-6b), and (1.5-7) are used in plate girder design to establish the maximum allowable flange stresses in bending. It is not likely the D/t_w (same as the d/t_w of rolled beams) ratio will ever be such that a compact section is produced, so that, in general, the maximum allowable flexural stress is

$$F_b \leq 0.6F_y$$

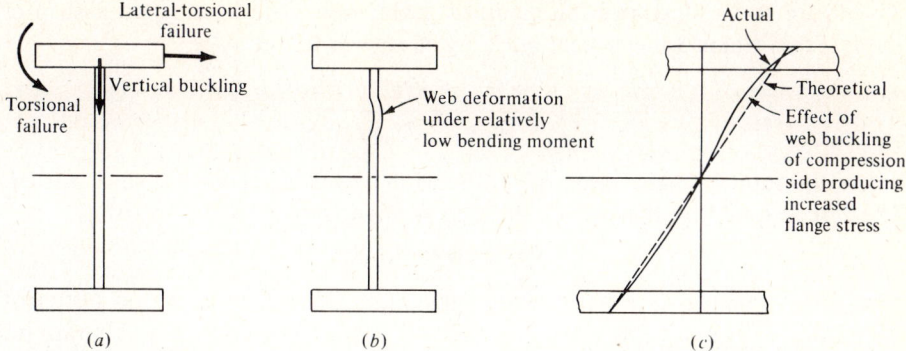

Figure 10-9 Modes of failure prevention required in a stability analysis for plate girders. (*a*) Mode of compression flange failure. (*b*) Web buckling. (*c*) Observed (qualitative) stress distribution on a girder cross section.

For $F_b \leq 0.6F_y$, the upper limit of the $b_f/2t_f$ ratio is

$$\text{fps:} \quad \frac{b_f}{2t_f} \leq \frac{95}{\sqrt{F_y}} \qquad \text{SI:} \quad \frac{b_f}{2t_f} \leq \frac{250}{\sqrt{F_y}} \qquad (10\text{-}3)$$

10-6.2 Web Plate Design

The web plate curvature in bending causes radial pressures to be developed at the interface of the web and flange plates (Fig. 10-10). If the web plate is too thin as measured by the h/t ratio (analogous to KL/r for a column), it will buckle.

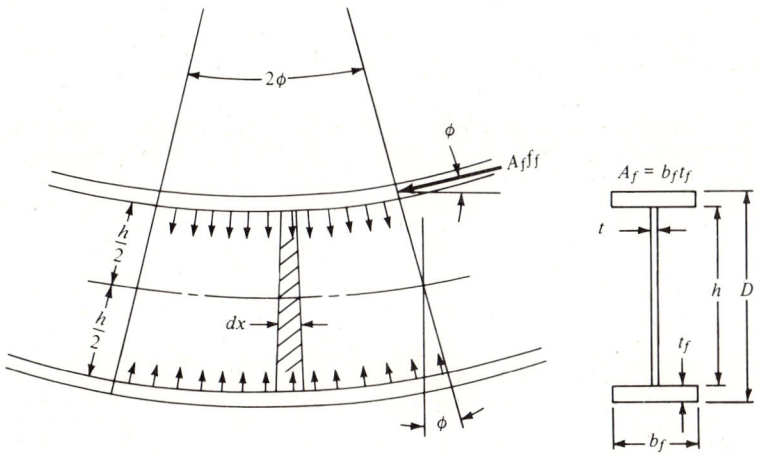

Figure 10-10 Compressive stresses developed in girder web due to bending.

472 STRUCTURAL STEEL DESIGN

Figure 10-10 illustrates the assumed girder web plate loading on a strip of $dx(h)$. From Chap. 3 the critical buckling stress was found to be

$$F_{cr} = \frac{k_c \pi^2 E}{12(1 - \mu^2)(h/t)^2} \qquad (3\text{-}5)$$

where the ratio h/t has been substituted for b/t as given in Eq. (3-5) earlier. The force exerted on the web of the web segment shown in Fig. 10-10 is

$$A_f f_f \sin \phi \cong A_f f_f \phi$$

since ϕ (in radians) is a very small angle. The stress f_f must be of sufficient magnitude to overcome any residual stress F_r in the web; thus the web strain (in units of FL^{-2}) at yield is

$$\epsilon_f = \epsilon_r + \epsilon_y = \frac{F_r + F_y}{E}$$

and the angle ϕ is

$$\phi = \epsilon \frac{dx}{fh/2} = \frac{2(F_r + F_y)}{E} \frac{dx}{h}$$

At web yield the applied force is

$$A_f f_f \tan \phi = A_f f_f \phi$$

for small angles and ϕ in radians, giving

$$A_f f_f \phi = 2 F_y (F_y + F_r) \frac{dx}{Eh} \qquad (a)$$

This value should not be larger than the critical web buckling force of

$$F_{cr} t \, dx = \frac{k_c \pi^2 E}{12(1 - \mu^2)} \left(\frac{t}{h}\right)^2 (t) \, dx \qquad (b)$$

Equating Eqs. (a) and (b) and solving for h/t, we obtain

$$\frac{h}{t} = \left[\frac{k_c \pi^2 E}{24(1 - \mu^2)} \frac{A_w}{A_f} \frac{1}{F_y(F_y + F_r)} \right]^{1/2}$$

When A-7 steel was used, it was assumed that the residual stress F_r could be adequately approximated as $F_r = f_y/2 = 33/2 = 16.5$ ksi. This value is currently being used for all values of steel F_y. It is also assumed that ratio $A_w/A_f \cong \frac{1}{2}$, Poisson's ratio for steel $\mu = 0.3$, $E = 29\,000$ ksi, and for $k_c = 1$, we obtain

$$\frac{h}{t} = \frac{13\,784}{[F_y(F_y + 16.5)]^{1/2}}$$

The AISC rounds this to obtain the limiting h/t ratio as

$$\text{fps:} \quad \frac{h}{t} \leq \frac{14\,000}{[F_y(F_y + 16.5)]^{1/2}} \quad (10\text{-}4)$$

$$\text{SI:} \quad \frac{h}{t} \leq \frac{97\,100}{[F_y(F_y + 114)]^{1/2}} \quad (10\text{-}4\text{m})$$

The AISC specification allows a somewhat larger h/t ratio if transverse stiffeners are used at a spacing ratio of $a/h \leq 1.5$ of

$$\text{fps:} \quad \frac{h}{t} \leq \frac{2000}{\sqrt{F_y}} \quad (10\text{-}5)$$

$$\text{SI:} \quad \frac{h}{t} \leq \frac{5270}{\sqrt{F_y}} \quad (10\text{-}5\text{m})$$

The maximum h/t ratios for several grades of steel are as follows:

F_y		h/t ratio	
		w/stiffeners	w/stiffeners, but
ksi	MPa	[by Eq. (10-4)]	$a/h \leq 1.5$
36	250	322	333
50	345	244	284
60	415	207	259
65	450	193	248

Experimental studies on full-size girders have shown that the web plate in the compression zone deflects laterally by small amounts at very early stages of bending, with the resulting transfer of stresses from the compression web portion to the compression flange. This results in an increase in the flange stress over that amount indicated by theory as shown in Fig. 10-10c. This increase in compression flange stress requires a reduction of the allowable compressive stress so that the stress actually developed does not cause a flange failure. The experimental studies indicated that this flange stress reduction could be expressed in terms of A_f, A_w, h/t, and F_y. A possible equation in terms of ultimate moment is

$$M_u = M_y \left[1 - 0.0005 \frac{A_w}{A_f} \left(\frac{h}{t} - \beta_0 \right) \right]$$

Since the ratio of $M_u/M_y = F_u/F_y = F_b'/F_b$, we may rewrite this equation in terms of stresses as long as the compression flange is sufficiently stable with respect to L/r_T and d/A_f. This is done by limiting the maximum allowable stress to F_b as defined by AISC Eqs. (1.5-6) and/or (1.5-7). The value of β_0 depends on the flange restraint effects on the web, and if we assume partial restraint, β_0 may be taken as $5.7\sqrt{E/F_{cr}}$ (see Basler and Thürlimann, "Strength of Plate Girders in Bending," *Journal of Structural Division, ASCE*, August 1961), where F_{cr} is the critical web buckling stress. If we replace $F_{cr} = 1.65 F_b$, we obtain (with slight rounding):

$$\text{fps:} \quad F_b' = F_b\left[1 - 0.0005\frac{A_w}{A_f}\left(\frac{h}{t} - \frac{760}{\sqrt{F_b}}\right)\right] \quad (10\text{-}6)$$

$$\text{SI:} \quad F_b' = F_b\left[1 - 0.0005\frac{A_w}{A_f}\left(\frac{h}{t} - \frac{1980}{\sqrt{F_b}}\right)\right] \quad (10\text{-}6\text{m})$$

Inspection of Eq. (10-5) indicates that if

$$\frac{h}{t} \le \frac{760}{\sqrt{F_y}}$$

or the SI equivalent, no flange stress reduction is necessary (i.e., $F_b' = F_b$).

10-6.3 Shear and Stiffener Requirements

Where the h/t ratio is sufficiently small, web buckling will not occur under shear before shear yielding occurs. Actually, a beam web as part of a flexural member subjected to a bending moment is carrying the shear in a "tension field" mode. Some researches have likened this to a Pratt truss, where the stiffeners (if used) are the compression members and the web segment between a pair of stiffeners is the tension element, as illustrated in Fig. 10-11a. Actual girder tests carried to yield so that mill scale forms stress lines displays a stress pattern similar to that idealized in Fig. 10-11b.

Girder webs are designed by AISC specifications assuming that the shear is carried by beam action shear—as for rolled shapes—until shear web buckling stresses are reached; then the additional shear is carried by tension field action. The ultimate shear resistance in a web panel (space between two stiffeners) is the sum of the two shear (beam, V_b, and tension field, V_t) contributions, or

$$V_u = V_b + V_t \quad (a)$$

The beam shear contribution is

$$V_b = F_{cr}(ht_w) = F_{cr}A_w \quad (b)$$

The ultimate shear force capacity of a girder for web plastification (plastic capacity) is

$$V_p = F_{ys}A_w \quad (c)$$

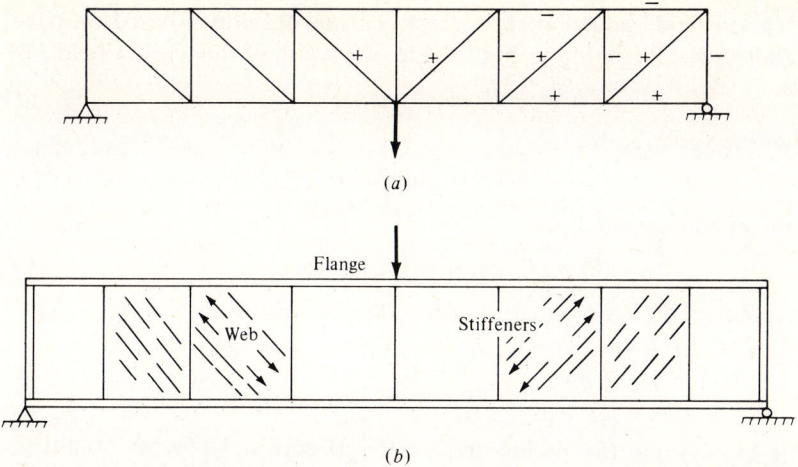

Figure 10-11 Analogy of plate girder to Pratt truss. (a) Pratt truss (−) = compression member. (b) Plate girder with tension fields.

but the yield shear stress F_{ys} according to the von Mises criteria is

$$F_{ys} = \frac{F_y}{\sqrt{3}} \qquad (d)$$

From Eqs. (c) and (d), we obtain A_w and substitution into (b),

$$V_b = \frac{\sqrt{3}\,(V_p)F'_{cr}}{F_y} \qquad (e)$$

But for values of $F'_{cr} \leq 0.8 F_{ys}$, the value of F'_{cr} is, from Eq. (3-5), repeated here

$$F'_{cr} = \frac{k\pi^2 E}{12(1-\mu^2)}\left(\frac{t_w}{h}\right)^2 \qquad (3\text{-}5)$$

If $F'_{cr} > 0.8 F_{ys}$ from using Eq. (10-6), the critical stress is taken as the mean of Eq. (10-6) and the value of $0.8 F_{ys}$, to obtain

$$F'_{cr} = \sqrt{0.8 F_{ys} \frac{k\pi^2 E}{12(1-\mu^2)}\left(\frac{t_w}{h}\right)^2} \qquad (10\text{-}7)$$

and where as previously defined (see Table 3-2):

$$\left.\begin{array}{ll} k = 4.00 + \dfrac{5.34}{(a/h)^2} & \dfrac{a}{h} \leq 1.0 \\[2mm] k = 5.34 + \dfrac{4.00}{(a/h)^2} & \dfrac{a}{h} > 1.0 \end{array}\right\} \qquad (10\text{-}8)$$

The "tension field" action shear force V_t can be developed based on panel geometry and statics. Referring to Fig. 10-12b, the width of the tension field s is

$$s = h \cos \phi - a \sin \phi \tag{f}$$

The web tension force T_w is

$$T_w = f_t s t_w \tag{g}$$

and the vertical component V_t is

$$V_t = T_w \sin \phi = f_t s t_w \sin \phi \tag{h}$$

The largest value of V_t is obtained by taking the derivative

$$\frac{dV_t}{d\phi} = 0$$

Substituting Eq. (f) into (h), taking derivative = 0, solving for ϕ, we obtain

$$\tan \phi = \left[1 + \left(\frac{a}{h}\right)^2\right]^{1/2} - \frac{a}{h} \tag{i}$$

If we limit a/h between 0.5 and 3.0, the corresponding values of ϕ are 9.2 to 31.75°, which is a reasonable range.

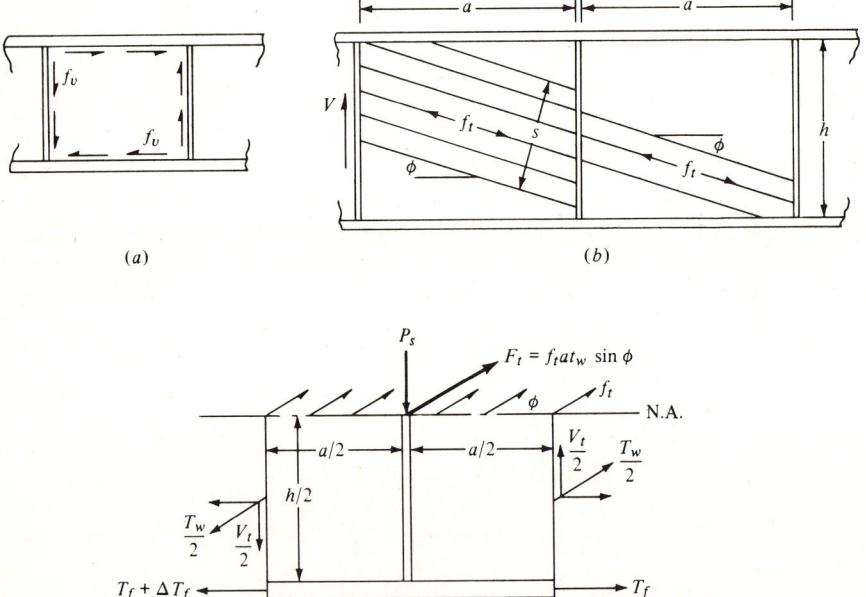

Figure 10-12 Shear stress distributions and "tension field" in girder panels for derivation of allowable shear stresses and stiffener size. (a) Girder panel under pure shear. (b) Girder panels with "tension field" action. (c) Stresses and free-body diagram out of base of girder panels shown in (b).

Using the free body of Fig. 10-12c, the change in flange force ΔT_f is

$$\Delta T_f = F_t \cos \phi = f_t a t_w \sin \phi \cos \phi \qquad (j)$$

Taking the $\Sigma M = 0$ about point 0 of Fig. 10-12c gives

$$\Delta T_f = V_t \frac{a}{h} \qquad (k)$$

Equating Eqs. (j) and (k), using the identity that

$$\sin \phi \cos \phi = \frac{1}{2} \sin 2\phi = \frac{1}{2} \frac{\tan \phi}{1 + \tan^2 \phi} = \frac{1}{2} \frac{1}{[1 + (a/h)^2]^{1/2}}$$

we obtain

$$V_t = f_t h t_w \frac{1}{2[1 + (a/h)^2]^{1/2}} \qquad (l)$$

Now we may substitute Eqs. (l) and (e) into Eq. (a) and, taking $ht_w = V_p/F_{ys} = \sqrt{3}\, V_p/F_y$, we obtain

$$V_u = V_p \left\{ \frac{F'_{cr}}{F_{ys}} + \frac{\sqrt{3}}{2} \frac{f_t}{F_y} \frac{1}{[1 + (a/h)^2]^{1/2}} \right\}$$

An approximation of

$$\frac{f_t}{F_y} = 1 - \frac{F'_{cr}}{F_{ys}} \qquad (m)$$

can be made to give

$$V_u = V_p \left\{ \frac{F'_{cr}}{F_{ys}} + \frac{1 - F'_{cr}/F_{ys}}{1.15[1 + (a/h)^2]^{1.2}} \right\}$$

Now dividing by $A_w = ht_w$, using a safety factor of 1.67, $F_{ys} = F_y/\sqrt{3}$, and taking $C_v = F'_{cr}/F_{ys}$, we obtain the AISC design equation [Eq. (1.10-2)]:

$$F_v = \frac{F_y}{2.89} \left\{ C_v + \frac{1 - C_v}{1.15[1 + (a/h)^2]^{1/2}} \right\} \qquad (C_v < 1) \qquad (10\text{-}9)$$

Note that the term

$$\frac{1 - C_v}{1.15[1 + (a/h)^2]^{1/2}}$$

is the tension field contribution; thus, if the actual shear stress

$$f_v \le \frac{F_y C_v}{2.89}$$

no stiffeners are required. Note also that when $C_v > 1$ [see Eq. (m)] only the first term of Eq. (10-9) is used to give the allowable shear stress F_v as

$$F_v = \frac{F_y C_v}{2.89} \leq 0.4 F_y \qquad [\text{AISC Eq. (1.10-1)}] \qquad (10\text{-}10)$$

The C_v terms are computed by substitution into Eq. (3-5) or (10-7) as appropriate to obtain:

For $C_v \leq 0.8$:

$$C_v = \frac{F'_{cr}}{F_{ys}} = \frac{k\pi^2 E}{12(1-\mu^2)(h/t_w)^2} \frac{\sqrt{3}}{F_y} = \frac{45398k}{F_y(h/t_w)^2}$$

The AISC uses

$$\text{fps:} \quad C_v = \frac{45\,000k}{F_y(h/t_w)^2} \qquad (10\text{-}11)$$

$$\text{SI:} \quad C_v = \frac{312\,500k}{F_y(h/t_w)^2} \qquad (10\text{-}11\text{m})$$

For $c_v > 0.8$:

$$C_v = \sqrt{\frac{0.8 F_{ys} k\pi^2 E}{12(1-\mu^2)(h/t_w)^2} \frac{\sqrt{3}}{F_y}}$$

The AISC uses

$$\text{fps:} \quad C_v = \frac{190h}{t(k/F_y)^{1/2}} \qquad (10\text{-}12)$$

$$\text{SI:} \quad C_v = \frac{500h}{t(k/F_y)^{1/2}} \qquad (10\text{-}12\text{m})$$

The AISC specifications further limit girder design and panel sizes as follows:

1. Bearing stiffeners for reactions (and generally under concentrated loads also) are always required.
2. Intermediate transverse stiffeners (AISC does not have a specification for longitudinal stiffeners) are not required if both of the following criteria are met:

$$\frac{h}{t_w} \leq 260 \quad \text{and} \quad f_v \leq \frac{F_y C_v}{2.89}$$

3. Intermediate stiffeners are required for any other shear stress condition. When stiffeners are required, the spacing is limited to

$$0.5 \leq \frac{a}{h} \leq \left(\frac{260}{h/t_w}\right)^2$$

but $a/h \leq 3.0$.

4. In girders designed on the basis of tension field action, the spacing of the first interior stiffener from the end bearing stiffener, or any stiffeners adjacent to large holes, will be based on limiting the shear stress to AISC Eq. (1.10-1).

10-6.4 Combined Bending and Shear (Interaction Check)

When a girder is simultaneously subjected to a large value of shear and bending simultaneously, the allowable shear stresses may have to be reduced. An interaction solution was developed by Basler (see "Strength of Plate Girders Under Combined Bending and Shear," *Journal of Structural Division, ASCE*, October 1961) based on girder tests at Lehigh University. AISC has slightly rounded the values to obtain

$$F_b \le \left(0.825 - \frac{0.375 f_v}{F_v}\right) F_y \le 0.6 F_y \qquad (10\text{-}13)$$

In general, no interaction check is required if

(a) $f_v \le 0.6 F_v$ and $f_b \le 0.6 F_y$
(b) $f_v = F_v$ but $f_b \le 0.75 F_b$.

When either one of these conditions is not met, the interaction using Eq. (10-13) must be checked for possibly reducing the allowable flange bending stress F_b by use of this equation.

10-6.5 Intermediate Stiffeners

Stiffener design is based on obtaining the vertical stiffener force based on $P_v = 0$ from Fig. 10-12c to obtain

$$P_s = f_t t_w a \sin \phi \sin \phi$$

Substitution for f_t, ϕ, and $A_{st} = P_s/F_{\text{allow}}$ gives

$$A_{st} = \frac{1 - C_v}{2} \left\{ \frac{a}{h} - \frac{(a/h)^2}{[1 + (a/h)^2]^{1/2}} \right\} YDht_w \qquad (10\text{-}14)$$

which is AISC Eq. 1.10-3. Where $Y = F_{ys}/F_{yw}$, to account for possible use of a different yield-grade steel for the stiffener than for the web, D = factor to account for reduced efficiency for a stiffener plate on one side of the web or both sides:

$D = 1.0$ for stiffener plates on both sides of web plate
$D = 1.8$ for angle used as stiffener on one side of web
$D = 2.4$ for stiffener plate on one side of web plate

The required area of stiffener is often very small. To ensure a stiffener sufficiently large and stiff to maintain the girder web shape, the moment of inertia must be at least

$$I_{st} \geq \left(\frac{h}{50}\right)^4 \qquad (10\text{-}15)$$

Use h in in or m to obtain $I_{st} = $ in^4 or m^4. Since the stiffener is a compression member the minimum b/t ratio should be maintained.

The stiffener must be fastened to the web to carry some vertical force. Partial differentiation of the equation for P_s with respect to a/h gives

$$P_s \cong 0.015 F_y h^2 \sqrt{\epsilon_y}$$

Using $\epsilon_y = F_y/E$, a safety factor of 1.67, and assuming that the stiffener force is developed in the distance $h/3$ from the compression flange, we obtain

$$\frac{3P_s}{h} = \frac{3h(0.015)}{1.67}\sqrt{F_y^3/E}$$

And with some additional manipulation and slight rounding the design equations are obtained as

$$\text{fps:} \quad f_{vs} = h\left[\left(\frac{F_y}{340}\right)^3\right]^{1/2} \quad \text{kips/in} \qquad (10\text{-}16)$$

$$\text{SI:} \quad f_{vs} = h\left[\left(\frac{F_y}{6.5}\right)^3\right]^{1/2} \quad \text{kN/m} \qquad (10\text{-}16m)$$

Use $h = $ in or m.

10-6.6 Bearing Stiffeners

Bearing stiffeners are always required in pairs over the reactions. Bearing stiffeners may be required beneath concentrated loads carried by plate girders. These stiffeners must extend the full flange to flange distance and have a close bearing against the flange delivering the load. The stiffener width must be such as to extend approximately to the outer edges of the flange or angles.

Bearing stiffeners are designed as columns with an area that includes the stiffeners and a central web area of $12t_w$ for end and $25t_w$ for interior bearing stiffeners (see Fig. 10-13). This area is used for computing the radius of gyration and for checking the column stresses. The effective length of the stiffener may be taken as $0.75h$ because of being securely connected to the web.

The effective bearing area A'_e is taken as the area outside the flange angle fillet or the flange-to-web welds.

The design requires computing L'/r to find the allowable column stress F_a and checking

$$P = AF_a \leq \text{applied load or reaction}$$

Also check using the effective bearing area $A_b = b'_s t_s$ (Fig. 10-13) to obtain

$$P = A_b F_{\text{brg}} \leq \text{applied load or reaction}$$

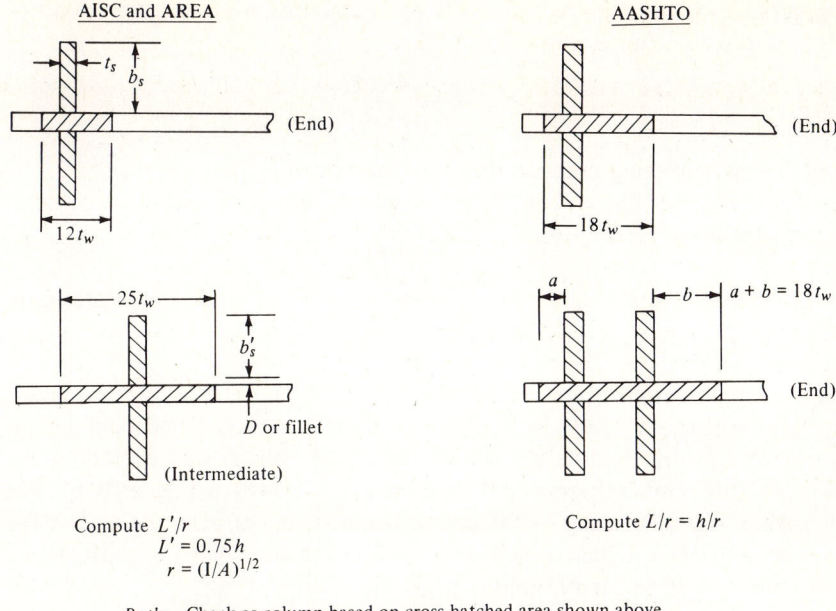

Figure 10-13 Bearing stiffener design requirements for specifications indicated.

The allowable bearing stress $F_{\text{brg}} = 0.90 F_y$ (AISC Sec. 1-5.1.5.1) based on the lesser F_y if the flange and stiffener are of different yield grades of steel.

10-6.7 Web Crippling

Webs of plate girders are required to be proportioned so that web crippling (same phenomena as for rolled beams) does not occur. This is accounted for at reactions by bearing stiffeners. Where the compression flange supports a uniform load or concentrated loads for which bearing stiffeners may not be required, the compressive stress delivered by the flange to the girder web must be sufficiently low that crippling (Fig. 10-9b illustrates buckling) does not occur. This is an edge-loaded plate stability problem, and again Eq. (3-5) is used (see Basler, "New Provisions for Plate Girder Design," *Proceedings* 1961 *AISC National Conference*, AISC, New York), SF = 2.65, E = 29 000, and some rounding, to obtain the allowable web compressive stress as

$$\text{fps:} \quad F = \frac{10\ 000 k_c}{(h/t_w)^2} \qquad (10\text{-}17)$$

$$\text{SI:} \quad F = \frac{68\ 900 k_c}{(h/t_w)^2} \qquad (10\text{-}17\text{m})$$

Basler suggested (AISC requires), for a flange delivering a compressive load to the web free to rotate (unrestrained),

$$k_c = 2 + \frac{4}{(a/h)^2}$$

and for a flange delivering compressive load restrained against rotation,

$$k_c = 5.5 + \frac{4}{(a/h)^2}$$

In the AISC specifications, Eq. (10-17) is directly combined with the appropriate k_c term for displaying the design equations.

Example 10-4 Design a welded plate girder to support two columns spanning an auditorium space in a high-rise building. Floor loads deliver an equivalent uniform load to the top flange of 2.8 kips/ft (not including the girder weight). General span and loading is as shown in Fig. E10-4a. We will assume lateral bracing of the compression flange at the ends and at the concentrated loads. Other design data: E60 electrodes, AISC specifications, A-36 steel, and girder depth limited to 84 in.

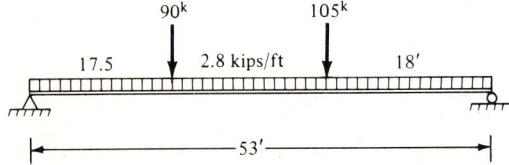

Figure E10-4a

SOLUTION Compute the girder reactions:

$$R_R = \frac{17.5(90 + 2 \times 105) + 53(2.8)53/2}{53} = 173.25 \text{ kips}$$

$$R_L = \frac{18(105) + 35.5(90) + 53(2.8)53/2}{53} = 170.15 \text{ kips}$$

$$\Sigma R = 343.4 \text{ kips}$$

Now draw shear and moment diagrams as shown in Fig. E10-4b.
 Step 1. Make a preliminary web plate design.
 Assume that web depth = 78 in. The limiting h/t_w for no reduction in flange bending stress is

$$\frac{h}{t_w} = \frac{760}{\sqrt{F_b}} \cong \frac{760}{\sqrt{22}} = 162 \text{ (approximately)}$$

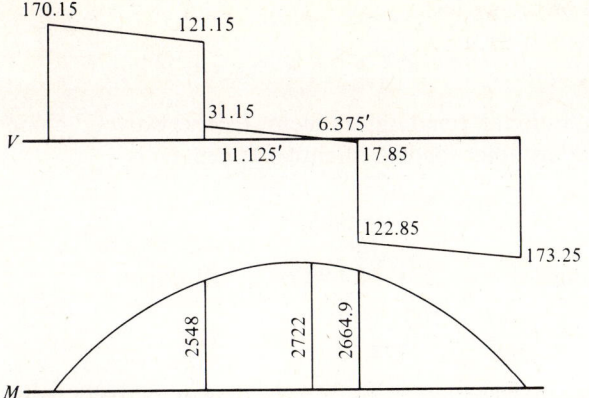

Figure E10-4b

For $h/t = 162$, obtain $t = 78/162 = 0.48$ in. Try tentative $t_w = 0.5$ in.

$$\text{Maximum } \frac{h}{t_w} = \frac{14\,000}{\sqrt{F_y(F_y + 16.5)}} = 322$$

$$t_{w(min)} = \frac{78}{322} = 0.242 \text{ in}$$

For $t_w = 0.5$ in, $h/t = 78/0.5 = 156$.

Step 2. Make a preliminary flange design.

$$A_f \cong \frac{M}{f_b h} - \frac{A_w}{6}$$

$$= \frac{2722(12)}{21(79)} - \frac{78(0.5)}{6} = 19.6 - 6.4 = 13.2 \text{ in}^2$$

(Assume that $f_b \cong 21$ ksi and an average distance to the center of the flange area of 79 in.) Try a flange plate $7/8 \times 18$ in.

$$\frac{b}{2t_f} = \frac{18}{2(0.875)} = 10.3 < \frac{95}{\sqrt{F_y}} \quad \text{(AISC Sec. 1-9.1.2)}$$

Step 3. Compute the actual moment of inertia and section modulus of the trial section and revise the dimensions as required.

$$I_{xx} = \frac{bh^3}{12} + 2Ad^2 \quad \text{(neglect } I_0 \text{ of flange plates about centroid axis)}$$

$$= \frac{0.5(78)^3}{12} + 2(0.875)(18)\left(\frac{78}{2} + \frac{0.875}{2}\right)^2$$

$$= 19\,773 + 48\,992 = 68\,765 \text{ in}^4$$

$$S_x = \frac{68\,765}{78/2 + 0.875} = \frac{68\,765}{39.875} = 1724.5 \text{ in}^3$$

Compute the weight of the girder:

$$w = \frac{0.490}{144}(2 \times 18 \times 0.875 + 0.5 \times 78) = 0.24 \text{ kip/ft}$$

The approximate additional bending moment (conservatively computed since the moment due to other loads is slightly off center) is

$$M = \frac{0.24(53)^2}{8} = 84.3 \text{ ft} \cdot \text{kips}$$

The total bending moment $= 2722 + 84.3 = 2806.3$ ft · kips.

$$f_b = \frac{M}{S} = \frac{2806.3(12)}{1724.5} = 19.53 \text{ ksi}$$

Since this is considerably less than 21 or 22 ksi, let us tentatively revise the web thickness t_w to $\frac{3}{8}$ in.

$$\frac{h}{t_w} = \frac{78}{0.375} = 208 < 322 \quad \text{O.K.}$$

Check the web shear so that the plate is not too thin (neglect beam weight at this point):

$$f_v = \frac{173.25}{78(0.375)} = 5.92 \ll 0.4 F_y \quad \text{O.K.}$$

Step 4. Recompute I and f_b.

$$I_{xx} = 14\,829.8 + 48\,992 = 63\,822 \text{ in}^4$$

$$S_x = \frac{63\,822}{39.875} = 1600 \text{ in}^3$$

$$\text{New girder weight} = \frac{0.490}{144}(31.5 + 29.3) = 0.207 \text{ kip/ft}$$

The approximate additional moment due to girder weight is

$$M = \frac{0.207(53)^2}{8} = 72.7$$

$$M_{\text{total}} = 2722 + 72.7 = 2795 \text{ ft} \cdot \text{kips}$$

$$f_b = \frac{2795(12)}{1600} = 20.96 \text{ ksi}$$

Continue the design. Use two flange plates $\frac{7}{8} \times 18$ in and a web plate $\frac{3}{8} \times 78$.

$$\frac{h}{t_w} = 208$$

$$D = 79.75 \text{ in} \quad A_w = 29.3 \text{ in}^2$$

$$S_x = 1600 \text{ in}^3 \quad A_f = 15.75 \text{ in}^2$$

Step 5. Check the allowable stress F_b.

Compute L/r_T with r_T computed using I_y of the compression flange. The radius of gyration is obtained using $A_f + A_w/6$:

$$r_T = \sqrt{\frac{I_y}{A_f + A_w/6}}$$

$$I_y = \frac{bh^3}{12} = \frac{0.875(18)^3}{12} = 425.25 \text{ in}^4$$

$$A_f + \frac{A_w}{6} = 0.875(18) + \frac{29.3}{6} = 20.63 \text{ in}^2$$

$$r_T = \sqrt{\frac{425.25}{20.63}} = 4.54 \text{ in}$$

We must investigate both the end panel and the interior (between two columns) panel since C_b is different for each location. For the end panel: $L = 18$ ft (largest value), and $m_1 = 0$ and $M_2 = 2664$ ft · kips

$$C_b = 1.75 + 1.05\left(\frac{M_1}{M_2}\right) + 0.3\left(\frac{M_1}{M_2}\right)^2 = 1.75$$

For the interior panel: $C_b = 1.0$ (since interior moment of $2722 > 2664$ ft · kips):

$$\frac{L}{r_T} = \frac{18(12)}{4.54} = 47.3 < 53\sqrt{C_b}$$

For this L/r_T ratio, $F_b = 0.6F_y = 22$ ksi for both the end and interior panels.

Now check AISC Eq. 1.10-5, since $h/t_w = 208 > 760/\sqrt{22}$:

$$F'_b = F_b\left[1 - 0.0005\frac{A_w}{A_f}\left(\frac{h}{t_w} - \frac{760}{\sqrt{F_b}}\right)\right]$$

$$= 22\left[1 - 0.0005\left(\frac{29.3}{15.75}\right)(208 - 162)\right] = 21.05 > 20.96 \text{ ksi} \quad \text{O.K.}$$

At this point the bending stress and girder proportions are adequate unless a later interaction check requires a revision of girder section.

Step 6. Compute the stiffener requirements.

AISC specifications require bearing stiffeners under reactions and the two concentrated column loads. In the end panel the actual shear stress at the reaction is

$$f_v = \frac{V}{A_w} = \frac{173.25 + 0.207(53/2)}{0.375(78)} = 6.11 \text{ ksi}$$

According to AISC Sec. 1-10.5.3, intermediate stiffeners (other than the end bearing stiffener) are not required if

$$\frac{h}{t} \leq 260 \quad \text{and} \quad f_v \leq F_v = \frac{F_y C_v}{2.89}$$

We will provide bearing stiffeners under the column loads, which leaves a clear distance in the 18-ft end panel of $18 \times 12 = 216$ in.

$$\frac{h}{t_w} = 208 < 260 \quad \text{O.K.}$$

$$\frac{a}{h} = \frac{216}{78} = 2.77 > 1$$

$$k = 5.34 + \frac{4}{(a/h)^2} = 5.34 + \frac{4}{(2.77)^2} = 5.986$$

Assuming that C_v will compute < 0.8, we will try Eq. (10-11):

$$C_v = \frac{45\,000k}{F_y(h/t_w)^2} = \frac{45\,000(5.86)}{36(208^2)} = 0.169 < 0.8$$

The allowable shear stress if no intermediate stiffeners are used is limited to

$$F_v = \frac{36(0.169)}{2.89} = 2.11 \text{ ksi} \ll 6.11 \quad \text{N.G. stiffeners required}$$

Try one stiffener at half the distance (noting that if this works for the 18-ft panel it will also work on other end in 17.5-ft panel):

$$a = \frac{216}{2} = 108 \text{ in}$$

$$\frac{a}{h} = \frac{108}{78} = 1.38 < 3 \quad \text{O.K. (less than maximum } a/h \text{ allowed)}$$

$$\frac{a}{h} \leq \frac{260}{(h/t_w)^2} = \left(\frac{260}{208}\right)^2 = 1.56 > 1.38 \quad \text{O.K.}$$

$$k = 5.34 + \frac{4}{1.38^2} = 7.44$$

$$C_v = \frac{45\,000(7.44)}{36(208)^2} = 0.215$$

With intermediate stiffeners and $C_v < 1$, we can use AISC Eq. (1.10-2) or Eq. (10-9):

$$F_v = \frac{36}{2.89}\left[C_v + \frac{1 - C_v}{1.15(1 + 1.38^2)^{1/2}}\right] = 7.67 \text{ ksi}$$

Since 7.67 ksi is greater than the actual shear stress at the bearing stiffener, 6.11 ksi, and the shear is less at the interior points, it is not necessary to check the shear stress further for stiffener analysis. Note that F_v could have been obtained from Table II-8 of SSDD but requires double interpolation.

Step 7. Check the interaction at concentrated loads [Eq. (10-13)].

At dx to the right of the right column, the shear V is

$$V = 178.4 - 18(2.80 + 0.207) = 124.6 \text{ kips (including girder weight)}$$

For $V = 124.6$ kips, the web shear stress is

$$f_v = \frac{124.6}{29.3} = 4.25 \text{ ksi}$$

We have just found $F_v = 7.67$ ksi in step 6.

$$F_b = \left[0.825 - 0.375\left(\frac{4.25}{7.67}\right)\right]36 = 22.2 > 22 \text{ ksi} \quad \text{O.K.}$$

Since the allowable stress in interaction is 22 ksi (maximum) and the actual and allowable values based on flange stability are less, the lesser values control.

$$F_b = 21.05 \text{ ksi} \qquad f_b = 20.96 \text{ ksi}$$

Step 8. Check the stiffener requirements of the 17.5-ft interior span between columns.

The maximum shear is obtained coming from the left:

$$V = 170.15 + 0.207\left(\frac{53}{2}\right) - 17.5(2.8 + 0.207) - 90 = 33.01 \text{ kips}$$

$$f_v = \frac{33.01}{29.3} = 1.13 \text{ ksi}$$

Try no stiffeners:

$$\frac{a}{h} = \frac{17.5(12)}{78} = 2.69 < 3.0$$

$$k = 5.34 + \frac{4}{2.69^2} = 5.89$$

$$C_v = \frac{45\ 000(5.89)}{36(208)^2} = 0.170$$

$$F_v = \frac{36(0.170)}{2.89} = 2.12 > 1.13 \text{ ksi} \quad \text{O.K.} \quad \text{no stiffeners required}$$

Step 9. Check the web crippling under the compression flange due to uniform load.

Assume that the flange is restrained against rotation (since it carries a uniform load). The load carried in compression to the web is 2.8 kips/ft + weight of top flange. We will neglect flange weight, so that the compressive stress is

$$f_c = \frac{2.8}{0.375(12)} = 0.622 \text{ ksi}$$

The allowable compressive stress (checking at the location where a/h is critical is

$$F = \frac{10\,000}{(h/t_w)^2}\left[5.5 + \frac{4}{(a/h)^2}\right] \quad [\text{Eq. (10-17) combined with } k_c]$$

$$= \frac{10\,000}{208^2}\left(5.5 + \frac{4}{2.69^2}\right) = 1.4 > 0.622 \text{ ksi} \quad \text{O.K.}$$

Note that if $F < f_c$, we would have to either add stiffeners to decrease a/h or increase t_w.

Step 10. Design bearing and intermediate stiffeners.

For bearing stiffeners at girder ends (we will use the same size under column loads) try two $8 \times 1/2$ in bars, for a width ($2 \times 8 + 0.375 = 16.375$ in) approaching the width of the flange plate of 18 in.

$$\frac{b}{t} = \frac{8}{0.5} = 16 \simeq \frac{95}{\sqrt{F_y}} \quad \text{O.K.}$$

$$I = \frac{0.5(16.35)^3}{12} = 182.9 \text{ in}^4$$

The "effective" column area for the radius of gyration (see Fig. 10-13):

$$A = 16 \times 0.5 + 12(0.375)(0.375) = 9.69 \text{ in}^2$$

$$r = \sqrt{\frac{182.9}{9.69}} = 4.34 \text{ in}$$

$$L = 0.75h = 0.75(78) = 58.5 \text{ in}$$

$$\frac{L}{r} = \frac{58.5}{4.34} = 13.5$$

From Table II-5, the allowable column stress is $F_a = 20.98$ ksi. The actual column stress is

$$f_a = \frac{178.74}{9.69} = 18.45 > 20.98 \text{ ksi} \quad \text{O.K.}$$

Check the bearing stress; assume a 5/16-in weld, so that the effective bearing area $= (8 - 0.3125)(0.5)(2) = 7.7$ in^2:

$$F_{brg} = 0.9F_y = 32.4 \text{ ksi}$$

$$P_e = 32.4(7.7) = 249.5 \gg 178.74 \text{ kips} \quad \text{O.K.}$$

For intermediate stiffeners, with only one intermediate stiffener use two plates.

$$A_{st} = \frac{1 - C_v}{2}\left\{\frac{a}{h} - \frac{(a/h)^2}{[1 + (a/h)^2]^{1/2}}\right\} YDht_w \quad [\text{Eq. (10-14)}]$$

For a plate on both sides of the beam web, $D = 1.0$ and $Y = 1$ (A-36 steel for web and stiffener); $C_v = 0.215$ and $a/h = 1.38$ from step 6.

$$A_{st} = \frac{1 - 0.215}{2}\left[1.38 - \frac{1.38^2}{(1 + 1.38^2)^{1/2}}\right](1)(1)(29.3) = 3.02 \text{ in}^2$$

In terms of $A_{st} \times 100/A_w$ this value could have been obtained from Table II-8 by using double interpolation. Try two bars $3/8 \times 6$ in:

$$A = 2(\tfrac{3}{8} \times 6) = 4.5 \text{ in}^2 > 3.02 \quad \text{O.K.}$$

$$\frac{b}{t} = \frac{6}{0.375} = 16 \quad \text{O.K.}$$

The minimum moment of inertia of the stiffeners is

$$\left(\frac{h}{50}\right)^4 = \left(\frac{78}{50}\right)^4 = 5.92 \text{ in}^4$$

$$I_{\text{furn}} = \frac{(0.375)(12 + 0.375)^3}{12} = 59.22 \text{ in}^4 \gg 5.92 \quad \text{O.K.}$$

Use a stiffener plate length

$$h - 4t_w = 78 - 4(0.375) = 76.5 \text{ in}$$

Step 11. Design welding to fasten the stiffeners to the web.

$$f_{vs} = h\left[\left(\frac{F_y}{340}\right)^3\right]^{1/2} \quad \text{kips/in} \quad [\text{Eq. (10-16)}]$$

$$= 78\left[\left(\frac{36}{340}\right)^3\right]^{1/2} = 2.7 \text{ kips/in}$$

For a pair of stiffeners, we have 2.7 kips/in; for each side this becomes $2.7/2 = 1.35$ kips/in.

For 1/2-in bearing stiffeners, use a 3/16-in weld and E60 electrodes.

$$F_w = 0.1875(0.70711)(0.3 \times 60) = 2.39 \text{ kips/in}$$

Use a 3/16-in fillet weld continuous for both bearing and intermediate stiffeners.

Rationale: Few stiffeners and the weld can be made in one pass. It is too difficult to measure and set up alternating weld distances and gaps.

Step 12. Design welding to fasten the flange plate to the web.
Check the end for maximum shear:

$$v = \frac{VQ}{I} = \frac{178.74(18 \times 7/8)[(78 + 0.875)/2]}{63\,822} = 1.74 \text{ kips/in}$$

Use a 5/16-in ($t_f > 3/4$ in) continuous weld on both sides at $F_w = 3.98$ kips/in. Note that the weld is considerably overdesigned, but for an important girder the use of intermittent welds is not worth the savings—particularly if the weld can be made in one pass, as here.

Figure E10-4c illustrates the design summary for the girder.

490 STRUCTURAL STEEL DESIGN

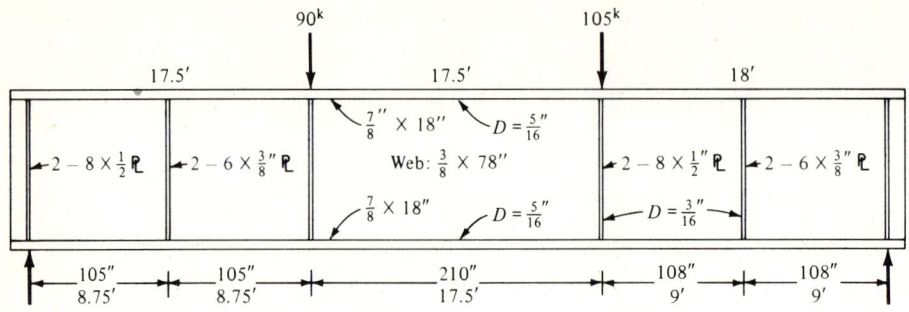

Figure E10-4c

///

10-7 PLATE GIRDER DESIGN THEORY—AASHTO AND AREA

Plate girder design using AASHTO and AREA specifications is very similar but more conservative than AISC because of the more hostile environment to which the girder will be subjected. In general, however, the same general considerations apply:

1. The girder is proportioned by the moment-of-inertia method.
2. No unique solution is possible.
3. Shear and stiffener requirements are more rigid. Bearing stiffeners are always required, but AASHTO allows use of longitudinal stiffeners.†

10-7.1 Girder Flanges (AASHTO and AREA)

In general, the basic allowable bending stress is

$$F_b \leq 0.55 F_y$$

However, if the compression flange is laterally unsupported in a length L, the stresses must be reduced as in Part III of SSDD and in the appropriate specification.

The $b/2t_f$ ratio for the flanges is also limited:

AASHTO:

fps	SI
$\dfrac{b}{2t_f} \leq \dfrac{51.4}{\sqrt{f_b}}$	$\dfrac{b}{2t_f} \leq \dfrac{135}{\sqrt{f_b}}$

The limiting $b/2t_f = 12$ for A-36 steel when $f_b = F_b = 0.55 F_y$.

† Also for AREA, but for continuous girders.

AREA:

When cross-ties rest directly on the girder flange:

$$\frac{b}{2t_f} \le \frac{60}{\sqrt{F_y}} \qquad \frac{b}{2t_f} \le \frac{160}{\sqrt{F_y}}$$

The limiting $b/2t_f = 10$ for A-36 steel

When cross-ties do not rest directly on the girder flange:

$$\frac{b}{2t_f} \le \frac{72}{\sqrt{F_y}} \qquad \frac{b}{2t_f} \le \frac{190}{\sqrt{F_y}}$$

The limiting $b/2t_f = 12$ for A-36 steel.

10-7.2 Web Plate Design

AASHTO also uses Eq. (3-5):

$$F_{cr} = \frac{k_c \pi^2 E}{12(1-\mu^2)(h/t_w)^2}$$

and solving for h/t_w, substituting $k_c = 23.9$, $E = 29\,000$, and $F_{cr} = 1.19 F_b$, we obtain

$$\text{fps:} \quad \frac{h}{t_w} = \frac{727}{\sqrt{f_b}} \qquad \text{SI:} \quad \frac{h}{t_w} = \frac{1920}{\sqrt{f_b}} \qquad (10\text{-}17)$$

This gives $h/t_w = 163$ for A-36 steel.

The AREA limiting h/t_w ratio is obtained from Eq. (3-5) by substitution of $k_c = 23.9$, $E = 29\,000$, and $F_{cr} = 0.6 F_y$ to obtain (with slight rounding)

$$\text{fps:} \quad \frac{h}{t_w} = \frac{1030}{\sqrt{F_y}} \qquad \text{SI:} \quad \frac{h}{t_w} = \frac{2700}{\sqrt{F_y}} \qquad (10\text{-}18)$$

For A-36 steel, the limiting $h/t_w = 171$. When the value of $f_b < F_b$ in the compression flange, AREA allows an increase in the h/t_w ratio:

$$\left(\frac{h}{t_w}\right)' = \left(\frac{h}{t_w}\right)\left(\frac{F_b}{f_b}\right)^{1/2}$$

with the ratio $(F_b/f_b)^{1/2} \le 2.0$.

AASHTO and AREA limits the web thickness of fabricated plate girders to:

AASHTO: $t_w \ge 5/16$ in (8.0 mm)   AREA: $t_w \ge 0.335$ in (8.5 mm)

AASHTO allows the limiting h/t_w to be increased with a longitudinal stiffener at 1/5 of the clear web depth from the compression flange. This value is based on Eq. (3-5) using $k_c = 129$ (since a more substantial edge fixity is obtained) and $F_{cr} = 1.6F_b$, to obtain

$$\text{fps:} \quad \frac{h}{t_w} \leq \frac{1455}{\sqrt{f_b}} \qquad \text{SI:} \quad \frac{h}{t_w} \leq \frac{3840}{\sqrt{f_b}} \qquad (10\text{-}19)$$

The limiting h/t_w for A-36 steel is 327.

10-7.3 Shear and Stiffener Requirements

The AASHTO requirement for stiffeners is also computed using Eq. (3-5) and the k factors given in Eq. (10-8) and with varying safety factors. If the plate is such that $a/h \to \infty$ (no stiffeners), $k_c = 5.34$ from Eq. (10-8), and taking $F_{cr}\text{-}2.5f_v$, we obtain

$$\text{fps:} \quad \frac{h}{t_w} \leq \frac{234}{\sqrt{f_v}} \qquad \text{SI:} \quad \frac{h}{t_w} \leq \frac{618}{\sqrt{f_v}} \qquad (10\text{-}20)$$

The limiting $h/t_w = 68$ for A-36 steel when $f_v = F_v = 0.33F_y$.

AREA requirements are slightly more conservative:

$$\text{fps:} \quad \frac{h}{t_w} \leq \frac{360}{\sqrt{F_y}} \qquad \text{SI:} \quad \frac{h}{t_w} \leq \frac{950}{\sqrt{F_y}} \qquad (10\text{-}21)$$

The limiting $h/t_w = 60$ for A-36 steel.

Intermediate stiffeners are not required by either specification if the h/t_w ratio is less than that given by Eq. (10-20) or (10-21) and may not be required for larger h/t_w ratios as given in the following section.

When h/t_w is larger than given by the preceding equations, the stiffener spacing is obtained from Eq. (3-5) by making some rearrangement to obtain

$$F_{cr} = \text{SF}(f_v) = \frac{k_c \pi^2 E}{12(1-\mu^2)} \left(\frac{t_w}{h}\right)^2 \left(\frac{a}{a}\right)^2$$

from which

$$a = \frac{a}{h} \left[\frac{k_c \pi^2 E}{12(1-\mu^2)\text{SF}}\right]^{1/2} \frac{t_w}{\sqrt{f_v}} = \beta \frac{t}{\sqrt{f_v}}$$

Using SF = 1.5, several a/h ratios from 0.5 to 1.0, and computing k_c from Eq. (10-8), we typically obtain

a/h	k	β
0.5	25.36	332.8
0.6	18.83	344.4
0.7	14.90	357.2
0.8	12.30	370.9
0.9	10.58	387.2
1.0	9.34	404.0

From an approximate averaging of β, we obtain 348, which if used to back-compute the SF gives 1.37 at $a/h = 0.5$ to 2.02 at $a/h = 1.0$. In earlier AASHTO specifications the stiffener spacing could be computed as

$$\text{fps:} \quad a = \frac{348t}{\sqrt{f_v}} \quad \text{(in)} \qquad \text{SI:} \quad a = \frac{914t}{\sqrt{f_v}} \quad \text{(mm)}$$

where t has units of in or mm. The previous two editions of the AISC specifications included this requirement for spacing the first interior stiffener from the bearing stiffener. Currently, the AASHTO specifications for intermediate stiffeners is somewhat more indirect and is given in Sec. 10-7.4 following under "Intermediate Stiffeners."

The AREA specifications reduce the 348 factor slightly to obtain

$$\text{fps:} \quad a = \frac{332t}{\sqrt{f_v}} \quad \text{(in)} \qquad \text{SI:} \quad a = \frac{875t}{\sqrt{f_v}} \quad \text{(mm)} \qquad (10\text{-}22)$$

In these equations t has units of in or mm. The stiffener spacing is limited to not more than 72 in or 1800 mm when using Eq. (10-22).

10-7.4 Stiffener Design

Longitudinal stiffeners The AASHTO value for longitudinal stiffeners in terms of moment of inertia (see Erickson and Eenam, "Application and Development of AASHTO Specifications to Bridge Design," *Journal of Structural Division, ASCE* July 1957) is

$$I_s = ht_w^3 \left[2.4\left(\frac{a}{h}\right)^2 - 0.13 \right] \qquad (10\text{-}23)$$

with a plate thickness limited to

$$\text{fps:} \quad \frac{b}{t} \leq \frac{71}{\sqrt{f_b}} \qquad \text{SI:} \quad \frac{b}{t} \leq \frac{188}{\sqrt{f_b}} \qquad (10\text{-}24)$$

where f_b = calculated bending stress in the compression flange. The stiffeners may be placed on one side of the web at the 1/5 point from the compression flange. The AREA allows longitudinal stiffeners for continuous girders in the negative-moment regions and the specifications are exactly the same as for AASHTO for I_s and b/t.

Bearing stiffeners Bearing stiffeners are always required for both AASHTO and AREA girders. They are designed as columns using an area as shown in Fig. 10-13. The allowable column stress depends on the L/r ratio computed as shown on Fig. 10-13. The bearing stiffeners must be checked for bearing as well as acting as columns. The bearing area is as shown on Fig. 10-13 and the allowable bearing stress is

$$\text{AASHTO:} \quad F_{\text{brg}} = 0.80 F_y \qquad \text{AREA:} \quad F_{\text{brg}} = 0.83 F_y$$

AASHTO specifications limit the b_s/t_s ratio of bearing stiffeners (using either angles or plates) to

$$\text{fps:} \quad \frac{b_s}{t_s} \leq \frac{69}{\sqrt{F_y}} \qquad \text{SI:} \quad \frac{b_s}{t_s} \leq \frac{182}{\sqrt{F_y}} \qquad (10\text{-}25)$$

AREA requirements for the b_s/t_s ratio of bearing stiffeners is the same as for any other compression member.

Intermediate stiffeners AASHTO does not require intermediate transverse stiffeners if

$$\frac{h}{t_w} \leq 150 \quad \text{and} \quad f_v \leq F_v$$

where

$$\text{fps:} \quad F_v = \frac{56\,250}{(h/t_w)^2} \qquad \text{SI:} \quad F_v = \frac{387\,830}{(h/t_w)^2} \qquad (10\text{-}26)$$

but $F_v \leq F_y/3$. This criterion gives the limiting $h/t_w = 68$ for $F_v = F_y/3$ for A-36 steel. Thus any h/t_w ratio that is less than 68 does not require transverse stiffeners if $f_v < F_y/3$. Any $h/t_w < 150$ does not require stiffeners if $f_v < F_v$ as given in Eq. (10-26).

When intermediate stiffeners are required or when $f_v > F_v$ of Eq. (10-26), the maximum spacing is limited to $a \leq 1.5h$ and the allowable shearing stress is limited to

$$F_v = \frac{F_y}{3} \left\{ C + \frac{0.87(1 - C)}{[1 + (a/h)^2]^{1/2}} \right\} \quad \text{(ksi or MPa)} \qquad (10\text{-}27)$$

where $C = \dfrac{B[1 + (h/a)^2]}{F_y(h/t_w)^2} \leq 1$

$B = 222\,000$ in fps units
$ = 1\,520\,000$ in SI units

It is often convenient to plot the limits of these stresses as in Fig. 10-14.

When transverse stiffeners are required, the first stiffener is placed such that $a/h \leq 0.5$ and the actual shearing stress, $f_v \leq F_v'$, is obtained from the following equation:

$$F_v' = \frac{B'[1 + (h/a)^2]}{(h/t_w)^2} \leq \frac{F_y}{3} \qquad (10\text{-}28)$$

where $B' = 70\,000$ in fps units
$ = 483\,000$ in SI units

The AASHTO specifications require a minimum moment of inertia of inter-

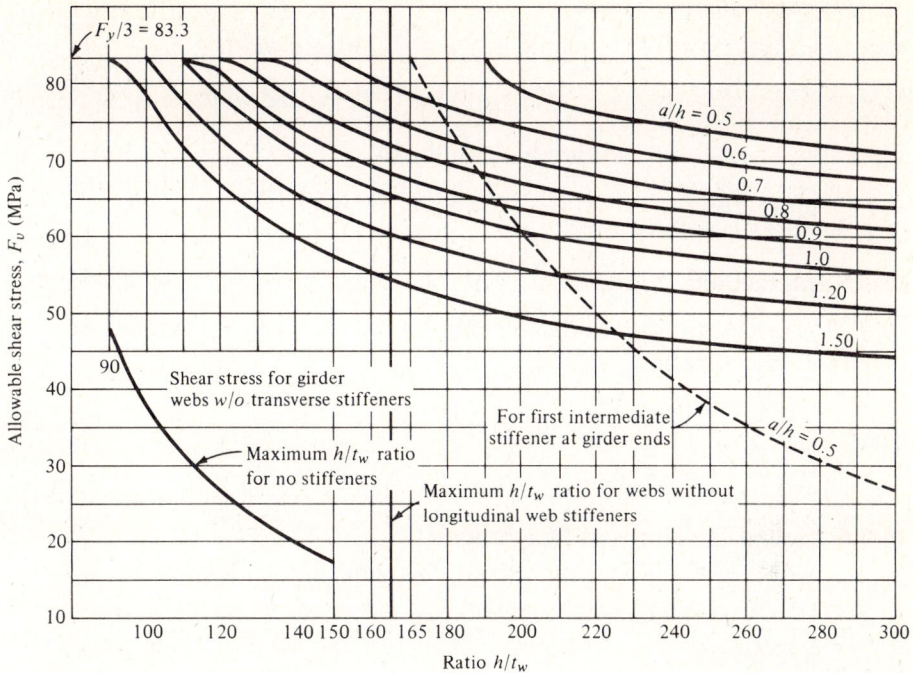

Figure 10-14 Allowable shear stress for h/t_w ratios shown and for stiffener spacing ratios a/h. Note the limiting shear stress is $F_y/3$.

mediate stiffeners (including other than reaction bearing stiffeners) of

$$I_s = \frac{at_w^3}{12(1-\mu^2)} J \qquad (10\text{-}29)$$

where $\mu = 0.3$ for steel
$a =$ actual stiffener spacing
$t_w =$ thickness of girder web plate
$J = 25(h/a)^2 - 20$ but $J \geq 5.0$

AREA requires intermediate stiffeners if the h/t_w ratio is greater than the values defined by Eq. (10-21). If stiffeners are required the spacing is limited to

$$a \leq \frac{332 t_w}{\sqrt{f_v}} \quad \text{(in)} \qquad a \leq \frac{0.875 t_w}{\sqrt{f_v}} \quad \text{(m)}$$

but $a \leq 72$ in (or 1.829 m),
where $t_w =$ girder web thickness, in or mm
$f_v =$ actual web shear stress $= V/dt_w$, ksi or MPa

The stiffener dimensions must be at least as follows:

AASHTO: width $\geq$ 2 in + $D/30$ and width $\geq b_f/4$

$\geq$ 51 mm + $D/30$ and width $\geq b_f/4$

thickness $\geq$ width/16

AREA: same as for AASHTO

10-7.5 Interaction

AASHTO specifications include a bending stress reduction if the shear stress is $f_v \leq 0.6 F_v$ according to

$$F_b' = \left(0.754 - \frac{0.34 f_v}{F_v}\right) F_y \leq 0.55 F_y \tag{10-30}$$

10-7.6 Lateral Bracing and Diaphragms

Figures 10-15 and 10-16 illustrate bracing and diaphragms in current practice. AASHTO specifications require either cross frames or diaphragms at the ends and at intervals along the span not to exceed 25 ft (7.62 m). Diaphragms (cross beams) must be at least one-third and preferably one-half or more of the girder depth.

When spans exceed 125 ft (38.1 m) for concrete deck bridges, an additional system of lateral bracing will be used parallel to and as close as practical to the bottom flange and in at least one-third of the bays (when multiple girders are used). The smallest angle permitted by AASHTO for this is the $3 \times 2\frac{1}{2}$ (76 $\times$ 64 mm) fastened with at least two bolts or the equivalent weld in each end connection of the angles.

(a)

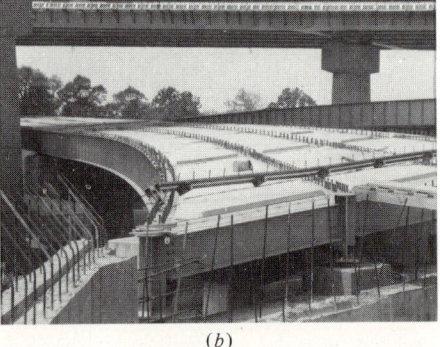
(b)

Figure 10-15 Bridge bracing using diaphragms. (a) Diaphragms used with five plate girders. (b) End diaphragm. Must be sufficiently set back from adjacent diaphragm or abutment so that it can be inspected and periodically painted.

PLATE GIRDERS **497**

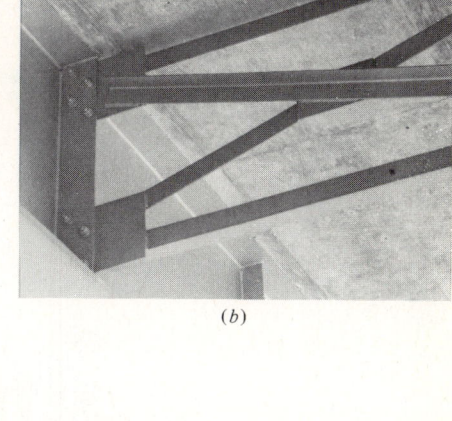
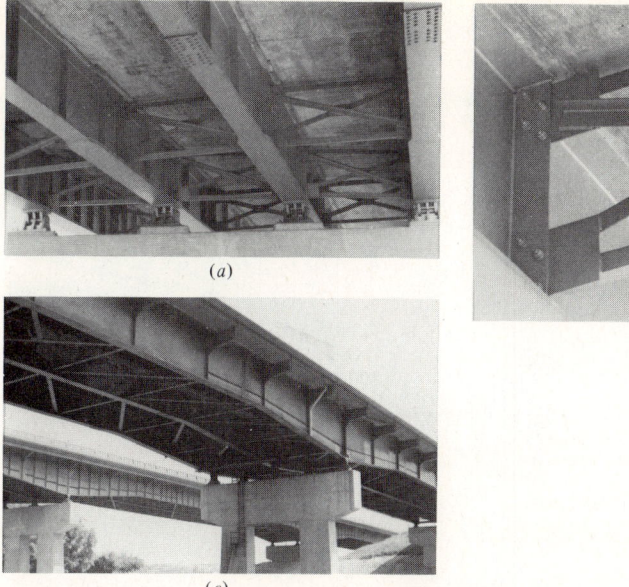

(a)

(b)

(c)

Figure 10-16 Longitudinal and cross bracing. Note use of vertical and longitudinal stiffeners. Bridge in background of (c) is older and uses riveted construction where bridge in foreground is all-welded. The three girders carry a four-lane roadway plus shoulders and walkway for both bridges. (a) Cross bracing. (b) Fabrication details. Note that the end fastener plates are welded to both web and flange. On continuous bridges these plates provide flange brace points. (c) Use of both cross bracing and longitudinal truss bracing. Note longitudinal stiffener along compression flange, where h/t_w is excessive.

AREA requires the following:

1. For beams or girders less than 42 in (1066 mm), use of I-shaped diaphragms (cross beams) as deep as practical and using double-angle beam framing connections.
2. For girders deeper than 42 in and spaced more than 48 in (1220 mm) on centers, a system of cross-frame bracing in which the angle of the cross-frame diagonals with the vertical is not greater than 60°.
3. Spacing of diaphragms or cross frames is
    18 ft (5.5 m) for open-deck construction
    12 ft (3.65 m) for ballasted-deck bridges
4. Where ballasted decking is carried on cross beams without stringers (as in Fig. 10-5), at least one line of longitudinal diaphragms is to be used, as shown in Fig. 10-3d.
5. Knee bracing is required to support the compression flange for through deck girders. This bracing (see Fig. 10-17) is usually placed at selected intermediate stiffeners for ease of attachment and ranges in slope from about $3V:1H$ to somewhat less. The maximum spacing is limited to 12 ft (3.66 m). It may be

498 STRUCTURAL STEEL DESIGN

Figure 10-17 Top flange bracing using web plate and bracket for the through-girder ballasted-deck railroad bridge shown in Fig. 10-4.

designed as a column for an axial force based on the horizontal component being $2\frac{1}{2}$ percent of the compression flange force

$$P_{kb} = \frac{0.025 f_b A_f}{H/V}$$

The approximation of the axial force using the sin defined by H and V rather than the actual length is sufficiently accurate for design.

Example 10-5 Design a welded railroad bridge girder for a 27.5-m-span single-track ballasted-through deck bridge. Use a Cooper E-110 loading, A-36 steel, E70 electrodes, and the AREA specifications. The general bridge configuration is as shown in Fig. E10-5a.

SOLUTION From Table 1-2, the design live load for one-half the track (one rail load to each girder) and adjusted for E-110 loading is:

$$M = 5339.1 \left(\frac{110}{80}\right) = 7341 \text{ ft} \cdot \text{kips} \times 1.36 = 9984 \text{ kN} \cdot \text{m}$$

The shear is

$$V = 274.5 \left(\frac{110}{80}\right) = 377.4 \text{ kips} \times 4.45 = 1680 \text{ kN}$$

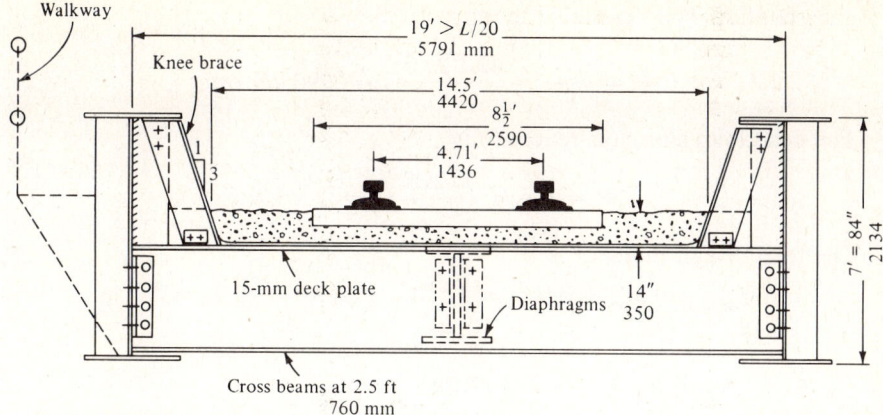

Figure E10-5a

The impact factor is computed (see Fig. E10-5a) based on $L > 25$ m as

$$I = \frac{30.5}{19} + 16 + \frac{185}{27.5 - 9} = 27.6 \quad \text{use 28 percent}$$

(Note that this impact factor is not the same as used in Example 4-15.) The design moment

$$M_d = 9984(1.28) = 12\,779.5 \text{ kN} \cdot \text{m}$$

The design shear

$$V_d = 1680(1.28) = 2150.4 \text{ kN}$$

We will neglect the live load on the pedestrian/service walkway. With this assumption the dead-load estimate is

Girder weight (including stiffeners and weld)	= 10.0 kN
Ballast, including ties at 350 mm depth and a reduced effective width: 120(9.807/62.4)(0.350)(4.42/2)	= 14.6 kN
Steel deck and cross beams and factor of 1.10 From Example 4-15, the beam is a W760 × 160.7/1.58 Deck: 0.015(77.0)(6.49)(1.10)/2.0	= 4.1 kN
Cross beams: 37(1.58)(5.79)/(27.5 × 2)	= 6.2 kN
Track at 200 lb/ft: 0.200 × 14.59 (AREA specification)	= 2.9 kN
Walkway: estimate 300 lb/ft = 0.300 × 14.59	= 4.4 kN
Total	= 42.2 kN/m

Note that some approximation is used for the steel deck to allow for forming the trough to hold the ballast to give an effective length of 6.49 m. Note also that an estimating factor of 10 percent is applied for other uncertainties. The dead load moment is

$$M = \frac{wL^2}{8} = \frac{42.2(27.5)^2}{8} = 3990 \text{ kN} \cdot \text{m}$$

The dead load shear (at reaction) is

$$V = \frac{wL}{2} = \frac{40.2(27.5)}{2} = 553 \text{ kN}$$

The total design moment is

$$M_{\text{design}} = M_L + M_{\text{dead}} = 12\,780 + 3990 = 16\,770 \text{ kN} \cdot \text{m}$$
$$V_{\text{design}} = V_L + V_{\text{dead}} = 2150.4 + 580 = 2730.4 \text{ kN}$$

Step 1. Find the girder proportions.

Take $h \simeq 2134 - 2(50) \simeq 2034$ mm. Also the web must be at least

$$t_w \simeq \frac{V}{hF_v} \simeq \frac{2730.4}{2.034(0.35 \times 250)} = 15.3 \text{ mm}$$

The maximum shear is at the reaction, but with a moving load contributing the major effects (with impact) will not change much for some distance along the girder, so we will take $t_w = 16$ mm.

$$\frac{h}{t_w} \simeq \frac{2034}{16} = 127 \ll 170 \quad \text{O.K.}$$

The tentative web area

$$A_w = 0.016(2034) = 32.54 \times 10^{-3} \text{ m}^2$$

Using Eq. (10-2), we find the tentative flange area as

$$A_f = \frac{16\,580}{2034 \times 0.5 F_y} - \frac{32.54}{6} = 65.21 - 5.42 = 59.79 \times 10^{-3} \text{ m}^2$$

Try two flange plates 70 × 815 mm (t available only in 10-mm increments):

$$h = 2134 - 2(70) = 1994 \text{ mm}$$

$$\frac{b}{2t_f} = \frac{815}{2 \times 70} = 5.82 \ll \frac{190}{\sqrt{F_y}} \quad \text{O.K.}$$

$$\frac{h}{t_w} = \frac{1994}{16} = 125 \ll 170 \quad \text{O.K.}$$

Step 2. Compute the moment of inertia of girder and determine the actual bending stress f_b so that we can check $f_b \leq F_b$.

$$I_x = I_{\text{web}} + 2A_f d^2 + \frac{b_f t_f^3}{12}$$

$$= \frac{10^6 (0.016)(1.994)^3}{12} + 2(0.070 \times 0.815)\left(997 + \frac{70}{2}\right)^2 + \frac{815(0.070)^3 10^3}{12}$$

$$= 10\,571 + 121\,510 + 24 = 132\,114 \times 10^{-6} \text{ m}^4$$

The section modulus is

$$S_x = \frac{I}{d/2} = \frac{132\ 114}{1067} = 123.81 \times 10^{-3}\ m^3$$

The corresponding maximum bending stress is

$$f_b = \frac{M}{S_x} = \frac{16\ 770}{123.81} = 135.4\ \text{MPa} < 137.5 \qquad \text{(O.K. so far)}$$

Step 3. Check the girder weight.

$$A_{\text{total}} = 2(0.070)(0.815) + 0.016(1.994) = 0.1460\ m^2$$

$$\text{Wt/m} = 77(0.1460) = 11.2\ \text{kN/m}$$

Try one web plate 16×1994 mm and two flange plates 70×815 mm.

$$I_x = 132\ 114 \times 10^{-6}\ m^4$$

$$S_x = 123.81 \times 10^{-3}\ m^3$$

Step 4. Compute the allowable bending stress F_b.

The compression flange will be laterally braced using knee brackets as in Fig. 10-17 at every fourth cross-beam supporting the deck. These intervals are approximately $4(0.76) = 3.04$ m < 3.66 (maximum by AREA). The radius of the gyration of the compression flange is

$$r_T = \left(\frac{I_f}{A}\right)^{1/2} = \left(\frac{I_f}{A_f + A_w/6}\right)^{1/2}$$

Flange area $= 0.070 \times 0.815 = 0.05705\ m^2$

$$\frac{A_w}{6} = \frac{1.994 \times 0.016}{6} = 0.0052\ m^2 \qquad \text{total} = 0.06237\ m^2$$

$$I_f = \frac{bt_f^3}{12} = \frac{0.070 \times 0.815^3}{12} = 0.003158\ m^4$$

$$r_T = \left(\frac{0.003158}{0.06237}\right)^{1/2} = 0.225\ m = 225\ mm$$

$$\frac{L}{r_T} = \frac{3.04 \times 10^3}{225} = 13.5 < \frac{282}{\sqrt{F_y}}$$

The allowable bending stress is

$$F_b = 0.55F_y\left[1.0 - \frac{F_y}{12.5 \times 10^6}\left(\frac{L}{r_T}\right)^2\right] = 137.0\ \text{MPa} > 135.4 \qquad \text{O.K.}$$

Step 5. Design the bearing stiffeners.

We will wrap the top flange around the end of the girder web to provide additional tension field resistance for the end post as shown in both Figs.

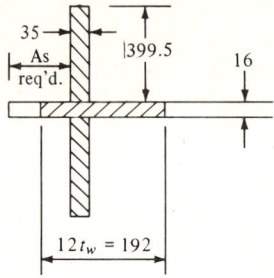

Figure E10-5b

10-1c and 10-5. The stiffener will be as in Fig. E10-5b.

$$\frac{b}{t_s} \le \frac{190}{\sqrt{F_y}} \qquad b = \frac{815 - 16}{2} = 399.5 \text{ mm}$$

Rearranging, $t_s = 399.5\sqrt{F_y}/190 = 32.2$; use 35 mm.

$$A_{st} = 2(0.3995)(0.035) + 0.192(0.016)$$
$$= 0.0310 \text{ m}^2$$

The moment of inertia of the stiffener is

$$I_{st} = \frac{bt_s^3}{12} = \frac{0.035(0.815)^3}{12} = 0.00158 \text{ m}^4$$

The radius of gyration is

$$r = \left(\frac{0.00158}{0.0310}\right)^{1/2} = 0.2257 \text{ m} = 226 \text{ mm}$$

$$\frac{L'}{r} = \frac{0.75(1994)}{226} = 6.62$$

The allowable column stress for this L/r is

$$F_a = 0.60F_y - \left(\frac{F_y}{316}\right)^{1.5} \frac{KL}{r} = 145.3 \text{ MPa} > 0.55F_y$$

The allowable column load is

$$P_{col} = 0.55F_y(31.0 \times 10^{-3} \text{ m}^2) = 4262 \text{ kN} \gg 2730.4 \qquad \text{O.K.}$$

Also, check the bearing stiffener for "bearing" and assume a 6-mm weld fillet for an area reduction.

$$A_{brg} = (2)35(0.3995 - 0.006) = 27.55 \times 10^{-3} \text{ m}^2$$
$$P_{brg} = 27.55(0.83 \times 250) = 5717 \text{ kN} \gg 2730.4 \qquad \text{O.K.}$$

Step 6. Design the intermediate stiffeners (both section and spacing).

The clear distance a is limited to

$$a = \frac{0.872 t_w}{\sqrt{f_v}} \leq 1.828 \text{ m (AREA 72-in limitation on spacing)}$$

At the bearing stiffener

$$f_v = \frac{V}{ht} = \frac{2730.4}{1994(0.016)} = 85.6 \text{ MPa} < 0.35 F_y \quad \text{O.K.}$$

This value of f_v results in an allowable spacing of

$$a = \frac{0.872(16)}{\sqrt{85.6}} = 1.51 \text{ m} < 1.828 \quad \text{O.K.}$$

We can now do one of two things:

1. Space all stiffeners (but so that they come out as an integer) at approximately this spacing.
2. Use this spacing for each end part of the bridge and a larger spacing for the interior part, since the live-load shear (without impact) is 701 kN (Table 1-2). Similarly, the dead-load shear is only 1/2 of the end value.

Let us investigate the spacing for the approximate center half of the span:

$$\text{Live load} = 764 \times 1.28 = 1234 \text{ kN}$$
$$\text{Dead load} = 580/2 = 290 \text{ kN}$$
$$\text{Total} = 1524 \text{ kN}$$

and

$$f_v = \frac{1524}{1994(0.016)} = 47.8 \text{ MPa}$$

$$a = \frac{.872(16)}{\sqrt{47.8}} = 2.01 \text{ m} > 1.828 \text{ maximum}$$

We will use the spacing as shown in Fig. E10-5c.

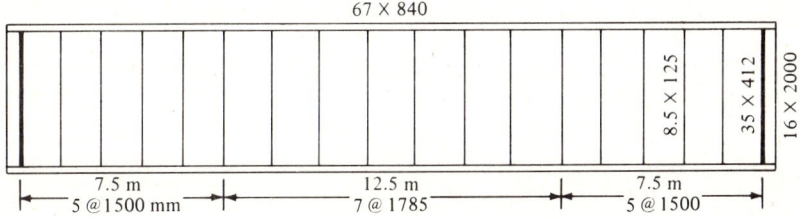

Figure E10-5c

Use a pair of stiffener plates for each intermediate stiffener.

$$b_s \geq 50 + \frac{D}{30} = 50 + \frac{2134}{30} = 121 \text{ mm} \quad \text{use } b_s = 125 \text{ mm}$$

$$t_s \geq \frac{b_s}{16} = \frac{125}{16} = 7.8 \quad \text{use } t_s = 8.5 \text{ mm (AREA minimum thickness)}$$

Step 7. Design the flange-to-web weld.

The AREA specifications allow either full-penetration groove welds or fillet welds; either must be continuous. We will arbitrarily use full-penetration groove welds. For this type of weld we will only have to check the shear produced and compare it to the maximum value allowed for E70 electrodes.

$$Q = A_f \bar{y} = 0.070(0.815)\left(997 + \frac{70}{2}\right)$$

$$= 58.88 \times 10^{-3} \text{ m}^3$$

$$v = \frac{VQ}{I} = \frac{2730.4(58.88 \times 10^{-3})}{0.132114} = 1216.6 \text{ kN/m}$$

The shear resistance is limited to

$$F_r = 0.35 F_y A_{\text{shear}} = 0.35(250)(0.016)(1 \text{ m}) \times 10^3$$

$$= 1400 \text{ kN/m} > 1216.6 \quad \text{O.K.}$$

Step 8. Check the deflection.

Assuming that the live-load moment is due to an equivalent uniform load, we obtain

$$w_L = \frac{8M}{L^2}$$

and the total equivalent uniform load w_e is

$$w_e = w_D + w_L = 42.2 + \frac{8(12\ 780)}{27.5^2} = 177.3 \text{ kN/m}$$

The deflection is approximately

$$\Delta = \frac{5wL^4}{384EI} = \frac{5(0.1773)(27.5)^4}{384(200\ 000)(0.132114)} = 0.04996 \text{ m (say 50 mm)}$$

$$\Delta_{\text{max}} = \frac{27.5(1000)}{640} = 43 \text{ mm} \simeq 50 \text{ mm} \quad \text{(take as O.K.)}$$

Step 9. Design the upper flange knee bracing.

We shall arbitrarily place a brace on every fourth floor beam for a spacing of $4(0.76) = 3.04$ m. Where this coincides with or is "close enough" to a stiffener, we will increase the stiffener thickness from 8.5 mm to 12 mm. For any bracing plate alone, we will use a 12-mm plate. Shop-weld the stiffener or brace plate to girder web and compression flange as shown in Fig. E10-5d. Field-weld the bottom of the stiffener brace plate to the cross beam. Shop-weld the angle for the bracket and punch for three 22-mm high-strength bolts at each end. Take the horizontal component of stiffener

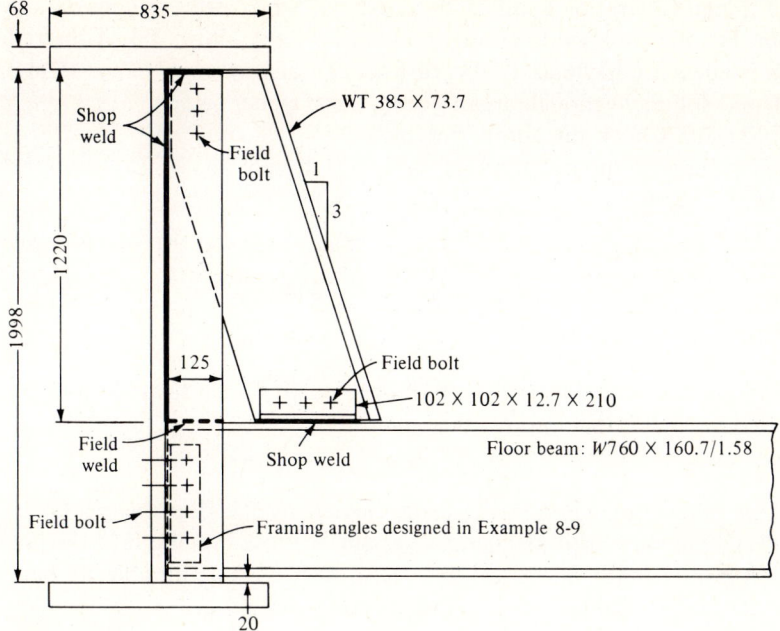

Figure E10-5d

force

$$F_h = 0.025 f_b A_f = 0.025(135.4)(0.070 \times 815) = 193 \text{ kN}$$

$$P_{kb} = \frac{193}{1220/(3)(1220)} = 579 \text{ kN}$$

The number of 22-mm A-325 bolts in double shear required for the horizontal component of stiffener force to attach stiffener to floor beam is

$$N = \frac{193}{(0.7854 \times 0.022^2)(138 \times 10^3)(2)} = 1.83$$

Use three fasteners (AREA minimum number of fasteners in a connection). For the stiffener try a WT 385 × 73.7

$$A = 9.39 \times 10^{-3} \text{ m}^2 \qquad t_f = 17.02 \text{ mm}$$

$$d = 376.6 \text{ mm} \qquad t_w = 13.21 \text{ mm}$$

$$r_x = 119.95 \text{ mm} \qquad r_y = 53.34 \text{ mm}$$

$$L = \left[1.220^2 + (1.220/3)^2\right]^{1/2} = 1.286 \text{ m}$$

$$L/r_y = 1286/53.34 = 24.1 > \frac{282}{\sqrt{F_y}}$$

$$F_a = 0.60(250) - \left(\frac{250}{316}\right)^{1.5}(24.1) = 133 \text{ MPa}$$

$$P_a = AF_a = 9.39 \times 133 = 1249 \text{ kN} \gg 579 \qquad \text{O.K.}$$

The flange b/t ratio is satisfactory. The web b/t ratio is considerably far from the allowable value of $16t$; however, AREA allows this if the reduced area is adequate for load. In this case we can discard most of the stem of the WT and still have enough area to carry the stiffener force

Step 10. Check the lateral bracing.

The lateral force/length of bridge is based on AREA requirements given in Sec. 1-9.

$$\text{Wind on girder: } 1.5(0.030 \text{ ksf})(47.88)(2.134 \text{ m depth}) = 4.6 \text{ kN/m}$$

$$\text{Wind on train: } 0.300 \text{ klf}(14.59) = 4.4 \text{ kN/m}$$
$$\text{Total} = 9.0 \text{ kN/m}$$

An additional concentrated force of one-fourth of the heaviest axle load is applied at each critical point; this gives a force of

$$110(4.45)(0.25) = 122.4 \text{ kN}$$

AREA allows this lateral load to be carried by the floor beams (as in Fig. 10-4d). One may arbitrarily add a system of cross bracing into the plane of the bottom flanges. If we use a system of cross bracing and six panels at $S = 27.5/6 = 4.58$ m, the panel loads are:

End panel: $\quad 122.4 + \dfrac{9(4.58)}{2} = 143.01$ kN

Interior panels: $\quad 122.4 + 9.0(4.58) = 163.6$ kN

The design of the cross-bracing members will not be carried out here, as it is simply a problem of finding the corresponding axial load and selecting a suitable compression member (probably an angle) to carry the load. ///

PROBLEMS

10-1 Design the cover plates for a rolled section to span 48 ft, carrying a single concentrated load of 225 kips. Limit the deflection to $L/360$. Consider both full- and partial-length cover plates. Assume lateral support at the ends and at the concentrated load. Use A-36 steel, E70 electrodes, AISC specifications and limit the depth to 36 in.

Answer: Try W30 × 211 and 1.25 × 24 in cover plates.

10-2 Design the cover plates for a rolled section to span 15.2 m and carrying a single concentrated load of 998 kN. Limit the deflection to $L/360$. Consider both full- and partial-length cover plates. Assume lateral support at the ends and at the concentrated load. Use A-36 steel, E70 (480 MPa) electrodes, AISC specifications and limit depth to 900 mm.

Answer: Try W760 × 314 and 36 × 600-mm cover plates.

10-3 What is the maximum bending moment capacity of a W36 × 194 with a C15 × 40 welded to the top flange to form a built-up crane runway girder as in Fig. P10-3? The span is 32 ft with no lateral support except at the girder ends. Use the AISC specifications, A-36 steel, and E60 electrodes.

Answer: 1492 ft · kips.

10-4 What is the maximum bending moment capacity for a W920 × 270.8 with a C380 × 50.45 welded to the top flange to form a built-up crane runway girder as in Fig. P10-3? The span is 9.75 m with lateral support only at the girder ends. Use the AISC specifications, A-36 steel, and E60 (480 MPa) electrodes.

Answer: 1638 kN · m.

Figure P10-3

10-5 Design the channel-to-beam weld for the girder of Prob. 10-3.

10-6 Design the channel-to-beam weld for the girder of Prob. 10-4.

The plate girders following are to be completely designed using the length or lengths furnished by the Instructor. You may work in groups using one length per group (and other data), or each person may have a separate set of design data of length depth and load. Be sure to use the following checklist of items to design: (1) shear and moment diagrams; (2) web thickness; (3) flange proportions; (4) moment of inertia and f_b; (5) allowable F_b and re-proportion as required; (6) bearing stiffeners; (7) intermediate stiffeners; (8) flange-to-web weld; (9) stiffener-to-web welds; (10) deflection check.

10-7 Design a welded plate girder to span _____ ft, carrying loads as shown in Fig. P10-7. Take $P = 175$ to 250 kips, $L = 40$ to 60 ft, and $D = 60$ to 96 in, as assigned. Use A-36 steel, E70 electrodes, and the AISC specifications. Assume lateral support at the ends and load points. (*Note:* If no specific data are assigned, take $P = 215$ kips, $L = 54$ ft, and $D = 80$ in.)

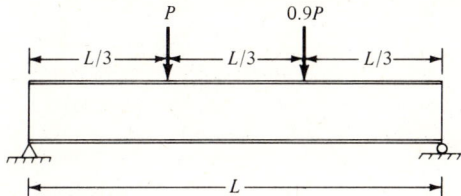

Figure P10-7

10-8 Design a welded plate girder to span _____ m, carrying loads as shown in Fig. P10-7. Take $P = 780$ to 1120 kN, $L = 12$ to 18 m, and $D = 1.52$ to 2.40 m, as assigned. Use A-36 steel, E70 electrodes, and the AISC specifications. Assume lateral support at the ends and load points. (*Note:* If no specific data are assigned, take $P = 950$ kN, $L = 15.3$ m, and $D = 2.14$ m.)

10-9 Design a welded plate girder for a crane runway for the conditions (maximum loads on one girder) shown in Fig. P10-9. Use the AISC specifications, A-36 steel, and E70 electrodes. Limit the

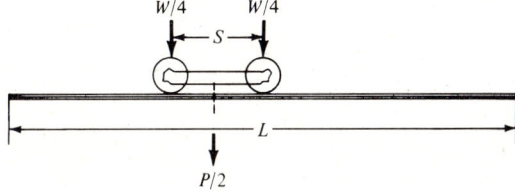

Figure P10-9

girder depth to 84 in. Lateral support is only at girder ends. Use $W/4 = 25$ kips, $P/2 = 200$ kips, $L = 80$ ft, and $S = 15$ ft.

10-10 Design a welded plate girder for a crane runway for the conditions (maximum loads to one girder) shown in Fig. P10-9. Use the AISC specifications, A-36 steel, and E70 (480 MPa) electrodes. Limit the girder depth to 2.15 m. Use $W/4 = 110$ kN, $P/2 = 950$ kN, $L = 26.2$ m, and $S = 4.6$ m.

10-11 Design a welded plate girder for a single-track open-deck railroad bridge. Assume that each track contributes 0.25 kip/ft loading, including rail, ties, and so on. Design for the Cooper E-80 or E-110 loading and span from 40 to 110 ft as assigned. Use A-36 steel, E70 electrodes, and the AREA specifications. Limit the overall girder depth to $L/10 \geq D \geq L/15$, but not under 4 ft and not over 8 ft. Take the girder spacing $S \geq L/15$. Design sway bracing and/or diaphragms in addition to the web, flanges, stiffeners, and welds. (*Note:* If specific problem data are not assigned, use Cooper E-110 loading, $L = 90$ ft, $D = 96$ in, and $S = 6.5$ ft.)

10-12 Design a welded plate girder for a single-track open-deck railroad bridge. Assume that each track contributes 3.65 kN/m of loading, including rail, ties, and so on. Design for Cooper E-80 or E-110 live loading and span from 14 to 34 m, as assigned. Use the data of Prob. 10-11 to obtain S and depth D but limit D to between 1.22 and 2.45 m. If specific problem data are not assigned, use Cooper E-80 loading, $L = 23$ m, $D = 2.3$ m, and $S = 1.6$ m.

APPENDIX
SELECTED COMPUTER PROGRAMS

A-1 FRAME ANALYSIS PROGRAM

This computer program will analyze any plane frame. The frame may be rigid, pinned, or a combination. It must have a constant modulus of elasticity, E. The program does not allow for support settlements. Support settlements can be accounted for by inputting by hand the equivalent in terms of fixed-end moments and resulting shears. The user is referred to Chap. 2 for typical coding for the several types of problems this program will analyze.

Generally, the user should place a node where deflections are wanted along a beam span and under concentrated loads, particularly if more than two concentrated loads are on the member. The program computes fixed-end moments for a uniform beam loading (and orientation from horizontal to vertical). It also computes fixed-end moments for up to two concentrated loads on the span. It also allows inputting column shears (loads from out of plane to the column, as for rigid one way and simple supported the other). These shears can be used to reduce the column load due to live-load reductions by appropriate values and signs.

The following steps are required to run the program, with terminology matching the computer program listing.

1. Code the structure according to rigid frame, truss, or combination. Take into account any hinges, as in Chap. 2.
2. Determine NP. The computer program computes $NPP1 = NP + 1$.

3. Determine the number of members, NM.
4. Determine the number of load conditions, NLC.
5. Determine NLC that uses wind the first time as NLW.
6. Determine NLC that uses $D + L$, which is to be later combined with wind $(D + L + W)$. Call this NLC = IDLPL. When this NLC comes up, the $D + L$ member forces are stored on disk file for recall with each subsequent NLC that uses wind.
7. Determine the NLC that has wind on any beams or columns. Call this NLC = IWINBE. This will generally be the same as NLW. When NLC = IWINBE comes up, the $D + L$ on disk is recalled and added to the current values, which should be only wind.
8. Find NBAND using the method outlined in the text.
9. Make sure that ISTIFF is dimensioned greater or at least equal to NBAND × NP.

Input

1. Punch cards for lines 4, 8, and 10 of computer listing.
2. Punch member data cards according to the FORMAT on line 31. FORMAT(7I4, 4G10.4, I2)

The G format allows use of either F10.4 or xxxE-3 or xxxE-6 for SI problems. NPE(7) uses the I2 format and goes in column 70.

Use NPE(7) = 1 for all members without transverse loads.

For any members with transverse loads:

NPE(7) = 0 for normal gravity beam loadings.
NPE(7) = −1 for wind loads on beam or transverse loads on vertical columns.

If NPE(7) ≤ 0, it is necessary to follow the member data card with NLC cards punched according to the FORMAT on line 51, where

w = uniform load on member (kips/ft or kN/m), direction of gravity = (+)
P1, P2 = maximum of 2 concentrated loads on member; use (+) = direction of gravity
X1, X2 = horizontal distance from origin end of member of P1 and P2 in ft or m
V1, V2 = column shears from alternate direction for a space frame, V1 = near end shear; use kips or kN and sign same as P1, P2

3. Put a blank card at the end of the member data.
4. If there are any concentrated node forces (and only loading for a pinned truss), go to DO 991 at line 222, which requires NNZP cards in the following

order:

M1 = NP value of the entry

PR(M1,L), L = 1,NLC This inputs the NLC nodal forces at NP = M1 for NLC load conditions using 8F10.4 format. Note that this requires at least two cards for each NNZP.

5. Add your system control cards.

Note that you may (but not recommended) read nodal moments in step 4 above. If you do read nodal moments, the output moments must be corrected by hand to get the final design moments as

$$F_{final} = F_{output} + M_{read}$$

with careful attention to signs. The output signs are interpreted using Fig. 2-6. When the program computes the fixed end moments, the output is automatically corrected for the FEMs to obtain the design values. Member forces do not have to be corrected for any node forces read in other than moments.

Output

If input units are as identified here, the output will be

X = displacements = in, mm, or rad

F = member forces

Truss members: kips or kN

Beam members: kips or kN for axial force

ft · kips or kN · m for moments

Identification of Program Variables

ZZZ any alphanumeric data using not more than 80 columns. Recommend using your name, problem number, or both.

UNIT1, UNIT2, UNIT3, UNIT4 units identification for fps and SI problems

UNIT1 FT or M (start in column 1 of card)

UNIT4 IN or MM (start in column 5)

UNIT2 K-IN or K-MM (start in column 11—unit used for P matrix)

UNIT3 K-FT or KN · M (start in column 21)

NP number of P's coded = 2 × number of nodes − number of reactions

NM	number of members (NM = NP for determinate trusses)
NLC	number of load conditions
LISTB	1 to list band matrix (part of total ASAT matrix used for finding X's)
NLW	first load condition with wind
NNZP	number of nonzero P-matrix entries to read in separately
IMET	0 for fps problems = 1 for SI problems
ITRUSS	1 for all truss members; = 0 if any beam members are present
ISTIFF	band matrix and must be dimensioned NP × NBAND at least
NBAND	band width used. Is difference between smallest and largest NP on ends of any member? Value may be larger than needed but not larger than NP. If value is only one or two numbers too small, tends to give answers almost right sometimes—other times gives overflow messages.
IDLPL	NLC with $D + L$ which is saved on disk file to add to wind NLC
IWINBE	NLC which uses $D + L$ from disk file with current forces
IWRITX	1 to write X (deflection) matrix
IWRITP	0 to obtain output with few members
	1 to obtain output when large number of NLC are used
E	modulus of elasticity, ksi or MPa
MEMNO	member number assigned to element during coding
NPE(I)	NP values for the element; truss has 1 through 4; beam 1 through 6
H	horizontal distance to far end of member including sign, ft or m
V	vertical distance to far end of member including sign, ft or m
A	cross-sectional area of member, in^2 or m^2
XI	moment of inertia of member, in^4 or m^4 (not needed for truss members)
NPE(7)	control switch for members with transverse loadings
	0 for members with transverse loading
	1 for members without transverse loading (also see COMMENT cards in computer program)
	BE SURE TO PUT BLANK CARD AT END OF MEMBER DATA
M1	NP value of P matrix entry where a concentrated force (or moment) is located
PR(I)	value of corresponding P-matrix entry. These last two cards are in a DO loop, so use as many pairs of cards as NNZP.

```
      PROGRAM JEB1(INPUT,OUTPUT,TAPE1=INPUT,TAPE3=OUTPUT,TAPE5)
C         J E BOWLES   STIFFNESS METHOD FOR SOLVING FRAMES AND TRUSSES  ***MODEL 4
C         ****************************************************************
C         CODE STRUCTURE WITH ALL REACTIONS OF ZERO DISPLACEMENT AS NPP1
C      (NP + 1).  FOR MEMBERS WTIH MORE THAN 2 CONC. LOADS SUBDIVIDE INTO
C      ADDITIONAL MEMBERS WITH NODES AT THE CONC. LOADS.  PUT NODES AT POINTS
C      CRITICAL DEFLECTION SO DEFLECTIONS (X(I)'S) ARE DIRECTLY OBTAINED.
C         *****************************
C         IWINBE = COUNTER TO ADD WIND LOAD OF BEAM TYPE ANALYSIS TO D + L
C         IDLPL  = COUNTER FOR DEAD + LIVE LOAD NLC
C
C      UNIT1 = K OR KN --- UNIT2 = IN-K OR KN-MM --- UNIT3 = FT-K OR KN-M
C      UNIT4 = IN OR MM --  *** NOTE CAREFULLY UNITS OF UNIT(I) FOR INPUT/OUTPUT
C
C
C  NP = NO OF NP IN CODING--NM = NO OF MEMBERS IN STRUCTURE--NLC = NO OF LOAD
C   CONDITIONS--LISTB = 1 TO LIST BAND MATRIX (USUALLY USE 0)----NLW = LOAD COND
C   WHERE WIND STARTS--NNZP = NO OF NON-ZERO P-MATRIX ENTRIES TO READ (READ ONLY
C   CONC LOADS)---IMET = 1 FOR METRIC PROB----ITRUSS = 1 FOR ALL TRUSS MEMBERS
C   ISTIFF = NO OF STIFF ENTRIES (NP*NBAND)--NBAND = BAND WIDTH---IDLPL = LOAD
C   COND OF DEAD + LIVE ONLY TO HOLD TO COMBINE WITH WIND----IWINBE = NO OF
C   LOAD CONDITION FOR WHICH D + L EFFECTS ARE ADDED TO OBTAIN DESIGN VALUES
C
C           IWRITX = 1 TO WRITE X-MATRIX
C           IWRITP = 1 TO WRITE BAR FORCE MATRIX WHEN NLC = 1
C                  = 0 TO WRITE MAX AND MIN BAR FORCES WHEN NLC = LARGE
C                  = - 1 TO WRITE BAR FORCE MATRIX WHEN NLC = LARGE
C
      DIMENSION NPE(7),EA(6,3),ZZZ(20),ES(3,3),ESAT(3,6),EASAT(6,6) ,
     1 F(3),FU(10),FM1(99),FM2(99),FFU(2,4)
      DIMENSION P(115,24),TF(115,24),PR(115),STIFF(1500),PMAX(115)
     1,PMIN(115)
C
      DATA FFU/12.,1000.,1.,1000.,1.,1000000.,12.,1./
C
C
C                                                           READ CARD
    2 READ(1,1000)ZZZ,UNIT1,UNIT4,UNIT2,UNIT3
C
 1000 FORMAT(20A4,/,A2,2X,A2,4X,4(A4,6X))
      IF(EOF(1))750,3
    3 WRITE (3,2000) ZZZ
 2000 FORMAT('1',//,1X,20A4//)
C
C                                                           READ CARD
      READ(1,1001)NP,NM,NLC,LISTB,NLW,NNZP,IMET,ITRUSS,ISTIFF,NBAND,
     1IDLPL,IWINBE,IWRITX,IWRITP
C
 1001 FORMAT(16I5)
C
C         ****** READ MODULUS OF ELASTICITY --KSI OR MPA
      READ(1,1002)E
C
 1002 FORMAT(8F10.4)
C
C         DEFINE UNIT CONVERSION FACTORS       ***********
      IUNIT = 1
      IF(IMET.GT.0)IUNIT = 2
      DO 40  I = 1,4
      FU(I) = FFU(IUNIT,I)
   40 CONTINUE
C
      NPP1 = NP + 1
      DO 102  I = 1,ISTIFF
  102 STIFF(I) = 0.
      DO 103  I = 1,NPP1
      DO 103  J = 1,NLC
      IF(I.EQ.1)PR(J) = 0.0
      TF(I,J) = 0.
```

```
      103 P(I,J) = 0.
          WRITE(3,2001)NP,NM,NLC,NLW,NNZP,IMET,ISTIFF,E,IWRITX,IWRITP,IDLPL,
         1IWINBE
     2001 FORMAT(/,       5X,'STIFFNESS METHOD OF RIGID FRAME AND TRUSS ANALY
         1SIS',//,6X,'NO OF NP = ',I4,5X,'NO OF MEMBERS = ',I4,5X,'NO OF LOA
         2D CONDIT = ',I4,/,  6X,'NO NLC TO WIND NLC = ',I4,   5X,'NO NON-ZER
         3O P = ',I4, 5X,'METRIC (1 = METRIC) = ',I2,/,
         4 6X,'NO ISTIFF ZEROED = ',I5,4X,'MOD ELASTICITY = ',F8.1,5X,'IWRIT
         5X = ',I2,3X,'IWRITP = ',I2,// 10X,
         6 'NLC FOR D + L = ',I2,3X,'NLC FOR WIND ON BEAM = ',I2,//)
          II = 0
          E = E*FU(2)
C
          JJ = 1
C                                                              READ CARD
C         READ MEMBER DATA AT 1 CARD/MEMBER AND FOLLOW WITH NLC CARDS OF
C    LOAD DATA (SEE ABOUT 20 LINES FOLLOWING).   READ NPE(7) = 1 FOR ALL TRUSS
C    OR OTHER MEMBERS WITH NO TRANSVERSE LOAD--NPE(7) = 0 FOR MEMBERS WITH
C       TRANSVERSE GRAVITY LOAD---NPE(7) = -1 FOR MEMBERS WITH UNIFORM LOAD
C    NORMAL TO MEMBER AXIS.
      106 READ(1,1007) MEMNO,(NPE(I),I=1,6),H,V,A,XI,NPE(7)
C
     1007 FORMAT (7I4,4G10.4,I2)
C
          H = H*FU(1)
          V = V*FU(1)
          XL=SQRT(H*H+V*V)
          IF(MEMNO.GT.0)GO TO 108
          XCOS=0.0
          XSIN=0.0
          GO TO 987
      108 XCOS=H/XL
          XSIN=V/XL
          DO 109  I = 1,NLC
          FM1(I) = 0.
      109 FM2(I) = 0.
C
          IF(NPE(7).GT.0)GO TO 430
C
          IF(JJ.GT.0)WRITE(3,2002)
     2002 FORMAT(1X, 'MEMNO', 4X,'W', 6X,'P1', 7X,'P2',7X,'X1',7X,'X2',
         1  7X,'V1',7X, 'V2',/)
          JJ = 0
C
C         COMPUTE FEM FOR P-MATRIX FOR ALL LOAD CONDITIONS---NOTE UNIFORM WIND
C    WIND LOADS CANNOT BE DIRECTLY COMBINED WITH UNIFORM GRAVITY BEAM LOADS
C    DUE TO HORIZ. COMPONENT--SOLVE SEPARATELY AND HAVE COMPUTER COMBINE
C    RESULTS.
C
          XZ = XL/FU(2)
          DO 988  I = 1,NLC
C                                                              READ CARD
          READ(1,1008)MEMNO,W,P1,P2,X1,X2,V1,V2
C
     1008 FORMAT(I5,7F10.4)
          WRITE(3,2005)MEMNO,W,P1,P2,X1,X2,V1,V2
     2005 FORMAT(I5,7F9.3)
          X1 = X1*FU(4)
          X2 = X2*FU(4)
          IF(ABS(XCOS).LE..000001)GO TO 42
          X1 = X1/XCOS
          X2 = X2/XCOS
C                        FOR CONCENTRATED GRAVITY LOADS ON MEMBER
       42 CONS = 12.*FU(4)
          XC = XCOS
          IF(ABS(XCOS).LT..00001)XC = 1.
          FG1 = ( P1*X1*(XZ-X1)**2/XZ**2 + P2*X2*(XZ-X2)**2/XZ**2)*XC
          FG2 = (P1*(XZ-X1)*X1**2/XZ**2 + P2*(XZ-X2)*X2**2/XZ**2)*XC
```

```
      IF(I.GE.NLW)GO TO 112
C                 BEAM WITH UNIFORM GRAVITY LOAD
  110 BMW = W*XCOS*XZ**2/CONS
      FEM1 = BMW + FG1
      FEM2 = -BMW - FG2
      GO TO 115
C  WIND OR OTHER LOADS PERPEND. TO BEAM AND/OR + OR - W ON COLUMN--
  112 BMW = W*XZ**2/CONS
      FEM1 = BMW + FG1
      FEM2 = -BMW - FG2
C
  115 FR1 = -(P1*(XZ-X1) + P2*(XZ-X2) + FEM1 + FEM2)/XZ
      FR2 = -(P1*X1 + P2*X2 - FEM2 - FEM1)/XZ
      XC = 1.
      IF(I.GE.NLW)XC = XCOS
      CON = 2.*FU(4)
      C = -W*XZ*XC/CON + FR1
      B = -W*XZ*XC/CON + FR2
      IF(I.LT.NLW)GO TO 118
      XS = XSIN
      P(NPE(2),I) = W*XZ*XSIN/CON + P(NPE(2),I) - FR1*XS
      P(NPE(5),I) = W*XZ*XSIN/CON + P(NPE(5),I) - FR2*XS
  118 IF(IMET.EQ.0)GO TO 119
      FEM1 = FEM1*FU(2)
      FEM2 = FEM2*FU(2)
  119 P(NPE(3),I) = C - V1 + P(NPE(3),I)
      P(NPE(6),I) = B - V2 + P(NPE(6),I)
   44 P(NPE(1),I)=+FEM1+P(NPE(1),I)
      P(NPE(4),I)=+FEM2+P(NPE(4),I)
      FM1(I) = FEM1
      FM2(I) = FEM2
  988 CONTINUE
  430 CONTINUE
C
C
  987 WRITE(5)MEMNO,(NPE(I),I=1,6),H,V,A,XI,XL,XCOS,XSIN,(FM1(I),FM2(I),
     1I=1,NLC)
      IF(II.LE.0.)GO TO 556
C                 ********************************
C       RE-ENTER HERE FOR COMPUTING MEMBER FORCES
C
  555 READ(5)MEMNO,(NPE(I),I=1,6),H,V,A,XI,XL,XCOS,XSIN,(FM1(I),FM2(I),I
     1=1,NLC)
  556 IF(MEMNO.GT.0)GO TO 201
C
C                 TRANSFER OF CONTROL DEPENDING ON ITRUSS AND MEMNO = 0
      IF(MEMNO.EQ.0.AND.ITRUSS.GT.0.AND.II.GT.0)GO TO 188
      IF(II.GT.0)GO TO 728
      WRITE(3,2006)
 2006 FORMAT(///,2X,'MEMBER NP1 NP2 NP3 NP4 NP5 NP6',6X,'H',7X,'V',7X,'A'
     1,10X,'I', 8X,  'L',6X,'COS', 4X,'SIN',10X,'FEM1',6X,'FEM2',/)
      REWIND 5
      DO 729 K=1,NM
      READ(5)MEMNO,(NPE(I),I=1,6),H,V,A,XI,XL,XCOS,XSIN,FEM1,FEM2
      WRITE(3,2007) MEMNO,(NPE(I),I=1,6),H,V,A,XI,XL,XCOS,XSIN,FEM1,FEM2
 2007 FORMAT(1X ,2I5,5I4,2X, 2F9.2,2(1X,G9.3),F9.2,3X,2F8.5,2F10.2)
  729 CONTINUE
      GO TO 728
  201 IF(NPE(6).GT.0)GO TO 185
      IF(II.GT.0.AND.NPE(6).EQ.0)GO TO 188
      EASAT(1,1) = E*A*XCOS*XCOS/XL
      EASAT(1,2) = E*A*XCOS*XSIN/XL
      EASAT(1,3) = -EASAT(1,1)
      EASAT(1,4) = -EASAT(1,2)
      EASAT(2,1) = EASAT(1,2)
      EASAT(2,2) = E*A*XSIN*XSIN/XL
```

```
      EASAT(2,3) = -EASAT(1,2)
      EASAT(2,4) = -EASAT(2,2)
      EASAT(3,1) = -EASAT(1,1)
      EASAT(3,2) = -EASAT(1,2)
      EASAT(3,3) = EASAT(1,1)
      EASAT(3,4) = EASAT(1,2)
      EASAT(4,1) = -EASAT(1,2)
      EASAT(4,2) = -EASAT(2,2)
      EASAT(4,3) = EASAT(1,2)
      EASAT(4,4) = EASAT(2,2)
      DO 179  I = 1,4
      IF(NPE(I).GE.NPP1)GO TO 179
      NS1 = (NPE(I)-1)*NBAND
      DO 178  J = 1,4
      IF(NPE(J).GE.NPP1)GO TO 178
      IF(NPE(J).LT.NPE(I))GO TO 178
      NS2 = NPE(J) - NPE(I) + 1
      STIFF(NS1 + NS2) = STIFF(NS1+NS2) + EASAT(I,J)
  178 CONTINUE
  179 CONTINUE
      GO TO 5500
  185 EAOL = E*A/XL
      EIOL = (E*XI/XL)*FU(3)
      SINOL = XSIN/XL
      COSOL = XCOS/XL
      EA(1,1) = 0.
      EA(1,2) = 1.
      EA(1,3) = 0.
      EA(2,1) = -XCOS
      EA(2,2) = SINOL
      EA(2,3) = SINOL
      EA(3,1) = -XSIN
      EA(3,2) = -COSOL
      EA(3,3) = -COSOL
      EA(4,1) = 0.
      EA(4,2) = 0.
      EA(4,3) = 1.
      EA(5,1) = XCOS
      EA(5,2) = -SINOL
      EA(5,3) = -SINOL
      EA(6,1) = XSIN
      EA(6,2) = COSOL
      EA(6,3) = COSOL
      ES(1,1) = EAOL
      ES(1,2) = 0.
      ES(1,3) = 0.
      ES(2,1) = 0.
      ES(2,2) = 4.*EIOL
      ES(2,3) = 2.*EIOL
      ES(3,1) = 0.
      ES(3,2) = ES(2,3)
      ES(3,3) = ES(2,2)
      DO 202  I = 1,3
      DO 202  J = 1,6
      ESAT(I,J) = 0.
      DO 186  K = 1,3
      ESAT(I,J) = ESAT(I,J) + ES(I,K)*EA(J,K)
  186 CONTINUE
  202 CONTINUE
C
      IF(II.GT.0)GO TO 605
  203 DO 204  I = 1,6
      DO 204  J = 1,6
      EASAT(I,J) = 0.
      DO 187  K = 1,3
      EASAT(I,J) = EASAT(I,J) + EA(I,K)*ESAT(K,J)
  187 CONTINUE
```

```
      204 CONTINUE
          DO 206   I = 1,6
          IF(NPE(I).GE.NPP1)GO TO 206
          NS1 = (NPE(I)-1)*NBAND
          DO 205  J = 1,6
          IF(NPE(J).GE.NPP1)GO TO 205
          IF(NPE(J).LT.NPE(I))GO TO 205
          NS2 = NPE(J) - NPE(I) + 1
          STIFF(NS1 + NS2) = STIFF(NS1+NS2) + EASAT(I,J)
      205 CONTINUE
      206 CONTINUE
     5500 GO TO 106
C
C         THE BAND MATRIX IS NOW FORMED FOR REDUCTION IN ISTIFF IN CORE
C         WRITE BAND MATRIX IF LISTB > 0
C
      728 IF(II.GT.0)GO TO 601
C
          M2 = NBAND*NP
C
          IF(LISTB.LE.0)GO TO 8889
          WRITE(3,2009)
     2009 FORMAT(  T10, 'THE BAND MATRIX WITH 1000 FACTORED',/)
          M1 = 1
          M2 = NBAND
          NCOU = NP*NBAND
          DO 305  I = 1,NP
          WRITE(3,2010)I,(STIFF(IZ),IZ=M1,M2)
     2010 FORMAT(1X,I4,1X,-3P(9F12.2),/, 5X,-3P(9F12.2),/, 5X,-3P(9F12.2))
          IF(M2.GE.NCOU)GO TO 8889
          M1 = M2 + 1
          M2 = M2 + NBAND
      305 CONTINUE
     8889 WRITE(3,2011)M2,NBAND
     2011 FORMAT(///,T5,'NO STIFF(I) ENTRIES = ',I6,10X,'BAND WIDTH =',I4,//)
C
C
C         ****   NOTE--DO NOT READ P-MATRIX ENTRIES FOR UNFORM LOADS ON BEAMS
C                DO NOT READ P-MATRIX ENTRIES FOR FEM--INPUT SO COMPUTER
C                    COMPUTES FEM SO FINAL MOMENTS ARE CORRECTED FOR FEM ********
C
          IF(NNZP.EQ.0)GO TO 426
          DO 991   NN = 1,NNZP
C                                                         READ CARD*********
          READ(1,1010)M1
C
     1010 FORMAT(16I5)
C                                                         READ CARD*********
          READ(1,1011)(PR(L),L=1,NLC)
C
     1011 FORMAT(8F10.4)
          DO 990  L = 1,NLC
          P(M1,L) = P(M1,L) + PR(L)
      990 CONTINUE
      991 CONTINUE
      426 CONTINUE
          DO 992   NS1 = 1,NP
          DO 992   NS2 = 1,NLC
      992 TF(NS1,NS2) = 0.
C                     *******************************************
C
      406 WRITE(3,2012)UNIT1,UNIT2
     2012 FORMAT(/,2X,'THE P-MATRIX, ',A2,' AND ',A4,/)
          NS1 = 1
          NS2 = 10
      427 IF(NS2.GT.NLC)NS2 = NLC
          DO 408   I = 1,NP
      408 WRITE(3,2014)I,(P(I,J),J=NS1,NS2)
```

518 STRUCTURAL STEEL DESIGN

```
 2014 FORMAT(T5, 'NP = ',I3,1X,10F11.2)
      IF(NS2.EQ.NLC)GO TO 428
      NS1 = NS2 + 1
      NS2 = NS2 + 10
      WRITE(3,2016)NS1
 2016 FORMAT(///,5X,'THE P-MATRIX CONTINUED BEGINNING WITH NLC = ',I3,/)
      GO TO 427
  428 CONTINUE
C
C     SUBROUTINE TO REDUCE THE BAND MATRIX
   60 N1 = 1
      DO 80  N = 1,NP
      I = N
      DO 70  L = 2,NBAND
      NL = (N-1)*NBAND + L
      I = I+1
      IF(STIFF(NL).EQ.0.)GO TO 70
      B = STIFF(NL)/STIFF(N1)
      J = 0
      DO 68  K = L,NBAND
      J = J + 1
      IJ = (I-1)*NBAND + J
      NK = (N-1)*NBAND + K
   68 IF(STIFF(NK).NE.0.)STIFF(IJ) = STIFF(IJ) - B*STIFF(NK)
      STIFF(NL) = B
C     INCLUDE LOAD MATRIX IN REDUCTION
      DO 67  M = 1,NLC
   67 P(I,M) = P(I,M) - B*P(N,M)
   70 CONTINUE
      DO 66  M = 1,NLC
      P(N,M) = P(N,M)/STIFF(N1)
   66 CONTINUE
   80 N1 = N1 + NBAND
C     COMPLETE SOLUTION BY BACK SUBSTITUTION
      N = NP
   85 N = N - 1
      IF(N.LE.0)GO TO 90
      L = N - 1
      DO 86  K = 2,NBAND
      NK = (N-1)*NBAND + K
      DO 86  M = 1,NLC
      IF(STIFF(NK).NE.0.)P(N,M) = P(N,M) - STIFF(NK)*P(L+K,M)
   86 CONTINUE
      GO TO 85
   90 CONTINUE
C
C     END OF MATRIX REDUCTION -- SOLUTION IS IN P(I,J)
C
      IF(IWRITX.LE.0)GO TO 431
      WRITE(3,2017)UNIT4
 2017 FORMAT(///,5X,'THE X-MATRIX, ',A2,' OR RADIANS',/)
      NS1 = 1
      NS2 = 10
  429 IF(NS2.GT.NLC)NS2 = NLC
      DO 503  I = 1,NP
  503 WRITE(3,2018)I,(P(I,J),J=NS1,NS2)
 2018 FORMAT(6X,'NX = ',I3,1X,10F11.5)
      IF(NS2.EQ.NLC)GO TO 431
      NS1 = NS2 + 1
      NS2 = NS2 + 10
      WRITE(3,2019)NS1
 2019 FORMAT(///,5X,'THE X-MATRIX CONT D BEGINNING WITH NLC = ',I3,/)
      GO TO 429
  431 CONTINUE
C                       INCREMENT COUNTERS II AND JJ--JJ COUNTS NLC
      WRITE(3,3003)
 3003 FORMAT(/////)
```

```
              JJ = 0
              II = 1
C
          601 JJ = JJ + 1
          237 IF(JJ.LE.NLC)GO TO 602
          238 REWIND 5
              GO TO 192
C
          602 IF(ITRUSS.EQ.0.OR.IWRITP.EQ.1)WRITE(3,2020)JJ,UNIT1,UNIT3
         2020 FORMAT(/, 4X, 'LOADING CONDITION NO = ',I3,/,
             1 37X, 'DESIGN END MOMENTS CORRECTED',/,    5X, 'MEMBER', 4X,'AXIAL
             2FORCE, ', A2, 8X,'FOR FEM AND WIND (NEAR END FIRST), ',A4,/)
              REWIND 5
              GO TO 555
          605 P(NPP1,JJ)=0.
              FEM1 = FM1(JJ)
              FEM2 = FM2(JJ)
C          NOTE IF YOU READ P-MATRIX ENTRIES FOR FEM THEN OUTPUT MUST BE
C          ADJUSTED BY HAND TO ACCOUNT FOR FIXED END MOMENTS ON BEAM ENDS
C
              DO 607  I = 1,3
              F(I) = 0.
              DO 607  K = 1,6
              NS3 = NPE(K)
              F(I)=F(I)+ESAT(I,K)*P(NS3,JJ)
          607 CONTINUE
              F(2)=(F(2)-FEM1)/FU(1)
              F(3)=(F(3)-FEM2)/FU(1)
C
C          REDUCE MOMENTS FOR WIND
C          ALSO NECESSARY TO REDUCE D + L MOMENTS BY 0.75 IF ADDED TO WIND
              IF(JJ.EQ.IDLPL)GO TO 972
C
              IF(JJ.LT.NLW.OR.NLW.EQ.0)GO TO 974
          972 F(1)=F(1)*3./4.
              F(2)=F(2)*3./4.
              F(3)=F(3)*3./4.
              IF(JJ.NE.IDLPL)GO TO 973
C                PUT D + L IN TF-MATRIX TO LATER ADD TO WIND LOAD CONDIT.
              DO 970  I = 1,3
          970 TF(MEMNO,I) = F(I)
          973 IF(JJ.LT.IWINBE.OR.IWINBE.EQ.0.OR.IDLPL.EQ.0)GO TO 974
              DO 971  I = 1,3
          971 F(I) = F(I) + TF(MEMNO,I)
          974 WRITE(3,2022) MEMNO,(F(I),I=1,3)
         2022 FORMAT(T5,I5,F15.2,5X,2F15.2)
              GO TO 191
          188 CONTINUE
              IF(ITRUSS.EQ.0)GO TO 189
              IF(MEMNO.EQ.0)GO TO 192
              PMAX(MEMNO) = 0.
              PMIN(MEMNO) = 1000000.
          189 F(1)  = E*A/XL*(XCOS*(P(NPE(3),JJ) - P(NPE(1),JJ)) + XSIN*(P(NPE(4
             1),JJ) - P(NPE(2),JJ)))
              IF(ITRUSS.LE.0)GO TO 190
              IF(NLC.EQ.1)GO TO 195
              IF(F(1).GT.PMAX(MEMNO))PMAX(MEMNO) = F(1)
              IF(F(1).LT.PMIN(MEMNO))PMIN(MEMNO) = F(1)
              TF(MEMNO,JJ) = F(1)
              IF(IWRITP.LE.0)GO TO 441
C                 WRITE BAR FORCES FOR NLC = 1 AND IWRITP = 1
              IF(MEMNO.EQ.1)WRITE(3,2025)JJ
         2025 FORMAT(5X,'BAR FORCES (KIPS OR KN) FOR NLC = ',I3,/)
C
          190 WRITE(3,2026)MEMNO,F(1)
         2026 FORMAT(T5,I5,F15.2, 10X,'----', 11X,'----')
          441 IF(ITRUSS.LE.0)GO TO 191
              IF(JJ.EQ.NLC)GO TO 195
```

```
          JJ = JJ + 1
          GO TO 189
      195 JJ = 1
C
      191 GO TO 555
      192 IF(ITRUSS.EQ.0)GO TO 193
          IF(NLC.EQ.1)GO TO 193
          REWIND 5
C         WHEN ITRUSS = 1 AND NLC IS LARGE USE IWRITP = 0 FOR MAX AND MIN
C                      BAR FORCES ONLY---USE IWRITP = -1 FOR COMPLETE LISTING OF
C                      BARFORCES WHEN NLC IS LARGER THAN 1
C
C
C
          WRITE(3,2033)
 2033 FORMAT(///,8X,'THE MAXIMUM LIVE LOAD BAR FORCES AND DEAD LOAD VALUE
     1S',/,5X,'MEM NO',5X,'MAX LL',9X,'MIN LL',8X,'DEAD LOAD OR LAST NLC
     2',/)
          DO 442  I = 1,NM
          WRITE(3,2035)I,PMAX(I),PMIN(I),TF(I,NLC)
 2035 FORMAT(6X,I5,4X,F10.2,5X,F10.2,6X,F10.2)
      442 CONTINUE
C              WRITE BAR FORCE MATRIX IF IWRITP = -1
          IF(IWRITP.NE.-1)GO TO 193
          NS1 = 1
          NS2=10
          IF(NS2.GT.NLC)NS2=NLC
          WRITE(3,2032)NS1
 2032 FORMAT(///,5X,'THE BAR FORCE MATRIX STARTING WITH NLC = ',I3,//)
      443 DO 444 I=1,NM
      444 WRITE(3,2028)I,(TF(I,J),J=NS1,NS2)
 2028 FORMAT(2X,I3,2X,10F11.3)
          IF(NS2.EQ.NLC) GO TO 193
          NS1=NS2+1
          NS2=NS2+10
          IF(NS2.GT.NLC)NS2=NLC
          WRITE(3,2032)NS1
          GO TO 443
      193 GO TO 2
C
      750 STOP
          END
```

A-2 LOAD MATRIX GENERATOR FOR AASHTO TRUCK LOADING ON A TRUSS BRIDGE

This program steps a moving load consisting in either two or three concentrated loads along a truss. At each step the panel loads are computed as simple beams with concentrated loads. These panel loads are the *P*-matrix entries for a weightless truss to use in the frame analysis program.
 The loads move from left to right. If the larger wheel loads are used first, the truck is "backed" across the truss. The limitations are:

1. The user must use at least two and not more than three loads.
2. The "step" must be an integer multiple of the panel length. Since this integer is computed, the step should be input sufficiently accurately and slightly smaller so that in computer truncation the correct integer is obtained.
3. All panels must be of the same length.

An *impact* factor is automatically computed and the panel loads include the impact factor in the output. MET is used to compute the impact factor in either fps or SI units.
 Output is automatically punched onto cards in the correct format for direct input as the *P* matrix in the frame analysis program. The user may wish to insert an optional punch control to inspect the output prior to having it punched.

Variable Identification

A sample set of data cards are listed with this program for both an fps and an SI bridge truss using three wheel loads, seven truss panels (25 ft or 7.5 m), and stepping the wheels 5 ft or 1.5 m. The coding is such that the panel NPs for M1(I) are 26, 2, 6, ... and NPPI = 26 here and is not punched as output.

NP	number of wheel loads
NS	number of truss panels
SL	step length, ft or m
STEP	increment of wheel movement, ft or m
MET	0 for fps; = 1 for SI bridges. This is used to properly compute the impact factor.
P(I)	wheel loads. Note the order of using loads on the sample cards. Using 8. 32. 32. runs the truck forward on the bridge.
X(I)	cumulative wheel spacing as 14. 28. for using 14 ft between each set of wheels. This spacing or the SI equivalent will usually be the most critical spacing.
M1(I)	NP numbers in order from left to right where the panel loads are placed. M1 has the same meaning here as in the frame analysis program. Note that NPP1 is not used. For a through deck truss, this section of the program requires a slight modification to get loads onto the truss where NPP1 may be, since the current method of use omits the first and last M1(I) values (they are input for identification).

522 STRUCTURAL STEEL DESIGN

```
C        NP IS THE NUMBER OF LOADS THAT ARE WANTED ON THE TRUSS      SL IS THE
C        LENGTH OF EACH PANEL IN THE TRUSS  NS IS THE NUMBER OF PANELS IN THE TRUSS
C        STEP   IS THE MOVEMENT THAT IS WANTED FOR THE LOADS BETWEEN EACH LOAD
C        CONDITION     P(I) IS THE WEIGHT OF EACH LOAD IN THE SERIES
C        X(I)IS THE DISTANCE THE LOAD IS BEHIND THE FIRST LOAD
C        MET=0.0 FBS =1.0 METRIC
         DIMENSION TEMP(20,70),REACT(20,70),P(5),X(5)  ,M1(20)
5000     READ(1,1,END=4000)NP,NS,SL,STEP,MET
    1    FORMAT(2I5,2F10.4,I5)
         WRITE(3,5)NP,SL,NS,STEP
    5    FORMAT('1NUMBER OF LOADS ON TRUSS',2X,I3,//,' SPAN LENGTH OF EACH
        1PANEL',F8.2 ,//,' NUMBER OF PANELS IN TRUSS',2X,I3,//,' LENGTH OF
        2MOVEMENT OF THE LOADS',F8.2 ,//)
         DO 20 I=1,20
         DO  20 J=1,70
         TEMP(I,J)=0.0
  20     REACT(I,J)=0.0
         DO 100 I=1,5
         X(I)=0.0
 100     P(I)=0.0
         NLS=SL/STEP
         NX=NP-1
         NSP1=NS+1
         NSM1=NS-1
         READ(1,2)(P(I),I=1,NP),(X(J),J=1,NX)
    2     FORMAT(10F8.2)
         WRITE(3,6)(I,P(I),I=1,NP)
    6    FORMAT(///,'   P',I1,' =',F8.3)
         WRITE(3,7)(J,X(J),J=1,NX)
    7    FORMAT(///,'   X',I1,' =',F8.3)
         SK=0.0
         SK2=0.0
         DO 110 M=1,NP
         DO 109 I=1,NS
         IP1=I+1
         DO 108 J=1,NLS
         RR=P(M)*(J*STEP-SK-SK2)/SL
         ALR=P(M)-RR
         IF (J*STEP+STEP .GT. SL)   SK=SL-J*STEP
         J2= (SL/STEP)*(I-1)
         J3=J+J2
C        TEMPERARY STORAGE OF REACTION VALUES
         TEMP(I,J3)=ALR
         TEMP(IP1,J3)=RR
 108     CONTINUE
 109     CONTINUE
C        PLACE CONSECUTIVE LOADS IN THEIR PROPER POSITION
         IX=X(M)/STEP
         IST=IX*STEP
         SK2=X(M)-IST
         INT=0
         IF (M.EQ.1) GOTO 105
         DO 107    I=1,NSP1
         DO 107  J=1,J3
         MM1= M-1
         INT=X(MM1)/STEP
         M2=J+INT
C        ADD TEMPERARY STORAGE TO PERMANENT STORAGE
 107     REACT(I,M2)=REACT(I,M2)-TEMP(I,J)
         GOTO 110
 105     DO   104 I=1,NSP1
         DO 104 J=1,J3
 104     REACT(I,J)=-TEMP(I,J)
 110     CONTINUE
         IF(MET.EQ.1)GOTO 200
         XIMPAT=50./(NS*SL+125.)
         GOTO 201
 200     XIMPAT=15.24/(NS*SL+38.10)
```

```
    201 CONTINUE
        IF (XIMPAT.GT.0.30) XIMPAT=0.30
        DO 103 I=1,M2
        DO 103 J=1,NSP1
    103 REACT(J,I)=REACT(J,I)*(1+XIMPAT)
        WRITE(3,4)
      4 FORMAT(/////,'       LOAD',/' CONDITION',T20,' THE REACTIONS')
        DO 106 J=1,M2
    106 WRITE(3,3) J, (REACT(I,J),I=1,NSP1)
      3 FORMAT(T5,I3,10(5X,F6.1))
C       PUNCH M1 FROM LEFT TO RIGHT ACROSS THE TRUSS
C          PUNCH FORMAT IS COMPATIBLE WITH ANALYSIS PROGRAM FOR DIRECT INPUT
        READ(1,8)(M1(I),I=1,NSP1 )
      8 FORMAT(20I4)
        DO 112 I=2,NS
        WRITE(2,9)M1(I)
      9    FORMAT(I5)
    112 WRITE(2,10)(REACT(I,J),J=1,M2)
     10  FORMAT(8F10.4)
        GOTO 5000
   4000 STOP
        END


C              **** FOLLOWING 3 CARDS ARE SET FOR FPS OUTPUT
      3    7  25.       5.          0
    8.      32.    32.     14.    28.
     26  2    6   10   14   18   22   26
C              **** FOLLOWING 3 CARDS ARE SET FOR SI OUTPUT
      3    7   7.5      1.5         1
    36.0   142.0  142.0    4.25    8.5
     26  2    6   10   14   18   22   26
/*
```

A-3 LOAD MATRIX GENERATOR FOR AREA COOPER'S E-80 LOADING ON A TRUSS BRIDGE

This program steps Cooper's E-80 train load across a truss bridge. At each step (DX) the panel loads are computed as a series of simple beams. These panel loads are the *P*-matrix entries for a weightless truss for use in the frame analysis program. Note that the program does not stop until all the wheels are off of the bridge and only the uniform load is on.

The only limitation to this program is that the panel lengths must be equal.

The Cooper's E-80 loads and wheel spacing have been permanently incorporated into the program as DATA FFP/ and DATA FFX/.

Variable Identification

(see sample data cards listed at end of program listing)

TITLE	up to 80 columns of alphanumeric data for problem identification
NOSPAN	number of panels
ISWIT	0 for fps; = 1 for SI problems; used to convert wheel loads and spacing to SI and to use correct form of impact factor
IPUNCH	0 for not punching output for initial inspection for correctness; = 1 to punch output in format for direct *P*-matrix input into the frame analysis program
IMP	0 for no impact factor; = 1 to include impact factor
IWRIT	1 to write selected intermediate computations for debugging program; = 0 when program working properly
DX	increment of wheel movement left to right, ft or m
SPAN	panel length, ft or m
WIDH	width between two trusses making up bridge and needed for computing the impact factor, ft or m
FAC	factor to convert E-80 to E-110, E-60, etc.; use FAC = 1.0 for E-80 loading
NPS(J)	NP numbers (same as M1) where *P*-matrix values are to be established (note that these entries do not include NPP1—modification may be required for a through deck bridge truss).

```
C     J E BOWLES      PROGRAM TO FIND TRUSS NODE FORCES FOR AREA BRIDGES
C
C        NOSPAN = NO OF SPANS OR GIRDER SEGMENTS;   ISWIT = 1 FOR METRIC
C        = 0 FOR FPS;   DX = INCR. OF LOAD MOVEMENT, FT OR M;   SPAN = LENGTH OF
C        TRUSS SPAN OR GIRDER SEGMENT;   FAC = LOAD RATIO (FFP = E80 LOADING) USE
C        FAC = 0.75 FOR E60--0.9 FOR E72--1.00 FOR E80--L.375 FOR E110.
C
C        WIDH = CENTER TO CENTER SPACING OF TRUSSES
C        IWRIT = 1 TO WRITE SELECTED COMPUTATIONS FOR DEBUGGING
C
C        IMP = 1 TO COMPUTE IMPACT TO APPLY TO GIVEN E-LOADINGAND/OR FOR
C           PUNCHED OUTPUT--PRINTED OUTPUT IS FOR CHECKING AND FOR USE IN
C           DESIGNING FLOOR BEAMS WITH USE OF SEPARATE IMPACT FACTOR
C        IPUNCH = 1 TO PUNCH OUTPUT IN FORMAT FOR ANALYSIS PROGRAM TO
C           DEVELOP BAR FORCES AS EQUIVALENT INFLUENCE LINE FOR DESIGN
C
      DIMENSION TITLE(20),P(200),X(200),FFX(18),FFP(19), PR(20,210)
     1,PT(50),NPS(40)
C
      DATA FFX/8.,5.,5.,5.,9.,5.,6.,5.,8.,8.,5.,5.,5.,9.,5.,6.,5.,5./
      DATA FFP/40.,80.,80.,80.,80.,52.,52.,52.,52.,40.,80.,80.,80.,80.,
     152.,52.,52.,52.,8./
C
 5000 READ(1,1000,END=150)TITLE,NOSPAN,ISWIT,IPUNCH,IMP,IWRIT
C
 1000 FORMAT(20A4,/,16I5)
C
      READ(1,1002)DX,SPAN,WIDH,FAC
 1002 FORMAT(8F10.4)
C                 NPS(J) = NP NUMBERS FOR NON-ZERO P-MATRIX ENTRIES
      NSPN = NOSPAN - 1
      READ(1,1006)(NPS(J),J=1,NSPN)
 1006 FORMAT(16I5)
      NP = 18
      DIS = 80.
      IF(ISWIT.GT.0)DIS = 25.
      B = NOSPAN
      TOTSPN = B*SPAN
C        SET WHEEL LOADS AND WHEEL SPACING
      SUMXL = 0.
      DO 5  I = 1,18
      X(I) = FFX(I)
      IF(ISWIT.GT.0)X(I) = X(I)*0.3048
      P(I) = FFP(I)*FAC
      IF(ISWIT.GT.0)P(I) = P(I)*4.44822
      SUMXL = SUMXL + X(I)
    5 CONTINUE
      WLOAD = FFP(19)*FAC
      IF(ISWIT.GT.0)WLOAD = WLOAD*14.59372
      PAC = 1.
C     COMPUTE IMPACT FACTOR FOR AREA 1976  (PAC)
      IF(IMP.LE.0)GO TO 7
      IF(TOTSPN.GT.DIS)GO TO 6
      IF(ISWIT.LE.0)PAC = (100./WIDH + 40. - 3.*TOTSPN**2/1600.)/100.
      IF(ISWIT.GT.0)PAC = (30.5/WIDH + 40. - 3.*TOTSPN**2/150.)/100.
      GO TO 7
    6 IF(ISWIT.LE.0)PAC = (100./WIDH + 16. - 600./(TOTSPN-30.))/100.
      IF(ISWIT.GT.0)PAC = (30.5/WIDH + 16. - 185./(TOTSPN-9.))/100.
C        NOW WRITE DATA FOR CHECKING
    7 PAC = 1. + PAC
      WRITE(3,2000)TITLE,NP,NOSPAN,DX,SPAN,WLOAD,SUMXL,PAC,FAC,WIDH
 2000 FORMAT(////,5X,20A4,/,5X,'NP = ',I5,3X,'NOSPAN = ',I5,3X,
     12X,'DX = ',F7.2,/,5X,'SPAN LENGTH = ',F8.3,3X,'UNIF LOAD/LENGTH =
     2',F6.1,    /, 5X,'SUM OF X(L) OF CONC. LOADS = ', F7.2,3X,'IMPACT F
     3ACTOR = ',F6.3,/,5X, 'FACTOR FOR E-LOADS--1.0 FOR E-80 = ',
     4F5.3, 3X,'DIST C-TO-C OF TRUSSES = ',F7.3,//)
      WRITE(3,2004)(X(I),I=1,NP)
 2004 FORMAT(5X,'X(I) = ',12F9.1,/,12X, 6F9.1,//)
```

```
              WRITE(3,2005)(P(I),I=1,NP)
 2005 FORMAT(5X,'P(I) = ',12F9.1,/,12X, 6F9.1,//)
              WRITE(3,2006)
 2006 FORMAT(10X,'*** LOADS W/O IMPACT--USE FOR FLOOR BEAM DESIGN AND CH
     1ECKING ***',/)
C
C
    9 ZSTOP = TOTSPN + SUMXL + DX
         DO 83   K = 1,20
         DO 73   L = 1,210
   73 PR(K,L) = 0.
   83 CONTINUE
         NY = 2*(NOSPAN+1)
C
C                                  ******       MAIN DO LOOP FOR INCREMENTING LOADS
         DO 140   I = 1,210
      A = I
      DIFF = 0.
      ADX = A*DX
      ADXH = ADX
C          TO STOP COMPUTATIONS WHEN ONLY UNIFORM LOAD ON BRIDGE
      IF(ADX.GT.ZSTOP)GO TO 141
      IF(ADX.GE.TOTSPN)ADX = TOTSPN
      SUM = 0.
      MM = 0
      N1 = 1
C
C
C         FIND NO OF SPANS TO USE
   85 SPUSD = ADX/SPAN
C
C          ZERO MATRICES ARBITRARY AMOUNTS
         DO 86   K = 19,200
      X(K) = 0.
   86 CONTINUE
C
         DO 88   K = 1,50
      PT(K) = 0.
   88 CONTINUE
      IM = SPUSD
      BM = IM
      IF(BM.LT.SPUSD)IM = IM + 1
      BM = IM
      IF(ADXH.GT.TOTSPN)DIFF = ADXH - TOTSPN
      SUM2 = 0.
      IF(ABS(DIFF).LE.0.005)GO TO 15
C
         DO 12   MM = 1,NP
      SUM2 = SUM2 + X(MM)
      IF(SUM2.GE.DIFF)GO TO 15
   12 CONTINUE
   15 NC = MM + 1
      AM = IM
      DZ = BM*SPAN - ADXH + SUM2
      NN = NC
      SUM1 = 0.
      SUM3 = 0.
      SUMP = 0.
      SUMPT = 0.
      IF(MM.EQ.NP.AND.DZ.LE.0.)DZ = 0.
C
C                    LOOP FOR NUMBER OF SPANS USED FOR CURRENT LOAD POSITIONS
   21    DO 102   JJ = 1,IM
      ZZX = 0.
      D1 = SPAN - DZ
      KC = 2*(IM-JJ) + 4
      Z = JJ
```

```
C                      FIND NO OF LOADS IN CURRENT SPAN
      DO 95   L = NC,200
      IF(L.EQ.NP+1.AND.ADXH.LE.TOTSPN)X(L) = Z*SPAN - SUMXL
      IF(L.EQ.NP+1.AND.ADXH.GT.TOTSPN)X(L) = DIFF + Z*SPAN - SUMXL
      IF(L.GT.NP+1.OR.X(L).GT.SPAN)X(L) = SPAN
      IF(X(L).LT.0.)GO TO 148
      IF(X(L).NE.SPAN)GO TO 22
      D1 = SPAN
      DZ = 0.
   22 SUM1 = SUM1 + X(L)
      IF(SUM1.GE.D1-0.0001)GO TO 96
      IF(L.EQ.NP+1)ZZX = SPAN - (DZ + 0.005)
      IF(L.EQ.NP+1.AND.SUM1.GE.ZZX)GO TO 96
   95 CONTINUE
   96 CONTINUE
      SUM3 = SUM3 + SUM1
      TOTSUM = SUM3 + SUM2
      NC = L + 1
      D2 = D1
      RR = 0.
      RL = 0.
      IF(IWRIT.LE.0)GO TO 550
      WRITE(3,3005)TOTSPN,NC,ADXH, IM, D2, D1, SUM1, DZ, NN, L,SPAN,SUM3
     1, SUM2, DIFF, TOTSUM,JJ,X(19),X(20),X(21),X(L)
 3005 FORMAT(3X,'TOTSPN = ',F8.1,3X,'NC=',I5,3X,'ADXXH = ',F7.3,
     2  3X,'IM = ',I5,   5X,'D2 = ',F7.3,2X,'D1 = ',F7.3,'  SUM1 =',F7.3,
     3/,5X,   'DZ = ',F7.3,3X,'NN=',I5,2X,'L=',I5,2X,'SPAN=',F7.3,
     43X, 'SUM3 = ',F7.3,3X,'SUM2 = ',F7.3,3X,'DIFF=',F7.3,3X, 'TOTSUM =
     5',F6.1,2X,'JJ=',I2, /,5X, 'X19 =',F7.3,3X,'X20=',F7.3,3X,'X21=',
     6F7.3,3X,'XL=',F7.3,/)
  550 CONTINUE
C                 ACCUMULATE EFFECTS OF LOADS IN ANY SPAN
      DO 99  LL = NN,L
      IF(LL.LE.NP)GO TO 97
      P(LL) = WLOAD*X(LL)
      IF(LL.EQ.NP+1.OR.NN.EQ.L)D2 = D2 - X(LL)/2.
   97 RR = D2*P(LL)/SPAN
      RL = P(LL) - RR
      PT(KC) = PT(KC) - RR
      PT(KC-2) = PT(KC-2) - RL
      D2 = D2 - X(LL)
        SUMP = SUMP + P(LL)
   99 CONTINUE
      SUMPT = SUMPT + PT(KC)
      IF(JJ.EQ.IM)SUMPT = SUMPT + PT(KC-2)
      Z = IM - JJ
      DZ = Z*SPAN + TOTSUM - ADXH
      NN = LL
      SUM1 = 0.
  102 CONTINUE
C        WRITE ONLY VERTICAL VALUES AT PANEL POINTS (HORIZ = 0.) SO CAN
C                    USE PRINTED OUTPUT TO DESIGN TRANSVERSE FLOOR BEAMS
C                    BASED ON LARGEST PANEL LOAD = FLOOR BEAM REACTIONS
C                    AND DEPENDING ON NUMBER OF TRACKS ON BRIDGE ** DOES NOT
C                    INCLUDE ANY IMPACT FACTOR *********
      WRITE(3,2008)I,(PT(J),J=2,NY,2)
 2008 FORMAT(//,2X,I5,12F10.2,/,  5X,12F10.2,/)
      WRITE(3,2009)SUMP,SUMPT
 2009 FORMAT(50X,'SUM TRUSS LOADS = ',F10.2,3X, 'SUM NODE LOADS = ',
     1F10.2,/)
      DO 138  J = 1,NY
      PR(J,I) = PT(J)
  138 CONTINUE
  140 CONTINUE
  141 WRITE(3,2011)I
 2011 FORMAT(//,5X,   'ONLY UNIFORM LOAD ON BRIDGE, I = ',I3,//)
      IF(IPUNCH.LE.0)GO TO 149
      IM1 = I - 1
```

```
C                       APPLY IMPACT FACTOR TO LOADS IF IMP > 0
      IF(IMP.LE.0)GO TO 25
          DO 143   J = 1,IM1
      DO 142   KK = 2,NY,2
  142 PR(KK,J) = PR(KK,J)*PAC
  143 CONTINUE
   25 KK = 2
          DO 145   J = 1,NSPN
      KK = KK + 2
      JJ = NPS(J)
      WRITE(2,1006)JJ
      WRITE(2,2014)(PR(KK,M),M=1,IM1)
 2014 FORMAT(8F10.4)
  145 CONTINUE
  149 GO TO 5000
  148 WRITE(3,2016)
 2016 FORMAT(////,5X,'*** PROBLEM TERMINATED--X(L) IS NEGATIVE ***',///)
  150 STOP
      END


C   ** FOLLOWING 4 CARDS TYPICAL SET OF FPS DATA FOR 9 PANEL TRUSS
    J E BOWLES  AREA BRIDGE FOR TEXT USING COOPER E-80 LOADING AND IMPACT
       9    0    0    1    0
 3.45      27.60     17.00      1.0
       4    8   12   16   20   24   28   32
C   ** FOLLOWING 4 CARDS TYPICAL SET OF SI   DATA FOR 9 PANEL TRUSS
    J E BOWLES  AREA BRIDGE FOR TEXT USING COOPER E-80 LOADING, IMPACT AND SI
       9    1    1    1    0
 1.05       8.40      5.20      1.0
       4    8   12   16   20   24   28   32
/*
```

INDEX

INDEX

AASHTO column formulas, 263
   with bending, 321
AASHTO minimum thickness of
   metal, 491
AISC column formula, 262
   with bending, 318
Angle tension members, 227
   net section with holes, 235
Approximate analysis of frames, 54
   cantilever method, 55
   portal method, 54
   torsion in, 56
AREA column formulas, 264
   with bending, 321
AREA minimum thickness of
   metal, 491
AREA track load distribution, 187

Base plates, column, 285
   bending stresses in, 287
   design criteria for, 287
   grouting of, 285
   welded, 439
BASLER, K., 474, 479, 481
Beam analysis, 57
   biaxial bending, 164

Beam analysis:
   differential equation, 58
   for unsymmetrical bending, 167
Beam columns, 297
   design formulas: AASHTO and
     AREA, 321
     AISC, 313, 318–319
     C_m for, 316, 320, 321
   design methods, 322
   effective length of, 303, 306
   G factors for, 306
   inelastic effect/reduction, 307
   K factor chart, 306
Beam framing connections, 385
Beam stresses, 146
   allowable: AASHTO and AREA,
     151
     AISC, 149, 174
     laterally unsupported, 174
   bending, 146
   biaxial bending, 164
   elastic design, 148
   shear, 146, 159
Beams:
   compact section criteria, 149
   deflections, 146
   laterally unsupported, 174

Beams:
   stress computations, 146
   types of, 57, 144, 154, 182, 183
BENDIGO, R., 360
Bending on columns, 298
   allowable, for beam columns, 313
BLEICH, F., 133
Bolts, high-strength, 355
   A-307, 355
     installation of, 357
     minimum tension, table of, 358
Bracing, sidesway control, 337
Bridge beams, 183
   AASHTO, 183
     wheel load distribution, table, 185
   AREA, 186
   composite, 189, 191
     shear studs for, 194
Bridge loads, table of coefficients, 185
Bridge stringers, 183
Bridges, terms used in, 46
Brittle fracture, 15
Buckling of plates, 133
   bracket plates, 380
Building loads, 18
Built-up shapes for columns, 272
   lacing for, 274, 276
   tabulated r for selected shapes, 272
   for tension members, 226
BUTLER, L. J., 421
Butt weld, 414, 422

C_c for columns, 262
   tabulated values of, 263
C_m for beam columns, 316, 320
Cables, 217, 222
   damping of, 244
   design data, tables, 224
   design equations for, 242
   modulus of elasticity of, 223
CHEN, W. F., 427

CHESSON, E., 364
COCHRANE, V., 233
Codes, structural list of, 17
Coefficient of thermal expansion, 5
Column design, steps in, 265
Column Research Council, 176
Columns, 255
   beam, 297
   C_m, 316, 320
   effective length, 303, 306
   formulas, 313, 318, 319, 321
   K factors for, 306, 331
   stepped, 329
Combined stresses, bending, 164, 300
   in columns, 312
   in connection fasteners, 396
   interaction in plate girders, 479, 496
   welded connections, 430, 432
Composite beam design, 189
   modular ratio tables for, 192
   shored construction, 190
Connection:
   welded corner, 445
   welded end plate, 442
Connection design, 237
   bearing connections, 351, 366
   edge distances, 361
   factor of safety in, 369
   friction connections, 349, 366
   minimum load, AISC, 237, 363
   type I, II, III, 347, 399
   welded, type I, 425
Constants, design: C_b, 176
   C_w, 174
   J, 171
Continuous beams, 154
   moments for design in, 154
   point of inflection, 155
Cover plates:
   built-up beam, 463
   perforated, 277

Deflections, beams, 125, 146, 161

Diaphragms for girders, 496
DISQUE, R., 308
DRANGER, T., 391
Ductility, 3

Earthquake loads, 33
　period for, 34
　zone map, 36
Eccentric loading on fasteners, 375
　AISC reduced, 382
Edge distance for fasteners, 361
EENAM, N., 493
Effective length of columns, 259
　control of, 337
　in frames, 303
　without sidesway, 259
　table of cases, 260
Elastic design, 114
Electrode, method of designation, 412
ERICKSON, E., 493
Euler column formula, 257
　stress F_e' defined, 308
Eyebar design, 222

Fabricator, steel, 5
Fahrenheit, conversion to Celsius, 7
Fasteners:
　allowable stresses, 353, 358, 366–368
　　ultimate strength of, 353
　　threads in shear plane, 366
　　　combined shear and tension, 396
　　　table of values for, 398
Fatigue, 39, 229
Fillet weld, 414, 416, 422
　LRFD used for, 451
Fire protection, 13
FISHER, J., 368
Framing connections, 385
　welded, 431

Gage distance, fasteners, 233, 362
GALAMBOS, T. V., 337
Girder, bending stresses: AASHTO allowable, 490
　AISC allowable, 470, 474
　AREA allowable, 490
Girder, built-up beam as, 464
　hybrid, 460
　partial length cover plate, 466
　　cut-off points, 466
　　　specifications for, 466
　properties of, 463, 469
Girder loads, 461
　AREA, 463
　location of AASHTO truck for, 461

Hanger connections, 390
　Dranger solution for, 391
HANSEN, R., 360
Hinge, length of plastic, 125
Hole diameter, effective, 223

Industrial buildings, 43

JOHNSTON, B., 133, 305
Julian and Lawrence chart, 306

K factor:
　for columns chart, 306
　stepped columns, 331
Knee bracing, girders, 497
　force in girders, 498
　industrial bents, 43
KRISHNAMURTHY, N., 442
KULAK, G. L., 421

Lacing:
　built-up sections for, 274
　shear force in, 275

Lamella tearing of welds, 420
Lateral bracing:
   for beams, 150, 155, 174
   for columns, 291
   for girders, 496
      knee bracing, 497
Live loads, 18
   reduction for, 19
Load conditions, 70
Load resistance factor design
      (LRFD), 132
   table of Φ factors, 132
Loads, 18
   bridge: AASHTO, 26, 27
      AREA, 28
      equations for shear and
         moment, 26
      impact, 32
      tabulated shear and moment,
         AREA, 29
   dead, 18
   earthquake, 33
   eccentric, on fasteners, 375
   impact, 32
   live, 18, 19
   ponding, 25
   snow, 24
      map for, 25
   wind, 21, 31
Low-temperature effects, 13
L/r ratio:
   compression members, 258, 260, 277
   tension members, 231, 237, 238, 242
LRFD, 132
   beam column design, 338
   beam design, 208
   column design, 292
   connections, 405
   tension member design, 248
Lug angles, 365

Mass density, steel, 7

Minimum thickness:
   AASHTO, 491
   AREA, 491
Modular ratio E_s/E_c, 192
Modulus of elasticity, 5
   in shear, 5
MUNSE, W., 364

Net area:
   column, 264
   in tension, 219, 225, 232
      effective hole diameter for, 233
      at thread root, 219
      use of $s^2/4g$, 233
NEWLIN, D. E., 427
NEWMARK, N. M., 185

Pitch distance, fastener, 233
Plastic design, 114, 117
Plastic section modulus, 120
   shape factor, 120
Plate girder:
   design of, 469, 470
      AASHTO, 490
      AISC, 470
      AREA, 490
      stresses: bending, 469, 470, 474, 490
         interaction, 479
         shear, web, 477, 492
      web plate, 473, 491
         crippling in, 481
         thickness of, 491
   use of, 457
      D/L ratios, typical, 457, 459
      lengths, typical, 457
Plug welds, 419
   AISC specifications for, 419
Principal axes for pending, 167
   product of inertia, 167
Prying action in connections, 390

Radius of gyration, 175, 237, 255, 277, 307
  built-up members, tables, 272
  compression flange of beams, 179
  Euler formula used in, 260
  lacing for, 276
Ratios:
  a/h, 473, 478, 494
  b/t, 150, 465, 471, 490, 493
  d/t, 150
  h/t, 456, 481, 491
  L/r, 263, 271, 275, 307
Reaction distance for beams, 157
  bracing for, 158
Reduced eccentricity of fasteners, 382
Residual stresses, 118, 472
RUMPF, J., 360

Safety factor, 123
  in connections, 369
    elastic design for, 123
    plastic design for, 124
Sag rods, 230
SANDHU, B., 331
Secondary stresses, 299
  P-Δ, 301
Shape factor, 120
Shapes, rolled, 7–10
Shear, computation: for beams, 146, 159
  for girders, 466
  for lacing bars: AASHTO and AREA, 275
    AISC, 275
Shear center, 171
  J constant, 171
Shear lag, 364
  joint length as factor, 360
Shear stress, allowable: in beams, 159
  in girders, 477
Shear studs, 194
Shear walls, 336

SI conversion factors, 49
Sidesway, control of, 337
Slot weld, 414, 419
  AISC specifications for, 419
Spacing of bolts, 363
Stepped columns, 329
  K factor for, 331
Stiffeners, bearing, 158, 478, 480, 493
  column flange, 427
    web, 426
  diagonal, for corner connections, 446
  girder, required for, 456
    intermediate, 479, 493, 495
    longitudinal, AASHTO, 493
    moment of inertia of, 480, 495
    spacing for, 493, 495
  seat angles for, 436
Stiffness method of analysis, 54
Stress, allowable: base plate design, 287
  bending, 149, 174, 469, 474, 490
  compression: AASHTO, 263
    AISC, 261
    AREA, 264
  fasteners, tables, 353, 358, 366–368
  interaction, 312
  shear, 146, 159
    in girder webs, 478, 492
  tension, 219
    table of values, 221
  weld, 416
Stress range:
  defined, 40
  table of values, 41
Stress-strain curves, 11, 113
Structural codes, list of, 17
Structural shape data, table, 6
Structural shapes, 7–10

Table values (*see specific tables*)
Tangent modulus, 135

Tapered flange beam, 182
Temperature coefficient, 5
Temperature versus strength, 13
Tension stresses, 219
    allowable values, 219
    effective net area for, 225
Threaded area in tension, 219
    table of design data for, 230
THÜRLIMANN, B., 474
TIMOSHENKO and GOODIER, 133
Torsion constant, J, 171, 174
Truss analysis, 59
    coding for computer, 60
Type 1 connections, 347, 425

Ultimate strength, 7
    table of values, 6
Unbraced length, 175
    L_c, 151, 323
    L_u, 151
Unit weight of steel, 7

WALLAERT, J., 368
Warping constant C_w, 174

Web crippling:
    beams, 156
    girder, 481
Weld capacity at cut-off, 467
Weld stresses, allowable, 416
    table, 417
Welded connections, AISC, 420
Welded framing angles, 431
    corner connections, 445
    end plates, design of, 442
    seat angles, 433
        stiffened, 436
Welding, types of, 312, 415
    electrodes used for, 414
Welds:
    lamella tearing of, 420
    maximum size, 418
    minimum size, table, 419
    orientation for strength, 421
Wind loads, 21

Yield point, 7
    table of values for steel, 6
Yield strength, 10
YURA, J., 307